BIOTIC FEEDBACKS IN THE
GLOBAL CLIMATIC SYSTEM

Biotic Feedbacks IN THE Global Climatic System

Will the Warming Feed the Warming?

Edited by

GEORGE M. WOODWELL
FRED T. MACKENZIE

New York   Oxford
OXFORD UNIVERSITY PRESS
1995

Oxford University Press

Oxford   New York   Toronto
Delhi   Bombay   Calcutta   Madras   Karachi
Kuala Lumpur   Singapore   Hong Kong   Tokyo
Nairobi   Dar es Salaam   Cape Town
Melbourne   Auckland   Madrid

and associated companies in
Berlin   Ibadan

Library of Congress Cataloging-in-Publication Data
Biotic feedbacks in the global climatic system : will the warming feed the
warming? / George M. Woodwell and Fred T. Mackenzie, editors.
   p.   cm.   Includes bibliographical references and index.
   ISBN 0-19-508640-6
   1. Global warming.   2. Climatic changes.   3. Ecosystem management.
   4. Biotic communities.   I. Woodwell, G. M.   II. Mackenzie, Fred T., 1934–
QC981.8.G56B47 1995   94-7773
551.5—dc20   CIP

Printing (last digit): 9 8 7 6 5 4 3 2 1

Printed in the United States of America
on acid-free paper

Foreword

One of the main successes of the Rio Conference on Environment and Development held in June 1992 was the Climate Convention signed by 161 countries. The objective of the convention as stated in Article 2 is the stabilization of greenhouse gases at levels and on a time scale that do not produce unacceptable damage to ecosystems and that allow for sustainable economic development. This objective of the Climate Convention presents a significant challenge to the scientific community. It means that scientists need to provide, with sufficient accuracy for the needs of policymakers, the following data:

1. Predictions of the future concentrations of greenhouse gases in the atmosphere for different scenarios of anthropogenic emissions, as well as the range of those scenarios that lead to the stabilization of greenhouse gas concentrations.
2. Predictions of the regional climate change likely to result from different concentrations of greenhouse gases.
3. Estimates of the impacts of likely climate change on natural and managed ecosystems.

The first and third of these items require a detailed understanding of the feedbacks in the biosphere. At the moment, our knowledge of these is very limited. In particular, there are substantial discrepancies to be resolved in our knowledge of the exchanges of carbon dioxide and methane between the atmosphere and the land surface and between the atmosphere and the ocean; these exchanges are important components that help to make up the global carbon cycle.

Because of the importance of these feedbacks, the Science Working Group of the Intergovernmental Panel on Climate Change (IPCC) was pleased to respond positively to the proposal made to us by Dr. George M. Woodwell that a workshop addressing biotic feedbacks be held at the Woods Hole Research Center in October 1992. Many experts from a number of countries attended. The discussions were lively, and it was interesting to note the considerable disagreement that exists regarding the likely scale, magnitude, and, in some cases, even the sign of the potential feedbacks. The papers that were prepared for the meeting and that serve as the basis for the chapters in this volume provide an important and valuable input into the IPCC assessment process; they have, of course, not received any formal IPCC approval.

All who attended agreed that it was a most useful meeting; in particular it has been of great assistance in setting the agenda for future work in this area. The IPCC is pleased to acknowledge the valuable contribution of all who submitted papers and who assisted in the development of the Conference Statement. Finally, we are particularly grateful to Dr. Woodwell and his colleagues for their organization of the workshop and their work in the preparation of this volume.

JOHN R. HOUGHTON

Preface

Progress over the past two decades in defining the details of global climate, its causes, and its effects has enabled not only prediction of weather in the short term with remarkable accuracy, but even the prediction of climate and climatic trends on the time scale of years. The current trends in climate are clearly toward a warming and are of particular importance because they have the potential, if the warming is as rapid as some anticipate, to bring about substantial disruption of major patterns of atmospheric and oceanic circulation and major changes in vegetation, in the distribution of arable land, and in the capacity of the earth for supporting people.

The particular advances in scholarship that affect the human prospect are those that have allowed an appraisal of the effect of the accumulation of the heat-trapping gases, including especially carbon dioxide and methane, in the atmosphere, as outlined in this volume. The warming from these gases is expected to exceed over the next few decades and into the indefinite future any warming that we know of in the earth's previous history. The topic has been explored in remarkable detail and with increasing intensity since an especially probing series of meetings held at Williams College during the summer of 1970 in preparation for the 1972 United Nations Conference on the Human Environment held in Stockholm. The Williams College meetings were followed almost immediately by a series of discussions of the biotic causes and effects of the accumulation of CO_2 and CH_4 in the atmosphere at a conference held at Brookhaven National Laboratory during 1972. Most recently the Intergovernmental Panel on Climate Change (IPCC), under international auspices, has marshaled international scientific talent, exposed earlier in more than a score of compelling reviews, to the challenge of defining a perspective on global warming that is as close to a global consensus among scientists as is possible. Meanwhile, the advances have forged ahead of the consensus to refine the predictions and to verify the trends in climate through a series of studies led by James Hansen and his colleagues at the Goddard Institute of Space Studies in New York. The most remarkable progress is the accumulating evidence that the warming is occurring and that it is modulated not only by the heat-trapping gases in the atmosphere but also by volcanic emissions of aerosols into the stratosphere and by anthropogenic emissions of sulfur dioxide into the troposphere. The effects are now being predicted with a remarkable degree of accuracy. The progress in prediction conveys confidence that we have at the moment a sound and increasingly accurate grasp of the factors that determine the broad details of global temperature and climate.

The warming, however, is expected to continue. This statement is based on the assumption that we shall continue to emit large quantities of CO_2 and CH_4 through reliance on fossil fuels for energy to drive industrialization and shall continue to reduce the area of forests globally. The warming is rapid, continuous, and open-ended. It will continue into the indefinite future as long as heat-trapping gases continue to accumulate in the atmosphere. The earth is not merely moving from one state of approximate equilibrium to another; it is moving rapidly from an approximate stability of temperature and climate to progressive instability. The next decades and the foreseeable future will probably be marked by rapid changes in climate globally. We may have the potential for changing that trend through massive reductions in releases of heat-trapping gases, but the world is not yet moving toward such reductions. Assumptions based on new climatic equilibria that are achievable only over centuries to millennia are of little use in appraising the changes likely to mark the next several decades.

The speed of the warming is important. It is rapid by any measure; over the last 15 years, it has been 0.26°C per decade. Forests, disturbed by whatever cause, take centuries to return to relative stability. Temperature changes on the order of tenths of a degree centigrade per century may be accommodated; such changes in a decade are unquestionably destructive, a further contribution to the wave of biotic impoverishment associated with the surge of human activities in these latter decades of the twentieth century.

The fact that the terrestrial vegetation globally contains a large amount of carbon in plants and soils and uses annually through the metabolism of plants as much as 15–20% of the carbon held in the atmosphere means that changes in the metabolism of plants or in the rates of decay of organic matter in soils have the potential for affecting the composition of the atmosphere significantly in a short time. That potential implies that climatic changes will have biotic feedbacks that may affect the temperature of the earth. The significance of such effects will clearly depend on the rate of climatic change, but the feedbacks are potentially large enough to be of concern.

The potential of the terrestrial vegetation for responding to an increase in the amount of CO_2 in the atmosphere has led to the assumption that the difficulty that scientists have had in determining details of the global cycle of carbon might be resolved by assuming that the rates of storage of carbon in forests globally are increasing in proportion to the changes in the atmospheric CO_2 burden. That possibility led to an emphasis on terrestrial ecosystems in this review: is there any basis for thinking that the land may be accumulating carbon in plants and soils at higher rates than in the past in response to the changing atmospheric and climatic circumstances?

To address this topic, not previously considered as a central part of the IPCC's review, the Woods Hole Research Center sought IPCC interest in a conference in Woods Hole. The summary presented here is the product of considerable discussion and advice from scientists and scholars around the world. Much of the advice has been centered on placing further emphasis on the oceans, so heavily has the role of the oceans been emphasized in earlier analyses. There is, however, little basis for anticipating strong biotic feedbacks in the oceans in response to a warming of the earth. The immediate issue appears to be the potential of terrestrial ecosystems for influencing the course of any rapid global warming by affecting, even determining, the composition of the atmosphere. Although we have explored the potential of the oceans for biotic feedbacks and the role of the oceans in resolving the "missing carbon" issue, we have persisted in keeping the emphasis on the land in this book.

Approximately 100 scientists from around the world met in Woods Hole on October 25–29, 1992, to explore the implications of the interactions between the biota globally and the warming of the earth now underway. The papers published here were prepared initially as background for the meeting and distributed to the participants in advance. They have been revised for publication, reviewed by sundry anonymous reviewers, and edited for publication. They and the comments of the editors reflect the advances made by individuals in their own analyses of the problem addressed by the conference.

G.M.W.
F.T.M.

Acknowledgments

Special thanks are due for the advice, encouragement, and active participation of the staff of the Intergovernmental Panel on Climate Change in Bracknell, U.K., under the leadership of Sir J. R. Houghton. Their persistent skepticism regarding the importance of the biotic interactions defined here has brought a continuous refinement of perspective to all and significant improvement to the insights presented in this volume.

We also acknowledge the continuing interest of several financial supporters who have had confidence in us over the years and who specifically contributed to this undertaking either directly by supporting the conference or indirectly by support of the center and those of its long-term programs that led to the conference. These supporters include the Andrew W. Mellon Foundation, the W. Alton Jones Foundation, and the Jessie Smith Noyes Foundation.

Specific grants were made for the conference by the U.S. Environmental Protection Agency, the United Nations Environment Programme, the Commission of the European Community, the U.S. Department of State, and the Intergovernmental Panel on Climate Change.

The conference was also supported by the Woods Hole Research Center, its staff, and its trustees.

The organizers and participants express appreciation to all these supporters and to other individual supporters too numerous to list here.

G.M.W.
F.T.M.

Contents

Contributors

LEON HARTWELL ALLEN, Jr.
U.S. Department of Agriculture
Agricultural Research Service,
Building 164
University of Florida
Gainesville, Florida 32611

JEFFREY S. AMTHOR
The Woods Hole Research Center
P.O. Box 296, 13 Church Street
Woods Hole, Massachusetts 02543

MICHAEL J. APPS
Northern Forestry Center
Canadian Forest Service
5320 122 Street
Edmonton, Alberta TGH 355
Canada

JACK G. BALDAUF
Department of Oceanography and
*   Ocean Drilling Program*
Texas A&M University
College Station, Texas 77843

DENNIS B. BALDOCCHI
Atmospheric Turbulence and
*   Diffusion Division*
National Oceanic and Atmospheric
*   Administration*
P.O. Box 2456
Oak Ridge, Tennessee 37831-2456

W. DWIGHT BILLINGS
Department of Botany
Duke University
Durham, North Carolina 27706

O. K. BORISOVA
Institute of Geography
Academy of Sciences
Staromenetry 29
Moscow 109017
Russia

R. J. CHARLSON
Department of Atmospheric Sciences,
*   AK-40*
Department of Chemistry and Institute
*   for Environmental Studies*
University of Washington
Seattle, Washington 98195

ROSANNE D. D'ARRIGO
Tree Ring Laboratory
Lamont-Doherty Geological Observatory
Palisades, New York 10964

ERIC A. DAVIDSON
The Woods Hole Research Center
P.O. Box 296, 13 Church Street
Woods Hole, Massachusetts 02543

IAN G. ENTING
Division of Atmospheric Research,
CSIRO
Private Bag No. 1
Mordialloc, Victoria 3195
Australia

INEZ Y. FUNG
School of Earth & Ocean Sciences
MS-4015
University of Victoria
Victoria, British Columbia V8W 2Y2
Canada

EVILLE GORHAM
Department of Ecology
University of Minnesota
318 Church Street SE
Minneapolis, Minnesota 55455

R. A. HOUGHTON
The Woods Hole Research Center
P.O. Box 296, 13 Church Street
Woods Hole, Massachusetts 02543

B. INGHAM
Department of Geology
University of Saskatchewan
Saskatoon, Saskatchewan S7N 0W0
Canada

GORDON C. JACOBY
Tree Ring Laboratory
Lamont-Doherty Geological
  Observatory
Palisades, New York 10964

A. W. KING
Environmental Sciences Division
Oak Ridge National Laboratory
P.O. Box 2008
Oak Ridge, Tennessee 37831-6034

WERNER A. KURZ
ESSA Technologies Ltd.
1765 West 8th Avenue, Third Floor
Vancouver, British Columbia
  V6J 5C6
Canada

ROBERT J. LUXMOORE
Environmental Sciences, MS 6038
Oak Ridge National Laboratory
P.O. Box 2008, Building 1505
Oak Ridge, Tennessee 37831-6038

FRED T. MACKENZIE
Department of Oceanography
University of Hawaii
1000 Pope Road, MSB525
Honolulu, Hawaii 96822

V. P. NECHAYEV
Institute of Geography
Academy of Sciences
Staromenetry 29
Moscow 109017
Russia

E. G. NISBET
Royal Holloway and Bedford
New College, University of London
Egham Hill
Egham, Surrey TW20 0EX
United Kingdom

W. M. POST
Environmental Sciences Division
Oak Ridge National Laboratory
P.O. Box 2008
Oak Ridge, Tennessee 37831-6034

I. COLIN PRENTICE
Department of Plant Ecology
Lund University
Ostra Vallgatan 14
S-223 61 Lund
Sweden

GILBERT T. ROWE
Department of Oceanography
Texas A&M University
College Station, Texas 77843

WILLIAM H. SCHLESINGER
Department of Botany
Duke University
Durham, North Carolina 27706

RAYMOND C. SMITH
University of California at
  Santa Barbara
Computer Systems Lab
6832 Ellison Hall
Santa Barbara, California 93106

STEPHEN V. SMITH
Department of Oceanography
University of Hawaii

1000 Pope Road, MSB510
Honolulu, Hawaii 96822

BRIAN J. STOCKS
Canadian Forest Service
Ontario Region
1219 Queen Street East
P.O. Box 490
Sault Ste. Marie, Ontario P6A 5M7
Canada

MARTIN T. SYKES
Department of Plant Ecology
Lund University
Ostra Vallgatan 14, S-223
61 Lund
Sweden

PIETER P. TANS
National Oceanic and Atmospheric
    Administration
325 Broadway
Boulder, Colorado 80303-3328

A. A. VELICHKO
Institute of Geography
Academy of Sciences

Staromenetry 29
Moscow 109017
Russia

W. JAN A. VOLNEY
Canadian Forest Service
Northwest Region
Edmonton, Alberta TGH 355
Canada

GEORGE M. WOODWELL
Woods Hole Research Center
P.O. Box 296, 13 Church Street
Woods Hole, Massachusetts 02543

STAN D. WULLSCHLEGER
Environmental Sciences Division
Oak Ridge National Laboratory
P.O. Box 2008
Oak Ridge, Tennessee 37831-6034

E. M. ZELIKSON
Institute of Geography
Academy of Sciences
Staromenetry 29
Moscow 109017
Russia

1

GLOBAL WARMING: PERSPECTIVES FROM LAND AND SEA

The topic addressed in this book is how the warming of the earth will affect the flow of carbon dioxide mobilized by human activities among the atmosphere, the land, and the oceans. The topic is intrinsically complicated and is not well understood by scientists, despite the 4-billion-year success of the biosphere in providing a habitat into which *Homo sapiens* could emerge and thrive. The topic is also important: the warming, if unchecked, promises a change in the human habitat of sufficient magnitude to threaten our future through drastic reductions in the capacity of the earth for support of the human race. Knowledge of how to deflect that threat becomes more important daily as the heat-trapping gases accumulate, as the size of the human population approaches its second doubling since 1950, and as pressures on all resources build.

The point that seems clear in the following analyses is that the rapid warming projected for the next several decades holds substantial possibility of causing changes on the land, especially in forests, that will release additional quantities of CO_2, CH_4, and other climatically important trace gases into the atmosphere. There is no salvation in the response of the oceans: the warming will probably reduce their capacity for absorbing CO_2. These conclusions emerge despite the lingering uncertainties surrounding explanations of the details of the current carbon fluxes among the three large pools of atmosphere, land, and sea. Oceanographers believe that the capacity of the oceans for absorbing the excess of CO_2 that is accumulating in the atmosphere is limited. Those who examine forests believe that deforestation and the warming of the earth are releasing substantial quantities of carbon as CO_2 into the atmosphere in addition to the amount emitted by combustion of fossil fuels. The analyses pose a dilemma: the total emissions projected by these calculations exceed the amount accumulating in the atmosphere plus the amount estimated as being absorbed into the oceans. The excess must be accumulating either on land or in the oceans, but intensive reviews of the analyses have failed to reveal the errors. The topic arises repeatedly in later discussions.

The two chapters of Part I address this issue from different perspectives. The first chapter examines data that appear to define the role of forests in determining the composition of the atmosphere now and as the earth warms. The second is written through the eyes of a geochemist and explores the coupled global carbon, nitrogen, phosphorus, and sulfur cycles over time with a special emphasis on the role of the oceans.

1

Biotic Feedbacks from the Warming of the Earth

GEORGE M. WOODWELL

THE PROBLEM: WHEN THE EARTH WARMS, WHAT HAPPENS?

The data from the Vostok Core (Barnola et al. 1987; Lorius et al. 1990; Raynaud et al. 1993) show that during the cycles of warming and cooling of the past 160,000 years temperature, CO_2 concentrations, and methane concentrations have marched together. The data do not reveal cause and effect, but they do indicate that the overall pattern of climatic change recorded in that period was dominated by a positive feedback: a warming was associated with an increase in CO_2 and CH_4 levels in the atmosphere, and a cooling, whatever the initial cause, was correlated with a removal of CO_2 and CH_4 from the atmosphere. During the cooling phases the temperature changes led the changes in the concentrations of gases. A 1°C change in temperature produced an approximately 7.5-ppm change in CO_2 or a change in the atmospheric burden of about 15 petagrams (Pg) (= 10^{15} g) carbon. The record is difficult to interpret in increments of less than a thousand years, but the pattern is strong enough to raise a general question as to whether a similar mechanism can be expected to dominate in the current warming.

A rapid and continuous warming is a serious threat to the human enterprise. The threat lends urgency to technical questions about how the world works. Answers lie in the details of the global carbon cycle. But the question of what to do to protect human interests now demands from science not a narrow technical proof but a reasoned appraisal of how the current and past warming will affect the course of the further warming.

Our interest is in the changes imminent in the time of our lives and our children's lives, on a time scale of years to decades, a century or so, not millennia. A rapid and continuous warming is expected as the concentrations of heat-trapping gases in the atmosphere rise: an increase in the average temperature of the earth of 0.2°C to as much as 0.5°C or greater per decade will be unevenly distributed latitudinally. Little change in temperature is expected in the tropics and as much as twice the average can be expected in the higher latitudes. Winter temperatures will rise more than summer temperatures by most appraisals. The warming will proceed into the indefinite future, unless definite steps are taken soon to avoid it. The evidence for this conclusion has been marshaled in at least a hundred papers in recent years and a score of international reviews, culminating most recently in the Villach-Bellagio Report (WMO/UNEP 1987), a review by the U.S. National Academy of Sciences (NAS 1991), another by the Office of Technology Assessment of the U.S. Congress (OTA 1991), and two reviews by the Intergovernmental Panel on Climate Change established by the UNEP and the WMO (J.T. Houghton et al. 1990, 1992). Several less formal reviews have been published as books, such as those edited by Abrahamson (1989) and Leggett (1990). These reviews leave little question as to the conclusions of the scientific community regarding the seriousness of the changes underway: the continued addition of heat-trapping gases to the atmosphere will warm the earth at rates expected to lie in the ranges just outlined. If the warming is allowed to proceed, devastating effects on terrestrial ecosystems, especially forests, can be anticipated (Peters and Lovejoy 1992). These effects will be large enough and may be rapid enough to add to the atmospheric burden of CO_2 and CH_4 and thereby speed the warming. The continued habitability of the earth is clearly in question, but the topic has not received the intensive scrutiny accorded the primary climatic changes.

The steps to correct the warming will be determined by how the global cycle of carbon release and absorption of carbon influences the heat-trapping gas content of the atmosphere. The analyses hinge heavily on the relative importance of terrestrial and oceanic influences. Answers are far from clear, as shown by the recent summary by Sundquist (1993) and by the other chapters in this book. The relative influence of land versus sea probably varies heavily with the length of the period of interest: in the shorter term of days to decades, there is reason to look to the terrestrial influences as dominant. They are the emphasis of the following discussion. In the longer term of centuries to millennia and more, the massive capacity of the oceans for holding CO_2 as part of the carbonate-bicarbonate system and CH_4 as hydrate (MacDonald 1990) has dominated.

WHAT WE KNOW

Warming from Heat-Trapping Gases

CO_2, CH_4, and a suite of other heat-trapping trace gases are currently accumulating in the atmosphere at a rate that will result in a doubling of the heat-trapping capacity of the atmosphere above what existed in the middle of the last century before the middle of the next century. CO_2 is emphasized because it is, except for water vapor, the most important of the heat-trapping gases and is the cause of more than half of the warming anticipated from current accumulations. The gas accumulation is open-ended: it will continue indefinitely as long as we continue to use large quantities of fossil fuels. Current use of fossil fuels is releasing about 6.0 Pg of carbon per year as CO_2 into an atmosphere that contains about 30% more CO_2 than it contained in the middle and late 1800s and at least 25% more than it has contained at any time in the last 160,000 years.

The most important data on the accumulation of CO_2 in the atmosphere are the records from Mauna Loa and the South Pole initiated by C. D. Keeling in 1958 (C. D. Keeling et al. 1989a,b) and continued most recently by the U.S. National Oceanic and Atmospheric Administration as part of a larger sampling program. They show the accelerating upward course of the CO_2 content of the atmosphere and the spectacular northern hemisphere annual oscillation in the concentration in response to the metabolism of plants. The oscillation is especially important because it shows clearly the potential of photosynthesis and respiration of terrestrial ecosystems for affecting the composition of the atmosphere on a time scale of weeks. It is this observation more than any other that calls attention to the possibility of short-term biotically controlled changes in the composition of the atmosphere. The changes may produce feedbacks, either positive or negative,[*] in the warming of the earth.

The Record from Glacial Time: Current Changes

The data from the Vostok Core, however, appear to set a limit on the patterns of global changes in temperature and CO_2 and CH_4 concentrations over millennia through several cycles of temperature changes (Figure 1.1) in a positive feedback system. The causes of the abrupt reversals in the record are not known. Possibilities include a recent suggestion that volcanic activity introduced energy-absorptive clouds high in the atmosphere that cooled the surface of the earth sufficiently to have started a positive feedback that led to the cooling and the glacial advance within the past 73,000 years (Rampino and Self 1993; Ramaswamy et al. 1993). A more widely held view is that variation in the output of the sun caused the reversals (Raynaud et al. 1993). Still more recent data from the Greenland ice cap have been interpreted to suggest that changes in the density of the surface water of the North Atlantic Ocean caused by the opening of the St. Lawrence drainage during the retreat of the glacial mass brought sudden, global changes in the venting of the oceans and equally sudden changes in the composition of the atmosphere and the temperature of the earth (Edwards et al. 1993). These analyses simply confirm the general view of the scientific community that the temperature and climates of the earth are vulnerable to substantial changes within a short time, perhaps as little as decades.

The records of temperature and CO_2 concentrations from more recent periods are consistent with the data from the Vostok Core. The correlation is conspicuous, for example, during the "Little Ice Age" of the earlier centuries of this millennium, before the surge of carbon emissions from the use of fossil fuels and the gross human distur-bance associated with the industrial revolution. Enting (Figure 18.3, page 319) has summarized data on atmospheric CO_2 from ice cores for the period 1300–1900, the period when temperatures in Europe declined and communication with Iceland and colonies on the coast of Greenland was lost because of ice in the northeastern Atlantic. These data also show a decline in the concentration of carbon dioxide coincident with the decline in temperature during that period. The observation confirms for this more recent period and for a time scale measured in centuries as opposed to millennia the positive feedback suggested by the data of the Vostok Core.

Contemporary data are also consistent with these observations (C. D. Keeling et al. 1989a; Oppenheimer et al. 1989; Kuo et al. 1990; Marston et al. 1991). In these studies (Figure 1.2), the gas concentration follows the temperature, but with a lag that is measured in weeks to months, not centuries to millennia.

---

[*]A *positive* feedback would accentuate a trend: the warming would speed the warming. A *negative* feedback would diminish a trend: a warming would trigger mechanisms that would tend to reduce the warming.

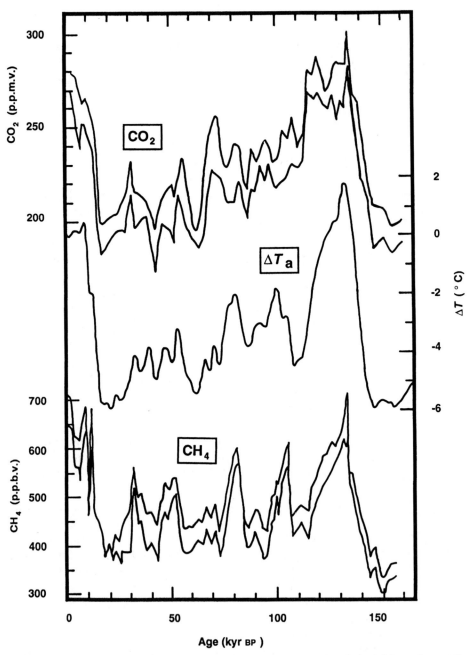

Figure 1.1. A summary of data from the Vostok Core, a sample of glacial ice taken at the Russian Station in the East Antarctic. The data provide a 160,000-year record of the range of carbon dioxide and methane concentrations in the atmosphere and the fluctuations in temperature over that period. For CO_2 and CH_4 the double traces take into account the various sources of uncertainty. Temperature is a smoothed curve. The current CO_2 concentration is about 360 ppm, approximately 80 ppm above the highest concentration over the past 160,000 years. (Adapted from Lorius et al. 1990. Used by permission.)

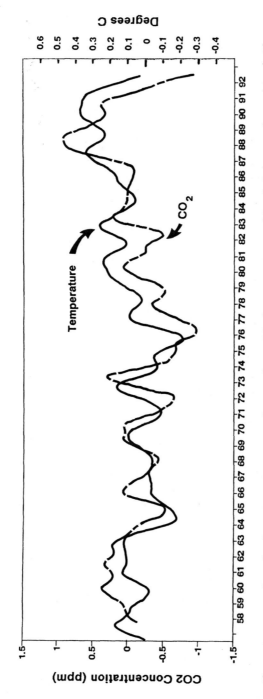

Figure 1.2. Anomalies of surface air temperature and atmospheric CO_2 for Mauna Loa and South Pole records combined. Observe the sharp decline in CO_2 emissions correlated with the decline in global temperature in 1991–1992. (From C. D. Keeling et al. 1989a. Modified with supplemental data for this publication by Keeling and Whorf.)

The dominant, global response is a positive feedback: a warming feeds more CO_2 and CH_4 into the atmosphere; a cooling, however it is caused, removes CO_2 and CH_4. Within those general limits there exists ample room for a series of subordinate positive and negative feedbacks. But their sum appears to be known: a positive feedback system of significant strength.

The Contemporary Carbon Budget: An Equation

Contemporary changes in concentrations of CO_2 and CH_4 in the atmosphere might be expected to be explained on the basis of flows of carbon among the major global pools of the atmosphere, the oceans, and the land and its release from combustion of fossil fuels. The relationship has been expressed as an equation (Woodwell et al. 1983) that is slightly modified here:

$$A = FF + NTE + NOE \qquad (1.1)$$

where A is the annual accumulation in the atmosphere in petagrams of carbon; FF is the release from combustion of fossil fuels; NTE is the net terrestrial effect, either positive or negative; and NOE is the net oceanic effect, also either positive or negative. In the present world there is a diffusion pressure gradient from the atmosphere into the oceans that drives a net flux from the atmosphere into the oceans.

Evaluating this equation has proven more difficult than most had expected. The uncertainty arises not only from the difficulties in making accurate global measurements of differences that may amount to only a few percent of large flows, but also from uncertainty as to what processes are active, especially in terrestrial ecosystems.

The annual increase in the amount of carbon in the atmosphere is measured with considerable accuracy. In 1989–1990 it was measured at Mauna Loa as 1.20 ppm (Keeling and Whorf in Marland and Boden 1991). There is also little question as to the use of fossil fuels globally and the amount of carbon release from that source. In 1989 it was estimated as 5.97 Pg carbon (Marland and Boden 1991). Yet the roles of sea and land are much in question.

The net oceanic effect (NOE) is universally considered at the present time to be a net absorption of carbon as CO_2 from the atmosphere. Data on the oceanic absorption of CO_2 from the atmosphere, accumulated over recent decades and summarized here by Mackenzie (Chapter 2), seem to confirm through several different routes that the net amount of carbon absorbed currently into the oceans is 2–2.5 Pg annually. The uncertainty surrounding the estimate varies greatly depending on the method used for its determination. Yet the estimate itself gains strength from the fact that it can be arrived at independently via several different methods. The most recent innovation is based on a very careful analysis of atmospheric concentrations of oxygen by R. F. Keeling and Shertz (1992). This estimate suggested an oceanic carbon uptake of 3.0 ± 2 Pg annually.

The contemporary relationships among these pools, especially the relationship between the atmosphere and the oceans, are the result of the rapid release of CO_2 from combustion of fossil fuels. Prior to this human influence, there was probably a small net flux of CO_2 from the oceans to the atmosphere as organic matter, washed from the land, decayed in the coastal oceans. If such a return flux had not existed, the continued erosion of carbon compounds from the land into the seas would have removed CO_2 from the atmosphere to the point of depletion (Garrels and Mackenzie 1972; see also Smith, Chapter 13).

Direct measurements of the role of forests are more difficult and the conclusions derived from them are thus less certain. The largest single human influence on forests has been thought to be progressive deforestation, a process that has followed the spread of humans over the earth and has gained momentum in recent years with the acceleration in growth of the human population (Woodwell and Pecan 1972; Woodwell and Houghton 1977; Woodwell 1983, 1990a,b; Woodwell et al. 1983; R. A. Houghton 1990a,b, 1991; R. A. Houghton and Skole 1990; R. A. Houghton et al. 1983, 1986). The process is currently estimated on the basis of "direct measurement" as releasing a net amount of 1.6 ± 0.5 Pg carbon annually (see Chapter 19). Stocks of carbon in forests are also being reduced by the continuing removal of the largest trees for timber, the process of "high grading," with its release of an unknown but significant additional quantity of carbon. Extensive areas of forest have been and continue to be affected by cumulative toxification, also not appraised to date but most conspicuous around smelters such as those in Sudbury, Ontario, Canada, at Mikel and Monchegorsk on the Kola Peninsula, and elsewhere in the former Soviet Union. Effects attributed to "acid rain" extend over such large areas that they may be considered general effects in forested regions of the northern hemisphere. In addition there are the effects of the warming itself on forests. These effects, too, are difficult to appraise, but they include the release of carbon from plants and soils through the stimulation of respiration, especially the respiration of decay. The tundra and the boreal forest may carry on one-third of the total respiration of terrestrial ecosystems. A warming that increased this respiration regionally by a net amount of 10% would produce a further release of carbon on the order of 3 Pg annually, a significant further contribution. This topic is discussed in detail later in this chapter.

Countering these trends is the possibility that the elevation of CO_2 in the atmosphere will increase photosynthetic rates (Strain and Cure 1985; see also Chapters 3 and 4) and, combined with the mobilization of nitrogen as nitrate and ammonia through accelerated decay, will stimulate carbon storage on land in the residual vegetation, as proposed recently in a "process model" by Melillo et al. (1993). Measurement of such a change in the short term of years would be difficult, even if it were large enough to result in a storage of carbon that was significant in the global budget.

Indeed both processes—accelerated carbon release via deforestation, toxification, and elevated respiration as well as carbon removal via increased photosynthesis—may be underway currently.

Evaluation of the Equation

The estimates available to evaluate the terms of equation 1.1 appear in Table 1.1.

Only at the extremes of the ranges of the data of Table 1.1, by pushing the oceanic estimates to a maximum and reducing terrestrial estimates, can we find a suite of data to "balance" this equation. In doing so, we raise further questions about what we know, how well we know it, and whether our analysis can be correct. The possibility of feedbacks, both positive and negative, lends further uncertainty.

These observations alone are enough to raise doubts about facile predictions of how a warming, triggered by the accumulation of heat-trapping gasses, will proceed. If

Table 1.1. Net Changes in Major Pools of Carbon in the Biosphere Approximated for the Year 1990

Evaluation of equation 1.1	Value (Gtons C/yr)	Reference
A (annual atmospheric increase)	(+) 3.8	Boden et al. (1991
FF (release from combustion of fossil fuels)	(+) 6.0	Marland and Boden (1991)
NTE (net terrestrial effect)		
NTE_d (deforestation)	(+) 1.6 ± 0.5	R. Houghton (1993)
NTE_{hg} (high grading)	(+) 0.1–0.5 (?)	—
NTE_t (toxic effects)	(+) 0.1–0.5 (?)	—
NTE_r (enhanced respiration)	(+) 1–3 (?)	Woodwell (1983)
NTE_p (enhanced photosynthesis)	(–) 1– (?)	—
NOE (net oceanic effect)	(–) 2–2.5	Chapter 2
		Sarmiento (1993)
	(–) 3 ± 2	R. F. Keeling and Shertz (1992)

increases in temperature cause an increase in the CO_2 and CH_4 concentrations in air, it is reasonable to assume that at some point the accumulation of heat-trapping gases might be expected to reach a level at which their influence becomes large in proportion to that of the other factors that affect the temperature of the earth. At that point, assuming that the inferences from the experience available now are correct and continue to apply, a further accumulation of CO_2 and CH_4 in the atmosphere might be expected. The circumstance has all the requirements, at least at first sight, of a self-perpetuating warming that would run as long as there were additional sources of carbon available to be mobilized by the warming. The CO_2 content of the atmosphere is now 25–30% higher than it has been at any time within the last 160,000 years. This change alone must be considered as raising the potential for unusual feedbacks.

What processes might be effective in producing an increase in CO_2 concentrations in the atmosphere as the earth warms and a decrease as the earth cools? The emphasis is, of course, on biotic processes, but there are physical and chemical processes as well that might be involved. The question lies at the core of this review.

Oceans and Land: Which Dominates?

There are further apparent constraints, which are even more puzzling. The most conspicuous influence on the CO_2 concentration in the atmosphere apart from the release of about 6 Pg carbon (Marland and Boden 1991) from burning fossil fuels is the annual oscillation in the CO_2 content of the atmosphere in the northern hemisphere, caused by the metabolism of forests (R. A. Houghton 1987a,b). The amplitude of the oscillation varies with latitude and place relative to the distribution of forests. At Mauna Loa the amplitude is about 5 ppm between the minimum, observed in the northern hemisphere fall, and the maximum, observed in late winter or spring. The important observation is that biotic processes have the potential for changing the composition of the atmosphere significantly within a short time. Any change in these processes can be expected to have a similarly large influence.

Much significance has been attributed to changes observed in the amplitude of the oscillation over the last two decades. R. A. Houghton (1987a,b) has shown that the oscillation is due to the integration in the atmosphere of two curves describing segments of the metabolism of terrestrial ecosystems. The first is the curve of gross photosynthesis; the second, that of total respiration of the ecosystem. These curves are out of phase and, when integrated by normal atmospheric mixing, produce the oscillation observed. A change in the amplitude could be caused by a change in gross photosynthesis, a change in total respiration, a change in the relative phasing of the two processes, or a combination of those changes (Woodwell 1983). One or more may be underway now as the earth warms.

The total pool of carbon held in plants and soils globally is large (probably three times the amount of carbon in the atmosphere), and any change in that large pool through, for instance, deforestation also affects the composition of the atmosphere (Woodwell 1983; see also Chapter 19). These changes affect the atmosphere in the short term of weeks to months.

There can be little question as to the capacity of the oceans, with their large reservoir of carbon and heat, for influencing, even determining, the composition of the atmosphere over periods of decades to millennia. But, apart from massive physical changes in the circulation of the oceans that change the distribution patterns of energy and water masses, the influence of the oceans on the atmospheric composition would seem to be in response to shorter-term changes in the atmosphere that are heavily influenced by terrestrial ecosystems. The flux between the atmosphere and the oceans hinges on the diffusion pressure gradient between the atmosphere and the surface water. The rate of exchange of the surface water with the abyssal water has a large influence on the characteristics of the surface water at any moment. That rate is determined by physical characteristics, especially density. Shorter-term changes in terrestrial carbon pools that influenced the magnitude and direction of the diffusion gradient appear to have resulted in massive long-term flows of carbon between the atmosphere and the oceans throughout glacial time, if the analysis by Adams et al. (1990) is correct.

A PLACE TO START: THE HAZARDS OF CLIMATIC CHANGE

Ecotypes Affect the Responses of Ecosystems

There is, unfortunately, little general agreement as to how the biotic components of the biosphere are assembled to constitute the earth's life-giving green mantle. On the one hand, we have the obvious complexity of the plant and animal communities of the tropics, where mutual dependencies among species are the rule and we are not surprised to discover that the complexities of mutualism are virtually infinite and unrelenting: the removal of one species may doom several others. A forest on ancient soils with little capacity to renew or hold nutrient elements that supports more than 400 species of trees per hectare, pollinated by such diverse mechanisms as insects, birds, and bats, is clearly a complicated system.

On the other hand, we have the flora and fauna of the Arctic, surviving on new soils developing on glacially churned mineral matter and comprising few species in comparison with the tropics—species that have been sorted and resorted and sorted again by countless changes in climate and circumstance to the point where mutualisms are few and virtually any combination of a small number of species will have existed in the past and can exist now. There is substance to support virtually any assumption about the importance of species or the structure or behavior of plant communities.

However loose the concept of community may be, there are clear, definitive, genetic principles that apply to natural communities, including the forests of the higher latitudes, and offer guidance in making inferences about the effects of a rapid warming of climate. One simple principle was elaborated in a series of classical experiments in transplanting species among different parts of the natural range of the species published by Clausen et al. (1940). That work and subsequent studies, including innumerable studies of the provenance of trees, have confirmed the fact that species are continuously variable throughout their ranges and that the variation is, not surprisingly, genetically fixed and related to the local environment. In May of 1992, for instance, I visited a transplant garden near Krasnoyarsk in central Siberia where there was a collection of larch from widely separated parts of Asia. The plantation was rich in lessons in genetics and ecology. Trees from regions that had a shorter growing season than occurs in Krasnoyarsk were active only for that shorter period, although the opportunity existed for a substantially longer period of growth (and, presumably, more growth) than in their native habitats. Trees brought as seedlings from Norilsk in the far north were, after 30 years, no more than 3 ft high. Others from other parts of the range of larch had reached 40 ft or more in height, although conspicuous differences existed among them. Such genetic differences are the product of selection for a particular set of morphological and physiological traits appropriate for a particular site. They can be set over very few generations, but they exist now for the current climatic regime and they exist throughout the range of a species. Although every species has not been tested for such variability, there is every reason to believe that it occurs in every species and that the speed with which changes in the genetic constitution of the species occur is related to the period required for reproduction, the generation time. The generation time for trees and most other perennials is long in comparison with periods of a few weeks for many annuals, some fungi, insects, and other smaller-bodied species.

The important point—long established in ecology but forgotten in the rash of contemporary analyses of the effect of warming the earth suddenly—is that a species, although it may have a wide range, is not genetically uniform throughout that range. The populations of any place have been selected for the circumstances of that place. Changing climate suddenly—for instance by changing the temperature and moisture regimes—moves the environment away from the set of circumstances that selected that particular subpopulation of the species, and the population becomes progressively maladapted to the new environment. The assumption that a warmer or wetter climate or a change in the nutrient regime is universally advantageous to trees or plants is not justified. The assumption that a species is sufficiently flexible to adapt to a new, especially a continuously changing, climatic regime is also unjustified. On the contrary, experience shows that such changes lead to impoverishment of the community: the replacement of longer-lived, later-successional species with shorter-lived, earlier-successional species, or of a forest by a shrub or grass community. The transition occurs systematically and follows patterns that have been defined for various types of vegetation around the world but are especially well defined for forests. The topic has been reviewed extensively by scholars experienced in work on various natural communities around the world (Woodwell 1990a,b).

Such considerations emphasize a conclusion that seems obvious, but also seems to be commonly ignored: the speed and continuity of the warming of the earth will be of greatest importance in determining the effects of the process. The severity of effects will be measured by the speed of change in temperature relative to the generation times of the dominant plants. A 1°C change in temperature in the middle latitudes is the equivalent of moving latitudinally 100–150 km. A 1°C change in temperature over one to several decades would constitute a rapid change for forests. Such rates of change are being predicted (J. T. Houghton et al. 1990, 1992). The rate of warming over recent decades has been about 0.2°C per decade globally. If the earth responds as expected, the warming in middle and higher latitudes will be greater than the average for the earth as a whole by a factor of two or more; in the lower latitudes the warming will be less than the mean for the earth as a whole. The greater warming will occur in winter. If the rates of change experienced so far are an indication of what is to be expected over the coming years, we might expect a warming on the order of 0.4–0.5°C per decade in the boreal forest zone and in the northern tundra. Such a warming would place a significant burden of increased respiration on existing forests; it would also start a series of successional changes in those forests as the accommodation to a new climatic regime began. The difficulty lies in the continuity of the warming: continuous warming is chronic disturbance, the classical cause of a shift from later-successional species to earlier-successional species, from larger, long-lived species to small-bodied, short-lived species, from forest to woodland to grassland—progressive biotic impoverishment.

The pattern of the changes expected in forests is well known. In the higher latitudes, where the climatic changes are expected to be most rapid, the changes will not be subtle but will be marked by increased morbidity and mortality of trees and other plants from diseases and pests, complicated by an increased frequency of fires. The latter will follow the drying of forests in the continental centers and more widely as temperatures rise. The impoverishment of forests leads to their transformation into shrubland and grassland, with the release of carbon stocks into the atmosphere through burning or accelerated decay.

The genetic adaptations of species extend to nutrient requirements as well as climate. The relationships are also subtle in one context and gross in another. There is little question that the presence or absence of certain mineral elements affects the species composition of communities on land and in the water. Ample experience with eutrophication has defined the patterns of impoverishment from that cause in both aquatic and terrestrial ecosystems. Far more subtle influences accrue from lesser shifts in nutrient ratios and moisture regimes. There is little question as to the potential of additions of nutrients, especially nitrogen and phosphorus, for causing the eutrophication of both aquatic and terrestrial ecosystems. The early stages may increase carbon storage on land and in the water, but the long-term effect is not necessarily an increase in storage.

Other Major Factors That Affect the Carbon Balance

The answer to the question of whether forests globally are accumulating or losing carbon depends on several factors. Two of the most important are the changes in the area of forests year by year and the relationship between net primary production and the respiration of the heterotrophs, especially the organisms of decay. A third factor is the partial harvesting of the largest trees, the process of high grading that goes on continuously in most forests around the world. This latter process is most difficult to appraise because the "forest" remains, although it is impoverished.

Changes in Area and Standing Stock of Forests

There is not much question whether the area of forests globally is increasing or decreasing. R. A. Houghton has shown that there is a continuous flow of land from forest into agriculture, thence to impoverishment (see Chapter 19).

Land that remains in forest supports over much of its area progressively smaller stocks of carbon as the process of high grading proceeds. The harvest takes two forms: clear-cutting of the best stands with the land left to recover as forest, and systematic removal of the largest and most valuable trees from areas left in forest. Once a forest has become accessible to harvest, it will be harvested again for timber when doing so appears financially advantageous. The effect is a progressive reduction in the mean standing stock of carbon in the residual stands (Woodwell et al. 1983). The reduction in standing stock results in a net release of carbon into the atmosphere. The magnitude globally is accounted for by R. A. Houghton (see Chapter 19), although the effect of the systematic partial harvest of timber through high grading is not easily estimated.

Metabolism of Forests

The terrestrial vegetation globally probably contains about 500 Pg of carbon; soils, including peats, contain approximately 1500 Pg. This total pool, widely thought to approximate 2000 Pg, is maintained by a flux of carbon through net primary production on the order of 100 Pg annually. It is approximately balanced by total respiration of about 100 Pg. Obviously, any change in this relationship on the order of 1% has the potential for altering the composition of the atmosphere by a significant amount. The relationships among the various segments of metabolism of ecosystems were defined initially by Woodwell and Whittaker (1968) as follows:

$$NEP = NPP - Rs(h)$$
$$= GPP - Rs(a) - Rs(h) \qquad (1.2)$$

NEP is net ecosystem production, the net change in carbon or energy storage over a period, usually a year, for a unit of area of the surface of the earth, an ecosystem. It can be positive or negative depending on the relationship between NPP and Rs(h) + Rs(a).

NPP is net primary production of a plant or a plant community. It is the net of gross primary production, GPP, less the total respiration of the plant or the plant community, Rs(h) + Rs(a). It is never negative.

Rs(h) is the respiration of the heterotrophs. It is the total respiration of the consumers of organic matter in an ecosystem, including insects, other grazers, and the organisms of decay.

Rs(a) is the respiration of the autotrophs, the plant or the plant community within an ecosystem.

GPP is gross primary production, the total amount of carbon (or organic matter or energy) fixed by the ecosystem, including the amount fixed but respired by the green plants.

NPP is difficult to measure directly in terrestrial ecosystems, not only because it requires a direct measurement of gross photosynthesis, but also because it requires measurement of the respiration of the plant or the plant community, including roots. Accurate measurement of root respiration is usually confounded by the decay of organic matter in soils. That decay is part of heterotrophic respiration. A partial correction is possible by subtracting the carbon influx in litter. The remaining respiration in soils is the respiration of roots plus the decay of any stored organic matter, whatever the source.

The question of how the relationship between NPP and Rs(h) changes as the earth warms in response to the accumulation of heat-trapping gases is difficult to address. On the one hand is the possibility that the increase in CO_2 in the atmosphere is causing a stimulation in rates of photosynthesis and therefore of GPP globally (Oechel and Strain 1985; Strain and Cure 1985; Oechel and Riechers 1986, 1987; Drake 1989; Oechel et al. 1989, 1991; Drake et al. 1990), a negative feedback. On the other hand is the threat that the increase in temperature will increase rates of respiration and speed the release of carbon into the atmosphere from plants and soils (Woodwell 1983, 1989; J. T. Houghton et al. 1992), a positive feedback. Both processes will, of course, work together, but the question of which will dominate under the conditions existing at any moment remains unresolved.

Negative Feedbacks

The question of negative feedbacks has been examined in detail previously without clear resolution. The central issue is whether the combined effects of warming the earth by adding heat-trapping gases to the atmosphere will, in conjunction with other changes in the environment such as the mobilization of fixed nitrogen (Melillo et al. 1993), increase NEP. Direct measurements of NEP in forests are few, and most methods of measurement are thought to have insufficient resolution to allow a definitive answer in any but a very few circumstances. A global increase in NEP of 3–5 Pg would be a highly significant factor in the global carbon cycle, but such an increase would be virtually unmeasurable in the short term by any technique currently available. It would involve an increase in the total carbon held on land on the order of 0.25%, or 5% or less of terrestrial NPP. Over the longer term of decades such a change would result in a significant, and presumably measurable, accumulation.

An index of the growth of trees is offered by measurements of diameter growth, discussed in detail in Chapter 5. Previous studies have not been definitive. The diameter growth of trees is influenced by many factors, including the history of the stand, and the effects of enhanced carbon dioxide concentrations and warmer temperatures may not appear as dominant influences until they are much larger than at present. Unfortunately, these effects may also continue to be masked by other factors, most of them deleterious, whose effects are also accumulating rapidly.

The changes in climate will, of course, shift the geographic distribution of climates favorable for forests and, presumably, open regions such as tundra to afforestation. The speed of such a transition, especially in regions with soils and climates only marginally suitable for forest, will be slow, with progress in the development of new forest requiring decades to centuries.

The best possibilities of accurate direct measurements of such transitions seem to lie at present in detailed long-term measurements of the distribution of species and the magnitude of major carbon stores, and in systematic improvements in carbon flux measurements.

Positive Feedbacks

Virtually any chronic disturbance of a forest can be expected to alter the relationship between gross photosynthesis, GPP, and total respiration, Rs(a) + Rs(h). The alteration is almost always a shift in favor of total respiration. If the disturbance is a systematic, open-ended warming, the effect on respiration can be expected to be large in proportion to the effect on GPP because of the higher sensitivity of respiration to temperature (Woodwell 1983; Raich and Schlesinger 1992).

It is this change, a differential stimulation of respiration by a rapid warming of the earth, that offers the greatest possibility of a positive feedback. The effects fall into three classes: effects on the distribution of forests; effects on the morbidity of forest plants, including vulnerability to disease and to fire; and direct effects on the metabolism of forests and other ecosystems.

Effects on the Distribution of Forests

A rapid warming can be expected to reduce forests rapidly at their warmer and drier limits of distribution. This transition is the well-known forest-steppe or forest-prairie transition, where the forest is classically vulnerable to fire and drought (Clements 1928; Curtis 1959). The change in climate in these regions will reduce the area of forest rapidly—far more rapidly than forests can be established elsewhere. The importance of this transition will hinge heavily on the speed of the warming.

Effects on the Morbidity of Forest Plants

The genetic structure of forests, discussed earlier, is well known and definitive. There is every reason to assume that a change in climate that is as general, directional, and definitive as the changes anticipated for the coming years will quickly exceed the ecological adaptability of individuals of many if not most species and render them increasingly vulnerable to fire and disease, including the effects of insects and other parasites. Highly suggestive evidence is now available that the frequency of forest fires in the boreal forest rose abruptly during the rapid warming of the 1980s and early 1990s (see Chapter 6). At the same time the warming provides conditions suitable for the spread of insect pests of both trees and agricultural crops into regions previously protected by cold winters. The recent lethal advances of the spruce budworm (*Choristoneura*) into the spruce forests of central and southern Alaska is attributed to the warming. The experience simply confirms for trees the projections of the spread of disease organisms for humans set forth previously and summarized so expertly by Dobson and Carper (1992).

Direct Effects on the Metabolism of Forests and Other Ecosystems

The differential warming of the earth that is expected to result in a warming in the high latitudes that is two times or more the average for the earth as a whole can also be expected to bring differential effects on natural ecosystems. The greater the warming the greater the differential effect on respiration as opposed to GPP and the more rapid the loss of carbon from forests and soils. A 1°C warming commonly increases rates of respiration by as much as 10% or more. If the boreal forest and tundra zones support one-quarter of the total respiration on land or 25 Pg carbon per year, such a warming might increase the annual release of carbon by 2.5 Pg, a significant further contribution to the global release of CO_2 into the atmosphere.

The issue is still more complicated. The period during which deciduous plants are in leaf is often, if not always, determined genetically through photoperiod, not simply by temperature. A lengthening of the frost-free season does not assure that primary production will increase; nor does it assure that additional carbon storage (NEP) will accrue to terrestrial ecosystems. What is assured is that rates of respiration will respond to the longer periods of warm weather. To the extent that total respiration rises above net primary production, forests will lose carbon to the atmosphere from plants and soil.

WHAT COULD BE DONE NOW TO LIMIT FURTHER WARMING?

The rate of accumulation of carbon in the atmosphere during the late 1980s was about 3.8 Pg/year. I have pointed out before (Woodwell 1989) that—because the current rate of diffusion of CO_2 into the oceans depends on the difference in partial pressure of CO_2 between the atmosphere and the water of the oceanic surface, which is exchanged only very slowly with that in the abyssal waters, stabilizing the composition of the atmosphere at the moment would require a reduction in releases by about 3.8 Pg/year, an amount equivalent to about 60% of the current release through fossil fuel use. The reduction might be divided between a reduction in fossil fuel use and a reduction in deforestation, but there is no possibility of stabilizing the composition of the atmosphere with respect to CO_2 without both a global reduction in use of fossil fuels and an effective plan for managing forests. The initial stabilization will be effective as long as the transfer into the oceans continues at the same rate and as long as the earth does not warm further. As the difference in partial pressure of CO_2 between the atmosphere and the oceans declines, maintaining a stable atmospheric composition will require further reductions in releases from human activities.

The longer this transition in human activities is delayed, the more difficult and disruptive it will be.

CONCLUSIONS

There will be no absolute proof that adding heat-trapping gases to the atmosphere will warm the earth, or that the effects of the warming will be as outlined here, until we have experienced the warming. Yet by that time, a retreat from the warming may well be impossible. There are, however, several points that are clear at the moment and offer a firm basis for proceeding, both with research and with steps toward stabilizing the temperature of the earth.

1. The climates globally are moving from a period of slow change into a period of accelerating change as heat-trapping gases accumulate in the atmsophere. Assumptions based on a new period of "stability" are misleading.
2. The changes are open-ended; there is no obvious mechanism that will slow or stop the warming apart from human action.
3. Positive feedbacks involving temperature and carbon dioxide and methane levels have dominated in such climatic changes over the past 160,000 years on a millennial scale and on the scale of centuries, as suggested by data from the Little

Ice Age. They continue in the contemporary world, as shown by recent correlations between contemporary temperature and atmospheric CO_2 concentrations.

4. The magnitude of these changes has been an approximately 1°C increase in temperature associated with a 7–10 ppm increase in CO_2 levels in the atmosphere, as estimated from the data of the Vostok Core.

5. In contemporary experience the correlation suggests a lag of days to weeks between a change in temperature and the resultant change in the CO_2 concentration.

6. These two observations are so general and overriding as to establish a context within which we can reasonably expect the world to continue to operate. Within that envelope of effects, there is a high probability that various mechanisms will induce positive and negative feedbacks, but the overall effect globally appears to be a positive feedback system: warming produces more CO_2 in the atmosphere, which in turn favors a further warming.

7. Scientists continue to have difficulty understanding how carbon flows are balanced globally. The difficulty arises from an inability to rationalize the evidence that human effects on forests, including the effects of warming the earth, seem to be adding carbon to the atmosphere at rates that exceed the transfer to the oceans and the current accumulation in the atmosphere.

8. The most satisfactory explanation at the moment appears to be that the net release from the land is less than the direct estimates suggest owing to unmeasured (and possibly unmeasurable) increases in NEP that reduce the net loss of carbon from forests.

9. There is reason, however, to question whether this relationship will continue as the warming accelerates. A systematic and continuous warming that follows the patterns projected can be expected to affect northern forests and tundra in particular by reducing their distribution at their warmer and drier margins; by increasing the morbidity of trees and other plants throughout their ranges in current ecosystems from insect damage, disease, and fire; and by increasing the rates of respiration of plants and soils throughout the forest and tundra zones.

10. Evidence suggests that this transition is underway now.

11. The effect can be expected to be an increase in the emissions of CO_2 and CH_4 of at least 1 Pg of carbon annually, and perhaps more.

12. Although this surge will be balanced in part by a parallel surge in the storage of carbon in the residual forests, there is no reason to believe that the net effect will be anything but a positive feedback: a net further accumulation of carbon in the atmosphere.

13. The probability that the warming of the earth will cause a positive feedback through acceleration of the release of CO_2 and CH_4 from forests and tundra of the northern hemisphere is high. The assumption that present rates of accumulation of CO_2 and CH_4 in the atmosphere will continue as the warming accelerates would be naive in the extreme.

14. Stabilization of the composition of the atmosphere will require both an immediate reduction in CO_2 emissions from human activities equivalent to 60% of the current use of fossil fuels and further reductions over time.

15. Delay in implementing these changes in releases increases the probability of accelerated releases of CO_2 and CH_4 from biotic sources and raises the probabil-

ity of increased emissions from biotic sources beyond the level at which reductions in human emissions will be able to stabilize the atmosphere.

REFERENCES

Abrahamson, D. E., ed. 1989. *The Challenge of Global Warming.* Island Press, Washington, D. C.

Adams, J. M., H. Faure, L. Faure-Denard, J. M. McGlade, and F. I. Woodward. 1990. Increases in terrestrial carbon storage from the last glacial maximum to the present. *Nature* 348:711–714.

Barnola, J. M., D. Raynaud, Y. S. Korotkevich, and C. Lorius. 1987. Vostok ice core provides 160,000-year record of atmospheric CO_2. *Nature* 329:408–414.

Clausen, J., D. D. Keck, and W. H. Heisey. 1940. *Experimental Studies on the Nature of Species.* Carnegie Institution of Washington Publication 520. Carnegie Institution of Washington, Washington, D.C.

Clements, F. E. 1928. *Plant Succession and Indicators.* H. W. Wilson, New York.

Curtis, J. T. 1959. *The Vegetation of Wisconsin.* University of Wisconsin Press, Madison.

Dobson, A., and R. Carper. 1992. Global warming and potential change in host-parasite and disease-vector relationships. In *Global Warming and Biological Diversity,* ed. R. L. Peters and T. E. Lovejoy, pp. 201–217. Yale University Press, New Haven, Conn.

Drake, B. 1989. Elevated atmospheric CO_2 concentration increases carbon sequestering in coastal wetlands. Carbon Dioxide Information Analysis Center, Environmental Sciences Division, Oak Ridge National Laboratory, Oak Ridge, Tenn.

Edwards, R. L., J. W. Beck, G. A. Burr, D. J. Donahue, J. M. A. Chappell, A. L. Bloom, E. R. M. Druffel, and F. W. Taylor. 1993. A large drop in atmospheric C^{14}/C^{12} and reduced melting in the Younger Dryas, documented with Th^{230} ages of corals. *Science* 260:962–968.

Garrels, R. M., and F. T. Mackenzie. 1972. A quantitative model of the sedimentary rock cycle. *Mar. Chem.* 1:22–41.

Grulke, N., G. H. Reichers, and W. C. Oechel. 1990. Carbon balance in tussock tundra under ambient and elevated atmospheric CO_2. *Oecologia.* 83:485–494.

Houghton, J. T., G. J. Jenkins, and J. J. Ephraums. 1990. *Climate Change. The IPCC Scientific Assessment.* Cambridge University Press, Cambridge.

Houghton, J. T., B. A. Callander, and S. K. Varney. 1992. *Climate Change 1992. The Supplementary Report to the IPCC Scientific Assessment.* Cambridge University Press, Cambridge.

Houghton, R. A. 1987a. Terrestrial metabolism and atmospheric CO_2 concentrations. *BioScience* 37:672–678.

Houghton, R. A. 1987b. Biotic changes consistent with the increased seasonal amplitude of atmospheric CO_2 concentrations. *J. Geophys. Res.* 92:4223–4230.

Houghton, R. A. 1990a. The long-term carbon flux between terrestrial ecosystems and the atmosphere as a result of changes in land use. Carbon Dioxide Information Analysis Center, Environmental Sciences Division, Oak Ridge National Laboratory, Oak Ridge, Tenn.

Houghton, R. A. 1990b. Emissions of greenhouse gases. In *Deforestation Rates in Tropical Forests and Their Climatic Implications,* ed. N. Myers, pp. 53–62. Friends of the Earth, London.

Houghton, R. A. 1991. Tropical deforestation and atmospheric carbon dioxide. *Climatic Change* 19:99–118.

Houghton, R. A., and D. L. Skole. 1990. Carbon. In *The Earth Transformed by Human Action,* ed. B. L. Turner, pp. 393–408. Cambridge University Press, New York.

Houghton, R. A., J. E. Hobbie, J. M. Melillo, B. Moore, B. J. Peterson, G. R. Shaver, and G. M. Woodwell. 1983. Changes in the carbon content of terrestrial biota and soils between 1860 and 1980: A net release of CO_2 to the atmosphere. *Ecol. Monogr.* 53:235–262.

Houghton, R. A., R. D. Boone, J. R. Frucci, J. E. Hobbie, J. M. Melillo, C. A. Palm, B. J. Peterson, G. R. Shaver, G. M. Woodwell, B. Moore, D. L. Skole, and N. Myers. 1986. The flux of carbon from terrestrial ecosystems to the atmosphere in 1980 due to changes in land use: Geographic distribution of the global flux. *Tellus* 39B:122–139.

Keeling, C. D., R. B. Bacastow, A. F. Carter, S. C. Piper, T. P. Whorf, M. Heimann, W. G. Mook, and H. Roeloffzen. 1989a. A three-dimensional model of atmospheric CO_2 transport based on observed winds. 1. Analysis of observational data. Geophysical Monograph 55. American Geophysical Union, Washington, D. C., pp. 165–236.

Keeling, C. D., S. C. Piper, and M. Heimann. 1989b. A three-dimensional model of atmospheric CO_2 transport based on observed winds. 4. Mean annual gradients and interannual variations. Geophysical Monograph 55. American Geophysical Union, Washington, D.C., pp. 305–363.

Keeling, R. F., and S. R. Shertz. 1992. Seasonal and interannual variations in atmospheric oxygen and implications for the global carbon cycle. *Nature* 358:723–727.

Kuo, C., C. Lindberg, and D. J. Thompson. 1990. Coherence established between atmospheric carbon dioxide and global temperature. *Nature* 343:709–713.

Leggett, J. 1990. *Global Warming: The Greenpeace Report*. Oxford University Press, Oxford.

Lorius, C., J. Jouzel, D. Raynaud, J. Hansen, and H. Le Treut. 1990. The ice-core record: Climate sensitivity and future greenhouse warming. *Nature* 347:139–145.

MacDonald, G. J. 1990. Role of methane clathrates in past and future climates. *Climatic Change* 16:247–281.

Marland, G., and T. A. Boden. 1991. CO_2 emissions—Global. In *Trends '91. A Compendium of Data on Global Change*, ed. T. A. Boden, R. J. Sepanski, and F. W. Stoss, pp. 386–389. ORNL/CDIAC-46. Oak Ridge National Laboratory, Oak Ridge, Tenn.

Marston, J. B., M. Oppenheimer, R. M. Fujita, and S. R. Gaffin. 1991. Carbon dioxide and temperature. *Nature* 349:573–574.

Melillo, J. M., A. D. McGuire, D. W. Kicklighter, B. Moore III, C. J. Vorosmarty, and A. L. Schloss. 1993. Global climate change and terrestrial net primary production. *Nature* 363:234–240. [Commentary by C. Prentice. 1993. *Nature* 363:209–210.]

Mooney, H. A., B. G. Drake, R. J. Luxmoore, W. C. Oechel, and L. F. Pitelka. 1990. Predicting ecosystem responses to elevated CO_2 concentrations. *Bioscience* 41:96–104.

National Academy of Sciences. 1991. *Policy Implications of Greenhouse Warming*. National Academy Press, Washington, D. C.

OTA. 1991. *Changing by Degrees: Steps to Reduce Greenhouse Gases*. Congress of the United States, Office of Technology Assessment.

Oechel, W. C., and G. H. Riechers. 1986. Impacts of increasing CO_2 on natural vegetation, particularly the tundra. In *Proceedings of the Climate-Vegetation Workshop*, ed. National Aeronautics and Space Administration, Goddard Space Flight Center, pp. 36–42. Greenbelt, Md.

Oechel W. C., and G. H. Riechers. 1987. *Response of a Tundra Ecosystem to Elevated Atmospheric Carbon Dioxide*. Greenbook No. 37. U.S. Department of Energy, Washington, D.C.

Oechel, W. C., and B. R. Strain. 1985. Native species responses to increased carbon dioxide concentration. In *Direct Effects of Increasing Carbon Dioxide on Vegetation*, ed. B. R. Strain and J. D. Cure, pp. 117–154. DOE/Er-238. U.S. Department of Energy, Washington, D.C.

Oechel, W. C., M. Jenkins, S. J. Hastings, G. Vourlitis, N. Grulke, and G. H. Reichers. 1991. Effects of recent and predicted global change on Arctic ecosystems. *Abstr. Bull. Ecol. Soc. Am.* 72:209.

Oppenheimer, M., J. B. Marston, and R. M. Fujita. 1989. Equatorial temperatures and fluctuations in atmospheric carbon dioxide. Unpublished preprint.

Peters, R. L., and T. E. Lovejoy, eds. 1992. *Global Warming and Biological Diversity*. Yale University Press, New Haven, Conn.

Raich, J. W., and W. H. Schlesinger. 1992. The global carbon dioxide flux in soil respiration and its relationship to vegetation and climate. *Tellus* 2:81–99.

Ramaswamy, V. 1992. Explosive start to last ice age. *Nature* 359:14.

Rampino, M. R., and S. Self. 1992. Volcanic winter and accelerated glaciation following the Toba super-eruption. *Nature* 359:50–52.

Raynaud, D., J. Jouzel, J. M. Barnola, J. Chappellaz, R. J. Delmas, and C. Lorius. 1993. The ice record of greenhouse gases. *Science* 259:926–934.

Sarmiento, J. L. 1993. Ocean carbon cycle. *Chemical and Engineering News,* May 31, pp. 30–40.

Strain, B. R., and J. D. Cure, eds. 1985. *Direct Effects of Increasing Carbon Dioxide on Vegetation.* U.S. Department of Energy, National Technical Information Service, Springfield, Va.

Sundquist, E. T. 1993. The global carbon dioxide budget. *Science* 259:934–941.

WMO/UNEP. 1988. *Developing Policies for Responding to Climatic Change: A summary of discussions and recommendations of workshops held in Villach and Bellagio.* WMO/TD No. 225. World Meteorological Organization–United Nations Environmental Programme, Geneva and Nairobi.

Woodwell, G. M. 1983. Biotic effects on the concentration of atmospheric carbon dioxide; A review and projection. In *Changing Climate.* pp. 216–241. NAS Press, Washington, D.C.

Woodwell, G. M. 1989. The warming of the industrialized middle latitudes 1985–2050: Causes and consequences. *Climatic Change* 15:31–50.

Woodwell, G. M., ed. 1990a. *The Earth in Transition: Patterns and Processes of Biotic Impoverishment.* Cambridge University Press, New York.

Woodwell, G. M. 1990b. The earth under stress: A transition to climatic instability raises questions about biotic impoverishment. In *The Earth in Transition: Patterns and Processes of Biotic Impoverishment,* ed. G. M. Woodwell, pp. 3–7. Cambridge University Press, New York.

Woodwell, G. M., and R. A. Houghton. 1977. Biotic influences on the world carbon budget. In *Global Chemical Cycles and Their Alterations by Man,* ed. W. Stumm, pp. 61–72. Dahlem Konferenzen, Berlin.

Woodwell, G. M., and E. V. Pecan, eds. 1972. *Carbon and the Biosphere.* Brookhaven National Laboratory, Upton, N.Y.

Woodwell, G. M., and R. H. Whittaker. 1968. Primary production in terrestrial ecosystems. *Am. Zool.* 8:19–30.

Woodwell, G. M., J. E. Hobbie, R. A. Houghton, J. M. Melillo, B. Moore, B. J. Peterson, and G. R. Shaver. 1983. Global deforestation: Contribution to atmospheric carbon dioxide. *Science* 222:1081–1086.

2

Global Climatic Change: Climatically Important Biogenic Gases and Feedbacks

FRED T. MACKENZIE

In the next century the earth will almost certainly experience climatic change. In the normal course of events, the planet would enter another glacial maximum about 23,000 years from now (Figure 2.1). However, the flywheels of population growth and fossil fuel burning are turning rapidly and will be difficult to slow within a century. The global population is growing at a rate of 1.6% per year, representing a doubling time of 44 years; the rate implies a population of at least nine billion people by 2040 (Figure 2.2A). All these people will require energy to sustain themselves; most scenarios of future global energy use project a continuous heavy reliance on fossil fuel into the 21st century (Figure 2.2B). Such reliance will result in continuous emissions of the greenhouse trace gases CO_2, CH_4 (indirectly stratospheric H_2O, a greenhouse gas), and N_2O, along with trace metals, nonmethane hydrocarbons (NMHCs), SO_x, and NO_x. These latter three substances react with other chemical components of the climatic system, particularly the OH radical, and have various effects on the radiative properties of the atmosphere. SO_x and NO_x are the principal constituents in acid deposition; NO_x and NMHCs are involved in the formation of tropospheric ozone, another greenhouse gas. All of these greenhouse gases are biogenic in origin (Moore and Schimel 1992). The continued accumulation of these gases in the atmosphere could lead to a "super interglaciation" (Figure 2.1).

Carbon, nitrogen, phosphorus, and sulfur, in addition to oxygen and hydrogen, are the principal elemental components of living systems, and the biogeochemical cycles of these elements are intimately coupled through biological productivity, according to two reactions:

Marine plankton:
$$106CO_2 + 16HNO_3 + 2H_2SO_4 + H_3PO_4 + 120H_2O \rightarrow C_{106}H_{263}O_{110}N_{16}S_2P + 141O_2$$

Terrestrial vegetation:
$$882CO_2 + 9HNO_3 + H_2SO_4 + H_3PO_4 + 890H_2O \rightarrow C_{882}H_{1794}O_{886}N_9SP + 901.5O_2$$

I consider subsequently the problem of biotic feedbacks in the global climatic system through some examples of how the coupled C-N-P-S system might react to the warming

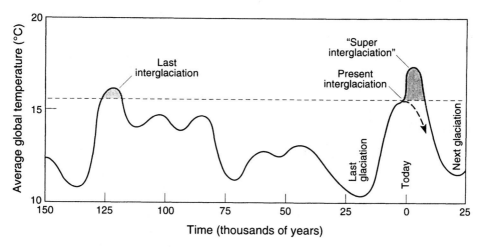

Figure 2.1. The earth's climate during the last 150,000 years and an interpretation of its future. (After Imbrie and Imbrie 1986.)

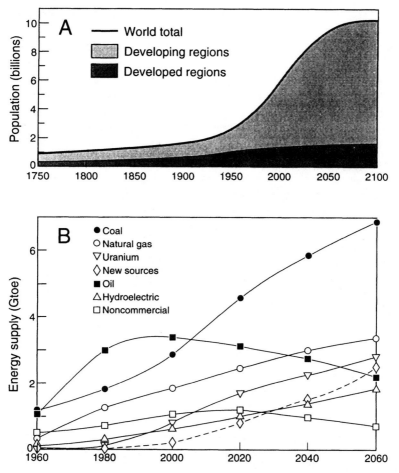

Figure 2.2. (A) The growth of the human population in the developed and developing regions of the world from 1750 to 2100 A.D. (World Resources Institute 1988). (B) Evolution of world energy supplies. (After Frisch 1986.)

envisioned for the next century. The analysis is qualitative, because the biogeochemical framework of the natural system is poorly known.

CONSTRAINTS ON ARGUMENTS

Biotic systems take energy from their surrounding environment and as a result produce organic matter, entropy on the planet external to the organic matter, and waste. The waste may act as a pollutant. The biogenic gases of carbon, nitrogen, and sulfur are a material consequence of this entropy production, and their fluxes maintain the planetary atmosphere in a state of disequilibrium. These gases have their natural sources at the earth's surface or through reactions involving precursor trace gases in the atmosphere. Processes involving the cycling of biogenic gases and other components of the coupled C-N-P-S system, although in some environments operating close to a thermodynamic threshold, are principally controlled by kinetic factors, that is, they are rate-controlled processes. Thus, the problem of biotic response and feedback is primarily one of how the kinetics of biological processes will change as the climate warms.

Because of the reductionist nature of science in the last half century, most global systems have been little studied or have only been studied in a piecemeal fashion. Only recently has some attention been paid to the coupled earth-surface system of atmosphere–hydrosphere–biosphere–shallow lithosphere. Basic information on global reservoir sizes (stocks) and fluxes (e.g., productivity) is lacking or of a qualitative or semiquantitative nature. Examples of this lack of data are estimates of tropical forest biomass, which vary by a factor of two to three for Amazonia alone (Brown et al. 1992), and of oceanic new production, which also varies by a factor of two to three (Knauer 1993). Thus, any attempt to evaluate the future of the global system, especially its biotic response to global warming, will be constrained by the poor data base available and is likely to be only qualitative, or at best semiquantitative.

As will be shown, fossil fuel combustion and forest and biomass burning are the principal human-induced fluxes of most biogenic gases to the earth's atmosphere. Fossil fuel burning and land use practices are also responsible for many of the global problems of "environmental change" (Table 2.1). The combustion of fossil fuels produces a significant fraction of the total emissions of SO_2 (80%), CO (50%), NO_x (50%), CH_4 (20%), NMHCs (20%), NH_3 (5%), and N_2O (4%) from the land surface to the atmosphere. Fossil fuel fluxes account for 70–90% of human-induced CO_2 emissions to the atmosphere, equivalent to about 10% of the terrestrial respiration-decay CO_2 flux. It is not likely that these emission percentages will decline substantially as we enter the 21st century. Population growth and our global reliance on fossil fuels as an energy source make such a scenario highly unlikely. Thus, because of fossil fuel burning and present land use practices, continuous global environmental change is virtually inevitable. Because global practices are unlikely to change greatly into the early 21st century, I see little likelihood that our planet will not experience a doubling of atmospheric CO_2 concentration and increases in concentrations of other greenhouse gases by the middle of the next century. Such a change in the gas composition of the atmosphere presents a strong probability of global warming.

Finally, it should be remembered that a feedback exists if the rate of change is affected by the change itself. In the context of this chapter, a feedback is a biotic process

Table 2.1. Some Problems of Global Environmental Change Owing to
Human Activities[a]

- Climatic changes from anthropogenic inputs to the atmosphere of CO_2 and other greenhouse gases, and SO_2 and its fate
- Disruptions in biogeochemical cycles of carbon, nitrogen, phosphorus, sulfur, trace metals, and other elements
- Acid precipitation
- Alterations in the ozone layer and associated effects on ultraviolet radiation
- Increasing rates of tropical deforestation and other large-scale destruction of habitat, with potential effects on climate
- Disappearance of biotic diversity through explosive rates of species extinctions
- Potential global consequences of distribution and application of xenobiotic chemicals and biotechnology
- Cultural eutrophication from agricultural runoff and municipal and industrial sewage disposal
- Exploitation of natural resources (e.g., metals and fossil fuels)
- Water quality and usage
- Waste disposal (radioactive, toxic and municipal)

[a]Population growth at 1.6–2.0% per year in the last 40 years is a factor common to all these problems.

that changes the rate (in a negative or positive sense) of climatic change. How will the biologically driven, coupled C-N-P-S system react to warming of the earth?

HISTORICAL FRAMEWORK: WHAT IS CHANGE?

How was the earth functioning prior to human interference in its biogeochemical cycles and climatic system? Table 2.2 shows some glacial-interglacial and present global environmental conditions of the earth's surface. Over the last century, nearly all of these components of the surface system have increased in magnitude. Exceptions are land runoff and the suspended load flux of rivers, which have probably decreased in magnitude because of damming of water courses. Also shown in Table 2.2 is one possible scenario for the changes in these conditions in the future. This scenario is predicated on the basis of two of the principal factors forcing future global environmental change— population growth and fossil fuel usage—continuing to increase in the 21st century. The scenario is necessarily qualitative and is, in part, a product of consideration of the coupled C-N-P-S system discussed briefly in the following sections.

THE COUPLED C-N-P-S SYSTEM

The coupled C-N-P-S system is multidimensional, consisting of numerous processes, reservoirs, and fluxes (see, e.g., Lerman et al. 1989; Mackenzie et al. 1993; Wollast, et al. 1993), and difficult to portray in detail. For simplicity, various components of the system have been isolated in Figures 2.3–2.14. These vignettes emphasize processes and magnitudes of present-day fluxes. The biogeochemical cycles of the various C-N-P-S system components are based on the recent summary of Mackenzie et al. (1993) and references therein, and on data in the IPCC volume on climate change (J. F. Houghton et al. 1990).

Table 2.2. Historical and Present Global Environmental Conditions of the Earth's Surface Environment and Recent and Future Changes in These Conditions

| Component | Conditions[a] | | | | |
	Glacial	Interglacial	Present	Last 100 yrs (avg)	Future
CO_2 concentration	180 ppmv	280 ppmv	356 ppm	↑ 0.3% y^{-1}	↑ Acceler
CH_4 concentration	0.3 ppmv	0.8 ppmv	1.7 ppmv	↑ 0.6% y^{-1}	↑ Acceler
N_2O concentration	?	285 ppbv	310 ppbv	↑ 0.02% y^{-1}	↑ Acceler
SO_2 emissions	—	—	90×10^9 kg y^{-1}	↑ 1% y^{-1}	↑
$(CH_3)_2S$ emissions	$60–400 \times 10^9$ kg y^{-1}	40×10^9 kg y^{-1}	40×10^9 kg y^{-1}	↑ ?	↑ ?
Temperature	284°K	288°K	288°K	↑	↑
Mineral aerosol flux	$5–10x(y)$	(y)	2×10^{12} kg y^{-1}	↑	↑
Land runoff (H_2O) flux	2×10^{16} kg y^{-1}	3.7×10^{16} kg y^{-1}	3.7×10^{16} kg y^{-1}	↓ 0.1% y^{-1}	↑ Acceler[b]
Particulate erosion products in runoff flux	$\sim 1 \times 10^{13}$ kg y^{-1}	$\sim 7 \times 10^{12}$ kg y^{-1}	1.5×10^{13} kg y^{-1}	↑	↑[b]
Dissolved salts in runoff flux	—	—	4×10^{12} kg y^{-1}	↑	↑
N_{tot} riverine flux	—	1.4×10^{10} kg N y^{-1}	3.5×10^{10} kg N y^{-1}	↑	↑ Acceler
P_{tot} riverine flux	—	1.4×10^9 kg P y^{-1}	3×10^9 kg P y^{-1}	↑	↑ Acceler
C_{org} riverine flux	—	4×10^{11} kg C y^{-1}	8×10^{11} kg C y^{-1}	↑	↑ Acceler
Total marine net primary production	$>(y)$	(y)	3.8×10^{13} kg y^{-1}	↑	↑ Acceler

[a](y), A quantity not accurately known at present; Acceler, accelerating trend; ↑, increase; ↓, decrease.
[b]Damming will decrease runoff and particulate discharge; temperature increase will increase runoff.
Source: Mackenzie et al. (1991).

A casual scan of Figures 2.3–2.14 reveals that most of the natural processes affecting mass transfer in the C-N-P-S system are biologically driven. In particular, the fluxes of the biogenic gases CH_4, CO, N_2O, NO_x, NMHCs, and reduced species of sulfur (CS_2, $[CH_3]_2S$, OCS) to the atmosphere are mediated by biochemical processes, commonly bacterial, occurring in the terrestrial and oceanic environments. In the atmosphere, these reduced gases are oxidized, in most cases by the OH radical, to species that are then returned to the earth's surface via biological productivity or in wet and dry depositions of oxidized species. It is also generally evident from the figures that forest and other biomass burning and fossil fuel combustion are the major sources of most biogenic gases released into the earth's atmosphere from human activities.

The natural processes and fluxes shown in Figures 2.3–2.14 are of concern in terms of biotic feedbacks in the global climatic system. Before we explore briefly and qualitatively these biotic feedbacks, a general and somewhat discouraging statement can be made. It is difficult with present knowledge of the global system to establish quantitatively negative biotic feedbacks in a scenario of future global warming. Even the arguments of enhanced terrestrial carbon storage owing to CO_2 stimulation, NO_3 and NH_4 eutrophication ("greening of the earth"; e.g., Broecker and Severinghaus 1992), and the dimethyl sulfide cloud condensation nuclei-cooling hypothesis (the so-called

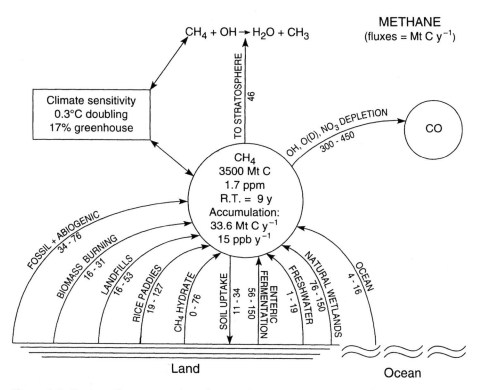

Figure 2.3. Earth surface–atmosphere biogeochemical cycle of CH_4. Climate sensitivity in this figure and those that follow refers to the temperature response to a doubling of gas concentration. The contribution of the gas to the potential of the enchanced greenhouse effect during the past 100 years is given in percent. All fluxes in this diagram and Figures 2.4–2.14 are in units of millions of tons of the substance per year, unless otherwise specified.

CLAW hypothesis; Charlson et al. 1987) are not sufficiently constrained by data to allow these hypotheses to qualify as proven negative biotic feedbacks on global warming. In addition, many of the potential biotic feedbacks identified are positive in nature, enhancing radiatively active biogenic gas fluxes to the atmosphere and the potential warming of the earth. Let us now explore briefly the potential feedbacks in various components of the global C-N-P-S system.

CH_4-CO-CO_2

The biogeochemical cycles of CH_4, CO, and CO_2 (Figures 2.3–2.5) are coupled through the process of oxidation of the reduced carbon gases of CH_4 and CO by atmospheric OH radical to CO_2. The CO_2 produced is then available for the production of organic matter, which, when it decays, releases CH_4 and CO from terrestrial and oceanic environments. The overall bacterial reactions leading to CH_4 production in anaerobic environments are the disproportionation of organic matter into CH_4 and CO, that is,

$$2CH_2O \rightarrow CO_2 + CH_4$$

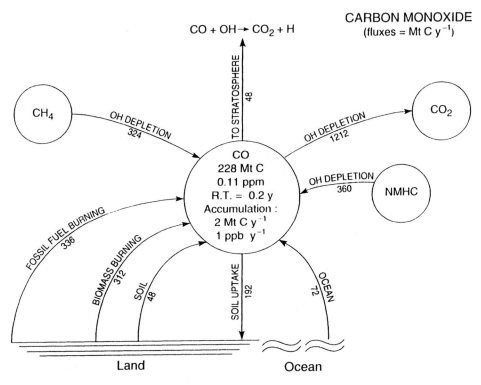

Figure 2.4. Earth surface–atmosphere biogeochemical cycle of CO.

and the reduction of CO_2 to CH_4 according to the reaction

$$CO_2 + 4H_2 \rightarrow CH_4 + 2H_2O$$

The natural CO flux comes from the bacterial decomposition of organic matter in soils, and from bacteria and algae in the ocean that actively generate CO from respiration processes according to the reaction

$$2CH_2O + O_2 \rightarrow 2CO + 2H_2O$$

Because microbial respiration and decomposition of organic matter are bacterially mediated, one would expect that in a warmer world, respiration fluxes of CH_4 and CO would be enhanced, a positive climatic feedback. However, because soils are also sinks for atmospheric CH_4 and CO and because emissions of these gases are sensitive to soil moisture, the situation is more complex. For tropical wetlands and rice paddies, soil moisture changes, resulting from changes in temperature and precipitation, could result in either increased or decreased CH_4 fluxes. An increase in soil moisture would enhance the flux; a decrease would result in a smaller flux. For high-latitude wetlands, where the temperature change owing to an enhanced greenhouse effect will be greater than that in low-latitude regions, an increase in soil moisture will lead to an increase in the flux of CH_4. Warmer, drier soils might give rise to decreased CH_4 fluxes. However, as pointed out in the 1990 IPCC volume (J. T. Houghton et al. 1990), higher temperatures in high

Table 2.3. Reaction Scheme for Conversion of CH_4 and CO to CO_2

$$CH_4 + OH \rightarrow CH_3 + H_2O$$
$$CH_3 + O_2 + M \rightarrow CH_3O_2 + M$$
$$CH_3O_2 + NO_x \rightarrow NO_x + CH_3O$$
$$CH_3O + O_2 \rightarrow H_2CO + HO_2$$
$$H_2CO \rightarrow CO + H_2$$

$$CH_4 + OH \rightarrow CO + \text{products}$$
$$CO + OH \rightarrow H + CO_2$$

northern latitudes will very likely lead to enhanced CH_4 fluxes from CH_4 trapped in permafrost, decomposable organic matter frozen in permafrost, and decomposition of CH_4 hydrates.

As seen in Figures 2.3–2.5, the major sink for the trace gases of CH_4 and CO is reaction with the OH radical in the atmosphere to form CO_2. The reaction scheme is given in Table 2.3. Thus, these gases, along with reduced nitrogen and sulfur gases, are important regulators of the oxidizing capacity of the atmosphere. An important potential climatic feedback involving CH_4, CO, and OH is suggested by the possibility that enhanced fluxes of CO to the atmosphere, because of its short residence time, could lead to depletion of the OH radical and consequently less effective CH_4 removal. Such a reaction sequence could lead to a faster rate of CH_4 accumulation in the atmosphere and an increased rate of global warming. Furthermore, an enhanced CH_4 flux will probably lead to increased production of stratospheric water vapor (Figure 2.3) according to the scheme

$$CH_4 + OH \rightarrow H_2O + CH_3$$

Water vapor is a greenhouse gas, and its increased production in the stratosphere is a positive, but minor (Lelieveld and Crutzen 1990), feedback in a warming-earth scenario.

In summary, although it is difficult to quantify, it is likely that an initial warming would lead to a net increase in fluxes of CH_4 and CO to the atmosphere and accumulation in that reservoir. This situation constitutes a positive feedback within the scenario of a warming earth.

Obviously, the most important trace gas contributing to the potential of an enhanced greenhouse is CO_2. For the period 1765–1990, CO_2 accounted for about 60% of the radiative forcing owing to accumulation of greenhouse gases in the atmosphere from human activities. Major processes affecting CO_2 in the atmosphere and estimates of associated fluxes are shown in Figure 2.5. CO_2 is coupled to the reduced carbon gases through the oxidation of CO by the OH radical. Figure 2.5 shows this flux to be on the order of 1.2 petagrams (Pg) (= 10^{15} g) carbon per year, much of which is derived from CH_4 and CO emissions to the atmosphere from human activities. Therefore, an initial warming of the planet and consequent enhanced emissions of CH_4 and CO from the earth's surface could potentially lead to an enhanced accumulation of CO_2 in the atmosphere, a positive climatic feedback.

As already noted in Chapter 1, the notorious problem with CO_2 today is the difficulty in balancing its fossil fuel and forest and biomass burning fluxes to the atmosphere with known sinks. The problem has been dealt with extensively in the recent literature and is

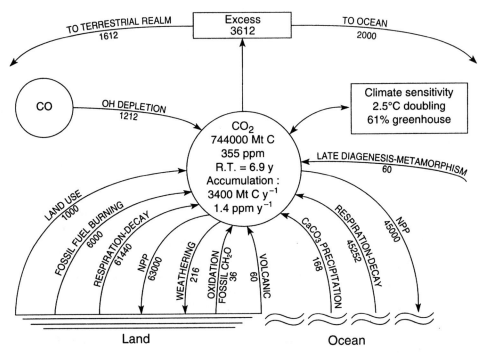

CARBON DIOXIDE
(fluxes = Mt C y^{-1})

Figure 2.5. Earth surface–atmosphere biogeochemical cycle of CO_2.

illustrated in Figure 2.5. Assuming a 1-Pg/year land use carbon flux and a 6-Pg/year flux to the atmosphere caused by fossil fuel burning and cement manufacturing, accumulation of 3.4 Pg carbon per year as CO_2 in the atmospheric reservoir leaves an imbalance (the "missing" flux) of 3.6 Pg carbon per year (rounded). It should be kept in mind, however, that the full range of estimates of the tropical deforestation flux for the years around 1990 is 1.6±1.0 Pg carbon per year (J. T. Houghton et al. 1990; R. A. Houghton 1991). Thus, the "missing" atmospheric CO_2 is on the order of 3–5 Pg carbon per year (rounded). The lower bound is certainly within the rather large error margins of recent estimates of the net oceanic sink strength (e.g., C. D. Keeling et al. 1989; R. F. Keeling and Shertz 1992; Quay et al. 1992), particularly if we include corrections to the synoptic air-to-sea CO_2 influx, as envisioned by Sarmiento and Sundquist (1992), or enhanced organic carbon sequestration fluxes, as suggested by Mackenzie and colleagues (Mackenzie 1981; Wollast and Mackenzie 1989; Sabine and Mackenzie 1991). The upper bound appears to be out of the range of current estimates of the sink strength of the ocean, suggesting a net terrestrial sink. Such a net sink implies fertilization of forest growth owing to increased atmospheric levels of CO_2, eutrophication of terrestrial ecosystems because of fallout of anthropogenic NO_3 and NH_4 nutrients, aggradation of previously disturbed or undisturbed terrestrial ecosystems, or some combination of these factors. Our failure to date to resolve the issue of the "missing carbon" makes it difficult to assess biotic feedbacks to climate change associated with CO_2 fluxes. However, some qualitative statements can be made.

Terrestrial Ecosystem

Perhaps one of the strongest potential positive biofeedbacks in the terrestrial environment is the effect of increasing temperature on photosynthesis and respiration. The rates of both processes involving vegetation and microbial life increase with increasing temperature. However, respiration rates are more sensitive to temperature change, and it is likely that in a warmer world there would be an enhanced flux of CO_2 to the atmosphere. The flux could be as large as a few petagrams of carbon per year (Woodwell 1983; R. A. Houghton and Woodwell 1989). This temperature-enhanced flux of CO_2 is a strong positive feedback on global warming.

There has been an increase in the availability of phosphorus, NO_3, and NH_4 nutrients in the terrestrial and oceanic environments. These excess nutrients enter the environment from such human activities as the application of fertilizers to the land surface, the burning of fossil fuels and biomass, and discharging nitrogen- and phosphorus-laden sewage. The nutrients are available to stimulate plant growth in the soil and aquatic environments and cause eutrophication of both land and marine environments. The eutrophication acts as a negative biofeedback on warming of the earth in that atmospheric CO_2 can be sequestered in plant production. Recently Kauppi et al. (1992) estimated from studies of western European forests that NO_3 stimulation of plant growth was sequestering 85–120 million tons of carbon per year in the late 1970s and 1980s. The effect of eutrophication and storage of organic matter in both coastal and open ocean marine systems may be on the order of 0.5 Pg carbon per year (Wollast and Mackenzie 1989; Sabine and Mackenzie 1991). Melillo (in J. T. Houghton et al. 1990) estimated that the effect of eutrophication, both on land and in the ocean, could be as large as 1 Pg carbon per year. Unless human-induced inputs of nutrients to the environment slow, it is likely that the cultural eutrophication sink for CO_2 will be enhanced in the future and act as a small negative feedback on global warming (Mackenzie et al. 1993). This statement must be qualified by the fact that increasing levels of acid deposition, air pollution, and degradation and erosion of soils caused by human activities can actually lead to a decrease in terrestrial plant growth. In addition, increased fluxes of organic matter to the coastal environment because of human activities and oxidation of that organic matter could lead to a net release of CO_2 from the coastal zone.

One of the more controversial of the potential negative feedbacks in terrestrial ecosystems is CO_2 fertilization. It is known from experimental simulations that certain agricultural crops and a few perennials, when subjected to increased CO_2 levels, will exhibit an increase in their photosynthetic and growth rates (e.g., Lemon 1977; Idso 1982; Strain and Cure 1985). If this enhancement occurred in large-scale terrestrial ecosystems, more carbon could be stored in vegetation or soil organic matter. Thus, CO_2 stimulation would lead to withdrawal of CO_2 from the atmosphere and its sequestration in plant organic matter, a potentially strong negative feedback on global warming. Although not a demonstrable fact, CO_2 stimulation has been invoked as a means of balancing the carbon cycle (e.g., Tans et al. 1990) and as part of the argument of several geochemists for a global greening of the earth (Broecker and Severinghaus 1992). It is difficult for many ecologists and others to accept this hypothesis of increased carbon storage in terrestrial ecosystems in recent decades because of the uncontroversial evi-

dence of degradation of terrestrial ecosystems supplied by both satellite images and
ground-truth data. Land use activities in tropical ecosystems alone may be responsible
for the release of nearly 3 Pg carbon per year into the atmosphere (R. A. Houghton
1991).

The role of terrestrial ecosystems today and in the future in terms of the carbon
balance awaits further clarification; however, the potential for strong biotic feedbacks
on global warming is large. Terrestrial ecosystem net primary production is about 63 Pg
carbon per year, or 60% of total world organic production (Figure 2.5). Total terrestrial
ecosystem (living plants, litter, soil organic matter) carbon biomass today is about 2100
Pg, 25 times all the organic matter presently stored in the ocean as living, particulate,
and dissolved organic carbon (Mackenzie et al. 1993). Obviously, biotic feedbacks
involving CO_2 and the terrestrial ecosystem, whether positive (enhanced respiration) or
negative (CO_2 stimulation and eutrophication), can represent large mass transfers and
substantially affect the course of a warming earth.

Changes in the quantity and distribution of soil water, in the geographical distribution
of vegetation, and in the UV-B flux to the land surface are other factors that may affect
CO_2 feedbacks in terrestrial ecosystems on global warming (J. T. Houghton et al. 1990).
It is difficult to predict the direction and magnitude of these effects, but it is likely that
CO_2 fluxes to the atmosphere would be enhanced on a rapidly warming earth experienc-
ing stratospheric ozone depletion and rapid environmental change, resulting in a positive
feedback on global warming.

Oceanic Ecosystem

For CO_2, several potential feedbacks in the oceanic system that involve changes in
oceanic temperature, circulation, and biogeochemical cycling warrant attention. These
feedbacks can also influence the oceanic fluxes of reduced carbon and sulfur and N_2O.

Oceanic temperature changes affect the amount of dissolved inorganic carbon in
seawater according to the reaction

$$CO_2(g) + H_2O + CO_3^{2-} \rightarrow 2HCO_3^-$$

For a warmer surface ocean, this reaction is driven to the left, and the P_{CO_2} of the water
will increase. Elevated P_{CO_2} causes gas evasion. For a temperature change of $+1°C$, the
change in P_{CO_2} is on the order of 10 µatm. This change would decrease the net uptake of
CO_2 by the ocean. This is a positive feedback and, although nonbiotic in nature, it should
be kept in mind because it could amplify a future CO_2 increase by about 5% (Lashof
1989).

Feedbacks involving oceanic circulation are strongly linked to biogeochemical cycling of
carbon and nutrients. On the one hand, as a result of an initial warming, the stability of the
oceanic water column could increase. Vertical mixing of nutrients across the thermocline
and into the euphotic zone may be decreased, leading to a decrease in biological productivity
and less organic carbon escaping the euphotic zone in the "biological pump." On the other
hand, it is completely conceivable that changes in wind patterns and intensity along the
western coastal margins of continents could lead to enhanced nutrient upwelling. Whatever
the case, upwelling waters transport dissolved carbon and nutrients in the approximate
Redfield C:N:P ratio of organic production of 106:16:1. All the carbon for production is
supplied by upwelling waters. Thus, changes in the intensity of upwelling because of
climatic change should have little effect on atmospheric CO_2. An exception to this statement

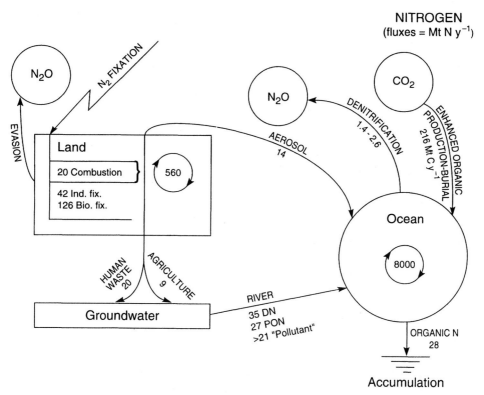

NITROGEN
(fluxes = Mt N y^{-1})

N$_2$O

N$_2$ FIXATION

EVASION

Land

20 Combustion

42 Ind. fix.
126 Bio. fix.

560

N$_2$O

AEROSOL
14

DENITRIFICATION
1.4 - 2.6

CO$_2$

ENHANCED ORGANIC
PRODUCTION-BURIAL
216 Mt C y^{-1}

Ocean

8000

HUMAN
WASTE
20

AGRICULTURE
9

Groundwater

RIVER
35 DN
27 PON
>21 "Pollutant"

ORGANIC N
28

Accumulation

Figure 2.6. Land-ocean biogeochemical cycle of nitrogen. In the present cycle, nitrogen is accumulating on land (see page 31).

is the possibility that climatic change results in a global change in the C:N:P ratio of marine communities. A higher C:N ratio of organic matter might result in a drawdown of atmospheric CO$_2$. However, there is little evidence supporting major changes in the C:N:P ratios of marine organisms owing to climatic change on time scales of decades to a century. At present this argument is highly speculative.

Some feeling for the magnitude of change in oceanic productivity and organic carbon efflux from the euphotic zone may be gained by looking at the global oceanic fluxes of nutrient nitrogen (Figures 2.6 and 2.7). Although the atmosphere and rivers are sources of nitrogen to fuel new production in both coastal margins and the open ocean, nutrient nitrogen fluxes involved with upwelling and vertical mixing in the ocean are most important. These fluxes of nitrogen (mostly NO$_3$) are on the order of 0.9 Pg nitrogen per year. The flux of riverine dissolved nitrogen plus particulate nitrogen today is about 7% of this flux (Figure 2.6). Now consider a scenario in which the organic riverine nitrogen fluxes are enhanced globally by a factor of 2—not an unlikely future situation, because this flux has about doubled since preindustrial time (Wollast and Mackenzie 1989; Wollast 1991, 1993; Meybeck 1993). Assume further that all the riverine nitrogen flux is reactive in the marine environment and forms organic matter. Such an enhanced nitrogen flux could lead to an increase in the accumulation of organic matter, with a molar C:N ratio of 12 in marine sediments representing about 0.7 Pg carbon per year. The ultimate

ORGANIC NITROGEN
(fluxes = Mt N y^{-1})

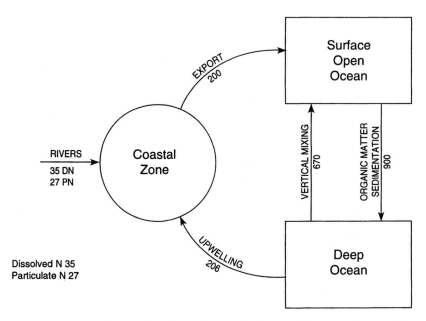

Figure 2.7. Biogeochemical cycling of organic nitrogen between the coastal and open oceanic realms.

source of this carbon is the atmosphere, and such a flux would thus represent a drain on atmospheric CO_2 equal to that amount.

Now, for argument's sake, let us increase the flux of nitrogen-bearing atmospheric aerosol into the ocean by a factor of 2 (Figure 2.6); about 25% of this enhanced flux falls on waters of the coastal margin and the rest falls on the open ocean (Wollast 1991). In the coastal margin, this nitrogen flux could lead to an annual burial of organic matter, assuming a C:N ratio of 12, containing 36 million tons of carbon. In the open ocean, the enhanced nitrogen flux could lead to production of about 60 million tons of organic carbon per year, using a C:N ratio of 6.6. Assuming that 10% of this enhanced flux escapes the euphotic zone and passes to the deep sea (Knauer 1993), 6 million tons of carbon per year in organic matter could be sequestered in the deep sea.

This scenario results in an enhanced organic carbon flux in the oceanic environment owing to increased autotrophic activity totaling about 740 million tons of carbon per year. The flux represents about 12% of today's fossil fuel flux and only 7% of the flux projected for the year 2020 in the business-as-usual case (R. A. Houghton et al. 1990). It is a possible negative biotic feedback on global warming, but the scenario does not take into account the possibility of enhanced denitrification rates from the ocean owing to warming. Furthermore, the scenario assumes that co-limiting nutrients, like phosphorus, are available to fuel bioproductivity. It is also possible that future increased riverine fluxes of organic matter into the ocean because of human activities, and its oxidation, could increase the net heterotrophy of the ocean.

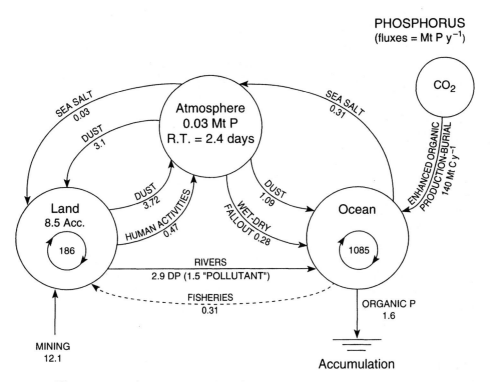

Figure 2.8. Land-ocean-atmosphere biogeochemical cycle of phosphorus.

There are several other potential feedbacks involving CO_2, the ocean, and climatic change; however, a quantitative assessment, or even a qualitative assessment like the one developed above, is fraught with difficulties. Increased UV-B radiation owing to stratospheric ozone depletion may affect the capacity of certain marine ecosystems, such as the rich waters of the Antarctic convergence, to store carbon. With warming, the composition and distribution of plants and animals (e.g., coccolithophoridae, salpas) and iron, molybdenum, and other bioessential trace metal fluxes to the ocean may change, and enhanced decomposition of dissolved organic carbon (of which there is a large oceanic reservoir, on the order of 1000 Pg carbon) may occur. The biological processes associated with these possibilities can change the P_{CO_2} of surface waters and hence the strength of the oceanic sink.

Phosphorus

The phosphorus biogeochemical cycle is shown in Figure 2.8. The major difference between this cycle and the carbon, nitrogen, and sulfur biogenic gas cycles is that there is no important biological process that generates a gas phase at the earth's surface. The phosphine (PH_{3g}) flux is small compared to other fluxes of phosphorus. Phosphine is also very reactive and oxidizes rapidly after emission from swamps. However, the phosphorus cycle is coupled to the cycles of carbon and nitrogen and is considered here for completeness.

Two major coupled feedbacks in the phosphorus cycle may be of importance in global warming. One involves enhanced eutrophication of aquatic systems because of inputs of fertilizer and sewage phosphorus to these systems. The global dissolved phosphorus flux of rivers to coastal marine margins has doubled because of these human activities. This process has resulted in eutrophication of coastal marine systems and a potential accumulation of organic carbon in these systems of 100–200 million tons of carbon per year (Figure 2.8). The flux is a negative feedback on accumulation of anthropogenic CO_2 in the atmosphere.

Notice from Figure 2.8 that the land appears to be accumulating phosphorus today because of our mining of this element and its utilization as a fertilizer and because of sewage inputs to the land surface. Because most mineral chemical weathering and biological decomposition rates increase with increasing temperature, with warming this reservoir of phosphorus may be more easily leached into aquatic systems, enhancing eutrophication processes in these systems and leading to enhanced accumulation of organic matter. The flux is a small negative feedback on CO_2 accumulation in the atmosphere and global warming. Provided that co-limiting nutrients in coastal marine systems are available, the total phosphorus presently stored on land, if leached and transported by rivers to those systems, would amount to only 0.8 Pg of organic carbon storage.

N_2O, NH_3, NO_x, and NMHCs

Vignettes of the biogeochemical cycles of N_2O, NH_3, NO_x, and NMHCs are shown in Figures 2.9–2.12. N_2O represents less than 5% of the global denitrification flux to the atmosphere; the rest is diatomic nitrogen. However, N_2O is an important greenhouse gas, accounting for about 4% of the radiative forcing between 1765 and 1990 (J. T. Houghton et al. 1990). The N_2O cycle provides another example of our difficulties in balancing global sources and sinks of elements. Given a 150-year lifetime, a present atmospheric concentration of 310 parts per million by volume ([ppmv] about 8% greater than that in preindustrial times), and a rate of growth of 0.2–0.3% per year, current global emissions exceed global sinks by about 30%.

With warming, the most important biotic feedbacks involving N_2O are changes in the denitrification (and nitrification) rates in soils and sediments and in the water column of the ocean, as well as changes in N_2O fluxes from nitrogen-bearing fertilizers applied to the land surface and sewage discharges into aquatic systems. Because the reactions involving N_2O are bacterially mediated, it is likely that an increase in temperature will lead to enhanced evasion rates of N_2O from the earth's surface. This is a positive biotic feedback on global warming that also could lead to enhanced destruction of stratospheric ozone (Figure 2.9).

The NH_3 cycle (Figure 2.10) is important in considerations of warming of the earth for two reasons. The first is that NH_3 interacts with the OH radical to produce NO_x; the second is that it reacts with NO_3 and SO_4 to produce ammoniated aerosols. The major natural source of NH_3 release into the atmosphere is bacterially mediated decomposition of organic matter and volatilization of NH_3 gas from soils and the ocean (Dawson 1977; Gammon and Charlson 1993; Clarke and Porter 1994). Once more it is likely that this biotic flux would be enhanced in a warmer world, leading to the potential for additional

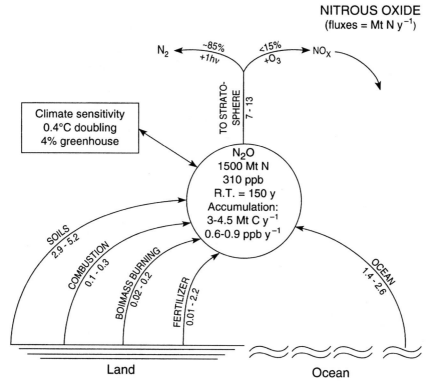

Figure 2.9. Earth surface–atmosphere biogeochemical cycle of N₂O.

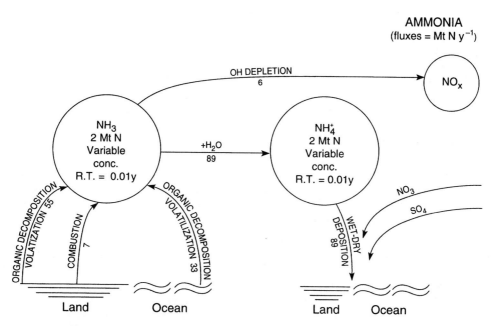

Figure 2.10. Earth surface–atmosphere biogeochemical cycle of NH₃.

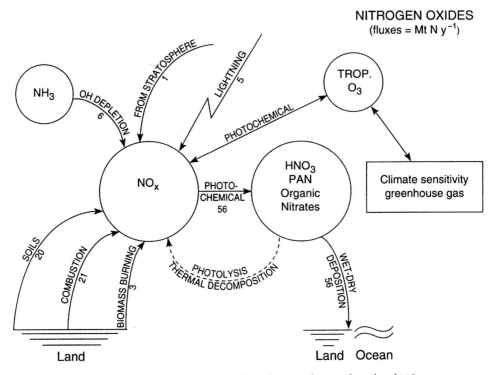

Figure 2.11. Earth surface–atmosphere biogeochemical cycle of NO_x.

stress on the OH radical concentration of the atmosphere and enhanced production of NO_x (Figure 2.11). The latter gas is involved in the complex process of formation of tropospheric ozone, a greenhouse gas.

The ammonia cycle is also informative in terms of the potential for cultural eutrophication of the terrestrial biosphere. The total land-to-atmosphere flux of 62 million tons of NH_3 per year (Figure 2.10) is principally a result of human activities. Only a few million tons of ammonia nitrogen per year have as their source natural bacterial decomposition processes in soils (Warneck 1988; Jenkinson 1990; Schlesinger and Hartley 1992; Isermann 1993). The anthropogenic NH_3 flux from the land to the atmosphere is about 50 million tons of nitrogen per year, of which about 25% is transported away from the continents to the oceanic atmosphere (e.g., Jenkinson 1990; Isermann 1993; Liss and Galloway 1993). The rest, about 37 million tons of nitrogen per year, falls back on the land surface in wet and dry deposition and may be available for terrestrial bioproductivity. If this nitrogen were to fuel land plant production with a C:N ratio of 100 (Delwiche and Likens 1977; Likens et al. 1981), 3 Pg carbon per year would be required. Interestingly, the phosphorus accumulating on land each year from agricultural fertilizers and sewage amounts to about 8.5 million tons (Figure 2.8), and this is about equal to the phosphorus requirement needed to sustain a land plant production of 3 Pg carbon per year.

The cycles of nitrogen oxides (NO_x) and NMHCs appear in Figures 2.11 and 2.12. Both gases are biologically produced—NO_x from the bacterial decomposition of organic matter in soils and NMHCs as by-products of plant productivity in terrestrial and marine

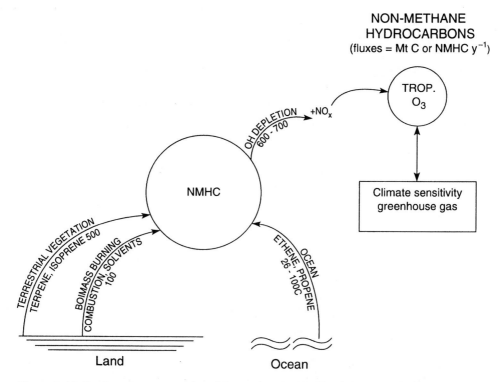

Figure 2.12. Earth surface–atmosphere biogeochemical cycle of NMHCs. Fluxes are in units of millions of tons of carbon and millions of tons of NMHCs per year.

environments. Increasing temperature alone would probably increase the flux of NO_x from soils, leading to potential OH radical depletion and enhanced formation of tropospheric ozone (Figure 2.11). However, the situation is much more complex with respect to tropospheric ozone. The concentration of tropospheric ozone depends in a very non-linear way on the atmospheric concentrations of other reduced biogenic gases, including CH_4, CO, NMHCs, and NO_x. In general, under atmospheric conditions in which NO_x concentrations are low (5–30 parts per trillion by volume [pptv]), increases in concentrations of the trace gases CH_4, CO, and NMHCs lead to a decrease in the concentration of tropospheric ozone. At high NO_x concentrations, increases in CH_4, CO, and NMHCs lead to an enrichment in ozone. The situation can be illustrated using CO as an example. Consider two of the main reaction mechanisms for CO oxidation (Cicerone and Oremland 1988):

$$CO + OH \rightarrow H + CO_2$$

$$H + O_2 + M \rightarrow HO_2 + M$$

NO_x-poor environment	NO_x-rich environment
$HO_2 + O_3 \rightarrow OH + 2O_2$	$HO_2 + NO \rightarrow NO_2 + OH$
	$NO_2 + h\nu \rightarrow O + NO$
	$O + O_2 + M \rightarrow O_3 + M$
Net: $CO + O_3 \rightarrow O_2 + CO_2$	$CO + 2O_2 + h\nu \rightarrow CO_2 + O_3$

The NO_x-rich scenario breaks even when background NO_x levels reach about 90 pptv (Warneck 1988). Generally, the OH radical behaves like ozone with changing NO_x concentrations, except to a lesser degree.

Clearly, oxidation of CO and other atmospheric trace gases has a strong effect on the oxidative capacity of the atmosphere, as outlined by Thompson (1992). The concentration of the OH radical, which is the main determinant of the oxidative capacity of the atmosphere, depends on concentrations of various trace gases, including tropospheric ozone and water vapor. Elevated concentrations of ozone, NO_x, and H_2O will result in increases in OH radical levels, whereas increases in the trace gases CH_4, CO, and NMHCs will lead to lower levels. One critical positive climatic feedback is that increases in CO concentrations in the atmosphere could lead to reduction in OH radical levels because NO_x has too short a lifetime to counteract that effect on a global scale (J. T. Houghton et al. 1990). Decreased concentrations of the OH radical lead to an increase in the lifetime of CH_4, a positive feedback.

Reduced and Oxidized Sulfur Gases

Biological emissions of sulfur-containing gases from the earth's surface are dominated by the reduced form of sulfur in $(CH_3)_2S$, H_2S, and OCS (Figure 2.13). Through oxidation these reduced sulfur-containing gases are coupled to the oxidized sulfur cycle (Figure 2.14), which is dominated by fluxes involving SO_2 gas and sulfate aerosol.

The sulfur biogenic gases released from the ocean and land surfaces differ in composition; $((CH_3)_2S$ and OCS characterize emissions from the ocean surface, whereas H_2S is the predominant form of sulfur released from decaying terrestrial vegetation. OCS is the most abundant reduced sulfur species in the remote marine atmosphere; because of its inert chemical behavior and consequently long residence time in the troposphere, it can enter the stratosphere (Andreae 1986). OCS and $(CH_3)_2S$ are important to the chemistry of the atmosphere and climate. In the stratosphere, OCS is destroyed by UV photolysis ($\lambda = 250$ nm) and atomic oxygen and is converted to SO_2 and on to sulfate aerosol (Warneck 1988). The lifetime of OCS is about 30 years, and this reduced sulfur gas supplies about half of the sulfate aerosol in the Junge layer of the lower stratosphere (Crutzen 1976). In the troposphere, $(CH_3)_2S$ that has entered the atmosphere from an aqueous phase is oxidized on a time scale on the order of a day by the OH radical to produce either SO_2 or methane sulfonic acid (MSA) in varying proportions. The SO_2 is then oxidized further to sulfate, and this aerosol and MSA are removed from the atmosphere to the earth's surface mainly in rain (Gammon and Charlson 1993).

The climatic connections to the chemistry of $(CH_3)_2S$ and OCS are as follows:

1. The major source of cloud condensation nuclei (CCN) in the remote marine atmosphere is most likely $(CH_3)_2S$. Charlson et al. (1987) have pointed out that because the reflectance (albedo) of clouds is sensitive to CCN density, there is a link between marine phytoplankton production and climate. $(CH_3)_2S$ is produced by bacterial degradation of a water-soluble chemical intermediate, dimethyl sulfonium propionate, produced by phytoplankton metabolism. The CLAW hypothesis argues that a warming of the earth's climate could lead to enhanced phytoplankton growth, hence enhanced excretion and emission of $(CH_3)_2S$ from the sea surface. The increased DMS flux could result in increased production of sulfate aerosol

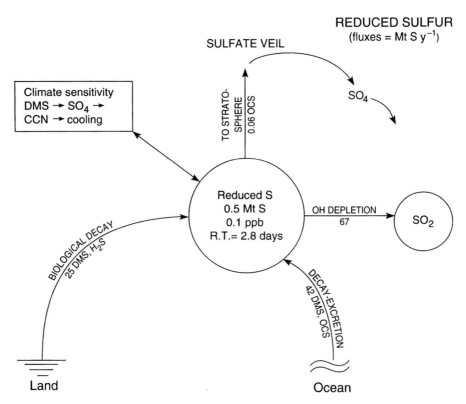

Figure 2.13. Earth surface–atmosphere biogeochemical cycle of reduced gaseous sulfur species.

and CCN in the remote marine atmosphere and increased cloud albedo, giving rise to a cooling of the troposphere, a negative feedback on global warming. Charlson et al. (1987) conclude that to counteract a warming owing to doubling of atmospheric CO_2 would require a doubling of CCN. The validity of the CLAW hypothesis is currently being argued in the literature (e.g., Schwartz 1988; Clarke 1992), but it suggests a potentially strong negative biofeedback on warming.

2. OCS is produced mainly by the photolysis of organic sulfur in the surface waters of the ocean and by photochemical oxidation of the biogenic gas CS_2 in the atmosphere. This atmospheric oxidation is accomplished by the OH radical, with a lifetime on the order of 15 days, to yield a one-for-one production of OCS (Warneck 1988). Any change in the flux of OCS into the stratosphere induced by global warming will have an effect on climate through the resulting change in the sulfur burden of the stratospheric sulfate veil. An increased stratospheric sulfate burden would give rise to cooling of the troposphere, a decreased burden to warming. The likely feedback to an initial warming is difficult to predict.

To complete the picture of the global cycle of sulfur gas species, the earth surface-atmosphere oxidized sulfur cycle is shown in Figure 2.14. Of importance here is the argument recently advanced by Charlson et al. (1992) that anthropogenic emissions of

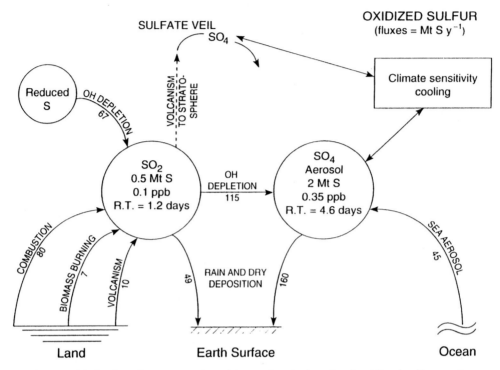

Figure 2.14. Earth surface–atmosphere biogeochemical cycle of oxidized sulfur species.

SO$_2$ to the atmosphere have led to an increased burden of Northern Hemisphere sulfate aerosol. The authors further argue that the enhanced SO$_4^{2-}$ aerosol burden through cooling of the atmosphere could be responsible for the discrepency between part of the observed temperature record of the last 100 years and that predicted on the basis of climate forcing owing to the accumulation of greenhouse gases in the atmosphere.

CONCLUSIONS

The processes discussed in this chapter lead us to the following conclusions regarding climatically important biogenic gases:

1. The natural transfer of CO$_2$ and reduced gases from the earth's surface to the atmosphere is driven by biological (often bacterial) reactions. These gases, or their reaction products in the atmosphere, can affect the radiative properties of the atmosphere. Most of these gases are greenhouse gases; important exceptions are (CH$_3$)$_2$S and OCS.
2. Most biotic feedbacks on global warming are positive. An important exception to this statement is the possibility of CO$_2$ stimulation of terrestrial bioproductivity and consequent storage of carbon in terrestrial ecosystems. However, with a warming earth and the greater sensitivity of respiration rather than productivity to

temperature change, net ecosystem metabolism will still probably be a positive feedback in the short term. This will be a strong feedback because of the large mass of carbon stored on land in above-ground living and dead vegetation, in humus, and in below-ground roots and dispersed organic matter.

3. Some identified potential negative feedbacks to global warming are actually due to inadvertent effects of human activities. The enhanced eutrophication of terrestrial and marine ecosystems is an important example.

4. Oxidized and reduced sulfur gases (SO_2, $[CH_3]_2S$, OCS) emitted into the atmosphere from natural sources and human activities potentially constitute an important negative feedback on global warming. The reduced sulfur gases are natural biogenic gases produced as a result of bacterial and plant metabolism at the earth's surface.

5. In recent years the rates of increase of the trace gas concentrations of CO_2, CH_4 and CO in the atmosphere have slowed. The reasons for this observation are complex. The eruption of Mt. Pinatubo in 1991 and the resultant dispersal of sulfate aerosol have certainly affected the physical climate system. The record of atmospheric temperature anomalies showed a decrease of about 0.5 °C during the period of late 1991 to late 1993. This cooling trend led to a negative anomaly in atmospheric CO_2 during this time interval, and the cooling and sulfate aerosol probably were responsible for the strong development of the Antarctic ozone hole during the austral spring and summer of 1992 and 1993. Furthermore, there is evidence that the fluxes related to the sources of CH_4 and CO may have recently decreased; these changes also may be responsible for decreasing atmospheric CO concentrations and significant slowing of atmospheric CH_4 growth. As emphasized in this chapter, these recent observations document the strong linkages between the biogeochemical cycles of trace gases and the physical climate system.

6. The strongest potential negative feedback on global warming during the next century is that related to human activities. The fluxes of radiatively active biogenic gases into the atmosphere or their radiatively active reaction products (oxone, water vapor) are being strongly modified by fluxes from human activities. It is obvious that slowing the growth of the human population and our exploitation of fossil fuels as an energy source and managing our forests in a sustainable manner would be the most effective steps in avoiding or alleviating a global warming in the next century.

ACKNOWLEDGMENTS

I thank Drs. Steve Smith and Tony Clarke of the University of Hawaii for their helpful comments on an earlier version of this chapter. Dr. George M. Woodwell, director of the Woods Hole Research Center, kindly provided space, support, and enthusiastic conversation while I prepared an initial draft. The research was partially supported by National Science Foundation grant EAR 89–13650 and EAR 93–16133. This chapter is School of Ocean and Earth Science and Technology Contribution No. 3222.

REFERENCES

Andreae, M. O. 1986. The ocean as a source of atmospheric sulfur compounds. In *The Role of Air-Sea Exchange in Geochemical Cycling,* ed. P. Buat-Menard, pp. 331–362. Reidel, Dordrecht, The Netherlands.

Broecker, W. S., and J. P. Severinghaus. 1992. Diminishing oxygen. *Science* 358:710–711.

Brown, I. F., D. C. Nepstad, I de O. Pires, L. M. Luz, and A. S. Alechandre. 1992. Carbon storage and land use in extractive reserves, Acre, Brazil. *Environ. Conserv.* 19:307–316.

Charlson, R. J., J. E. Lovelock, M. O. Andreae, and S. G. Warren. 1987. Oceanic phytoplankton, atmospheric sulfur, cloud albedo, and climate. *Nature* 326:655–661.

Charlson, R. J., S. E. Schwartz, J. M. Hales, R. D. Cess, J. A. Coakley, Jr., J. E. Hansen, and D. J. Hoffman. 1992. Climate forcing by anthropogenic aerosols. *Science* 255:423–430.

Cicerone, R. J., and R. S. Oremland. 1988. Biogeochemical aspects of atmospheric methane. *Global Biogeochem. Cycles* 2:299–327.

Clarke, A. D. 1992. Atmospheric nuclei in the remote free-troposphere. *J. Atmos. Chem.* 14:479–488.

Clarke, A. D., and J. N. Porter. 1994. Pacific marine aerosol part II: Equatorial gradients, ammonium and chlorophyll during SAGA 3. *J. Geophys. Res.* 98:16997–17010.

Crutzen, P. J. 1976. The possible importance of COS for the sulfate layer of the stratosphere. *Geophys. Res. Lett.* 3:73–76.

Dawson, G. A. 1977. Atmospheric ammonia from undisturbed land. *J. Geophys. Res.* 82:3125–3133.

Delwiche, C. C., and G. E. Likens. 1977. Biological response to fossil fuel combustion products. In *Global Chemical Cycles and Their Alterations by Man,* ed. W. Stumm, pp. 73–88. Dahlem Konferenzen, Berlin.

Frisch, J. R. 1986. Future stresses for energy resources. World Energy Conference—Conservation Commission. Graham and Trotman, London.

Gammon, R. H., and R. J. Charlson. 1993. Origins, atmospheric transformations and fate of biologically exchanged C, N and S gases. In *Interactions of C, N, P and S Biogeochemical Cycles and Global Change,* ed. R. Wollast, F. T. Mackenzie, and L. Chou, pp. 283–304. Springer-Verlag, Berlin.

Houghton, J. T., G. J. Jenkins, and J. J. Ephraums, eds. 1990. *Climate Change. The IPCC Scientific Assessment.* Cambridge University Press, Cambridge.

Houghton, R. A. 1991. Tropical deforestation and atmospheric carbon dioxide. *Climate Change* 19:99–118.

Houghton, R. A., and G. M. Woodwell. 1989. Global climate change. *Sci. Am.* 260:36–44.

Idso, S. B. 1982. Carbon dioxide: Friend or foe. Institute for Carbon Dioxide Research Press, Tempe, Ariz.

Imbrie, J., and K. P. Imbrie. 1986. *Ice Ages: Solving the Mystery.* Harvard University Press, Cambridge, Mass.

Isermann, K. 1993. Territorial, continental and global aspects of C, N, P and S emissions from agricultural ecosystems. In *Interactions of C, N, P and S Biogeochemical Cycles and Global Change,* ed. R. Wollast, F. T. Mackenzie, and L. Chou, pp. 79–121. Springer-Verlag, Berlin.

Jenkinson, D. S. 1990. An introduction to the global nitrogen cycle. *Soil Use Man.* 6:56–61.

Kauppi, P. E., K. Mielikäinen, and K. Kuusela. 1992. Biomass and carbon budget of European forests, 1971 to 1990. *Science* 256:70–74.

Keeling, C. D., R. B. Bacastrow, A. F. Carter, S. C. Piper, T. P. Whorf, M. Heimann, W. G. Mook, and H. Roeloffzen. 1989. A three-dimensional model of atmospheric CO_2 transport based on observed winds. 1. Analysis of observational data. In *Aspects of Climate Variability in the Pacific and the Western Americas,* Geophysical Monograph 55, ed. D. H. Peterson, pp. 165–235. American Geophysical Union, Washington, D.C.

Keeling, R. F., and S. R. Shertz. 1992. Seasonal and interannual variations in atmospheric oxygen and implications from the global carbon cycle. *Nature* 358:723–727.

Knauer, G. 1993. Productivity and new production of the oceanic system. In *Interactions of C, N, P and S Biogeochemical Cycles and Global Change,* ed. R. Wollast, F. T. Mackenzie, and L. Chou, pp. 211–231. Springer-Verlag, Berlin.

Lashof, D. A. 1989. The dynamic greenhouse: Feedback processes that may influence future concentrations of atmospheric trace gases and climatic change. *Climatic Change* 14: 213–242.

Lelieveld, J., and P. J. Crutzen. 1990. Influence of cloud photochemical processes on tropospheric ozone. *Nature* 343:227–233.

Lemon, E. 1977. The land's response to more carbon dioxide. In *The Fate of Fossil Fuel CO₂ in the Oceans*, ed. N. R. Andersen and A. Malahoff, pp. 97–130. Plenum Press, New York.

Lerman, A., F. T. Mackenzie and R. J. Geiger. 1989. Environmental chemical stress effects associated with carbon and phosphorus biogeochemical cycles. In *Ecotoxicology: Problems and Approaches,* ed. S. A. Levin, M. A. Hartwell, J. R. Kelly, and K. D. Kimball, pp. 315–350. Springer-Verlag, New York.

Likens, G. E., F. T. Mackenzie, J. Richey, J. R. Sedwell, and K. K. Turekian, eds. 1981. *Flux of Organic Carbon by Rivers to the Sea.* U.S. Department of Energy Conference Report 8009140, Washington, D.C.

Liss, P. S., and J. N. Galloway. 1993. Air-sea exchange of sulfur and nitrogen and their interaction in the marine atmosphere. In *Interactions of C, N, P and S Biogeochemical Cycles and Global Change,* ed. R. Wollast, F. T. Mackenzie, and L. Chou, pp. 259–281. Springer-Verlag, Berlin.

Mackenzie, F. T. 1981. Global carbon cycle: Some minor sinks for CO₂. In *Flux of Organic Carbon by Rivers to the Sea,* ed. G. E. Likens, F. T. Mackenzie, J. Richey, J. R. Sedwell, and K. K. Turekian, pp. 360–384. U.S. Department of Energy Conference Report 8009140, Washington, D.C.

Mackenzie, F. T., J. M. Bewers, R. J. Charlson, E. E. Hoffman, G. A. Knauer, J. C. Kraft, E.-M. Nöthig, B. Quack, J. J. Walsh, M. Whitfield, and R. Wollast. 1991. Group report: What is the importance of ocean margin processes in global change? In *Ocean Margin Processes in Global Change,* ed. R. F. C. Mantoura, J.-M. Martin, and R. Wollast, pp. 433–454. Wiley, New York.

Mackenzie, F. T., L. M. Ver, C. Sabine, M. Lane, and A. Lerman. 1993. C, N, P, S global biogeochemical cycles and modeling of global change. In *Interactions of C, N, P and S Biogeochemical Cycles and Global Change,* ed. R. Wollast, F. T. Mackenzie, and L. Chou, pp. 1–61. Springer-Verlag, Berlin.

Meybeck, M. 1993. C, N, P and S in rivers: From sources to global inputs. In *Interactions of C, N, P and S Biogeochemical Cycles and Global Change,* ed. R. Wollast, F. T. Mackenzie, and L. Chou, pp. 163–193. Springer-Verlag, Berlin.

Moore, III, B., and D. Schimel, eds. 1992. *Trace Gases and the Biosphere.* UCAR/Office for Interdisciplinary Earth Studies, Boulder, Colo.

Quay, P. D., B. Tilbrook, and C. S. Wong. 1992. Oceanic uptake of fossil fuel CO₂: Carbon-13 evidence. *Science* 256:74–79.

Sabine, C., and F. T. Mackenzie. 1991. Oceanic sinks for anthropogenic CO₂. *Int. J. Energy Environ. Econ.* 1:119–127.

Sarmiento, J. L., and E. T. Sundquist. 1992. Revised budget for the oceanic uptake of anthropogenic carbon dioxide. *Nature* 356:589–593.

Schlesinger, W. H., and A. E. Hartley. 1992. A global budget for atmospheric NH₃. *Biogeochemistry* 15:191–211.

Schwartz, S. E. 1988. "Are global cloud albedo and climate controlled by marine phytoplankton?" *Nature* 347:372–373.

Strain, B. R., and J. D. Cure, eds. 1985. *Direct Effect of Increasing Carbon Dioxide on Vegetation.* DOE/ER-0238. U.S. Department of Energy, Washington D.C.

Tans, P. P., I. Y. Fung, and T. Takahashi. 1990. Observational constraints on the global atmospheric CO₂ budget. *Science* 247:1431–1438.

Thompson, A. M. 1992. The oxidizing capacity of the Earth's atmosphere: Probable past and future changes. *Science* 256:1157–1165.

Warneck, P. 1988. *Chemistry of the Natural Atmosphere*. Academic Press, London.

Wollast, R. 1991. The coastal organic carbon cycle: Fluxes, sources and sinks. In *Ocean Margin Processes in Global Change*, ed. R. F. C. Mantoura, J.-M. Martin, and R. Wollast, pp. 365–381. Wiley, New York.

Wollast, R. 1993. Interactions of carbon and nitrogen in the coastal zone. In *Interactions of C, N, P and S Biogeochemical Cycles and Global Change*, ed. R. Wollast, F. T. Mackenzie, and L. Chou, pp. 195–210. Springer-Verlag, Berlin.

Wollast, R., and F. T. Mackenzie. 1989. Global biogeochemical cycles and climate. In *Climate and Geo-Sciences*, ed. A. Berger, S. Schneider, and J.-Cl. Duplessy, pp. 453–510. Kluwer, Dordrecht, The Netherlands.

Wollast, R., F. T. Mackenzie, and L. Chou, eds. 1993. *Interactions of C, N, P and S Biogeochemical Cycles and Global Change*. Springer-Verlag, Berlin.

Woodwell, G. M. 1983. Biotic effects on the concentration of atmospheric carbon dioxide: A review and projection. In *Changing Climate*, pp. 216–241. National Academy Sciences Press, Washington, D.C.

World Resources Institute. 1988. *World Resources 1988–89*. Basic Books, New York.

II

BIOTIC PROCESSES AND POTENTIAL FEEDBACKS

A

Plants and Plant Communities

A large body of experimental data on the effects of increased concentrations of carbon dioxide on the metabolism of plants supports the assumption that one of the biotic effects of the current changes in the atmosphere will be an acceleraton of carbon fixation through photosynthesis. The assumption has gained further, indirect support through the uncertainty (explained by Woodwell and Mackenzie in Chapters 1 and 2) of defining the current flows of carbon globally among the three major pools of atmosphere, land, and oceans. The least satisfactory quantitative data on global net metabolism on a decadal time scale, and therefore the greatest uncertainty, exist for the terrestrial pools of carbon. The uncertainty is taken as one reason for assuming that additional carbon is being not only fixed on land but also stored there as net ecosystem production.

The topic is not quite so easily dismissed, as the chapters by Allen and Amthor and by Wullschleger and colleagues show. Although carbon fixation in photosynthesis may increase with increasing carbon dioxide levels, the increase is not universal, may not be persistent in perennials over years, and the fate of that carbon in plants and in ecosystems is not necessarily increased storage in plants. The fate hinges heavily on rates of respiration, which also are affected by carbon dioxide concentrations, temperature, and other factors.

One of the most promising techniques for ascertaining whether there has been an increase in the growth of plants as a result of the approximately 30% increase in atmospheric carbon dioxide is to examine the marks of growth that exist in perennial plants whose life extends back before the 20th-century changes in the atmosphere. Study of tree growth, especially the growth in diameter of stems, offers this possibility. There have been various explorations of this topic over recent years, none more comprehensive than the studies of trees growing at the limits of tree growth around the world carried out by Jacoby and D'Arrigo. None of the evidence to date shows an effect of carbon dioxide on the diameter growth of trees. Yet it is difficult to believe that, if there is a general stimulation of plant growth attributable to the increase in carbon dioxide and temperature, such an increase would not be reflected in any degree in diameter growth.

But the topic does not end there. There is a question as to whether forests, especially temperate zone and northern forests, have ever in the few thousand years since the glacial retreat reached a stable point at which gross photosynthesis and total ecosystem respiration were approximately equal. It is possible that the recovery from the glacial retreat continues, albeit slowly, and that we have overlooked a flow of carbon from the atmosphere into northern forests and tundra. Kurz and Apps and colleagues point to succession and stability in forests as a major consideration and call attention as well to the possibility that the changes in climate anticipated for the next decades will introduce sufficient disruptions to release significant new quantities of carbon as carbon dioxide into the atmosphere—a clearly important positive feedback.

Velichko et al. have applied a still different approach, using paleoclimatic records from the glacial periods to infer the changes that can be anticipated as the earth warms. They consider the rates of migration of species as determined from pollen records from glacial time and compare them to anticipated rates of migration of climate over the next decades. They also explore the distribution of frozen ground in the past, currently, and in a hypothesized series of warmer climates. They add a further perspective to the view that a warming will lead to the decay of organic matter in soils of the north and a further release of carbon into the atmosphere. Proof of these analyses, as of most others, will have to await the primary experience. Nevertheless there is abundant insight here into rates of migration of vegetation as climates fluctuated in the glacial past.

3

Plant Physiological Responses to Elevated CO₂, Temperature, Air Pollution, and UV-B Radiation

LEON HARTWELL ALLEN, JR., AND JEFFREY S. AMTHOR

Measurements of the CO_2 and CH_4 content of entrapped gas bubbles in the Vostok Core show a clear correspondence with deuterium content–derived temperatures (Lorius et al. 1990). It has not, however, been established whether changes in these two greenhouse gases have been the primary drivers of the global climatic cycles that have occurred over the last 160,000 years of ice core records. The Milankovich orbital variations of the earth appear to be primary factors in governing global climates (Lorius et al. 1990). Recent analyses of the effects of frequency variations of the orbital obliquity of the earth, superimposed on records of the eccentricity of its orbit, indicate that the frequency variations may induce climatic changes by modulating the seasonal distribution of solar radiation (Liu 1992).

The sharp rises in global temperature indicated by the Vostok Core record, beginning about 13,000 years B.P. and also occurring during the previous interglacial period about 116,000 to 140,000 years B.P. were accompanied by sharp rises in CO_2 concentration from about 200 to about 270 ppm. The lead or lag time in these records cannot be separated by the scientific reader; the changes in the greenhouse gases and the temperature records appear to have occurred simultaneously during the rapid global warming events. During the several stepwise global cooling events that occurred between the two icesheet minima, however, the decreases in temperature apparently occurred several thousand years before decreases in concentrations of CO_2 and CH_4. Setting aside the earlier mindset that rising levels of radiatively active trace gases cause global warming, an equally plausible premise is that climatic changes govern the concentration of CO_2 and CH_4 through effects on fluxes to and from the biosphere and hydrosphere.

This volume focuses on important questions that have been raised again by ice core data. Does the biota govern climatic systems, or does climate respond to physical processes generated without regard to the biosphere? Or are the processes entwined? Or, more likely, are there components of the complex of global climate wherein physical processes govern biotic processes, and others wherein biotic processes govern climate through effects on atmospheric composition or aspects of surface energy exchange, such as albedo, transpiration, and convection?

This chapter examines the responses of plants to rising atmospheric CO_2 and to the various climate change factors that have been investigated within the last 10–15 years. The question underlying this analysis is "Will plant-environment interactions during the coming decades result in a net negative or positive feedback to atmospheric change, i.e., will the plant responses to atmospheric changes speed (positive feedback) or retard (negative feedback) those changes?" More specifically, "Will the storage of carbon in plants and soil be affected by plant physiological responses to the continuing increase in atmospheric CO_2 and change in climate?" Although most physiological research is conducted on relatively small spatial and temporal scales, it is long-term regional plant and soil metabolism that is significant to global carbon cycles.

RESPONSES OF PLANTS TO CO₂

Within the last 15 years, there have been numerous reports and reviews on responses of plants to elevated CO_2 (e.g., Kramer 1981; Gates et al. 1983; Kimball 1983; Lemon 1983; Pearcy and Björkman 1983; Strain and Cure 1985; Bolin et al. 1986; L. H. Allen 1990, 1993; Bazzaz 1990; Amthor 1991; Drake and Leadley 1991; Lawlor and Mitchell 1991; Stitt 1991; Field et al. 1992; Mousseau and Saugier 1992; Bowes 1993; Rogers et al. 1994; Rozema et al. 1993). Studies have been conducted in facilities ranging from leaf chambers and phytotrons to sunlit outdoor exposure systems (i.e., closed-cycle controlled environments and open flow-through exposure systems) to free-air CO_2 enrichment systems (L. H. Allen et al. 1992). With regard to biotic feedbacks of rising CO_2 in the global climate system, the most fundamental questions to be raised are these:

1. Will the net photosynthetic rates of individual leaves and plant canopies be increased by increasing levels of CO_2? If so, by how much?
2. How will photoassimilate be partitioned among the tissues and organs of the terrestrial autotrophs? What will be the projections for short- and middle-term sequestration of carbon?
3. How will the answers to (1) and (2) affect competition among species?
4. How will the answers to (1), (2), and (3) affect grazers?
5. How will the answers to (1), (2), and (3) affect the heterotrophic processes of fungi and microorganisms?
6. How will all of the foregoing processes affect the sequestration of carbon in short-term pools (annual carbon fixation and rapid heterotrophic respiration), middle-term pools (those associated with life cycles of trees and other perennial plants), and long-term terrestrial pools (those associated with the generation of stable soil organic matter)?
7. How will all of these factors affect the net exchange of carbon with the atmosphere and the atmospheric concentrations of CO_2 and CH_4?

This string of questions is also valid when considering each of the presumed climatic change factors (e.g., temperature, precipitation, and solar radiation), either individually or in combination with one another and with CO_2 concentration increases. Other factors, such as soil fertility, can certainly affect the outcome of each of the questions.

Photosynthetic Responses to CO_2

It is common knowledge that each species of terrestrial plants exhibits one of three modes of incorporating CO_2 into plant tissues (see reviews by, e.g., Osmond et al. 1982; Tolbert and Zelitch 1983): the C_3 (95% of known plant species), C_4 (1% of plant species), and crassulacean acid metabolism (CAM) (4% of plant species) metabolic pathways. The C_3 species fix carbon by the classic Calvin cycle. The carboxylating enzyme of the Calvin cycle, RuP_2 carboxylase/oxygenase (rubisco), functions at CO_2 concentrations below its K_M (CO_2) under present atmospheric conditions (Jordan and Ogren 1984). The C_4 plants have evolved an anatomy and biochemistry that, in effect, concentrate CO_2 into the bundle sheath cells of leaves for subsequent assimilation of carbon by the C_3 pathway. That is, CO_2 concentration is already "elevated" in C_4 species, and they should respond less than C_3 species to increasing levels of ambient CO_2 (Collatz et al. 1992); indeed, this has been the case in a number of experiments (Lawlor and Mitchell 1991). CAM species respond positively to elevated CO_2 (e.g., Nobel and Hartsock 1986) but generally display lower photosynthetic rates than either the C_3 or C_4 plants. CAM plants fix CO_2 (actually HCO_3^-) into organic acids such as malate at night. During the day, malic acid is decarboxylated by the mechanisms also present in C_4 bundle sheath cells and the resulting CO_2 is assimilated by the C_3 pathway. Thus, all three plant types use the C_3 pathway during some stage of photosynthesis, but C_3 plants show the greatest potential for response to elevated CO_2 levels. Because most plants are C_3 species, we would expect that on a global scale photosynthesis would be likely to increase with increasing atmospheric CO_2.

Many of the early studies of the effects of elevated CO_2 concentration on plants reported initial increases in leaf or canopy photosynthetic rates, followed by a decrease after exposure for a number of days or weeks as photosynthetic rates acclimated (downward) to high CO_2 levels (Aoki and Yabuki 1977; Kramer 1981; Cure 1985; Oechel and Strain 1985; Peet et al. 1986; Stitt 1991). Other studies, however, have shown no decreases or even increases in photosynthetic rates during long-term exposures to elevated CO_2 levels (Jones et al. 1984, 1985a,b; Valle et al. 1985a,b; Drake and Leadley 1991; Idso and Kimball 1991; Curtis and Teeri 1992). In cases in which decreases in photosynthesis have been observed, starch tended to accumulate in leaves (e.g., Guinn and Mauney 1980; DeLucia et al. 1985) rather than being translocated to sinks of photoassimilate such as growing leaves, roots, and fruits. Root growth restriction might have played a role in several observed declines in photosynthetic capacity following long-term CO_2 enrichment (R. B. Thomas and Strain 1991). Starch accumulation is not, however, always associated with inhibited leaf photosynthesis (L. H. Allen et al. 1988), and the mechanism responsible for potential feedback inhibition of photosynthesis is likely to involve insufficient organic phosphate for sucrose formation and export, rather than direct effects of starch accumulation in leaf chloroplasts (Herold 1980; Sharkey 1990; Stitt 1991).

L. H. Allen (1993) illustrated several types of photosynthetic responses of leaves to elevated CO_2 based on the data of Radin et al. (1987), Campbell et al. (1988), and Sage et al. (1989). Figure 3.1 illustrates these differing responses of leaf photosynthetic CO_2 uptake rate (A or assimilation) versus intercellular CO_2 concentration (C_i) (i.e., A-C_i curves) to long-term CO_2 enrichment:

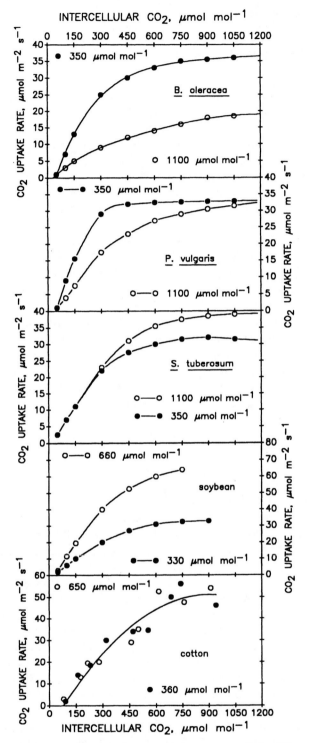

Figure 3.1. Leaf photosynthetic CO_2 uptake rates versus intercellular CO_2 concentration for five types of responses to elevated CO_2 concentrations (cabbage [*Brassica oleracea* L.], kidney bean [*Phaseolus vulgaris* L.], potato [*Solanum tuberosum* L.], soybean [*Glycine max* L. Merr.], and cotton [*Gossypium hirsutum* L.]). Different symbols refer to the CO_2 concentrations during plant growth. (After Allen 1994.)

1. Both the initial slope of the photosynthetic rate and the CO_2-saturated photosynthetic rate were decreased by elevated CO_2 concentration.
2. The initial slope was decreased, but the CO_2-saturated photosynthetic rate was affected little.
3. The initial slope was not affected, but the CO_2-saturated photosynthetic rate was increased.
4. Both the initial slope and the extrapolated CO_2-saturated photosynthetic rate were increased.
5. Neither the initial slope nor the CO_2-saturated photosynthetic rate was changed.

There are two other possible combinations of leaf-level responses to elevated CO_2 concentration. The main message is that there may be no single type of acclimation by leaves to elevated levels of CO_2. The response of leaves to elevated CO_2 concentration may depend on the inherent sink strength of a species (Stitt 1991). Pretreatment conditions and even the conditions under which measurements are made may predestine the outcome of the experiment. The up- or down-regulation of photosynthesis may involve anatomical as well as biochemical responses. For instance, photosynthetic capacity per unit leaf area may be increased as a result of increasing leaf thickness accompanied by production of an additional layer of palisade cells, as observed in soybean (*Glycine max* L.) (Hofstra and Hesketh 1975; J. F. Thomas and Harvey 1983; Vu et al. 1989). On the whole, present data show that CO_2 enrichment enhances net photosynthesis in C_3 plants, and this situation might contribute to a *negative* feedback on atmospheric CO_2 increase, at least in the short term.

Stomatal Responses to CO_2

Stomatal conductance is inversely related to CO_2 concentration (Morison and Gifford 1983; Morison 1987). Atmospheric water vapor is the most important greenhouse gas (Lorius et al. 1990), and the effects of stomatal closure elicited by CO_2 on climate via hydrologic cycles are unclear. Effects of stomatal closure on regional and global transpiration can be much smaller than effects of stomatal closure on individual leaf and plant transpiration because of negative feedbacks in the surface layer and planetary boundary layer (Jarvis and McNaughton 1986; Jacobs and de Bruin 1992).

The link between CO_2 and stomatal conductance appears to be at the level of intercellular CO_2 (Mott 1988). Because an increase in CO_2 level accelerates photosynthesis, slows photorespiration, and decreases leaf surface conductance, the instantaneous water use efficiency (CO_2 assimilated per unit water transpired) on a leaf area basis is positively related to CO_2 concentration (L. H. Allen et al. 1985). Although decreased stomatal conductance owing to elevated CO_2 concentration can decrease transpiration on a leaf area basis, increased leaf area resulting from elevated CO_2 concentration may offset the leaf-level transpiration decline when transpiration is considered at the level of the canopy (L. H. Allen 1991; Eamus 1991). Increased water use efficiency under elevated levels of CO_2 may be due more to increased CO_2 assimilation than to decreased transpiration, especially for C_3 species (L. H. Allen et al. 1985; Jones et al. 1985a). Decreased stomatal conductance and leaf-level transpiration will tend to increase leaf and canopy temperature, and this elevation will reduce somewhat the leaf level decreases in transpiration resulting from stomatal closure caused by an increased vapor pressure gradient from the leaf to the atmosphere (L. H. Allen et al. 1985; L. H. Allen 1991, 1993). The increase in canopy temperature has the potential to accelerate developmental rate at a given air temperature. Finally, decreased stomatal conductance under elevated CO_2 levels may contribute to a small *positive* feedback to increasing surface temperatures.

Partitioning

Following photosynthesis, both temporally and quantitatively, the most important physiological processes with respect to carbon storage and cycling in higher plants are the coarse partitioning of photosynthate among organs and the subsequent fine partitioning of carbon among storage, biosynthesis (growth), and respiration in those organs. Controls on partitioning of carbon are presumably largely genetic, with enough flexibility to respond adequately to changes in the environment likely to be encountered by a specific plant. Effects of elevated CO_2 concentration and nonstructural carbohydrate status on partitioning are poorly understood from a mechanistic perspective, for the state of the science is essentially phenomenological (e.g., Farrar 1988). Thus, broad and firm predictions of effects of increasing CO_2 levels on carbon partitioning are premature; we must rely on rather scattered empirical observations and models.

Evidence exists suggesting that elevated CO_2 levels, i.e., greater photosynthesis and elevated nonstructural carbohydrate levels, can result in a relatively greater partitioning of carbon to roots versus shoots (e.g., Rogers et al. 1992, 1994), although generalizations based on this evidence are premature (H. H. Rogers, personal communication). Increased root growth under elevated CO_2 conditions might permit water extraction from a greater soil volume, which could be important in a future warmer and drier environment. A relative increase in partitioning to roots is also of potential significance to enhanced storage of carbon below ground as levels of atmospheric CO_2 increase. The lack of a strong whole-plant growth response to elevated CO_2 concentration observed by, e.g., Norby et al. (1992) and Körner and Arnone (1992), in spite of markedly increased net CO_2 assimilation by leaves, may have been related to increased partitioning of carbon to roots where it entered the soil (as exudates or during fine root turnover) and accelerated the *cycling* of carbon through plants and the soil. Increased relative partitioning of carbon to roots or no change in relative partitioning but an increase in plant size under conditions of elevated CO_2 increases carbon transport to belowground tissues. To the extent that this situation leads to increased soil carbon content in the short or long term, it might contribute to a *negative* feedback on short- or long-term atmospheric CO_2 concentration increase, respectively. Direct measurements pertaining to this point for plants in the field are rare and inconclusive.

Respiration

Half or more of the carbon assimilated in photosynthesis can be released as CO_2 during subsequent plant respiration, with the remaining phytomass subject to eventual oxidation by decomposers. Plant respiration is not a simple oxidation of the products of photosynthesis, but is the primary means of supplying plant cells with carbon skeletons (biosynthetic precursors), usable energy, and reductant (Beevers 1961, 1970; Amthor 1989). Respiration is linked to growth of new phytomass and to processes maintaining

Figure 3.2. Effects of ambient CO_2 level on apparent respiration rate in leaves of *Rumex crispus* at 15°C. Data are from a single leaf attached to a plant grown in a greenhouse at a daytime CO_2 concentration of ~350 ppm. Measurements were made in the dark in a leaf cuvette in which CO_2 level was cycled between 350 (o) and 650 (•) ppm. Sufficient time was allowed at each CO_2 concentration to attain a steady CO_2 efflux rate. In all cases, apparent respiration rate increased when CO_2 level was decreased from 650 to 350 ppm and apparent respiration rate decreased when CO_2 level was increased from 350 to 650 ppm. Similar results were obtained with other leaves from other plants. (After Amthor et al. 1992.)

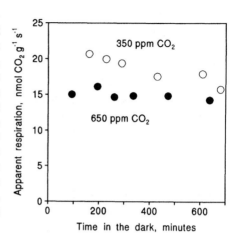

existing phytomass so that factors that influence growth or maintenance processes will in turn influence respiration (Ryan 1991; Amthor 1994).

Growth of plants under conditions of elevated CO_2 often results in a decrease in *specific* (per unit dry mass) respiration rate (Bunce 1992; Poorter et al. 1992; Amthor 1994). Because of increased growth resulting from elevated CO_2 concentration, plant respiration on a ground area basis can be less influenced than specific respiration rate by CO_2 concentration (Amthor 1991; Baker et al. 1992). In fact, plant respiration can be increased on a ground area basis by CO_2 enrichment owing to increased photosynthesis and growth, but on a nitrogen or protein basis respiration may be relatively unaffected by CO_2 level. This finding suggests that the long-term effects of CO_2 on respiration are mediated through effects on plant composition, and in particular by an increase in the C:N ratio as CO_2 concentration increases (Wullschleger et al. 1992; Amthor 1994; but see Ziska and Bunce 1993). Concomitant measures of growth rate, tissue composition, and respiration rate are required in order to unravel the links between long-term elevated CO_2 levels and respiration and growth. A decline in whole-plant specific respiration rate as a result of elevated CO_2 levels can contribute to a *negative* feedback on atmospheric CO_2 increase and thus augment the effects of increased photosynthesis on carbon storage.

Over and above the effects of long-term daytime CO_2 level on subsequent respiration, it appears that CO_2 has a more direct, inhibitory effect on nighttime CO_2 efflux or apparent respiration (Figure 3.2) (Amthor, et al. 1992; Amthor 1994). The mechanism(s) underlying an immediate, reversible slowing of apparent respiration by ambient CO_2 level is unknown. A direct inhibition of respiratory metabolism by CO_2 has the potential to increase susceptibility to various stresses (Gale 1982) because it is respiration that supplies much of the energy, reductant, and carbon skeletons used in repair and detoxification processes. A simple direct inhibition of respiration by CO_2 could contribute to a *negative* feedback on increasing atmospheric CO_2, but to the extent that stress tolerance is reduced it might contribute to a *positive* feedback.

Growth and Biomass Yield

Crop biomass accumulation and grain yield are positively related to CO_2 concentration (e.g., Kimball 1983; Rogers et al. 1983; Acock and Allen 1985; Cure 1985; L. H. Allen et al. 1987; L. H. Allen 1990; Baker and Allen 1993) with a $34 \pm 4\%$ increase in biomass or yield indicated for a doubling of ambient CO_2 concentration (Kimball 1983; L. H. Allen et al. 1987). When a number of tree species were exposed to a doubling of CO_2 level (mostly for short periods of less than one year), dry mass increases averaged 32% but showed a wide range of responses (Chapter 4, this volume). Recently, Poorter (1993) compiled selected information from the literature and reported that the average growth stimulation of 156 plant species was 37% for a doubling of atmospheric CO_2 concentration. The dry weight growth stimulation was 41%, 22%, and 15% for C_3, C_4, and CAM species, respectively. Within the herbaceous C_3 types, crop plants responded more than wild species (58% versus 35%, respectively). Fast-growing wild species responded more than slow-growing species (54% versus 23%, respectively). Poorter (1993) also reported that nitrogen-fixing species and species with large sink strength were more responsive to a doubling of CO_2 concentration and that herbaceous dicots tended to be more responsive than monocots. These responses to CO_2 level should be considered "best case" responses, however, since (1) only vegetative stage responses (before flowering) were analyzed, (2) plants growing in competition were not included, and (3) data were selected from the most favorable environmental conditions when a range of environmental factors was used in the original studies.

The best way to obtain quantitative data on plant responses to CO_2 level is to conduct studies across a wide range of subambient and superambient CO_2 concentrations. For example, L. H. Allen et al. (1991) grew soybean throughout its life cycle at six levels of CO_2 (160, 220, 280, 330, 660, and 990 ppm). An expression to relate total shoot dry mass accumulation per unit ground area (Y) during the linear phase of growth of soybean to CO_2 treatment was derived. This relationship accounted for linear growth owing to effects of the passage of time (after canopy closure) from 24 to 66 days after planting (DAP) and nonlinear growth owing to CO_2 concentration:

$$Y = \beta_0 + \beta_1(DAP-D)(C-G)/(C + K_c) \qquad (3.1)$$

where β_0 is the y-axis intercept, β_1 is the maximum asymptotic value of the daily dry mass accumulation rate as the parameter carbon level increases, D is a DAP offset parameter to remove the time lag of seedling emergence and nonlinear early crop growth, C is the CO_2 concentration, G is the apparent CO_2 compensation point, and K_c is a hyperbolic function shape factor. This equation can also be applied to other components of dry mass or leaf area accumulation (Table 3.1). For the experiment of L. H. Allen et al. (1991), the values of β_0, β_1, D, G, and K_c for total dry mass accumulation were 4.24 g m^{-2}, 26.9 g m^{-2} d^{-1}, 24.0 d, 99 ppm, and 168 ppm, respectively (Table 3.1). The total shoot dry mass accumulation rates were 5.0, 8.4, 10.9, 12.5, 18.2, and 20.7 g m^{-2} d^{-1} for CO_2 treatments of 160, 220, 280, 330, 660, and 990 ppm, respectively (Table 3.2). The value of 12.5 g m^{-2} d^{-1} is comparable to common growth rates of many C_3 crops (Loomis 1983; Loomis and Connor 1992).

L. H. Allen et al. (1987) used a nonlinear model (Figure 3.3) to predict photosynthetic rate, final biomass, and seed yield responses of soybean to CO_2 relative to a CO_2 concentration of 330 ppm. The rise of CO_2 level from preindustrial values to present values was predicted to cause a 13% increase in seed yield (Table 3.3) (see also Polley et al. 1993). Furthermore, a doubling from 315 to 630 ppm could cause a 32% increase in seed yield, an increase that is in close agreement with that found in the survey of plant

Table 3.1. Estimated Parameters of the Soybean Growth Model Fitted by Nonlinear Regression Statistical Procedures to CO_2 Level and DAP over the 24- to 66-DAP Linear Growth Phase[a]

Dependent variable	β_0 (g m^{-2})	β_1 (g m^{-2} DAP^{-1})	D (DAP)	G (ppm)	K_c (ppm)[b]
		Parameter of growth model			
Total shoot dry weight	4.243	26.881	23.97	98.7	167.8
Stem dry weight	2.951	7.682	22.96	120.6	214.1
Petiole dry weight	0.987	3.930	22.14	116.3	183.8
Leaf dry weight	4.556	7.206	19.53	101.6	84.2
Leaf area	0.493[c]	0.214[d]		20.41	119.1

[a]Computed originally on a per-plant basis and then adjusted to g m^{-2} based on an average of 28 plants m^{-2} over the 24- to 66-DAP interval.
[b]The apparent Michaelis constant is $K_c + G$. See the text.
[c]In m^2 m^{-2}.
[d]In m^2 m^{-2} DAP^{-1}.

responses to doubled CO_2 (34 ± 4%) by Kimball (1983). The photosynthetic rate responses to CO_2 concentration increase were measured under midday conditions and thus these relative responses were greater than the long-term biomass accumulation responses. The seed yield responses were less than biomass responses because of greater vegetative growth and less efficient conversion to seed under high levels of CO_2. Advances in crop breeding and selection may be able to provide more efficient conversion to seed yield in the future.

Fruit trees and some vegetable crops appear to be just as responsive as soybean to increases in atmospheric CO_2 levels (e.g., Zimmerman et al. 1970; Krizek et al. 1971, 1974; Koch et al. 1983, 1986, 1987; Brakke 1989; Brakke and Allen 1994). For example, enrichment of citrus trees with 300 ppm of CO_2 above ambient level in field open-top chambers has maintained a growth advantage over a period of 3 years. Leaf photosynthetic rates and whole-plant carbon sequestration (wood accumulation) rates have been reported to be 2.2- and 2.8-fold greater for individual sour orange trees (*Citrus aurantium* L.) when they are exposed continuously to elevated CO_2 levels (Idso and Kimball 1991; Idso et al. 1991). The plants were amply irrigated and heavily fertilized. The large increases in biomass accumulation were undoubtedly due to the incremental effects of

Table 3.2. Growth Rates during the Linear Phase of Growth of Soybeans Grown at Subambient and Superambient Concentrations of CO_2 in 1984[a]

Plant component	160	220	280	330	660	990
	CO_2 concentration (ppm)					
Total shoot dry weight (g m^{-2} day^{-1})[b]	5.0	8.4	10.9	12.5	18.2	20.7
Stem dry weight (g m^{-2} day^{-1})	0.8	1.7	2.4	2.9	4.7	5.5
Petiole dry weight (g m^{-2} day^{-1})	0.5	1.0	1.4	1.6	2.5	2.9
Leaf dry weight (g m^{-2} day^{-1})	1.7	2.7	3.4	3.9	5.3	5.8
Leaf area (m^2 m^{-2} day^{-1})	0.070	0.116	0.140	0.153	0.185	0.195

[a]Computed during 24 to 66 DAP from equation 3.1.
[b]Includes seed and podwalls, and stems, petioles, and leaves.

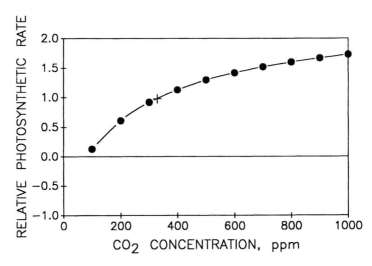

Figure 3.3. Relative canopy photosynthetic rates obtained from a nonlinear model of responses of soybean to elevated CO_2 concentration.

growth of larger crowns under elevated CO_2 conditions in conjunction with higher leaf photosynthetic rates. No competition for light between trees occurred since the trees were widely spaced. Nevertheless, all the experiments with citrus trees show that these species have a large photosynthetic capacity for responding to elevated CO_2 concentrations (Brakke 1989; Idso et al. 1991; Brakke and Allen 1994).

The large, linear increase in leaf photosynthetic rate with increasing ambient CO_2 concentration for citrus trees may be related to their low conductances for CO_2 diffusion from intercellular spaces to chloroplasts (mesophyll conductance) in comparison to other species that have been studied (e.g., Lloyd et al. 1992; Loreto et al. 1992). Thus, *Citrus*—especially widely spaced, irrigated, and fertilized plants—may not be a good example from which to make extrapolations of general plant responses, and especially not those of forest ecosystems, to elevated atmospheric CO_2 levels. These results, which imply a very strong *negative* feedback to increasing atmospheric CO_2 levels, are not necessarily relevant to global carbon cycling since irrigated and fertilized *Citrus* is a minute component of the total global phytomass.

Forest tree leaf photosynthesis responds positively to increases in CO_2 concentration (e.g., Jurik et al. 1984; Curtis and Teeri 1992). Rather than repeating the details of

Table 3.3. Soybean Responses to Rising CO_2 Predicted by a Modified Michaelis-Menten Model[a]

Assumed period (years)	CO_2 level change (ppm)	Percentage increase		
		Photosynthetic rate	Final biomass	Seed yield
1800–1958	276–315	12	10	8
1800–1986	276–345	20	17	13
1958–2058	315–630	53	43	32

Source: Allen et al. (1987).

Table 3.4. Comparisons of CO_2 Enrichment Effects on Photosynthetic CO_2 Exchange Rates and Biomass Growth Modification Ratios

CO₂ concentration (ppm)	CO₂ exchange rate (μmol m^{-2} s^{-1})			Biomass growth modification ratio	
	Average daily	Early morning	Afternoon	Measured	Calculated
		Sour orange			
350	3.1 [1.00]	7.1 [1.00]	1.2 [1.00]	1.00	
650	6.8 [2.20]	11.1 [1.56]	4.8 [4.00]	2.79	
		Yellow poplar			
355	7.4 [1.00]			1.00	(1.00)
505	10.8 [1.46]			1.18	(1.46)^a
655	12.3 [1.66]			1.15	(1.48)^a
	Artificial tropical ecosystem				
340	7.5 [1.00]			1.00	(1.00)^b
610	15.0^c [2.00]			1.11	(1.21)^b

^{a}BGMR computed from photosynthetic rate and LAI measurements if carbon were converted directly to biomass only by the CO_2 enrichment treatments.
^{b}BGMR computed from the statement by Körner and Arnone (1992) that the initial biomass was 700 g m^{-2}.
^{c}Assumed canopy-only photosynthetic rate based on qualitative description of Körner and Arnone (1992).
Sources: Sour orange: Idso et al. (1991), Idso and Kimball (1991); yellow poplar: Norby et al. (1992); artificial tropical ecosystem: Körner and Arnone (1992).

several earlier reviews (e.g., Eamus and Jarvis 1989; Field et al. 1992), we will analyze the results of two recent experiments on exposure of trees and ecosystems to elevated CO_2 levels (Körner and Arnone 1992; Norby et al. 1992). This analysis addresses the second question on the list presented earlier.

The biomass accumulation of yellow poplar trees after 28 months (2.7 growing seasons) of exposure to CO_2 enrichment of +150 and +300 ppm was relatively small (Norby et al. 1992). The biomass growth modification ratio (BGMR) reflected a tree dry mass only 18% and 15% greater than that of ambient CO_2–grown trees for the +150- and +300-ppm CO_2 treatments, respectively (Table 3.4). Midday net leaf CO_2 uptake rates, however, were 7.4, 10.8, and 12.3 μmol (CO_2) m^{-2} s^{-1} for the ambient control (355-ppm) and the +150- and +300-ppm CO_2 treatments, respectively (Table 3.4). That is, the +150- and +300-ppm treatments enhanced leaf-level midday CO_2 uptake rates by 46% and 66%, respectively, compared to ambient CO_2 levels. Based on individual tree leaf areas of 8.1, 8.1, and 7.2 m^2, the potential whole-tree photosynthetic rates under saturating light were 59.9, 87.5, and 88.6 μmol (CO_2) s^{-1} per tree. (In addition, leaf nighttime CO_2 efflux was slowed by elevated CO_2 concentration.) If that potential photosynthesis (ignoring the diurnal patterns of light) had been converted into standing biomass at a constant efficiency, the elevated CO_2 level would have enhanced biomass accumulation by 46% and 48% for the +150- and +300-ppm treatments, respectively (Table 3.4). The discrepancy between the potential and actual biomass accumulation might have been due to increased CO_2 efflux from the roots and soil, since soil CO_2 efflux rate was 24% and 22% greater for the +150- and +300-ppm CO_2 treatments, respectively. Carbon storage in the soil under the trees was also possibly greater with elevated CO_2 levels. But neither whole-day, whole-season, nor whole-experiment carbon flux and biomass budgets can be obtained from the data provided. Only long-term, closed-cycle, controlled environment chamber data similar to those reported by Jones et al. (1985b) can be used to determine such carbon balances. In any case, the accumulation of carbon in plant material was not as greatly stimulated by elevated CO_2 as was leaf net CO_2 uptake. It is noteworthy that no supplemental water or nutrients were supplied to these forest trees.

Another study of biomass accumulation, for an artificial tropical ecosystem in closed air-circulation polyethylene chambers, was reported by Körner and Arnone (1992). After 100 days of growth in either 340- or 610-ppm CO_2, total biomass was 11% greater ($P = 0.102$) in the elevated-CO_2 chambers (Table 3.4). Since these chambers contained 15 species of dicots and monocots, part of the population might have been C_4 plants (this is not stated in Körner and Arnone 1992) with a smaller response to elevated CO_2 levels than would be expected in a pure C_3 plant community. Furthermore, only final biomass was reported, with no specific measurement of incremental biomass over the treatment period. If the initial total ecosystem biomass was 700 g m^{-2} (Körner and Arnone 1992), then the biomass accumulation enhancement by the elevated CO_2 was about 21% (Table 3.4). Canopy CO_2 uptake near the end of the 100-day experiment, however, was almost twice as high in the elevated-CO_2 treatments (no quantitative values were given for elevated CO_2) as in the ambient-CO_2 treatments (Table 3.4).

When Körner and Arnone (1992) exposed both treatments to ambient CO_2 levels, they reported identical canopy CO_2 uptake rates of 7.5 μmol m^{-2} s^{-1}, after correcting for CO_2 evolution from the soil. Thus, they saw no evidence of loss of photosynthetic capacity of the artificial tropical ecosystem after long-term exposure to elevated CO_2. Since no mention is made of either actual photosynthetic photon flux densities or the time of year of the experiment (both serious omissions for any plant growth experiment), a strict interpretation is not possible. Körner and Arnone (1992) reported a large loss of carbon and nutrients from the rooting medium system. The "soil" was a mixture of sand and vermiculite, only 0.2 m thick, overlain by a thin layer of leaf and bark compost. This rooting medium may not be conducive to retention of mineral nutrients. Greater rooting depths in natural soils will likely contribute to less nutrient leaching.

The experiments by both Norby et al. (1992) and Körner and Arnone (1992) resulted in a doubling of leaf or canopy photosynthetic CO_2 uptake rates with a near doubling of CO_2 level. The dry mass accumulation, however, was affected to a markedly smaller degree; much of the additional carbon uptake is not accounted for in long-term biomass accumulation. Possibly, some of the increased photoassimilate is lost through fine root sloughing or through root exudation and oxidation by soil microorganisms. Fine root penetration into the leaf litter and exudation from roots may have contributed to the very large loss of organic matter from the "soil" system of Körner and Arnone (1992). Fine root production and mortality can differ noticeably between slightly different sites (e.g., Hendrick and Pregitzer 1993), and these patterns may differ further when elevated CO_2 concentration enhances the carbohydrate supply to roots.

Few long-term (multiple-year) experiments studying the effect of CO_2 enrichment on whole-plant woody species have been conducted, although a number of short-term studies have been reported (Poorter 1993; Chapter 4, this volume). Moreover, these studies have varied greatly in their methods as well as their results. We cannot tell whether these results reflect differences in the experimental methods, inherent differences in responses by different species and systems to elevated CO_2 concentration, or a combination of the two. In any case, the differences are nearly as dramatic as the individual leaf responses illustrated in Figure 3.1. Nevertheless, these studies indicate that growth and biomass accumulation of woody species under conditions of increasing atmospheric CO_2 will contribute to a *negative* feedback on CO_2 accumulation.

RESPONSES OF PLANTS TO TEMPERATURE AND INTERACTIONS BETWEEN TEMPERATURE AND CO_2

The effects of temperature on plant physiology have been well studied (e.g., Long and Woodward 1988), but only a limited amount of work concerning interactions between rising temperature and CO_2 has been conducted (Eamus 1991; Rawson 1992).

Effects on Photosynthesis

The temperature optimum for photosynthesis varies depending on the ecological niche occupied by the plant, as well as on physiological adjustments. The temperature optimum of many temperate-region species may be 25–30°C at ambient CO_2 levels (Berry and Björkman 1980), but this optimum may increase by 5–10°C as the CO_2 concentration is doubled (Farquhar et al. 1980; Pearcy and Björkman 1983; Long 1991). Species adapted to hotter and drier climates have higher temperature optima for photosynthesis. For some perennial species the optimum temperature for photosynthesis and the upper temperature limit for photosynthesis may increase by up to 10°C from spring to summer and then decrease again in the fall (Lange et al. 1974; Berry and Björkman 1980). If this acclimation potential exists in crop plants, or if it can be introduced, then impacts of increasing temperature on agriculture may be less severe than previously assumed (Adams et al. 1990; Hall and Allen 1993). If this type of genetic adaptability exists in part in components of natural ecosystems, then ecosystems may evolve, rather than collapse, in response to gradual climatic changes. The long replacement time of individuals in forests, however, may not allow evolution or genetic adaptation processes to come into play in forests soon enough to avoid serious ecological disruption if the temperature increases as rapidly as predicted.

Photosynthetic rates of C_4 plants tend to increase with temperature to a greater extent than those of C_3 plants at current levels of atmospheric CO_2. For C_3 and C_4 plants adapted to similar climates, leaves of C_4 plants generally have a higher temperature optimum for photosynthesis, as well as a higher overall photosynthetic rate at the temperature optimum (e.g., see Pearcy and Björkman 1983). High concentrations of CO_2 can give C_3 plants a photosynthesis-versus-temperature response that is similar to that of C_4 plants (Long 1991) (Figure 3.4). Thus, CO_2 enrichment in combination with higher temperature enhances net photosynthesis in C_3 plants, and this relationship might contribute to a *negative* feedback on atmospheric CO_2 increase.

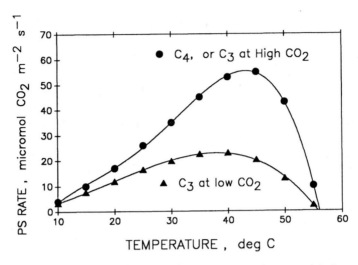

Figure 3.4. Photosynthetic rate versus temperature for C_3 and C_4 leaves.

Partitioning

Generally, a moderate increase in temperature results in an increase in the ratio of whole-plant growth to photosynthesis, i.e., low temperature results in enhanced storage of nonstructural carbohydrates at the apparent expense of growth. The mechanisms underlying the accumulation of nonstructural carbon at low temperature are not well understood, but it is apparent that growth, translocation, or perhaps respiration itself is more sensitive to low temperature than is photosynthesis. Warming can overcome this imbalance in cool habitats. Warming also tends to result in relatively greater partitioning of carbon to shoots compared to roots (Farrar and Williams 1991), and this process can enhance photosynthesis by providing increased leaf area and promoting interception of photosynthetically active radiation.

Farrar and Williams (1991) point out that increasing CO_2 concentration and increasing temperature may have opposite effects on partitioning. Warm plants have lower levels of nonstructural carbohydrates and lower ratios of root to shoot than cool plants. Plants in high-CO_2-level environments have large nonstructural carbohydrate stores and increased ratios of root to shoot. The combined effects of warming and increased CO_2 level on partitioning are less clear. Warming can increase specific respiration rate, whereas an elevated CO_2 level can slow it. Increased photosynthesis resulting from an elevated CO_2 level in combination with the increased sink metabolism (growth) allowed by warming has the potential to produce larger plants and increase carbon storage in terrestrial ecosystems as CO_2 concentration and temperature increase. That this is not clearly the case in experiments conducted to date points to our ignorance of important controls on growth, productivity, partitioning, respiration, litter production, and carbon sequestration processes. The processes involved in partitioning are too complex to permit speculation on net positive or negative feedbacks on atmospheric CO_2 concentration.

Respiration

In the short term, an increase in temperature is accompanied by an increase in respiration rate, at least up to the very high (and usually nonphysiological) temperatures at which structural damage to metabolic machinery occurs (Forward 1960). For individual plant organs, a Q_{10} of about 3 may apply at low temperatures (0–10°C), with the Q_{10} decreasing as the temperature increases (James 1953). The rate of respiration at any given temperature is generally higher in plants adapted to a colder as compared to a warmer climate (Forward 1960; Billings and Mooney 1968). Most available evidence also indicates that plants acclimate to a change in temperature within a few days with respect to respiration rate (Amthor 1994). We assume that this adjustment is a manifestation of acclimation by various growth and maintenance processes, with respiration responding in turn. (It appears, however, that there are some more or less "set" temperature limits to developmental, and therefore growth, processes.) Acclimation to a change in temperature thus diminishes the long-term effects of changing temperature on respiration compared to the responses expected from observations of short-term changes in temperature. Because respiration may be tightly linked to growth (Amthor 1994), an increase in temperature above the optimum for growth will slow growth and, concomitantly, may decrease respiration. The slowing of specific respiration rate with long-term CO_2 enrichment may be greatest at low compared to high temperatures (Ziska and Bunce 1993).

The Q_{10} of ecosystem respiration may be much lower than the Q_{10} for individual plants or plant organs, suggesting overriding controls at the system level. For "balanced" benthic, freshwater microecosystems, Beyers (1962) noted that short-term (24-h) changes in temperature of circa 7°C above and below the normal temperature had a negligible effect on ecosystem respiration. (For a Q_{10} of 2, a 7°C increase in temperature would cause a 62% increase in respiration.) From forest respiration rates reported by Woodwell and Dykeman (1966), it can be calculated that ecosystem Q_{10}s were about 1.34 and 1.39 from 0 to 10°C (winter) and 10 to 20°C (summer), respectively, as opposed to the much greater values expected for, say, individual leaves. These observations raise the important question of the respiratory responses to temperature, and to a change in climate, by whole ecosystems. Both baseline respiration rates and whole-ecosystem respiratory response to temperature may be governed more by system photosynthesis and growth than by temperature itself.

There is a need for measurements of growth and respiratory responses by plants and ecosystems to long-term changes in temperature. Work conducted to date has generally focused on herbaceous species and seedlings of woody plants. Most measurements have been of single organs or plants; little is known about the effects of temperature on the metabolism of ecosystems in nature. The few measures that are available are simple correlations between temperature and CO_2 exchange rates. Knowledge of cause-and-effect relationships is missing, and observations made to date could reflect the effects of some factor that normally covaries with temperature but that might be uncoupled from temperature as a result of changes taking place in the atmosphere and biosphere. Although information is scanty and variable, the effects of increases in global temperature on plant and ecosystem respiration are likely to contribute to *positive* feedbacks on increasing atmospheric CO_2 levels.

Biomass Growth Modification Ratio

Response curves of photosynthetic rate to temperature at ambient and elevated CO_2 concentrations indicate that the photosynthesis and biomass accumulation of many C_3

plant species could increase with both increasing CO_2 concentration and increasing temperature (Rawson 1992), at least up to a threshold value or temperature optimum. Idso et al. (1987) defined a plant growth modification factor, herein called the biomass growth modification ratio (BGMR) after L. H. Allen (1993), to express growth responses at an elevated level of CO_2 relative to growth responses at an ambient or control level of CO_2 (cf. BGMR to β, the biota growth factor of Bacastow and Keeling [1973] discussed by Gates [1985]). Idso et al. (1987) found that BGMR was somewhat linearly related to mean daily temperature. BGMR as a function of temperature is defined herein as

$$\text{BGMR} = (G_{\text{elevated}}/G_{\text{control}}) = 1 + m(T - T_1) \tag{3.2}$$

where G_{elevated} and G_{control} are growth responses at elevated and control levels of CO_2, respectively; m is the slope of BGMR with respect to temperature ($°C^{-1}$); T is the mean daily temperature ($°C$); and T_1 is the temperature at which BGMR is unity. Using data for vegetative growth of several species of plants—cotton (*Gossypium hirsutum*), water hyacinth (*Eichhornia crassipes* [Mart.] Solms), water fern (*Azolla pinnata* var. *pinnata*), carrot (*Daucus carota* L. var. *sativa* cv. Red Cored Chantenay), and radish (*Raphanus sativa* L. cv. Cherry Belle)—BGMR ranged from about 0.39 at 12°C to about 2.31 at 34°C, with a slope of $0.087°C^{-1}$. This relationship, derived from the data of Idso et al. (1987), showed a negative effect of elevated CO_2 at mean daily temperatures below $T_1 = 19°C$ (i.e., BGMR values were less than unity). Baker et al. (1989) observed that higher temperatures promoted more rapid biomass accumulation rates of soybean under conditions of CO_2 enrichment during early vegetative stages of growth. They did not, however, see this type of BGMR response for final seed yields of soybean grown in a CO_2 concentration (330 versus 660 ppm) × temperature (22.8, 27.8, and 32.8°C) interaction experiment. L. H. Allen (1991) calculated slopes of the soybean BGMR at final harvest to be -0.026 and $-0.031°C^{-1}$ for total dry mass and seed yield, respectively.

We do not know how universal the enhancement of photosynthesis and vegetative growth may be when plants are exposed to the combination of elevated CO_2 levels and high temperatures. The large BGMR at high temperatures, steep slope of BGMR, and concomitant relatively high T_1 reported by Idso et al. (1987) and S. G. Allen et al. (1990) may be partially confounded by the experimental conditions. Several of the data sets with low mean temperatures were associated with intervals of short photoperiod, low total daily solar radiation, and low nighttime minimum temperature. Moreover, BGMR is likely to vary among species (Idso and Kimball 1989). Nevertheless, higher temperature in combination with CO_2 enrichment appears to enhance vegetative biomass accumulation in C_3 plants, and this relationship might contribute to a *negative* feedback on atmospheric CO_2 increase.

Reproductive growth of plants appears to respond quite differently to temperature increases than does vegetative growth. For example, Baker et al. (1990) and Baker and Allen (1993) reported that in a cultivar of tropical lowland rice (cv. IR-30) grain yield decreased linearly about 10% for each 1°C increment across the range of 26 to about 36°C under CO_2 treatments of both 330 and 660 ppm. Linear regressions of grain yield (Y) were expressed as $Y = m(T - T_0)$; where the units of Y, m, and T are Mg ha^{-1}, Mg ha^{-1} $°C^{-1}$, and °C, respectively. Grain yield extrapolated to zero at $T = T_0 = 37°C$ for both CO_2 treatment levels. The values of the slopes (m) were -0.61 and -0.83 Mg ha^{-1} $°C^{-1}$ for both

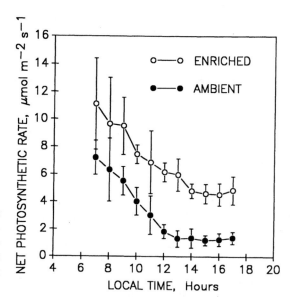

Figure 3.5. Leaf photosynthetic rates of sour orange trees exposed to ambient CO_2 concentrations and ambient + 300 ppm CO_2 concentrations in Phoenix, Arizona. (After Idso et al. 1991.)

CO_2 treatments. The ratio of the slopes (1.36) indicates a constant 36% increase in grain yield resulting from a doubling of CO_2 concentration across the same temperature range. Thus, there was apparently no change in BGMR with increasing temperature for this rice cultivar.

The steady decline in rice grain yield with increasing temperatures was accompanied by sharp declines in number of filled grains per panicle, small declines in grain mass per seed, and small increases in the number of panicles per plant (Baker et al. 1990; Baker and Allen 1993). If this elevated temperature effect on seed growth is a general phenomenon, it may completely negate the elevated temperature benefits for vegetative growth at elevated CO_2 levels, at least for seed-producing agricultural crops.

The large biomass accumulation response to elevated CO_2 of the sour orange trees of Idso and Kimball (1991) seems inexplicable at first glance. The results are reported in several brief overlapping papers rather than as one comprehensive treatise. The response is related in large part to the diurnal course of leaf photosynthesis under summer conditions at Phoenix, Arizona, reported by Idso et al. (1991) and summarized here in Figure 3.5. Under ambient CO_2 concentrations, leaf CO_2 uptake rates decreased linearly from about 7.1 μmol m^{-2} s^{-1} at 0700 hours local time (Mountain Standard Time) to about 1.2 μmol m^{-2} s^{-1} at 1300 hours, where they remained until at least 1700 hours. On the other hand, the sour orange trees enriched to 650 ppm CO_2 had leaf CO_2 uptake rates that decreased from about 11.1 μmol m^{-2} s^{-1} at 0700 hours to about 4.8 μmol m^{-2} s^{-1} at 1400 hours, where they remained to at least 1700 hours (Table 3.4; Figure 3.5). Idso et al. (1991) point out that the mean morning enhancement of leaf net photosynthetic rates was 83% and the mean afternoon enhancement was 284%. Over the hours of measurement, the mean leaf photosynthetic rate was about 2.2-fold greater for the elevated CO_2 treatment in comparison with the ambient CO_2 treatment.

Citrus trees (e.g., *Citrus sinensis* L. and *Citrus paradisi* L.) frequently show "midday depression" of leaf photosynthetic rates when exposed to high temperatures and high vapor pressure deficits (Sinclair and Allen 1982; Sinclair et al. 1983). Brakke (1989) showed that canopy photosynthetic rates of citrus (*Poncirus trifoliata* × *C. sinensis*, *P. trifoliata* × *C. paradisi*, and *C. sinensis*) can show a drastic midday depression when exposed to high temperatures (37°C) and CO_2 concentrations of 330 ppm. Exposure of the same plants to 840 ppm of CO_2, however, effectively eliminated midday depression of photosynthesis and also doubled CO_2 uptake rates. Thus, the large enhancement of citrus biomass accumulation (Idso and Kimball 1991) is quite likely due to interaction effects of temperature (direct or indirect, e.g., vapor pressure deficit) with CO_2 concentration on leaf photosynthesis. The 3-year ensemble BGMR was about 2.8 for citrus grown under the high-temperature, high-vapor-pressure-deficit conditions at Phoenix, Arizona (Table 3.4).

Brakke (1989) pointed out that progenitors of cultivated species of the genus *Citrus* originated as a mesophytic, understory vegetation in southeastern Asia (Webber 1948; Zukovskij 1962; Cameron and Soost 1976; Reuther 1977). Being derived from this ecological niche, it is not surprising that citrus may show a large CO_2 concentration × temperature interaction response. The very large enhancement of citrus photosynthetic rate and growth by elevated CO_2 level under Phoenix, Arizona, conditions illustrates the inability of this species to function at top photosynthetic capacity under current ambient CO_2 concentrations, even under irrigation and high nutrient availability. The large CO_2 enrichment effect could be alternatively viewed as a maladaptation of this mesophytic species to a xerophytic aerial environment.

PLANT RESPONSES TO AIR POLLUTANTS AND UV-B RADIATION

The three major phytotoxic air pollutants are ozone, SO_2, and NO_2 because of their widespread dispersion across the industrialized world (Heck 1984). Most NO produced in combustion of fossil fuels is rapidly oxidized to NO_2 in the atmosphere and thus is not of major concern, although NO may be fourfold more inhibitory to photosynthesis than NO_2 (Saxe 1986).

Leaf cuticles generally have low permeabilities for toxic gases, but these permeabilities are not as low as those for CO_2 and water vapor (Lendzian 1984). Most of the penetration of phytotoxic gases is presumed to occur through stomata, but experiments by Taylor and Tingey (1983) suggest that SO_2 may have a shorter mean deposition pathlength than evaporated water within the substomatal cavity. Furthermore, surface deposition of highly reactive, water-soluble gaseous HNO_3 may constitute 80% of total leaf deposition, in contrast to about 10% for NO_2, 20% for ozone, 25% for H_2S, and 30% for SO_2 (Taylor et al. 1988). Therefore, the main deposition sites may be the leaf surface and substomatal cavity for HNO_3, the substomatal cavity and spongy mesophyll for SO_2, and the whole intercellular surface for ozone, a gas with low water solubility (Taylor et al. 1988).

Several mechanisms of ozone injury to plant leaves have been proposed. The plasma membrane may be the initial site of injury, with inhibition of K^+-stimulated ATPase of the plasmalemma leading to osmotic or ionic imbalances within the cells via changes in plasmalemma permeability (Dominy and Heath 1985). However, this is unlikely to be the only effect of ozone (see Dugger and Ting 1970; Evans and Ting 1973; De Kok et al. 1983; Mudd et al. 1984).

Daytime background ozone concentration in the unpolluted atmosphere is about 25 ppb, but it may range from 40 to 70 ppb during the summer over much of the United States (Davis and Dean 1972; Heck et al. 1982). Such ozone levels can cause decreased photosynthetic rates and earlier senescence of leaves (Coyne and Bingham 1982; Reich and Amundson 1985; Reich et al. 1986), reduced stomatal conductance (Unsworth and Black 1981; Coyne and Bingham 1982; Amiro and Gillespie 1985; Temple 1986), increased maintenance respiration (Amthor 1988, 1994), reduced growth of trees (Pye 1988), and lowered crop yields (Davis and Dean 1972; Heck et al. 1982; Heagle et al. 1983). Although exposure levels are typically used to express ozone treatment effects, computed ozone flux density into leaves or time-integrated ozone uptake may be more appropriate bases for assessing injury (Amiro et al. 1984; Amiro and Gillespie 1985; Lefohn and Tingey 1985). There is no evidence of a light-driven detoxification of ozone (Olszyk and Tingey 1984).

Although SO_2 may react with plasma membrane proteins, it does penetrate the membrane and affect metabolic processes (Mudd et al. 1984). Low concentrations of SO_2 may cause stomata to open and transpiration rates to increase (Unsworth and Black 1981), but at high concentrations, or during prolonged exposure, stomata usually close (Mansfield and Freer-Smith 1984). Increasing SO_2 concentrations may reduce photosynthetic rates to a greater extent than stomatal conductance (Muller et al. 1979), which suggests that the adverse effects of SO_2 are due to direct action on the photosynthetic system rather than on stomata. Low concentrations of SO_2 have been observed to cause more damage to adjacent epidermal cells than to stomatal guard cells (C. R. Black and Black 1979). Unlike those exposed to ozone, leaves exposed to SO_2 in the light showed much less necrotic damage than when exposed in darkness with stomata remaining open (Olszyk and Tingey 1984).

Oxides of nitrogen usually have short residence time in the atmosphere because they tend to be oxidized. A more frequent concern than direct effects of NO_x in many areas has been the production of ozone or other oxidants by photochemical reactions in the presence of hydrocarbons (Heck 1984). To the extent that pollutants associated with fossil fuel combustion limit photosynthesis and carbon storage in vegetation and soils, pollutant-generating processes can contribute to a *positive* feedback on the accumulation of atmospheric CO_2.

Interactions among Air Pollutants

When plants are exposed to air pollutants in combination, additivity, synergism, or antagonism of responses can occur (Ormrod 1982). Stomatal conductance may increase in response to low levels of SO_2 (Unsworth and Black 1981; Winner et al. 1988) but decrease in response to ozone and combinations of ozone with other air pollutants (Olszyk and Tingey 1986). Soybean stomatal conductance was decreased sharply by the combination of SO_2 and NO_2, slightly by SO_2 alone, and not at all by NO_2 alone (Amundson and Weinstein 1981). Only the SO_2-treated soybeans developed necrosis. Exposure to NO_2 may cause further yield reductions when SO_2 is above a threshold concentration (Amundson 1983).

CO_2 exchange rates of soybean leaves were decreased by 5-day exposures to either SO_2 or ozone, with the combination of the two causing the greatest decrease (Le

Suer-Brymer and Ormrod 1984). Using the Weibull model, Heagle et al. (1983) predicted that soybean yields at 67 ppb ozone would be reduced to 60% and that yields at 200 ppb SO_2 would be reduced to 71% compared to yields for pollutant-free air.

Overdoses of ozone, SO_2, and NO_2 may cause different responses among various clones of trees. Sensitive clones of eastern white pine (*Pinus strobus* L.) showed considerably more needle injury in response to various exposures to ozone, SO_2, and NO_2 than did insensitive clones (Yang et al. 1982, 1983). The effects of NO_2 were much less than those of ozone or SO_2.

Leaf area, leaf dry mass, and whole-plant dry mass growth were markedly inhibited by ozone up to 200 ppb in silver maple (*Acer saccharimum* L.) seedlings (Jensen 1983). Furthermore, with the addition of 100 ppb SO_2 to 200 ppb ozone, leaf area, leaf dry mass, and total dry mass growth were inhibited about 90% compared to control plants.

Interactions between CO_2 and Air Pollutants

Increased atmospheric CO_2 levels have the potential to mitigate some air pollution injury by virtue of increased photosynthesis (hence increased supply of photoassimilate to be used in various repair and detoxification processes) as well as reduced stomatal conductance (hence slower uptake of the pollutants). However, continuing increases in air pollutant levels may counteract otherwise beneficial effects of increasing CO_2 on plant physiology and growth.

Few definitive experiments on air pollutant $\times$ CO_2 concentration interactions on plants have been conducted (Hällgren 1984; Acock and Allen 1985; Carlson and Bazzaz 1985; Kimball 1986; Saxe 1986; Morison 1988; Krupa and Kickert 1993). Some earlier studies showed that leaf damage from ozone was decreased by 66% for tobacco (*Nicotiana tabacum* L.) and by 14% for pinto bean (*Phaseolus vulgaris* L.) with doubled CO_2 concentration (Heck and Dunning 1967). Doubled CO_2 concentration decreased alfalfa leaf injury from SO_2 by 60% (Hou et al. 1977). Photosynthetic rates of C_4 plants appear to be less sensitive than rates of C_3 plants to SO_2 at ambient CO_2 levels (Winner and Mooney 1980; Carlson and Bazzaz 1982). Increasing CO_2 concentration and concomitant stomatal closure reduced sensitivity to SO_2 of three species (Carlson and Bazzaz 1982, 1985). Photosynthetic rates were inhibited progressively less by SO_2 when plants were exposed to CO_2 at concentrations increasing from 330 to 680 ppm (V. J. Black 1982). Stomatal closure with increasing CO_2 concentration could have afforded protection against SO_2. However, enhanced photosynthetic rates at elevated CO_2 levels could also provide protection through metabolic detoxification of SO_2 or repair of SO_2 damage.

Heagle et al. (1983) developed two empirical functions describing soybean seed yield response to air pollutants. One function was a polynomial expressing yield as a function of ozone and SO_2 concentrations. With this function, seed yield was calculated to be 339.6 g m^{-2} with the assumption that ozone and SO_2 concentrations were 55 and 26 ppb, respectively. Based on the stomatal conductance data of Rogers et al. (1983), L. H. Allen (1990) assumed that ozone and SO_2 penetration into the leaves would be only 69% at a CO_2 concentration of 680 ppm, compared with its value at 340 ppm, because of decreased leaf conductance. Therefore, the effective ozone and SO_2 concentrations would be 38 and 18 ppb, respectively. With these pollutant concentrations, soybean yield would be 391.9 g m^{-2}, a 15.4% increase due to reduction of pollutant fluxes into the leaves (especially ozone).

The other function used by Heagle et al. (1983) was the Weibull model. This model predicts exponentially decreasing soybean yields as ozone and SO_2 levels increase. The seed yield predicted at 340 ppm CO_2 was 340.5 g m^{-2}. With the expected reduction in pollutant entry at 680 ppm CO_2, the predicted yield was 390.6 g m^{-2}, or a yield increase of 14.7%.

Each model provided by Heagle et al. (1983) gave a predicted yield increase of about 15% owing to the decreased stomatal conductance calculated by L. H. Allen (1990) that would be associated with a doubling of the CO_2 concentration. There are a number of simplifying assumptions and unknown interactions embedded in these calculations, but as far as the effects of the air pollutants ozone and SO_2 are concerned, this exercise suggests that their concentration could increase by 1.45 (the reciprocal of 0.69) with a doubling of CO_2 level before the flux of ozone and SO_2 into C_3 plant leaves would be the same as current values. Nevertheless, any increase in the concentrations of these air pollutants associated with fossil fuel use might contribute to a *positive* feedback on atmospheric CO_2 concentration in comparison to the effects if no change in ozone and SO_2 concentrations occurred. A more precise estimate of gaseous air pollutant effects can only be provided by future CO_2 concentration × air pollutant interaction experiments.

UV-B Radiation

A decrease in stratospheric ozone level caused by, e.g., chlorine from chlorofluoro-carbons will increase the flux of UV-B radiation (280–320 nm) at the earth's surface. However, there are some skepticism and misunderstanding of the meteorological transport processes and atmospheric chemistry affecting ozone depletion in the stratosphere (Rowland 1993). The ozone depletion may now exceed 60% over the South Polar region in the Antarctic spring, creating the so-called "ozone hole." Recent research continues to demonstrate the relationship between the concentration of free ClO and ozone depletion during the Antarctic spring and, to a lesser extent, the Arctic spring (Waters et al. 1993). Furthermore, the daily global average ozone amount (area-weighted, 65°S to 65°N) was 2–3% lower in 1992 than in any other year observed (1979–1991) by the Total Ozone Mapping Spectrometer (Gleason et al. 1993). Global ozone levels appear to be even lower in 1993 and 1994. These decreases in ozone were probably caused by the eruption of Mount Pinatubo in June 1991. Thus, volcanic eruptions could add further concerns about stratospheric ozone depletion and increased UV-B radiation.

There are also many conflicting or inconclusive reports concerning the impact of increased UV-B levels on terrestrial vegetation. Although there are a number of technical difficulties in treating plants with radiation of the spectral quality expected in the solar UV waveband, there is much evidence that enhanced UV-B irradiance reduces plant photosynthesis and growth in many species, especially under growth chamber and greenhouse conditions (Baker and Allen 1994). For example, Vu et al. (1984) reported decreased rubisco activity of greenhouse-grown pea (*Pisum sativum* L.) and soybean when the plants were exposed to a range of UV-B treatments. The effect was also accompanied by decreased height, fresh mass, dry mass, and levels of leaf soluble protein, chlorophyll, and carotenoids, and by an increase in acetone leaf extract absorption below 400 nm. Dai et al. (1992) found differences in sensitivity of rice cultivars to UV-B radiation, but their control was no UV-B exposure rather than typical field UV-B exposure. The effects of UV-B on the physiology of field-grown plants under high solar irradiance are less dramatic than effects on greenhouse- or chamber-grown plants. Beyschlag et al. (1988) found no effects of UV-B on field-grown *Triticum aestivum* L. or *Avena fatua* L. photosynthesis, and Sinclair et al. (1990) found no effects of UV-B on the seed yield of field-grown soybean. In addition to its physiological effects, enhanced levels of UV-B can lead to changes in plant developmental rate and partitioning of photosynthate among organs that can alter community structure (Caldwell 1981; Beyschlag et al. 1988). The positive effects of increasing CO_2 concentration on photosynthesis and plant growth can be reduced, or perhaps eliminated, by a concomitant increase in UV-B radiation (Ziska and Teramura 1992; van de Staaij et al. 1993). In other cases, however, UV-B may not affect the growth response to elevated CO_2 levels (Teramura et al. 1990). Much additional research concerning the interactions among increasing UV-B irradiance, increasing CO_2 concentration, and changing climate on plant physiology and carbon storage in terrestrial ecosystems is needed. Nevertheless, to the extent that UV-B radiation may limit photosynthesis and biomass accumulation in vegetation, stratospheric ozone depletion with increased UV-B radiation can contribute to a *positive* feedback on the accumulation of atmospheric CO_2.

CONCLUSIONS

The anticipated negative and positive feedbacks on the earth's atmosphere that may result from plant physiological responses to rising CO_2 levels are summarized in Table 3.5. The background plant responses leading to our conclusions about these feedbacks are discussed in this section.

CO_2 is the substrate of photosynthesis and inhibits photorespiration. Thus, an increase in atmospheric CO_2 level increases leaf and canopy CO_2 assimilation in C_3 plant communities, other factors being equal. Moreover, the temperature optimum for C_3 net photosynthesis generally increases with an increase in CO_2 level. Acclimation of the photosynthetic process to long-term exposure to elevated CO_2 has spanned the range of possibilities in previous experiments, from a strong downregulation to a marked upregulation of photosynthetic *capacity* at any given CO_2 level.

As well as increasing photosynthesis, elevated CO_2 levels induce stomatal closure. Therefore, leaf-level water use efficiency is positively related to CO_2 concentration. The elevated leaf temperatures that can result from reduced stomatal aperture and the increased leaf area that can result from greater leaf carbon gain under conditions of elevated CO_2 can lead to more or less comparable amounts of transpiration on a ground area basis across CO_2 levels. Thus, the increase in water use efficiency is often largely a result of enhanced photosynthesis rather than reduced transpiration. Increasing air temperature will also tend to increase transpiration, at least as long as soil water is available. The interactions of CO_2 and soil water availability (or water stress) are uncertain. Under high CO_2 levels, stress effects may be delayed for a few days in some crops because of the higher level of leaf starch buildup and lower stomatal conductance to water loss. This interaction with CO_2 concentration may be highly dependent on other

Table 3.5. Anticipated Atmospheric Change Feedback Effects of Plant Physiological Responses to Rising Atmospheric CO_2 Concentration, Temperature, Air Pollutant Concentration, and UV-B Radiation

Atmospheric change and physiological process	Anticipated feedback
Elevated CO_2 concentration enhances photosynthetic CO_2 concentration uptake, especially by C_3 plants.	Decreases rate of CO_2 concentration increase
Elevated CO_2 concentration decreases stomatal conductance.	Increases vegetation surface temperature
Elevated CO_2 concentration results in increased partitioning to roots, or no change in partitioning but increased plant size, which may increase soil carbon.	Decreases rate of CO_2 concentration increase
Elevated CO_2 concentration results in decreased whole-plant specific respiration rate.	Decreases rate of CO_2 concentration increase
Elevated CO_2 concentration directly inhibits plant respiration.	Decreases rate of CO_2 concentration increase
Direct inhibition of respiration by elevated CO_2 concentration may reduce stress tolerance.	Increases rate of CO_2 concentration increase
Elevated CO_2 concentration enhances biomass accumulation of woody plants.	Decreases rate of CO_2 concentration increase
Elevated CO_2 concentration at higher temperature enhances C_3 plant photosynthesis.	Decreases rate of CO_2 concentration increase
Elevated CO_2 concentration and temperature effects on partitioning are too complex to draw conclusion on feedbacks.	?
Increased temperature is likely to increase plant and ecosystem respiration.	Increases rate of CO_2 concentration increase
Higher temperature in combination with elevated CO_2 concentration promotes vegetative biomass accumulation.	Decreases rate of CO_2 concentration increase
Increased air pollution can limit photosynthesis and carbon accumulation in plants.	Increases rate of CO_2 concentration increase
Increased UV-B radiation may decrease photosynthesis and biomass accumulation.	Increases rate of CO_2 concentration increase
Rising CO_2 concentration and increasing temperature are likely to result in a net carbon sequestration by plants, at least in early stages of atmospheric change.	Decreases rate of CO_2 concentration increase

climate factors, such as temperature. Regional and global atmospheric feedbacks on transpiration will also be likely to reduce the effectiveness of increased CO_2 as a global antitranspirant (Eamus 1991).

In addition to changes in photosynthesis and stomatal conductance, long-term growth under conditions of elevated CO_2 often causes a reduction in the *specific* (per unit dry mass) respiration rate of plants. CO_2 also appears to inhibit leaf and shoot respiration directly or increase dark CO_2 fixation into organic acids. To the extent that increased CO_2 concentration results in more rapid photosynthesis and reduced respiration, and that these effects in turn lead to larger plants and increased carbon storage in ecosystems, plant physiological responses to elevated CO_2 levels are expected to contribute to *negative* feedbacks on atmospheric CO_2 concentration.

In general, the *relative* enhancement of photosynthesis and plant productivity owing to elevated CO_2 is greatest under stressful conditions, whereas the *absolute* enhancement is greatest under favorable conditions, i.e., for plants well supplied with water and nutrients at temperatures favorable for growth. Thus, increasing CO_2 concentration may aid the survival of plants in extreme environments but may have its greatest quantitative effects on biotic carbon storage in environments favorable for plant growth. Elevated CO_2 increases crop productivity and the amount of carbon cycled through agricultural systems, but inasmuch as annual crop production is equal to annual crop consumption (carbon release) in human and livestock feeds, enhanced crop productivity does not represent either a *positive* or a *negative* long-term feedback on atmospheric CO_2 level. Tree crops are a potential exception, with the possibility of increased carbon storage in perennial wood (e.g., Chapter 4, this volume). Enhanced crop productivity resulting from elevated CO_2, however, could contribute to greater annual inputs to the soil carbon reservoirs (e.g., Lekkerkerk et al. 1990).

In unmanaged plant systems, leaf and canopy photosynthesis generally increases with increasing atmospheric CO_2 levels, as expected, but in the few systems studied to date *annual carbon storage* in the system has not been greatly enhanced by elevated CO_2. Instead, the cycling of carbon through the system, and especially through the root-soil component, is apparently accelerated. An increased area of vegetation could result from increased CO_2 concentration because of increased water use efficiency or increased growing season length in currently dry and cold environments. This situation could lead to increased carbon storage in currently nonproductive regions. The potential for increased long-term storage of carbon in terrestrial ecosystems as a result of increased atmospheric CO_2 concentration and photosynthesis clearly exists, and this implies the possibility of a *negative* feedback on atmospheric CO_2 levels by terrestrial ecosystems. But, as emphasized by Field et al. (1992), the productivity of natural ecosystems is often constrained by limiting resources—other than CO_2—that will make many natural systems less responsive to increased CO_2 levels than most agricultural ecosystems or plants in controlled environments.

Atmospheric CO_2 concentration is not increasing in isolation from other environmental factors. The same and related processes responsible for increasing CO_2 concentration are also leading directly or indirectly to increasing temperature, increasing UV-B irradiance at the earth's surface, and increasing levels of several phytotoxic air pollutants. Better quantitative assessments of interactions among these factors are needed.

Nearly all plant physiological and developmental processes are sensitive to temperature. Both photosynthesis and respiration, however, become acclimated to a change in temperature over the long term. Thus, a moderate change in temperature need not result in a significant change in the rate of plant CO_2 exchange processes. Moreover, physiological responses of individual plant organs to a change in temperature may differ from the temperature response of plant communities and ecosystems. In particular, ecosystem CO_2 exchange with the atmosphere is apparently less sensitive to temperature than CO_2 exchange by isolated components of that ecosystem. Yet developmental processes may be more tightly set by the plant genome. It is unclear whether climatic warming will cause a net expansion or contraction of the areal extent of, and carbon storage in, terrestrial ecosystems. An expansion would represent a *negative* feedback on atmospheric CO_2 levels, but the effects on the global hydrologic cycle and atmospheric water vapor levels and therefore on the greenhouse effect are much less clear.

Because CO_2 induces stomatal closure, gaseous pollutant uptake may be reduced as CO_2 concentration increases. However, pollution levels may increase concomitantly with levels of atmospheric CO_2. Because CO_2 increases photosynthesis and thus plant carbohydrate levels, the repair of air pollution injury may be increased under conditions of elevated CO_2. The same is true for UV-B damage repair. On the whole, increasing CO_2 concentration has the potential to reduce plant injury from environmental pollutants, or accelerate repair of injury that does occur by increasing the carbohydrate supplies available for repair metabolism. Stated the other way around, the increasing levels of air pollution and UV-B radiation expected in the future may diminish the positive effects of increasing CO_2 on plant physiology and growth.

ACKNOWLEDGMENTS

We thank J. T. Baker, J. P. Syvertsen, and S. D. Wullschleger for reviewing this chapter. The contributions of the senior author were made possible in part by support from U.S. Department of Energy Interagency agreements DE-AI01-81ER60001 and DE-AI05-88ER69014 and U.S. Environmental Protection Agency Interagency agreement DW12934099-0 with the U.S. Department of Agriculture, Agricultural Research Service, at the University of Florida. This chapter is Florida Agricultural Experiment Station Journal Series Contribution No. R-03304.

REFERENCES

Acock, B., and L. H. Allen, Jr. 1985. Crop responses to elevated carbon dioxide concentration. In *Direct Effects of Increasing Carbon Dioxide on Vegetation,* ed. B. R. Strain and J. D. Cure, pp. 53–97. DOE/ER-0238. U.S. Department of Energy, Washington, D.C.

Adams, R. M., C. Rosenzweig, R. M. Peart, J. T. Ritchie, B. A. McCarl, J. D. Glyer, R. B. Curry, J. W. Jones, K. J. Boote, and L. H. Allen, Jr. 1990. Global climate change and U.S. agriculture. *Nature* 345:219–224.

Allen, L. H., Jr. 1990. Plant responses to rising carbon dioxide and potential interactions with air pollutants. *J. Environ. Qual.* 19:15–34.

Allen, L. H., Jr. 1991. Effects of increasing carbon dioxide levels and climate change on plant growth, evapotranspiration, and water resources. In Colloquium on Managing Water Resources in the West under Conditions of Climate Uncertainty, Nov. 14–16, 1990, Scottsdale, Arizona, pp. 101–147. National Research Council, National Academy Press, Washington, D.C.

Allen, L. H., Jr. 1994. Carbon dioxide increase: Direct impacts on crops and indirect effects mediated through anticipated climatic changes. In *Physiology and Determination of Crop Yield,* ed. K. J. Boote, J. M. Bennett, T. R. Sinclair, and G. M. Paulsen, pp. 425–459. American Society of Agronomy, Madison, Wis.

Allen, L. H., Jr., P. Jones, and J. W. Jones. 1985. Rising atmospheric CO_2 and evapotranspiration. In *Advances in Evapotranspiration,* Proc. Nat. Conf. Adv. Evapotranspiration, pp. 13–27. ASAE Publication 14–85. American Society of Agricultural Engineers, St. Joseph, Mich.

Allen, L. H., Jr., K. J. Boote, J. W. Jones, P. H. Jones, R. R. Valle, B. Acock, H. H. Rogers, and R. C. Dahlman. 1987. Response of vegetation to rising carbon dioxide: Photosynthesis, biomass, and seed yield of soybean. *Global Biogeochem. Cycles* 1:1–14.

Allen, L. H., Jr., J. C. V. Vu, R. R. Valle, K. J. Boote, and P. H. Jones. 1988. Nonstructural carbohydrates and nitrogen of soybean grown under carbon dioxide enrichment. *Crop Sci.* 28:88–94.

Allen, L. H., Jr., E. C. Bisbal, K. J. Boote, and P. H. Jones. 1991. Soybean dry matter allocation under subambient and superambient levels of carbon dioxide. *Agron. J.* 83:875–883.

Allen, L. H., Jr., B. G. Drake, H. H. Rogers, and J. H. Shinn. 1992. Field techniques for exposure of plants and ecosystems to elevated CO_2 and other trace gases. *Crit. Rev. Plant Sci.* 11:85–119.

Allen, S. G., S. B. Idso, B. A. Kimball, J. T. Baker, L. H. Allen, Jr., J. R. Mauney, J. W. Radin, and M. G. Anderson. 1990. Effects of air temperature on atmospheric CO_2–plant growth relationships. Technical Report. TRO48. DOE/ER-0450T. U.S. Department of Energy Office of Energy Research, Office of Health and Environmental Research, Carbon Dioxide Research Program, and United States Department of Agriculture, Agricultural Research Service, Washington, D.C.

Amiro, B. D., and T. J. Gillespie. 1985. Leaf conductance response of *Phaseolus vulgaris* to ozone flux density. *Atmos. Environ.* 19:807–810.

Amiro, B. D., T. J. Gillespie, and G. W. Thurtell. 1984. Injury response of *Phaseolus vulgaris* to ozone flux density. *Atmos. Environ.* 18:1207–1215.

Amthor, J. S. 1988. Growth and maintenance respiration in leaves of bean (*Phaseolus vulgaris* L.) exposed to ozone in open-top chambers in the field. *New Phytol.* 110:319–325.

Amthor, J. S. 1989. *Respiration and Crop Productivity.* Springer-Verlag, New York.

Amthor, J. S. 1991. Respiration in a future, higher-CO_2 world. *Plant Cell Environ.* 14:13–20.

Amthor, J. S. 1994. Plant respiratory responses to the environment and their effects on the carbon balance. In *Plant-Environment Interactions,* ed. R. E. Wilkinson, pp. 501–554. Marcel Dekker, New York.

Amthor, J. S., G. W. Koch, and A. J. Bloom. 1992. CO_2 inhibits respiration in leaves of *Rumex crispus* L. *Plant Physiol.* 98:757–760.

Amundson, R. G. 1983. Yield reduction of soybean due to 14-day exposure to sulfur dioxide and nitrogen dioxide. *J. Environ. Qual.* 12:454–459.

Amundson, R. G., and L. H. Weinstein. 1981. Joint action of short-term sulfur dioxide and nitrogen dioxide on foliar injury and stomatal behavior in soybean. *J. Environ. Qual.* 10:204–206.

Aoki, M., and K. Yabuki. 1977. Studies on the carbon enrichment for plant growth. VII. Changes in dry matter production and photosynthetic rates of cucumber during carbon dioxide enrichment. *Agric. Meteorol.* 18:475–485.

Bacastow, R., and C. D. Keeling. 1973. Atmospheric carbon dioxide and radiocarbon in the natural carbon cycle. II. Changes from A.D. 1700 to 2070 as deduced from a geochemical model. In *Carbon and the Biosphere,* ed. G. M. Woodwell and E. V. Pecan, pp. 86–135. National Technical Information Service, Springfield, Va.

Baker, J. T., and L. H. Allen, Jr. 1993. Contrasting crop species responses to CO_2 and temperature: Rice, soybean, and citrus. *Vegetatio* 104/105:239–260.

Baker, J. T., and L. H. Allen, Jr. 1994. Assessment of the impact of rising carbon dioxide and other potential climate changes on vegetation. *Environ. Pollut.* 83:223–235.

Baker, J. T., L. H. Allen, Jr., K. J. Boote, P. Jones, and J. W. Jones. 1989. Response of soybean to air temperature and carbon dioxide concentration. *Crop Sci.* 29:98–105.

Baker, J. T., L. H. Allen, Jr., K. J. Boote, A. J. Rowland-Bamford, R. S. Waschmann, J. W. Jones, P. H. Jones, and G. Bowes. 1990. Response of vegetation to carbon dioxide, series no. 060. Temperature effects on rice at elevated CO_2 concentration (1989 Progress Report). Plant Stress and Protection Research Unit, United States Department of Agriculture, Agricultural Research Service, and Institute of Food and Agricultural Science, University of Florida, Gainesville. Joint Program of the U.S. Department of Energy and United States Department of Agriculture, Agricultural Research Service, Washington, D.C.

Baker, J. T., F. Laugel, K. J. Boote, and L. H. Allen, Jr. 1992. Effects of daytime carbon dioxide concentration on dark respiration in rice. *Plant Cell Environ.* 15:231–239.

Bazzaz, F. A. 1990. The response of natural ecosystems to the rising global CO_2 levels. *Ann. Rev. Ecol. Syst.* 21:167–196.

Beevers, H. 1961. *Respiratory Metabolism in Plants.* Row, Peterson, Evanston, Ill.

Beevers, H. 1970. Respiration in plants and its regulation. In *Prediction and Measurement of Photosynthetic Productivity,* ed. I. Setlik, pp. 209–214. Pudoc, Wageningen, The Netherlands.

Berry, J., and O. Björkman. 1980. Photosynthetic response and adaptation to temperature in higher plants. *Ann. Rev. Plant Physiol.* 31:491–543.

Beyers, R. J. 1962. Relationship between temperature and the metabolism of experimental ecosystems. *Science* 136:980–982.

Beyschlag, W., P. W. Barnes, S. D. Flint, and M. M. Caldwell. 1988. Enhanced UV-B irradiation has no effect on photosynthetic characteristics of wheat (*Triticum aestivum* L.) and wild oat (*Avena fatua* L.) under greenhouse and field conditions. *Photosynthetica* 22:516–525.

Billings, W. D., and H. A. Mooney. 1968. The ecology of arctic and alpine plants. *Biol. Rev.* 43:481–529.

Black, C. R., and V. J. Black. 1979. The effects of low concentrations of sulphur dioxide on stomatal conductance and epidermal cell survival in field bean (*Vicia faba* L.) *J. Exp. Bot.* 30:291–298.

Black, V. J. 1982. Effects of sulphur dioxide on physiological processes in plants. In *Effects of Gaseous Air Pollution in Agriculture and Horticulture,* ed. M. H. Unsworth and D. P. Ormrod, pp. 67–91. Butterworths, London.

Bolin, B., B. R. Döös, J. Jäger, and R. A. Warrick, eds. 1986. *The Greenhouse Effect, Climate Change, and Ecosystems.* Scientific Committee on Problems of the Environment (SCOPE); 29. Wiley, Chichester, U.K.

Bowes, G. 1993. Facing the inevitable: Plants and increasing atmospheric CO_2. *Annu. Rev. Plant Physiol. Plant Mol. Biol.* 44:309–332.

Brakke, M. 1989. Gas exchange and growth responses of citrus trees to partial irrigation, soil water, and atmospheric conditions. Ph.D. dissertation, University of Florida, Gainesville.

Brakke, M., and L. H. Allen, Jr. 1994. Gas exchange response of citrus seedlings to temperature, vapor pressure deficit, and soil water content. *J. Am. Soc. Hortic. Sci.* 119 (in press).

Bunce, J. A. 1992. Stomatal conductance, photosynthesis and respiration of temperate deciduous tree seedlings grown outdoors at an elevated concentration of carbon dioxide. *Plant Cell, Environ.* 15:541–549.

Caldwell, M. M. 1981. Plant response to solar ultraviolet radiation. In *Physiological Plant Ecology I: Responses to the Physical Environment,* ed. O. L. Lange, P. S. Nobel, C. B. Osmond, and H. Ziegler, pp. 169–197. Springer-Verlag, Berlin.

Cameron, J. W., and R. K. Soost. 1976. Citrus. In *Evolution of Crops,* ed. N. W. Simmonds, pp. 261–265. Longman, London.

Campbell, W. J., L. H. Allen, Jr., and G. Bowes. 1988. Effects of CO_2 concentration on rubisco activity, amount, and photosynthesis in soybean leaves. *Plant Physiol.* 88:1310–1316.

Carlson, R. W., and F. A. Bazzaz. 1982. Photosynthetic and growth responses to fumigation with SO_2 at elevated CO_2 for C_3 and C_4 plants. *Oecologia* 54:50–54.

Carlson, R. W., and F. A. Bazzaz. 1985. Plant response to SO_2 and CO_2. In *Sulfur Dioxide and Vegetation,* ed. W. W. Winter, pp. 313–331. Stanford University Press, Stanford, Calif.

Collatz, G. J., M. Ribas-Carbo, and J. A. Berry. 1992. Coupled photosynthesis–stomatal conductance model for leaves of C_4 plants. *Aust. J. Plant Physiol.* 19:519–538.

Coyne, P. I., and G. E. Bingham. 1982. Variation in photosynthesis and stomatal conductance in an ozone-stressed ponderosa pine stand: Light response. *Forest Sci.* 28:257–273.

Cure, J. D. 1985. Carbon dioxide doubling responses: A crop. In *Direct Effects of Increasing Carbon Dioxide on Vegetation,* ed. B. R. Strain and J. D. Cure, pp. 99–116. DOE/ER-0238. U.S. Department of Energy, Carbon Dioxide Research Division, Washington D.C.

Curtis, P. S., and J. A. Teeri. 1992. Seasonal responses of leaf gas exchange to elevated carbon dioxide in *Populus grandidentata*. *Can. J. For. Res.* 22:1320–1325.

Dai, Q., V. P. Coronel, B. S. Vergara, P. W. Barnes, and A. T. Quitos. 1992. Ultraviolet-B radiation effects on growth and physiology of four rice cultivars. *Crop Sci.* 32:1269–1274.

Davis, D. R., and C. E. Dean. 1972. Meteorological aspects of atmospheric ozone as a potential threat to the forest industry of north Florida. *J. Environ. Qual.* 1:438–441.

De Kok, L. J., C. R. Thompson, J. B. Mudd, and G. Kats. 1983. Effect of H_2S fumigation on water soluble sulfhydryl compounds in shoots of crop plants. *Z. Pflanzenphysiol.* 111:85–89.

DeLucia, E. H., T. W. Sasek, and B. R. Strain. 1985. Photosynthetic inhibition after long-term exposure to elevated levels of atmospheric carbon dioxide. *Photosyn. Res.* 7:175–184.

Dominy, P. J., and R. L. Heath. 1985. Inhibition of the K^+-stimulated ATPase of the plasmalemma of pinto bean leaves by ozone. *Plant Physiol.* 77:43–45.

Drake, B. G., and P. W. Leadley. 1991. Canopy photosynthesis of crops and native plant communities exposed to long-term elevated CO_2. *Plant Cell Environ.* 14:853–860.

Dugger, W. M., and I. P. Ting. 1970. Air pollution oxidants—Their effects on metabolic processes in plants. *Annu. Rev. Plant Physiol.* 21:215–234.

Eamus, D. 1991. The interaction of rising CO_2 and temperatures with water use efficiency. *Plant Cell Environ.* 14:843–852.

Eamus, D., and P. G. Jarvis. 1989. The direct effects of increase in the global atmospheric CO_2 concentration on natural and commercial temperate trees and forests. *Adv. Ecol. Res.* 19:1–55.

Evans, L. S., and I. P. Ting. 1973. Ozone induced permeability changes. *Am. J. Bot.* 60:155–162.

Farquhar, G. D., S. von Caemmerer, and J. A. Berry. 1980. A biochemical model of photosynthetic CO_2 assimilation in leaves of C_3 plants. *Planta* 149:78–90.

Farrar, J. F. 1988. Temperature and the partitioning and translocation of carbon. *Symp. Soc. Exp. Biol.* 42:203–235.

Farrar, J. F., and M. L. Williams. 1991. The effects of increased atmospheric carbon dioxide and temperature on carbon partitioning, source-sink relations and respiration. *Plant Cell Environ.* 14:819–830.

Field, C. B., F. S. Chapin, III, P. A. Matson, and H. A. Mooney. 1992. Responses of terrestrial ecosystems to the changing atmosphere: A resource-based approach. *Ann. Rev. Ecol. Syst.* 23:201–235.

Forward, D. F. 1960. Effect of temperature on respiration. In *Encyclopedia of Plant Physiology,* Vol. XII, Part 2, ed. W. Ruhland, pp. 234–258. Springer-Verlag, Berlin.

Gale, J. 1982. Evidence for essential maintenance respiration of leaves of *Xanthium strumarium* at high temperature. *J. Exp. Bot.* 33:471–476.

Gates, D. M. 1985. Global biospheric response to increasing atmospheric carbon dioxide concentration. In *Direct Effects of Increasing Carbon Dioxide on Vegetation,* ed. B. R. Strain and J. D. Cure, pp. 171–184. U.S. Department of Energy, Washington, D.C.

Gates, D. M., B. R. Strain, and J. A. Weber. 1983. Ecophysiological effects of changing atmospheric CO_2 concentration. In *Physiological Plant Ecology IV,* ed. O. L. Lange, P. S. Nobel, C. B. Osmond, and H. Ziegler, pp. 503–526. Springer-Verlag, Berlin.

Gleason, J. F., P. K. Bhartia, J. R. Herman, R. McPeters, P. Newman, R. S. Stolarski, L. Flynn, G. Labow, D. Larko, C. Seftor, C. Wellemeyer, W. D. Komhyr, A. J. Miller, and W. Planet. 1993. Record low global ozone in 1992. *Science* 260:523–526.

Guinn, G., and J. R. Mauney. 1980. Analysis of CO_2 exchange assumptions: Feedback control. In *Predicting Photosynthesis for Ecosystem Models,* Vol. II, ed. J. D. Hesketh and J. W. Jones, pp. 1–16. CRC Press, Boca Raton, Fla.

Hall, A. E., and L. H. Allen, Jr. 1993. Designing cultivars for the climatic conditions of the next century. In *International Crop Science I.,* ed. D. R. Buxton, R. Shibles, R. A. Forsberg, B. L. Blad, K. H. Asay, G. M. Paulsen, and R. F. Wilson, pp. 291–297. Crop Science Society of America, Madison, Wisc.

Hällgren, J. E. 1984. Photosynthetic gas exchange in leaves affected by air pollutants. In *Gaseous Air Pollutants and Plant Metabolism*, ed. M. J. Koziol and F. R. Whatley, pp. 147–159. Butterworths, London.

Heagle, A. S., W. W. Heck, J. O. Rawlings, and R. B. Philbeck. 1983. Effects of chronic doses of ozone and sulfur dioxide on injury and yield of soybeans in open-top field chambers. *Crop Sci.* 23:1184–1191.

Heck, W. W. 1984. Defining gaseous pollution problems in North America. In *Gaseous Air Pollutants and Plant Metabolism*, ed. M. J. Koziol and F. R. Whatley, pp. 35–48. Butterworths, London.

Heck, W. W., and J. A. Dunning. 1967. The effects of ozone on tobacco and pinto bean as conditioned by several ecological factors. *J. Air Pollut. Control Assoc.* 17:112–114.

Heck, W. W., O. C. Taylor, R. Adams, G. Bingham, J. Miller, E. Preston, and L. Weinstein. 1982. Assessment of crop loss from ozone. *J. Air Pollut. Control Ass.* 32:353–361.

Hendrick, R. L., and K. S. Pregitzer. 1993. Patterns of fine root mortality in two sugar maple forests. *Nature* 361:59–61.

Herold, A. 1980. Regulation of photosynthesis by sink activity—The missing link. *New Phytol.* 86:131–144.

Hofstra, G., and J. D. Hesketh. 1975. The effects of temperature and CO_2 enrichment on photosynthesis in soybean. In *Environmental and Biological Control of Photosynthesis*, ed. R. Marcelle, pp. 71–80. Dr. W. Junk, The Hague, The Netherlands.

Hou, L., A. C. Hill, and A. Soleimani. 1977. Influence of CO_2 on the effects of SO_2 and NO_2 on alfalfa. *Environ. Pollut.* 12:7–16.

Idso, S. B., and B. A. Kimball. 1989. Growth response of carrot and radish to atmospheric CO_2 enrichment. *Environ. Exp. Bot.* 29:135–139.

Idso, S. B., and B. A. Kimball. 1991. Downward regulation of photosynthesis and growth at high CO_2 levels. *Plant Physiol.* 96:990–992.

Idso, S. B., B. A. Kimball, M. G. Anderson, and J. R. Mauney. 1987. Effects of atmospheric CO_2 enrichment on plant growth: The interactive role of air temperature. *Agr. Ecosys. Environ.* 20:1–10.

Idso, S. B., B. A. Kimball, and S. G. Allen. 1991. Net photosynthesis of sour orange trees maintained in atmospheres of ambient and elevated CO_2 concentration. *Agric. For. Meteorol.* 54:95–101.

Jacobs, C. M. J., and H. A. R. de Bruin. 1992. The sensitivity of regional transpiration to land-surface characteristics: Significance of feedback. *J. Climate* 5:683–698.

James, W. O. 1953. *Plant Respiration*. Oxford University Press, Oxford.

Jarvis, P. G., and K. G. McNaughton. 1986. Stomatal control of transpiration: Scaling up from leaf to region. *Adv. Ecol. Res.* 15:1–49.

Jensen, K. F. 1983. Growth relationships in silver maple seedlings fumigated with O_3 and SO_2. *Canad. J. For. Res.* 13:298–302.

Jones, P., L. H. Allen, Jr., J. W. Jones, K. J. Boote, and W. J. Campbell. 1984. Soybean canopy growth, photosynthesis, and transpiration responses to whole-season carbon dioxide enrichment. *Agron. J.* 76:633–637.

Jones, P., L. H. Allen, Jr., J. W. Jones, and R. Valle. 1985a. Photosynthesis and transpiration responses of soybean canopies to short- and long-term CO_2 treatments. *Agron. J.* 77:119–126.

Jones, P., J.W. Jones, and L.H. Allen, Jr. 1985b. Seasonal carbon and water balances of soybeans grown under stress treatments in sunlit chambers. *Trans. Am. Soc. Agr. Eng.* 28:2021–2028.

Jordan, D. B., and W. L. Ogren. 1984. The CO_2/O_2 specificity of ribulose 1,5-bisphosphate carboxylase/oxygenase. *Planta* 161:308–313.

Jurik, T. W., J. A. Webber, and D. M. Gates. 1984. Short-term effects of CO_2 on gas exchange of leaves of bigtooth aspen (*Populus grandidentata*) in the field. *Plant Physiol.* 75:1022–1025.

Kimball, B. A. 1983. Carbon dioxide and agricultural yield: An assemblage and analysis of 430 prior observations. *Agron. J.* 75:779–788.

Kimball, B. A. 1986. CO_2 stimulation of growth and yield under environmental restraints. In *Carbon Dioxide Enrichment of Greenhouse Crops*, Vol. II, ed. H. Z. Enoch and B. A. Kimball, pp. 53–67. CRC Press, Boca Raton, FLa.

Koch, K. E., D. W. White, P. H. Jones, and L. H. Allen, Jr. 1983. CO_2 enrichment of Carrizo citrange and Swingle citrumelo rootstocks. *Fla. State Hortic. Soc. Proc.* 96:37–40.

Koch, K. E., P. H. Jones, W. T. Avigne, and L. H. Allen, Jr. 1986. Growth, dry matter partitioning, and diurnal activities of RuBP carboxylase in citrus seedlings maintained at two levels of CO_2. *Physiol. Plant.* 67:477–484.

Koch, K. E., L. H. Allen, Jr., P. Jones, and W. T. Avigne. 1987. Growth of citrus rootstock (Carrizo citrange) seedlings during and after long-term CO_2 enrichment. *J. Am. Soc. Hortic. Sci.* 112:77–82.

Körner, C., and J. A. Arnone, III. 1992. Responses to elevated carbon dioxide in artificial tropical ecosystems. *Science* 257:1672–1675.

Kramer, P. J. 1981. Carbon dioxide concentrations, photosynthesis and dry matter production. *BioScience* 31:29–33.

Krizek, D. T., R. H. Zimmerman, H. H. Klueter, and W. A. Bailey. 1971. Growth of crabapple seedlings in controlled environments: Effects of CO_2 level and time and duration of CO_2 treatment. *J. Am. Soc. Hortic. Sci.* 96:285–288.

Krizek, D. T., W. A. Bailey, H. H. Klueter, and R. C. Liu. 1974. Maximizing growth of vegetable seedlings in controlled environments at elevated temperature, light, and CO_2. *Acta Hortic.* 39:89–102.

Krupa, S. V., and R. N. Kickert. 1993. The greenhouse effect: The impacts of carbon dioxide (CO_2), ultraviolet-B (UV-B) radiation and ozone (O_3) on vegetation (crops). *Vegetatio* 104/105:223–238.

Lange, O. L., E.-D. Schulze, M. Evenari, L. Kappen, and U. Buschbom. 1974. The temperature-related photosynthetic capacity of plants under desert conditions. I. Seasonal changes of the photosynthetic response to temperature. *Oecologia* 17:97–110.

Lawlor, D. W., and R. A. C. Mitchell. 1991. The effects of increasing CO_2 on crop photosynthesis and productivity: A review of field studies. *Plant Cell Environ.* 14:807–818.

Lefohn, A. S., and D. T. Tingey. 1985. Injury response of *Phaseolus vulgaris* to ozone flux density. *Atmos. Environ.* 19:206–207.

Lekkerkerk, L. J. A., S. C. van de Geijn, and J. A. van Veen. 1990. Effects of elevated atmospheric CO_2-levels on the carbon economy of a soil planted with wheat. In *Soils and the Greenhouse Effect*, ed. A. F. Bouwman, pp. 423–429. Wiley, Chichester; U.K.

Lemon, E. R., ed. 1983. *CO_2 and Plants: The Response of Plants to Rising Levels of Atmospheric Carbon Dioxide*. Westview Press, Boulder, Colo.

Lendzian, K. J. 1984. Permeability of plant cuticles to gaseous air pollutants. In *Gaseous Air Pollutants and Plant Metabolism*, ed. M. J. Koziol and F. R. Whatley, pp. 77–81. Butterworths, London.

Le Suer-Brymer, N. M., and D. P. Ormrod. 1984. Carbon dioxide exchange rates of fruiting soybean plants exposed to ozone and sulphur dioxide singly or in combination. *Can. J. Plant Sci.* 64:69–75.

Liu, H. S. 1992. Frequency variations of the Earth's obliquity and the 100-kyr ice-age cycles. *Nature* 358:397–399.

Lloyd, J., J. P. Syvertsen, P. E. Kriedemann, and G. D. Farquhar. 1992. Low conductances for CO_2 diffusion from stomata to the sites of carboxylation in leaves of woody species. *Plant Cell Environ.* 15:873–899.

Long, S. P. 1991. Modification of the response of photosynthetic productivity to rising temperature by atmospheric CO_2 concentrations: Has its importance been underestimated? *Plant Cell Environ.* 14:729–739.

Long, S. P., and F. I. Woodward; eds. 1988. *Plants and Temperature*. Company of Biologists, Cambridge.

Loomis, R. S. 1983. Productivity of agricultural systems. In *Physiological Plant Ecology IV*, ed. O. L. Lange, P. S. Nobel, C. B. Osmond, and H. Ziegler, pp. 151–172. Springer-Verlag, Berlin.

Loomis, R. S., and D. J. Connor. 1992. *Crop Ecology*. Cambridge University Press, Cambridge.

Loreto, F., P. C. Harley, G. Di Marco, and T. D. Sharkey. 1992. Estimation of mesophyll conductance to CO_2 flux by three different methods. *Plant Physiol.* 98:1437–1443.

Lorius, C., J. Jouzel, D. Raynaud, J. Hansen, and H. Le Treut. 1990. The ice-core record: Climate sensitivity and future greenhouse warming. *Nature* 347:139–145.

Mansfield, T. A., and P. H. Freer-Smith. 1984. The role of stomata in resistance mechanisms. In *Gaseous Air Pollutants and Plant Metabolism,* ed. M. J. Koziol and F. R. Whatley, pp. 131–145. Butterworths, London.

Morison, J. I. L. 1987. Intracellular CO_2 concentration and stomatal response to CO_2. In *Stomatal Function,* ed. E. Zeiger, G. D. Farquhar, and I. Cowan, pp. 229–252. Stanford University Press, Stanford, Calif.

Morison, J. I. L. 1988. Effect of increasing atmospheric CO_2 on plants and their responses to other pollutants, climate, and soil factors. *Asp. Appl. Biol.* 17:113–122.

Morison, J. I. L., and R. M. Gifford. 1983. Stomatal sensitivity to carbon dioxide and humidity. *Plant Physiol.* 71:789–796.

Mott, K. A. 1988. Do stomata respond to CO_2 concentration other than intercellular? *Plant Physiol.* 86:200–203.

Mousseau, M., and B. Saugier. 1992. The direct effect of increased CO_2 on gas exchange and growth of forest tree species. *J. Exp. Bot.* 43:1121–1130.

Mudd, J. B., S. K. Banesjee, M. M. Dooley, and K. L. Knight. 1984. Pollutants and plant cells: Effects on membranes. In *Gaseous Air Pollutants and Plant Metabolism,* ed. M. J. Koziol and F. R. Whatley, pp. 106–116. Butterworths, London.

Muller, R. N., J. E. Miller, and D. G. Sprugel. 1979. Photosynthetic response of field-grown soybeans to fumigations with sulphur dioxide. *J. Appl. Ecol.* 16:567–576.

Nobel, P. S., and T. L. Hartsock. 1986. Short-term and long-term responses of crassulacean acid metabolism plants to elevated CO_2. *Plant Physiol.* 82:604–606.

Norby, R. J., C. A. Gunderson, S. D. Wullschleger, E. G. O'Neill, and M. K. McCracken. 1992. Productivity and compensatory responses of yellow-poplar trees in elevated CO_2. *Nature* 357:322–324.

Oechel, W. C., and B. R. Strain. 1985. Native species responses to increased atmospheric carbon dioxide concentrations. In *Direct Effects of Increasing Carbon Dioxide on Vegetation,* ed. B. R. Strain and J. D. Cure, pp. 117–154. DOE/ER-0238. U.S. Department of Energy, Carbon Dioxide Research Division, Washington, D.C.

Olszyk, D. M., and D. T. Tingey. 1984. Phytotoxicity of air pollutants. *Plant Physiol.* 74:999–1005.

Olszyk, D. M., and D. T. Tingey. 1986. Joint action of O_3 and SO_2 in modifying plant gas exchange. *Plant Physiol.* 82:401–405.

Ormrod, D. P. 1982. Air pollutant interactions in mixtures. In *Effects of Air Pollution in Agriculture and Horticulture,* ed. M. H. Unsworth and D. P. Ormrod, pp. 307–331. Butterworths, London.

Osmond, C. B., K. Winter, and H. Ziegler. 1982. Functional significance of different pathways of CO_2 fixation in photosynthesis. In *Physiological Plant Ecology II. Water Relations and Carbon Assimilation,* ed. O. L. Lange, P. S. Nobel, C. B. Osmond, and H. Ziegler, pp. 479–547. Springer-Verlag, Berlin.

Peet, M. M., S. C. Huber, and D. T. Patterson. 1986. Acclimation to high CO_2 in monoecious cucumbers. II. Carbon exchange rates, enzyme activities, and starch and nutrient concentrations. *Plant Physiol.* 80:63–67.

Pearcy, R. W., and O. Björkman. 1983. Physiological effects. In *CO₂ and Plants: The Response of Plants to Rising Levels of Atmospheric Carbon Dioxide,* ed. E. R. Lemon, pp. 65–105. Westview Press, Boulder, Colo.

Polley, H. W., H. B. Johnson, H. S. Mayeux, and S. R. Malone. 1993. Physiology and growth of wheat across a subambient carbon dioxide gradient. *Ann. Bot.* 71:347–356.

Poorter, H. 1993. Interspecific variation in the growth response of plants to an elevated ambient CO_2 concentration. *Vegetatio* 104/105:77–97.

Poorter, H., R. M. Gifford, P. E. Kriedemann, and S. C. Wong. 1992. A quantitative analysis of dark respiration and carbon content as factors in the growth response of plants to elevated CO_2. *Aust. J. Bot.* 40:501–513.

Pye, J. M. 1988. Impact of ozone on the growth and yield of trees: A review. *J. Environ. Qual.* 17:347–360.

Radin, J. W., B. A. Kimball, D. L. Hendrix, and J. R. Mauney. 1987. Photosynthesis of cotton plants exposed to elevated levels of carbon dioxide in the field. *Photosyn. Res.* 12:191–203.

Rawson, H. M. 1992. Plant responses to temperature under conditions of elevated CO_2. *Aust. J. Bot.* 40:473–490.

Reich, P. B., and R. G. Amundson. 1985. Ambient levels of ozone reduce net photosynthesis in tree and crop species. *Science* 230:566–570.

Reich, P. B., A. W. Schoettle, R. M. Raba, and R. G. Amundson. 1986. Response of soybean to low concentrations of ozone. I. Reductions in leaf and whole plant net photosynthesis and leaf chlorophyll content. *J. Environ. Qual.* 15:31–36.

Reuther, W. 1977. Citrus. In *Ecophysiology of Tropical Crops,* ed. P. de T. Alvim and T. T. Kozlowski, pp. 409–439. Academic Press, New York.

Rogers, H. H., G. E. Bingham, J. D. Cure, J. M. Smith, and K. A. Surano. 1983. Responses of selected plant species to elevated carbon dioxide in the field. *J. Environ. Qual.* 12:569–574.

Rogers, H. H., C. M. Peterson, J. N. McCrimmon, and J. D. Cure. 1992. Response of plant roots to elevated atmospheric carbon dioxide. *Plant Cell Environ.* 15:749–752.

Rogers, H. H., G. B. Runion, and S. V. Krupa. 1994. Plant responses to atmospheric CO_2 enrichment with emphasis on roots and the rhizosphere. *Environ. Pollut.* 83:155–189.

Rowland, F. S. 1993. President's lecture: The need for scientific communication with the public. *Science* 260:1571–1576.

Rozema, J., H. Lambers, S. C. van de Geijn, and M. L. Cambridge, eds. 1993. *CO₂ and Biosphere.* Kluwer, Dordrecht, The Netherlands.

Ryan, M. G. 1991. Effects of climate change on plant respiration. *Ecol. Appl.* 1:157–167.

Sage, R. F., T. D. Sharkey, and J. R. Seemann. 1989. Acclimation of photosynthesis to elevated CO_2 in five C_3 species. *Plant Physiol.* 89:590–596.

Saxe, H. 1986. Effects of NO, NO_2, and CO_2 on net photosynthesis, dark respiration and transpiration of pot plants. *New Phytol.* 103:185–197.

Sharkey, T. D. 1990. Feedback limitation of photosynthesis and the physiological role of ribulose bisphosphate carboxylase carbamylation. *Bot. Mag. Tokyo Special Issue* 2:87–105.

Sinclair, T. R., and L. H. Allen, Jr. 1982. Carbon dioxide and water vapor exchange of leaves on field-grown citrus trees. *J. Exp. Bot.* 33:1166–1175.

Sinclair, T. R., L. H. Allen, Jr., and M. Cohen. 1983. Citrus blight effects on carbon dioxide assimilation. *J. Am. Soc. Hortic. Sci.* 108:503–506.

Sinclair, T. R., O. N'Diaye, and R. H. Biggs. 1990. Growth and yield of field-grown soybean in response to enhanced exposure to ultraviolet-B radiation. *J. Environ. Qual.* 19:478–481.

Stitt, M. 1991. Rising CO_2 levels and their potential significance for carbon flow in photosynthetic cells. *Plant Cell Environ.* 14:741–762.

Strain, B. R., and J. D. Cure, eds. 1985. *Direct Effects of Increasing Carbon Dioxide on Vegetation,* DOE/ER-0238. U.S. Department of Energy, Carbon Dioxide Research Division, Washington, D.C.

Taylor, G. E., Jr., and D. T. Tingey. 1983. Sulfur dioxide flux into leaves of *Geranium carolinianum* L. *Plant Physiol.* 72:237–244.

Taylor, G. E., Jr., P. J. Hansen, and D. D. Baldocchi. 1988. Pollution deposition to individual leaves and plant canopies: Sites of regulation and relationship to injury. In *Assessment of Crop Loss from Air Pollutants,* ed. W. W. Heck, pp. 227–257. Elsevier, London.

Temple, P. J. 1986. Stomatal conductance and transpirational responses of field-grown cotton to ozone. *Plant Cell Environ.* 9:315–321.

Teramura, A. H., J. H. Sullivan, and L. H. Ziska. 1990. Interaction of elevated ultraviolet-B radiation and CO_2 on productivity and photosynthetic characteristics in wheat, rice, and soybean. *Plant Physiol.* 94:470–475.

Thomas, J. F., and C. N. Harvey. 1983. Leaf anatomy of four species grown under continuous CO_2 enrichment. *Bot. Gaz.* 144:303–309.

Thomas, R. B., and B. R. Strain. 1991. Root restriction as a factor in photosynthetic acclimation of cotton seedlings grown in elevated carbon dioxide. *Plant Physiol.* 96:627–634.

Tolbert, N. E., and I. Zelitch. 1983. Carbon metabolism. In *CO_2 and Plants: The Response of Plants to Rising Levels of Atmospheric Carbon Dioxide,* ed. E. R. Lemon, pp. 21–64. Westview Press, Boulder, Colo.

Unsworth, M. H., and V. J. Black. 1981. Stomatal responses to pollutants. In *Stomatal Physiology,* ed. P. G. Jarvis and T. A. Mansfield, pp. 187–203. Cambridge University Press, Cambridge.

Valle, R., J. W. Mishoe, W. J. Campbell, J. W. Jones, and L. H. Allen, Jr. 1985a. Photosynthetic responses of "Bragg" soybean leaves adapted to different CO_2 environments. *Crop Sci.* 25:333–339.

Valle, R., J. W. Mishoe, J. W. Jones, and L. H. Allen, Jr. 1985b. Changes in transpiration rates and water use efficiency of soybean leaves adapted to different CO_2 environments. *Crop Sci.* 25:477–482.

van de Staaij, J. W. M., G. M. Lenssen, M. Stroetenga, and J. Rozema. 1993. The combined effects of elevated CO_2 level and UV-B radiation on growth characteristics of *Elymus athericus* (=*E. pycnanathus*). *Vegetatio* 104/105:433–439.

Vu, J. C. V., L. H. Allen, Jr., and L. A. Garrard. 1984. Effects of enhanced UV-B radiation (280–320 nm) on ribulose-1,5-bisphosphate carboxylase in pea and soybean. *Environ. Exp. Bot.* 24:131–143.

Vu, J. C. V., L. H. Allen, Jr., and G. Bowes. 1989. Leaf ultrastructure, carbohydrates, and protein of soybeans grown under CO_2 enrichment. *Environ. Exp. Bot.* 29:141–147.

Waters, J. W., L. Froidevaux, W. G. Read, G. L. Manney, L. S. Elson, D. A. Flower, R. F. Jarnot, and R. S. Harwood. 1993. Stratospheric ClO and ozone from the Microwave Limb Sounder on the Upper Atmosphere Research Satellite. *Nature* 362:597–602.

Webber, H. J. 1948. Plant characteristics and climatology. In *The Citrus Industry,* Vol. I: *History, Botany and Breeding,* ed. H. J. Webber and L. D. Batchelor, pp. 41–70. University of California Press, Berkeley and Los Angeles.

Winner, W. E., and H. A. Mooney. 1980. Ecology of SO_2 resistance. II. Photosynthetic changes of shrubs in relation to SO_2 adsorption and stomatal behavior. *Oecologia* 44:296–302.

Winner, W. E., C. Gillespie, W.-S. Shen, and H. A. Mooney, H.A. 1988. Stomatal responses to SO_2 and O_3. In *Air Pollution and Plant Metabolism,* ed. S. Schulte-Hostede, N. M. Darrall, L. W. Blank, and A. R. Wellburn, pp. 255–271. Elsevier, London.

Woodwell, G. M., and W. R. Dykeman. 1966. Respiration of a forest measured by carbon dioxide accumulation during temperature inversions. *Science* 154:1031–1034.

Wullschleger, S. D., R. J. Norby, and C. A. Gunderson. 1992. Growth and maintenance respiration in leaves of *Liriodendron tulipifera* L. exposed to long-term carbon dioxide enrichment in the field. *New Phytol.* 121:515–523.

Yang, Y.-S., J. M. Skelly, and B. I. Chevone. 1982. Clonal response of eastern white pine to low doses of O_3, SO_2, and NO_2, singly and in combination. *Can. J. For. Res.* 12:803–808.

Yang, Y.-S., J. M. Skelly, and B. I. Chevone. 1983. Sensitivity of eastern white pine clones to acute doses of ozone, sulfur dioxide, or nitrogen dioxide. *Phytopathology* 73:1234–1237.

Zimmerman, R. H., D. T. Krizek, H. H. Klueter, and W. A. Bailey. 1970. Growth of crabapple seedlings in controlled environments: Influence of seedling age and CO_2 content of the atmosphere. *J. Am. Soc. Hortic. Sci.* 95:323–325.

Ziska, L. H., and J. A. Bunce. 1993. Inhibition of whole plant respiration by elevated CO_2 as modified by growth temperature. *Physiol. Plant.* 87:459–466.

Ziska, L. H., and A. H. Teramura. 1992. CO_2 enhancement of growth and photosynthesis in rice (*Oryza sativa*). Modification by increased ultraviolet-B radiation. *Plant Physiol.* 99:473–481.

Zukovskij, P. M. 1962. *Cultivated Plants and Their Wild Relatives.* Commonwealth Agricultural Bureaux.

4

On the Potential for a CO₂ Fertilization Effect in Forests: Estimates of the Biotic Growth Factor Based on 58 Controlled-Exposure Studies

STAN D. WULLSCHLEGER, W. M. POST, AND
A. W. KING

There is little question that terrestrial ecosystems have a potentially pivotal role in determining both the direction and the rate of future changes in atmospheric CO_2 concentrations. There is some question, however, whether terrestrial ecosystems will contribute to global warming by releasing additional CO_2 into the atmosphere as a result of increasing respiration in a warming climate, or whether they will instead sequester additional carbon in response to the enhancing effects of atmospheric CO_2 on plant growth (Soloman and Cramer 1993). This latter response, the so-called CO_2 fertilization effect, has been advanced as an important negative feedback that has the potential for limiting increases in atmospheric CO_2 concentrations. The mechanism has often been invoked as an explanation of the 1.6-petagram ($Pg = 10^{15}$ g)-carbon-per-year imbalance or "missing sink" in calculations of the global carbon budget.

Despite the frequent incorporation of a CO_2 fertilization effect into ecosystem and global carbon models, questions concerning the magnitude of any additional storage of carbon in terrestrial biomass remain. On the one hand, direct experimental evidence shows that whole-plant carbon storage increases in response to elevated CO_2 concentrations (Kimball 1983); on the other, results from modeling exercises indicate that this enhanced short-term productivity of individual plants may not be sustained given limitations imposed on plant growth by the biogeochemical cycles of carbon and nitrogen (Rastetter et al. 1991). Historical tree ring chronologies are not definitive: in some instances tree growth has increased during the past century (LaMarche et al. 1984), whereas other chronologies indicate no such stimulation (Kienast and Luxmoore 1988). The issue is made more puzzling by the possibility that young or aggrading forests may be more likely than mature, closed-canopy forests to display increased productivity in response to elevated CO_2 concentrations (Graham et al. 1990).

There appears to be little hope that the uncertainty surrounding the magnitude of a CO_2 fertilization effect will be resolved soon. Progress is possible, however, using the results of controlled CO_2 exposure studies for species representative of ecosystems that store large amounts of carbon. Because forests are important reservoirs of carbon, a data base of over 390 prior observations on the response of trees to atmospheric CO_2 enrichment was compiled from 58 controlled-exposure studies. The data describe the growth of 73 boreal, temperate, and tropical tree species exposed to an approximate doubling of atmospheric CO_2. This information was used to explore the magnitude of a CO_2 fertilization effect, to highlight species-specific differences in this response, and to evaluate whether a growth response to increasing CO_2 concentrations in forest trees might be constrained by limiting water and nutrients. Furthermore, we explored how the growth in controlled-exposure studies might be represented, applied, and interpreted within the context of global carbon models. Finally, we examined the potential consequences of a CO_2-induced increase in plant productivity on litter decomposition and storage of carbon in soils.

LITERATURE SEARCH, DATA PRESENTATION, AND ANALYSIS

Literature describing the response of trees to CO_2 enrichment was identified through a computerized search of the AGRICOLA and BIOSIS data bases and by searching scientific journals. Information was also requested from investigators currently conducting CO_2 enrichment studies. The data base is a comprehensive if not exhaustive representation of past and current investigations.

Rather than simply reiterate the results for each of the studies, as has been the approach in previous reviews (Eamus and Jarvis 1989; Mousseau and Saugier 1992), we adopted instead the procedures of Kimball (1983), who, in documenting the growth response of agricultural crops to elevated CO_2 concentrations, looked not at the results of individual studies per se, but at responses characteristic of broad plant groups in a data base of over 430 observations. Although this type of analysis tends to ignore subtle differences between studies, it does offer the hope that the larger picture will be clearer and that generalities will appear that might otherwise remain obscure.

The response of plant growth to increasing CO_2 concentration can be measured in many ways. A common technique is to calculate the relative change in total dry weight for plants grown at elevated CO_2 concentrations compared to that for their controls (Kimball 1983). In this approach a ratio of 1 would indicate no effect of elevated CO_2 on plant growth, whereas a ratio of 2 would indicate a doubling of dry mass in response to CO_2 enrichment. For the purposes of this presentation, relative growth responses were calculated for each observation. The data were transformed to logs to normalize the distribution and plotted as a frequency distribution showing the number of times that a given response was observed. The mean response for all observations, for the species-specific and biome-specific observations, and for the CO_2-induced growth response of trees given limiting nutrient and water resources was calculated (Kimball 1983). In addition to the mean, 95% and 99.9% confidence intervals were also computed.

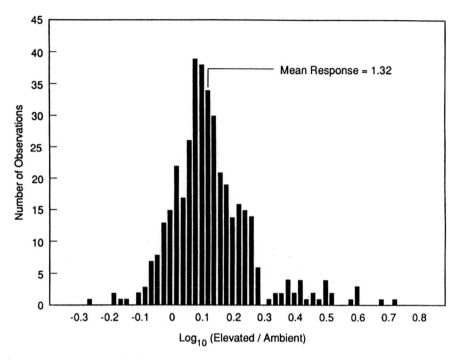

Figure 4.1. Frequency distribution of relative growth responses as calculated from the total dry mass of trees grown at ambient and elevated CO_2 concentrations. An average growth response for the entire data set was determined by averaging the log-transformed data to normalize the distribution.

RESPONSE OF FOREST TREES TO CO_2 ENRICHMENT

Based on the results of 58 controlled-exposure studies, the growth response of 73 tree species to elevated CO_2 concentrations was overwhelmingly positive (Figure 4.1). Of the 398 observations compiled in this study, 51 showed a relative growth response less than 1 (i.e., a value less than 0 on a log-transformed scale) and 31 showed values greater than 2 (i.e., $\log_{10} > 0.30$), indicating a twofold increase in plant dry mass with CO_2 enrichment. For the entire data set the relative growth response averaged 1.32, with a 95% confidence interval of between 1.28 and 1.36 and a 99.9% confidence interval of between 1.25 and 1.39. A relative growth response of 1.32 is equivalent to a 32% increase in total plant dry mass (i.e., above and below ground) in response to an approximate doubling of atmospheric CO_2 concentration. From a smaller number of observations, this stimulation of whole-plant dry mass was found to be more or less equally partitioned between leaves (1.33), stems (1.29), and roots (1.38), but it was not a fair representation of canopy leaf area, which increased only marginally (1.13) in response to CO_2 enrichment.

Separating the results by species indicated that the relative growth response varied fourfold across all 73 tree species, from below 1 to above 3 (Figure 4.2). The cause of the differences was not clear, except that species vary in response to CO_2 enrichment.

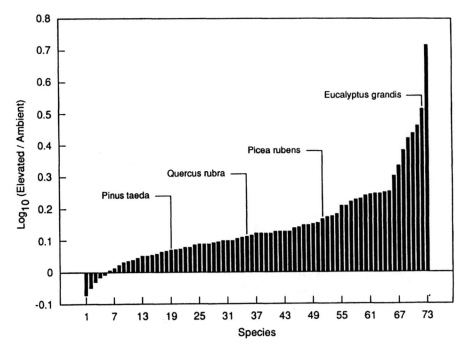

Figure 4.2. Species-specific estimates for the relative growth response as calculated from the total dry mass of trees grown at ambient and elevated CO_2 concentrations. Average growth responses were calculated by averaging the log-transformed data to normalize the distribution. From left to right the species are: *Populus trichocarpa, Quercus prinus, Betula* sp., *Myriocarpa longipes, Cecropia obtusifolia, Fraxinus lanceolata, Pseudotsuga menziesii, Populus euramericana, Carya ovata, Pinus virginiana, Senna multijuga, Liriodendron tulip-ifera, Platanus occidentalis, Picea sitchensis, Fraxinus americana, Acer saccharinum, Piper auritum, Nothofagus fusca, Pinus taeda, Quercus alba, Pinus strobus, Castanea sativa, Pinus radiata, Betula populifolia, Pinus ponderosa, Betula papyrifera, Populus trichocarpa × P. deltoides, Pinus echinata, Alnus rubra, Gliricidia sepium, Betula alleghaniensis, Picea mariana, Betula pendula, Tsuga canadensis, Pinus nigra, Pentaclethra macroloba, Prunus serotina, Robinia pseudoacacia, Pinus sylvestris, Acer rubrum, Trichospermum mexicanum, Pinus* sp., *Acer pensylvanicum, Salix × dasyclados, Picea glauca, Populus grandidentata, Acacia mangium, Pinus banksiana, Quercus rubra, Liquidambar styraciflua, Picea rubens, Alnus glutinosa, Picea pungens, Populus tremuloides, Elaeagnus angustifolia, Fagus syl-vatica, Cedrus atlantica, Populus deltoides, Juglans nigra, Fagus grandifolia, Picea* sp., *Abies fraseri, Acer saccharum, Picea abies, Ochroma lagopus, Pinus caribaea, Eucalyptus camaldulensis, Eucalyptus cypellocarpa, Tabebuia rosea, Eucalyptus pauciflora, Eucalyptus pulverulenta, Eucalyptus grandis, Pinus contorta.*

Those species most responsive to a doubling of atmospheric CO_2 were among the Eucalypts; estimates of the growth response for *Eucalyptus grandis* exceeded 3 and for the genus averaged over 2 (see the Appendix). No other genus-specific distinction in the growth response of forest trees to elevated CO_2 concentrations was apparent. Further separation of these species-specific results into those representative of three forest biomes indicated a trend: the relative growth response was greatest for trees of boreal

Table 4.1. Biome-Specific Estimates for the Relative Growth Response as Calculated from the Total Dry Mass of Trees Grown at Ambient and Elevated CO$_2$ Concentrations

Forest biome	Lower limit		Mean	Upper limit	
	−99.9%	−95%		+95%	+99.9%
Boreal ($n = 74$)	1.24	1.30	1.38	1.46	1.53
Temperate ($n = 305$)	1.24	1.27	1.31	1.35	1.39
Tropical ($n = 18$)	0.98	1.10	1.25	1.41	1.54

regions, intermediate for forests characterized by temperate zone broadleaf and coniferous species, and lowest for trees inhabiting tropical regions (Table 4.1). However, the differences were small and of questionable significance.

Limitations imposed on the relative growth response by the availability of nitrogen and water were minimal (Table 4.2). Although the distinction between adequate and deficient as used here and in many of the experiments themselves is purely subjective (note the exception of Conroy et al. [1990a]), the growth response for forest trees exposed to CO$_2$ enrichment and grown under adequate supplies of nitrogen or water was 1.32, a value only slightly higher than that for trees deficient in nitrogen or water (Table 4.2). In contrast, phosphorus deficiency did result in a lower, albeit still positive, relative growth response as compared with plants supplied with adequate phosphorus.

The relative growth responses as estimated from the 58 controlled-exposure studies are representative of experiments in which CO$_2$ exposures lasted between 8 and 798 days, with the average length of exposure being 136 days. The majority of these studies were conducted in growth chambers or greenhouses; a few were field studies using open-top chambers; most used transplanted seedlings grown in pots. Rarely were seedlings grown from seed under elevated CO$_2$ concentrations. Day and night temperatures ranged from 37/21°C to 20/10°C, with the 24-h average temperature being about 22°C when pooled across all studies. No study considered the interactive effects of temperature on the growth response of trees to CO$_2$ enrichment. Ambient CO$_2$ concentrations ranged from 300 to 400 ppm, with elevated concentrations of CO$_2$ ranging from 415 to 2000 ppm (see the Appendix). With the possible exception of the work by Silvola

Table 4.2. Estimates for the Relative Growth Response as Calculated from the Total Dry Mass of Trees Grown at Ambient and Elevated CO$_2$ Concentrations Given Limiting Nutrient or Water Resources

Resource	Lower limit		Mean	Upper limit	
	−99.9%	−95%		+95%	+99.9%
Nitrogen					
Adequate ($n = 295$)	1.24	1.27	1.32	1.36	1.40
Deficient ($n = 82$)	1.16	1.22	1.30	1.38	1.45
Phosphorus					
Adequate ($n = 314$)	1.25	1.28	1.33	1.37	1.40
Deficient ($n = 61$)	1.09	1.14	1.21	1.29	1.35
Water					
Adequate ($n = 359$)	1.26	1.28	1.32	1.37	1.40
Deficient ($n = 30$)	1.16	1.21	1.28	1.35	1.41

and Ahlholm (1992), no study explicitly included a subambient CO_2 treatment. In the case of the Silvola and Ahlholm study, however, there was no ambient control.

ESTIMATING A GLOBAL RESPONSE TO CO_2 ENRICHMENT

Although the majority of controlled-exposure studies indicate that forest trees have the capacity to increase whole-plant carbon storage in response to enhanced concentrations of CO_2, many questions concerning the magnitude of such effects in natural ecosystems remain. Favoring the hierarchical development of models for describing ecosystem and global carbon dynamics, Reynolds et al. (1992) argue that the growth results obtained in controlled-exposure studies with plants must be interpreted with caution because they may not result in additional carbon storage in an ecosystem (see Körner 1993). There are, of course, many processes that can intervene between the initial acquisition of carbon in photosynthesis, for which the response to rising CO_2 concentration is expected to be the greatest, and the ultimate expression of this in net ecosystem production. Integration of single-plant responses to heterogenous canopies and landscapes; interactions with limiting light, water, and nutrient resources; effects of plant life history and demography; competitive interactions among multiple species and life forms; and many more factors all need to be considered in extrapolating experimental results to the global carbon cycle.

    Although few would argue with the suggestion that intervening processes might indeed lessen the magnitude of a CO_2 fertilization effect as estimated from controlled-exposure studies (note the exception of Idso [1991]), it remains a difficult task to quantify the magnitude of a CO_2 fertilization effect at the ecosystem level, let alone on a global scale. Nevertheless, applying a CO_2 fertilization effect to ecosystem and global carbon models seems reasonable to mimic the assumed response of terrestrial vegetation to rising atmospheric CO_2 concentrations (Goudriaan and Ketner 1984; Kienast 1991; Gifford 1992; Polglase and Wang 1992). Here we describe, in some detail, how growth results obtained from controlled exposure studies can be cautiously used to derive estimates of the global CO_2 response by the terrestrial biosphere.

Experimental Studies and a Global Biotic Growth Factor

Keeling (1973) introduced a "biota growth factor," β, in his global carbon cycle model. The model represents carbon flux from the atmosphere to terrestrial biota and its response to increasing atmospheric CO_2 concentration. Keeling argued that "according to several studies in glass houses, the assimilation by plants at high CO_2 concentrations is roughly proportional to the logarithm of the ambient CO_2 concentration." In the form later used by Bacastow and Keeling (1973), this biota growth factor and its logarithmic dependency on CO_2 concentration were defined as

$$F_{ab} = F_{b0}[1 + \beta ln(N_a/N_{a0})](N_b/N_{b0}) \tag{4.1}$$

where F_{ab} is the flux from the atmosphere to the terrestrial biota, F_{b0} is the initial preindustrial equilibrium flux from atmosphere to biota (and equivalently biota to atmosphere), N_a and N_b are the carbon contents of the atmosphere and terrestrial biota, respectively; and N_{a0} and N_{b0} are their initial contents. The deficiencies of this logarithmic representation of CO_2 saturation have been frequently noted (e.g., Gifford 1980; Gates 1985; Kohlmaier et al. 1987). Nevertheless, because of its historical precedence and its simplicity, Keeling's formulation has become one of the most common representations of terrestrial biospheric response to elevated CO_2 levels in global carbon cycle models.

Keeling's biota growth factor applies to the global terrestrial biosphere as a whole, and its value has often, and appropriately, been determined by tuning for correspondence between observed and simulated changes in atmospheric CO_2 concentration. In its purest form, the global β of equation 4.1 cannot be determined experimentally. Nevertheless, the parameter β is explicitly related to plant response to CO_2 concentration, and it is reasonable to seek experimental evidence for appropriate a priori values of the biota growth factor. Within this context, some experimentalists have adopted β (often recoined the "biotic growth factor") as a means of characterizing plant response to increasing CO_2 concentration (e.g., Gifford 1979a,b; Acock and Allen 1985; Allen et al. 1987), and some global carbon cycle modelers have explicitly recognized an experimentally measured biotic growth factor that approximates the global parameter of the Keeling formulation (e.g., Kohlmaier et al. 1987).

As noted in equation 4.1, Keeling's β applies to the global flux of carbon from atmosphere to biosphere (i.e, F_{ab}). To estimate β from experimental studies of plant response to elevated CO_2 levels, we must translate between this global flux and experimental measurements. Net primary productivity (NPP) is normally used to approximate F_{ab} and literature estimates of NPP are used to quantify NPP_0. Thus we can rewrite equation (4.1) as

$$NPP = NPP_0[1 + \beta\ln(C_a/C_{a0})](C_b/C_{b0}) \tag{4.2}$$

where NPP_0 is the initial equilibrium NPP and C_a and C_b are carbon contents of the atmosphere and biosphere, respectively. From equation 4.2 we can define β as

$$\beta = \frac{(NPP - NPP_0) / NPP_0}{\ln (C_a / C_{a0})} \tag{4.3}$$

If we assume that NPP is independent of the mass of the terrestrial biosphere (e.g., Goudriaan and Ketner 1984; Kohlmaier et al. 1987) then

$$NPP = NPP_0[1 + \beta_1\ln(C_a/C_{a0})] \tag{4.4}$$

and

$$\beta_1 = \frac{(NPP - NPP_0) / NPP_0}{\ln (C_a / C_{a0})} \tag{4.5}$$

For small changes in the terrestrial biosphere $\beta_0 \approx \beta_1$, and for a monotonically increasing terrestrial biosphere $\beta_0 \leq \beta_1$.

For small changes in atmospheric CO_2 concentration,

$$\beta_1 = \frac{(NPP - NPP_0) / NPP_0}{(C_a - C_{a0}) / C_{a0}} \tag{4.6}$$

We identify this approximation as β_2. Thus for small changes in the terrestrial biosphere and atmospheric CO_2, $\beta_2 \approx \beta_1 \approx \beta_0$, and β is approximately the fractional increase in NPP for a unit increase in atmospheric CO_2 concentration. β_2 of equation 4.6 is often described as the biotic growth factor and ascribed to Bacastow and Keeling (1973). We note that it is in fact an approximation of the β (equation 4.3) described in that paper.

Net primary productivity can be defined as

$$NPP = \Delta B + L + C \tag{4.7}$$

where ΔB is the change in vegetation biomass over some period of time and L and C are litter production and the loss of vegetation biomass to herbivore consumption, respectively, over the same period of time. If we assume that herbivory is absent or negligible in experimental CO_2 exposure studies, and that any litter produced is accounted for as a change in biomass, then we can define P as ΔB over the period of the experimental exposure and equate NPP with P. Then, if we assume that the functional relationship of equation 4.2 among global NPP, biospheric carbon, and atmospheric CO_2 changing over tens to hundreds of years also characterizes, at least approximately, the relationship among plant biomass production, plant biomass, and relatively short-term exposure to elevated CO_2 levels under controlled experimental conditions, we can define the relationship

$$P_e = P_c[1 + \beta\ln(C_e/C_c)](B_e/B_c) \tag{4.8}$$

where P_c is plant biomass production at control CO_2 concentrations C_c, P_e is production at experimental or elevated CO_2 concentrations C_e, and B_c and B_e are plant biomass at control and elevated CO_2 concentrations, respectively.

From equation 4.8 and the corresponding translations of equations 4.5 and 4.6, we can define the following experimental growth factors:

$$\beta E_0 = \frac{[P_e (B_c / B_e) - P_c] / P_c}{\ln (C_e / C_c)} \tag{4.9a}$$

$$\beta E_1 = \frac{(P_e - P_c) / P_c}{\ln (C_e / C_c)} \tag{4.9b}$$

$$\beta E_2 = \frac{(P_e - P_c) / P_c}{(C_e - C_c) / C_c} \tag{4.9c}$$

Upon equating P with NPP, βE_2 of equation 4.9c is equivalent to the experimental fertilization factor $\hat{\beta}$ of Kohlmaier et al. (1987).

Controlled CO_2 exposure studies often report only the biomass of treatments at the end of the experiment and not the explicit biomass production P, analogous to NPP, called for by equation 4.9. We have defined P as ΔB over the period of the CO_2 exposure experiment, or

$$P_c = B_{ct} - B_{c0} \tag{4.10a}$$

$$P_e = B_{et} - B_{e0} \tag{4.10b}$$

where t is the length of exposure and the subscript 0 indicates initial values at $t = 0$ such that B_{ct}, for example, is the biomass of the control exposure at the end of the experiment. B_c and B_e of equation 4.8 are defined as B_{ct} and B_{et}, respectively.

If the experimental treatment begins with seed germination, as might be the case with annual crop plants, the B_0s are essentially zero, $P = B_t$. βE of equation 4.9a is ill defined as zero by this substitution. However, βE_1 and βE_2 of equations 4.9b and 4.9c are unaffected by substituting B_t for P. If the initial biomass of the experimental material is nonzero, but appreciably the same for each treatment, as is the case for most studies with woody perennials, then the experimental growth factors can be evaluated as

$$\hat{\beta E}_0 = \frac{[(B_{ct} / B_{et})(B_{et} - B_{e0}) - (B_{ct} - B_{c0})] / B_{ct} - B_{c0}}{\ln (C_e / C_c)} \tag{4.11a}$$

$$\hat{\beta E}_1 = \frac{(B_{et} - B_{ct}) / (B_{ct} - B_{c0})}{\ln (C_e / C_c)} \tag{4.11b}$$

$$\hat{\beta E}_2 = \frac{(B_{et} - B_{ct}) / (B_{ct} - B_{c0})}{(C_e - C_c) / C_c} \tag{4.11c}$$

To the extent that $B_{c0} \neq B_{e0}$, the $\hat{\beta E}$s of equations 4.11 are approximations of the βE of equations 4.9. If $B_{e0} > B_{c0}$ they are overapproximations; they are underapproximations if $B_{e0} < B_{c0}$.

The relative growth response used earlier in this chapter to summarize the controlled exposure studies was defined as

$$\rho = B_{et}/B_{ct} \tag{4.12}$$

where, for the entire data set, $\rho = 1.32$. For B_{ct} much greater than B_{c0}, we can make the following approximations:

$$\hat{\beta E}_0 \approx \frac{[\rho^{-1}(B_{et} - B_{e0}) / B_{ct}] - 1}{\ln (C_e / C_c)} \tag{4.13a}$$

$$\hat{\beta E}_1 \approx \frac{\rho - 1}{\ln (C_e / C_c)} \tag{4.13b}$$

$$\hat{\beta E}_2 \approx \frac{\rho - 1}{(C_e - C_c) / C_c} \tag{4.13c}$$

and estimate a corresponding biotic growth factor for each approximation (Table 4.3). Since $B_{e0} = B_{c0} < B_{ct} \leq B_{et}$ βE_0 is again poorly approximated as being close to zero.

Table 4.3. Estimates of a Global Biotic Growth Factor Based
on the Results of Controlled-Exposure Studies

i	$N\hat{\beta E}_i a$	$\hat{\beta E}_i b$	$\beta E_i c$	$\beta_i d$
1	≈0.46	0.46ε	0.46ε	0.46γε ≤ 0.46
2	≈0.32	0.32ε	0.32ε	0.32γε ≤ 0.32

[a]Estimates of $N \hat{\beta E}$ based on equations 4.14.

[b]Estimates of $\hat{\beta E}$ of equations 4.11 based on the approximation of equations 4.14, where $0 < \varepsilon \leq 1$, and probably close to 1.

[c]Estimates of βE of equations 4.9 from estimates of $\hat{\beta E}$, assuming initial biomass of control and elevated CO₂ treatments is equal.

[d]Estimates of the global β of equations 4.3, 4.5, and 4.6 by equations 4.15, where $0 \leq \gamma \leq 1$.

The control and elevated CO_2 treatments vary among the studies summarized in the Appendix, but we can normalize for a doubling of CO_2 concentration ($C_e = 2C_c$) such that

$$N\hat{\beta}E_0 \approx \frac{[\rho^{-1}(B_{et} - B_{e0}) / B_{ct}] - 1}{\ln (2)} \tag{4.14a}$$

$$N\hat{\beta}E_1 \approx \frac{\rho - 1}{\ln (2)} \tag{4.14b}$$

$$N\hat{\beta}E_2 \approx \rho - 1 \tag{4.14c}$$

Once again, $N\hat{\beta}E_0$ is poorly approximated as close to zero. When the elevated CO_2 treatment is greater than twice the control, these normalized approximations $N\hat{\beta}E_1$ and $N\hat{\beta}E_2$ are overestimates of the $\hat{\beta}E$. We also note that for a doubling of atmospheric CO_2 concentration, the β_2 approximation of β_1 underestimates β_1 by roughly 30%.

Without data on rates of production (P) or the initial biomass (B_0) of control and elevated treatments (data absent from most reports of controlled CO_2 exposure studies, including those summarized here), the experimental growth factor $\hat{\beta}E_0$ (or βE_0) cannot be appropriately evaluated. However, estimates of the $2\times CO_2$ normalized $N\hat{\beta}E_1$ and $N\hat{\beta}E_2$ for the mean relative growth response of the entire data set presented here ($\rho = 1.32$) are given in Table 4.3. As noted previously, these estimates are approximations of the $\hat{\beta}E$ of equations 4.11 and are probably overestimates. The correction for this approximation is indicated by ε (Table 4.3), where $0 < \varepsilon \leq 1$. Because we believe B_{c0} to be much less than B_{ct} in most of the studies summarized here, and most elevated CO_2 concentrations are at least approximately a doubling of the control concentrations, we believe that ε is close to 1. We assume that $B_{c0} = B_{e0}$ and accept the approximations of $\hat{\beta}E$ as estimates of the βE of equation 4.9.

In translating from the experimental growth factors to an estimate of the global β, we make the following definitions:

$$\beta_0 = \gamma\beta E_0 \tag{4.15a}$$

$$\beta_1 = \gamma\beta E_1 \tag{4.15b}$$

$$\beta_2 = \gamma\beta E_2 \tag{4.15c}$$

where γ is a proportionality coefficient that incorporates the multitude of assumptions, errors, and uncertainties involved in translating from the measurements of short-term experimental exposures of small (generally young) plants in controlled chambers or small field exposures to a model parameter of long-term CO_2 response by the terrestrial biosphere. The value of γ is highly uncertain. Experimental evidence is mixed, but there is some indication that as the level of hierarchical system organization increases (e.g., from leaf photosynthetic rates to whole-plant growth), the CO_2 response declines (e.g., Allen et al. 1987) and that with increased exposure time acclimation can occur with a decline in the magnitude of the CO_2 response (Oechel and Strain 1985). Additional experiments, modeling, analysis, and theoretical treatment are required to substantiate or refute these trends, but for now we assume that $0 \leq \gamma \leq 1$. The resulting estimates of global β as defined in equations 4.3, 4.5, and 4.6 are given in Table 4.3.

Carbon Sequestration and Soil Organic Matter

The capacity of forest trees to increase whole-plant carbon storage is only one of many components that will dictate whether similar increases in carbon sequestration will be observed in natural ecosystems, where carbon will be stored not only in biomass but also in soils. In many terrestrial ecosystems, for example, the amount of carbon sequestered in soil organic matter greatly exceeds that in living plants. Similarly, on the global scale over twice as much carbon is stored within soil organic matter and its component fractions as in plant biomass (Post et al. 1990). Given these distinctions, if we assume for the moment that plant growth will increase in response to rising CO_2 concentrations, then the emphasis shifts away from one that focuses strictly on vegetation toward one of understanding how this increased biomass, once shed from the plant, decays and becomes soil organic matter. These are the processes that will largely govern the rate and magnitude of long-term carbon storage under changing atmospheric CO_2 concentrations.

The amount of carbon stored in soils is determined by the rate of organic matter input through litterfall (i.e., leaves, branches, boles), root exudates, fine root turnover, and the rate of decay. However, a large fraction of soil organic matter has a long turnover time (around 20–50 years), and this introduces a considerable time response for sequestering in soils any additional carbon that may result from increasing plant productivity owing to CO_2 enrichment. If, for example, the rate of organic matter input into soil increases by 30% in response to a doubling of atmospheric CO_2 concentration, then, all other factors remaining the same, soil carbon should also increase by an equivalent amount. An important question to ask in this regard, however, is how long it will take for this new equilibrium to become established.

Using the Rothamsted turnover model (Jenkinson 1990), it is possible to explore the time lags involved in soil carbon sequestration and to link CO_2-induced effects on plant productivity with decomposition. Plant residues are divided into two components. The first includes simple compounds subject to rapid uptake, transformation, and mineralization by decomposers. This pool is decomposable plant matter (DPM) and has a turnover time of 0.1 year. The second pool contains components such as lignin that are not readily attacked; it is identified as recalcitrant plant matter (RPM), with a turnover time of 3 years. Soil humus is divided into three components with turnover times differing by an order of magnitude. The soil biomass (BIO, turnover time 1.5 years) component is mostly live microbes and microbial products such as extracellular enzymes, whereas the active soil humus (HUM, turnover time 50 years) component consists of organic compounds that are protected physically or that exist in chemical forms that are biologically resistant to decomposition. Finally, the inert soil organic matter (IOM) component consists of chemically refractory and physically protected forms with turnover times in excess of 1000 years. In the simulations presented here, the IOM is truly inert, neither increasing nor decreasing during the simulations. Following Jenkinson et al. (1991) we set this pool at 3 tons/hectare.

The fate of soil carbon as estimated from the Rothamsted model was investigated for a hypothetical 200-year period during which organic matter inputs to the soil were gradually increased by 30% (Figure 4.3). Modeled results using Rothamsted Experiment Station climatic conditions (mean annual temperature 9.3°C) showed that the litter pools (DPM and RPM) and decomposer biomass (BIO) responded quickly, because of their rapid turnover times, and tracked closely the projected increases in organic matter inputs

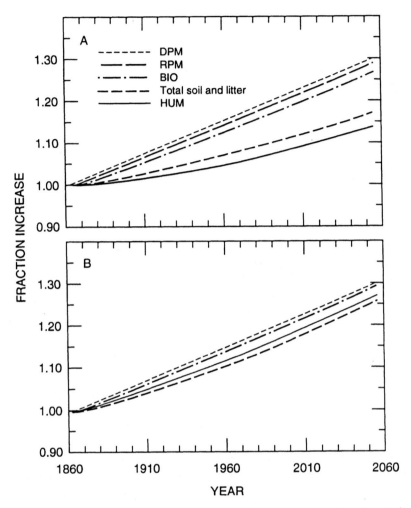

Figure 4.3. Soil carbon responses to increasing inputs as simulated by the Rothamsted turnover model. Two scenarios are presented, one using the climatic conditions of Rothamsted Experiment Station (A), where annual temperatures average 9.3°C, and another using the warmer conditions of Manaus, Brazil (B), where annual temperatures average 26.9°C. Organic matter inputs were assumed to increase linearly by 30% over a 200-year period beginning in 1860. The DPM pool increased at the same rate as total inputs. The RPM and BIO pools lagged closely behind the organic matter increases, although RPM in the Manaus simulations did equilibrate as readily as total inputs and overlay the DPM line in this subplot. The HUM pool lagged behind the other organic matter components and resulted in the response of the litter and active soil pools (DPM + RPM + BIO + HUM) not reaching equilibrium during the simulation period. The magnitude of the lag was related to the rate of decomposition, which was considerably higher with Manaus climate than with Rothamsted climate. The IOM pool of the Rothamsted model did not change during the simulation, is not plotted here, and was not used in the calculation of the total litter and soil response.

(Figure 4.3A). Increased inputs because of enhanced plant productivity were immediately reflected in these component fractions. These pools, however, contained less than 30% of total soil organic matter. The much larger active soil humus (HUM) pool, which comprises 70% of the soil organic matter, increased at a much slower rate. As a result, the total increase in soil organic matter after 200 years was only 17%, substantially less than one might expect from a 30% increase in litter inputs. Nonetheless, such an increase resulted in incorporating 15% of the enhanced productivity over this 200-year period into soil organic matter, mostly as active humus. Model runs using the warmer climatic conditions for Manaus, Brazil (mean annual temperature 26.9°C) resulted in higher decomposition rates (Figure 4.3B). The HUM pool, which in this case comprised 34% of the total soil carbon, increased faster under these conditions than for the Rothamsted scenario. As a result, the total soil carbon increased by 25% over the 200-year period, although much less of the enhanced biomass production during this time was sequestered, with only 4% retained in soil carbon pools—30% in RPM and 67% in HUM.

From these modeled results we observe that the storage of carbon in biomass and in soils may not be tightly coupled: an increase in whole-plant productivity resulting from rising CO_2 concentrations may not necessarily lead to a proportional increase in soil carbon storage, despite an equivalent increase in litterfall. There are two reasons for this. First, the turnover times of soil organic matter and its component fractions are sufficiently long that any increased input in the form of litterfall and its subsequent incorporation into soil organic matter will not occur immediately, but will be delayed depending on the turnover times of individual soil carbon pools and decomposition rates. Second, as indicated by the contrasting results of the Rothamsted and Manaus climatic effects, increasing decomposition rates will increase the speed at which the slower organic matter pools approach equilibrium, but as temperatures increase less carbon will be sequestered in soil carbon pools and will be returned to the atmosphere as CO_2.

GLOBAL CARBON MODELS AND A CO₂ FERTILIZATION EFFECT

As alluded to earlier, global carbon models frequently mimic the CO_2 fertilization effect in terrestrial vegetation by incorporating a biotic growth factor. In an interesting application of the biotic growth factor, Polglase and Wang (1992) divided the globe into 10 representative biomes and then used the Rothamsted soil carbon model (Jenkinson 1990)—along with estimates of areal distribution; net primary productivity; allocation of biomass to leaves, branches, stems, and roots (each with a characteristic residence time); and a reference-state biotic growth factor of 0.3—to calculate biome-specific and global estimates of CO_2-enhanced carbon storage. The biotic growth factor was made a function of annual air temperature, with the growth factor being lower in cooler and higher in warmer climates. They applied this growth factor to an NPP function like that of equation 4.4. Their results indicated that the total CO_2-enhanced carbon storage in 1990 was 1.14 Pg/year. The most important biomes for sequestering carbon were tropical forests, savannas, and temperate zone forests. Of the globally sequestered carbon, 42% was stored in vegetation, 19% in litter, and 39% in soil organic matter. Our estimate of a larger biotic growth factor, $\beta_1 \leq 0.46$, would presumably result in even greater CO_2-enhanced carbon storage.

Calculating probable rates of global carbon sequestration caused by land-use change, primarily deforestation in the tropics, and a CO_2 fertilization effect, Taylor and Lloyd (1992)

Table 4.4. Summary of Studies That Have Incorporated a Biotic Growth
Factor into Models of Global Carbon Sequestration

Reference	Biome divisions	Biotic growth factor	Rate of carbon storage (Pg/year)
Oeschger et al. (1975)	1	0.20	—
Gifford (1980)	9	0.60	1.0
Goudriaan and Ketner (1984)	6	0.50	—
Kohlmaier et al. (1987)	4	0.25–0.50	1.4
Kohlmaier et al. (1989)	—	0.38	—
Gifford (1993)	1	0.36	2.3
Polglase and Wang (1992)	10	0.30	1.1
	10	0.02–0.54	1.5
	10	0.02–0.54	1.3
Taylor and Lloyd (1992)	14	0.36	2.4

Note: Because of the many assumptions used in each model the rate of carbon storage cannot be ascribed solely to a CO_2 fertilization effect.

similarly invoked a biotic growth factor of 0.37 (i.e., a 25% increase in NPP associated with a doubling of atmospheric CO_2 concentration) for 14 major ecosystem types. They suggested that the total net sink effect in 1985 for carbon was 2.4 Pg/year globally, with 19% of the terrestrial carbon sink in the boreal zone, 37% in the temperate zone, and 44% in the tropics. Their biotic growth factor corresponds to our β_2 (i.e., 0.32) shown in Table 4.3. Our estimate of this biotic growth factor is only slightly smaller than that used by Taylor and Lloyd and would presumably reduce the estimates of a net sink only slightly.

The examples cited, along with those obtained by Gifford (1993) using the C-Quest model, and a sampling of other studies that include a biotic growth factor in global carbon models (Table 4.4) indicate that through a CO_2 fertilization effect on net primary productivity, terrestrial ecosystems can sequester vast amounts of carbon and that this process might indeed function as an important negative feedback on future increases in global CO_2 concentrations. For example, in simulating the concentrations of atmospheric CO_2 for the year 2030, Goudriaan and Ketner (1984) indicated that by reducing the biotic growth factor from 0.50 to 0.25, and thereby lessening the negative feedback potential of terrestrial vegetation, atmospheric CO_2 concentrations would increase by an estimated 21 ppm. Likewise, Polglase and Wang (1992) also point out the sensitivity of carbon sequestration to the biotic growth factor, suggesting that a twofold increase in β gives rise to a doubling of ecosystem and global CO_2-enhanced carbon storage. Given this sensitivity of model estimates to the often subjective choice of a biotic growth factor, it is important to consider carefully and to understand fully the assumptions and translations involved in using controlled-exposure studies to estimate a global biotic growth factor, and how that estimate is applied in global carbon models.

CONCLUSIONS

Where do the results obtained from controlled-exposure studies begin to provide answers for questions of truly global proportions? So far the link between results obtained from experimental studies and the application of such results on the ecosystem or global scale has been a limited one. This has not, however, been the case because of a lack of effort. Rather, it has occurred because of the realization that answers to ecosystem and global concerns may not always lend themselves to direct experimentation, and as a result modelers have been left, in many cases, with little more than intuition to guide them. This comment is not intended to be derogatory; instead it suggests that experimental studies, among all the important questions that they are addressing, may not be providing the simple information needed to verify or refute even the most basic of assumptions required in global carbon models. Within the context of the CO_2 fertilization effect and the data summarized herein, four points are illustrative:

1. Despite the fact that boreal, temperate, and tropical forests occupy regions of the globe with vastly different temperature regimes, these differences are not reflected in the growth conditions chosen for CO_2 enrichment studies. Of the 58 controlled-exposure studies considered here, virtually all were conducted at temperatures of 20–25°C, temperatures that apparently were chosen with little thought to the species' natural range.

2. Even if one wanted to scale the CO_2 fertilization effect for changing temperatures, we know very little about what such a correction curve would look like. Although Polglase and Wang (1992) derive a solution whereby β can be corrected for temperature, there has to date been no attempt to examine the potential for a temperature × CO_2 level interaction in forest trees. Because of the presumed association between rising CO_2 concentrations and global warming, there is an obvious need to pursue this area of research.

3. It is clear that to predict future atmospheric concentrations of CO_2 and hence to evaluate the potential for global warming, modelers must first match the predictive capabilities of a carbon model against the historical record. The period considered in this chapter was characterized by CO_2 concentrations below present-day levels, but no study has examined the growth response of long-lived woody perennials to subambient CO_2 concentrations.

4. Given the need to equate experimentally derived estimates of biomass or biomass productivity with net primary productivity, it is unfortunate that so few studies report an initial plant dry mass at the beginning of the CO_2 exposure. Doing so would help eliminate some of the assumptions, approximations, and uncertainties in estimating a global β from controlled-exposure studies.

Clearly, experimental studies have done much to address the needs of modeling the global carbon cycle, yet it is equally clear that much work remains ahead. In assessing the capacity of terrestrial vegetation to serve as an important biospheric feedback to rising CO_2 concentrations and associated climate change, it is apparent that we have only begun to meet the challenges. By working together, experimentalists and modelers must strive to complement each others' strengths and weaknesses to define and interpret empirical and mechanistic descriptions of terrestrial biota and subsequently incorporate them into ecosystem and global carbon models.

APPENDIX: SPECIES-SPECIFIC ESTIMATES FOR THE RELATIVE GROWTH RESPONSE AS CALCULATED FROM THE TOTAL DRY MASS OF TREES GROWN AT AMBIENT AND ELEVATED CO$_2$ CONCENTRATIONS

Ambient and elevated concentrations of CO$_2$ used in each study varied, as did the duration of each study, and this information is provided for the reader's convenience. NA denotes data not available.

Species	Growth CO$_2$ concentration (ppm)	Duration of experiment (days)	Relative growth response	Reference
Abies fraseri	362, 713	365	1.76	Samuelson and Seiler (1992)
Acacia mangium	354, 711	95	1.40	Ziska et al. (1991)
Acer pensylvanicum	350, 700	165	1.36	Bazzaz and Miao (1993)
Acer rubrum	400, 700	60	1.05	Bazzaz et al. (1990)
	350, 700	70	1.58	Bunce (1992)
	350, 700	66	1.36	Miao et al. (1992)
	350, 700	165	1.34	Bazzaz and Miao (1993)
Acer saccharinum	300, 600, 1200	35	1.75	Carlson and Bazzaz (1980)
	350, 500, 700	90	0.87	Williams et al. (1986)
	350, 700	55	1.66	Bunce (1992)
Acer saccharum	400, 700	100	2.52	Bazzaz et al. (1990)
	400, 800	85	1.55	Nobel et al. (1992)
Alnus glutinosa	350, 700	98	1.49	Norby (1987)
Alnus rubra	350, 650	15–47	1.27	Arnone and Gordon (1990)
Betula alleghaniensis	350, 700	165	1.27	Bazzaz and Miao (1993)
Betula papyrifera	400, 700	60	1.23	Bazzaz et al. (1990)
Betula pendula	350, 700	28–42	1.31	Pettersson and McDonald (1992)
Betula populifolia	350, 700	66	1.30	Miao et al. (1992)
	350, 700	165	1.17	Bazzaz and Miao (1993))
Betula sp.	350, 2000	40	0.93	Hardh (1966)
Carya ovata	350, 500, 700	90	1.08	Williams et al. (1986)
Castanea sativa	350, 700	365	1.23	El Kohen et al. (1992)
	350, 700	NA	1.14	Mousseau (1992)
	350, 700	730	1.14	Mousseau and Saugier (1992)
	350, 700	365	1.20	El Kohen et al. (1993)
Cecropia obtusifolia	350, 525, 700	111	1.00	Reekie and Bazzaz (1989)
Cedrus atlantica	350, 800	274	1.66	Kaushal et al. (1989)
Elaeagnus angustifolia	350, 700	84	1.61	Norby (1987)
Eucalyptus camaldulensis	330, 660	84	2.21	Wong et al. (1992)
Eucalyptus cypellocarpa	330, 660	84	2.48	Wong et al. (1992)
Eucalyptus grandis	340, 660	42	3.34	Conroy et al. (1992)
Eucalyptus pauciflora	330, 600	100	2.75	Wong et al. (1992)
Eucalyptus pulverulenta	330, 660	98	2.88	Wong et al. (1992)
Fagus grandifolia	400, 700	60	1.74	Bazzaz et al. (1990)
Fagus sylvatica	350, 700	365	1.62	El Kohen et al. (1993)
Fraxinus americana	350, 700	165	1.19	Bazzaz and Miao (1993)
Fraxinus lanceolata	350, 500, 700	90	1.03	Williams et al. (1986)
Gliricidia sepium	350, 650	31–71	1.26	Thomas et al. (1991)

Species	Growth CO_2 concentration (ppm)	Duration of experiment (days)	Relative growth response	Reference
Juglans nigra	325, 1500	NA	1.46	Tinus (1976)
Liquidambar styraciflua	340, 910	91	1.60	Rogers et al. (1983)
	350, 675, 1000	8–113	1.48	Tolley and Strain (1984a)
	350, 675, 1000	54–78	1.35	Tolley and Strain (1984b)
	350, 500, 700	224	1.44	Sionit et al. (1985)
Liriodendron tulipifera	350, 500, 700	90	0.95	Williams et al. (1986)
	367, 692	168	1.72	O'Neill et al. (1987)
	371, 493, 787	168	1.15	Norby and O'Neill (1991)
Myriocarpa longipes	350, 525, 770	111	0.96	Reekie and Bazzaz (1989)
Nothofagus fusca	340, 640	120	1.17	Hollinger (1987)
Ochroma lagopus	350, 675	60	1.79	Oberbauer et al. (1985)
Pentaclethra macroloba	350, 675	123	1.30	Oberbauer et al. (1985)
Picea abies	350, 900	21	1.78	Yeatman (1970)
Picea glauca	350, 900	21	1.61	Yeatman (1970)
	350, 700	30–100	1.37	Brown and Higginbotham (1986)
Picea mariana	340, 1000	21–42	1.27	Campagna and Margolis (1989)
Picea pungens	325, 1200	243–365	1.50	Tinus (1972)
Picea rubens	350, 700	122	1.12	Shipley et al. (1992)
	374, 713	365	1.66	Samuelson and Seiler (1994)
	362, 711	152	1.49	Samuelson and Seiler (1993)
Picea sitchensis	350, 700	213–426	1.14	Lee et al. (1992)
Picea sp.	350, 2000	40	1.75	Hardh (1966)
Pinus banksiana	350, 900	21	1.40	Yeatman (1970)
Pinus caribaea	340, 600	343	2.07	Conroy et al. (1990b)
Pinus contorta	300, 1000	152	5.20	Higginbotham et al. (1985)
Pinus echinata	368, 695	238–287	1.15	Norby et al. (1987)
	360, 700	42–168	1.36	O'Neill et al. (1987b)
Pinus nigra	350, 800	274	1.29	Kaushal et al. (1989)
Pinus ponderosa	325, 1200	243–365	1.35	Tinus (1972)
	355, 530, 705	122–243	1.28	Ball et al. (1992)
Pinus radiata	330, 660	63–154	1.21	Conroy et al. (1986a)
	330, 660	63–154	1.20	Conroy et al. (1986b)
	340, 640	120	1.27	Hollinger (1987)
	330, 660	154	1.24	Conroy et al. (1988)
	340, 660	798	1.19	Conroy et al. (1990a)
Pinus strobus	400, 700	100	1.19	Bazzaz et al. (1990)
Pinus sylvestris	350, 900	21	1.32	Yeatman (1970)
Pinus taeda	350, 675, 1000	8–84	1.15	Tolley and Strain (1984a)
	350, 675, 1000	54–82	1.08	Tolley and Strain (1984b)
	350, 500, 700	224	1.58	Sionit et al. (1985)
	350, 500, 650	155–512	1.24	Ball et al. (1992)
	350, 700	181	1.42	T. J. Tschaplinski et al. (1993)
Pinus virginiana	340, 415, 640, 940	122	1.09	CDIC (1985)
Pinus sp.	350, 2000	40	1.34	Hardh (1966)
Piper auritum	350, 525, 700	111	1.16	Reekie and Bazzaz (1989)

(Table continued on next page)

Species	Growth CO_2 concentration (ppm)	Duration of experiment (days)	Relative growth response	Reference
Platanus occidentalis	300, 600, 1200	35	1.22	Carlson and Bazzaz (1980)
	350, 500, 700	90	1.09	Williams et al. (1986)
Populus deltoides	300, 600, 1200	35	1.70	Carlson and Bazzaz (1980)
Populus euramericana	350, 700	92	1.72	Radoglou and Jarvis (1990)
	350, 700	228	0.65	R. Ceulemans (unpublished data)
Populus grandidentata	361, 707	70	1.27	Curtis and Teeri (1992)
	342, 692	152	1.50	Zak et al. (1993)
Populus tremuloides	350, 750	30–100	1.68	Brown and Higginbotham (1986)
Populus trichocarpa	350, 700	92	1.12	Radoglou and Jarvis (1990)
	350, 700	228	0.64	R. Ceulemans (unpublished data)
Populus trichocarpa × *P. deltoides*	350, 700	92	1.24	Radoglou and Jarvis (1990)
	350, 700	228	1.23	R. Ceulemans (unpublished data)
Prunus serotina	400, 700	60	1.32	Bazzaz et al. (1990)
Pseudotsuga menziesii	340, 640	120	1.03	Hollinger (1987)
Quercus alba	360, 700	42–210	1.62	O'Neill et al. (1987a)
	386, 496, 793	35–268	1.07	Norby and O'Neill (1989)
Quercus rubra	350, 500, 700	90	1.09	Williams et al. (1986)
	350, 700	165	1.98	Bazzaz and Miao (1993)
Quercus prinus	350, 700	64	0.89	Bunce (1992)
Robinia pseudoacacia	350, 700	105	1.32	Norby (1987)
Salix × *dasyclados*	300, 500, 700	122	1.36	Silvola and Ahlholm (1992)
Senna multijuga	350, 525, 700	111	1.10	Reekie and Bazzaz (1986)
Tabebuia rosea	354, 711	95	2.63	Ziska et al. (1991)
Trichospermum mexicanum	350, 525, 700	111	1.35	Reekie and Bazzaz (1989)
Tsuga canadensis	400, 700	100	1.28	Bazzaz et al. (1990)

ACKNOWLEDGMENTS

Research reported in this chapter was sponsored by the Carbon Dioxide Research Program, Office of Health and Environmental Research, U.S. Department of Energy, under contract DE-AC05–84OR21400 with Martin Marietta Energy Systems, Inc. This chapter is Publication No. 4280 from the Environmental Sciences Division, Oak Ridge National Laboratory.

REFERENCES

Acock, B., and L. H. Allen, Jr. 1985. Crop responses to elevated carbon dioxide concentrations. In *Direct Effects of Increasing Carbon Dioxide on Vegetation,* ed. B. R. Strain and J. D. Cure, pp. 54–97. DOE/ER-0238 U.S. Department of Energy Carbon Dioxide Research Division, Washington, D.C.

Allen, L. H., Jr., K. J. Boote, J. W. Jones, P. H. Jones, R. Valle, B. Acock, H. H. Rogers, and R. C. Dahlman. 1987. Response of vegetation to rising carbon dioxide: Photosynthesis, biomass, and seed yield of soybean. *Global Biogeochem. Cycles* 1:1–14.

Arnone, J. A., and J. C. Gordon. 1990. Effect of nodulation, nitrogen fixation and CO_2 enrichment on the physiology, growth and dry mass allocation of seedlings of *Alnus rubra* Bong. *New Phytol.* 116:55–66.

Ball, J. T., D. W. Johnson, P. Ross, B. R. Strain, R. Thomas, and R. B. Walker. 1992. Forest response to CO_2—Annual Report for 1991. Electric Power Research Institute, Palo Alto, Calif.

Bacastow, R., and C. D. Keeling. 1973. Atmospheric carbon dioxide and radiocarbon in the natural carbon cycle. II. Changes from A.D. 1700 to 2070 as deduced from a geochemical model. In *Carbon and the Biosphere,* ed. G. M. Woodwell and E. V. Pecan, pp. 86–135. U.S. Atomic Energy Commission, Washington, D.C.

Bazzaz, F. A., and S. L. Miao. 1993. Successional status, seed size and responses of tree seedlings to CO_2, light and nutrient availability: A search for patterns. *Ecology* 74:104–112.

Bazzaz, F. A., J. S. Coleman, and S. R. Morse. 1990. Growth responses of seven major co-occurring tree species of the northeastern United States to elevated CO_2. *Can. J. For. Res.* 20:1479–1484.

Brown, K., and K. O. Higginbotham. 1986. Effects of carbon dioxide enrichment and nitrogen supply on growth of boreal tree seedlings. *Tree Physiol.* 2:223–232.

Bunce, J. A. 1992. Stomatal conductance, photosynthesis and respiration of temperate deciduous tree seedlings grown outdoors at an elevated concentration of carbon dioxide. *Plant Cell Environ.* 15:541–549.

Campagna, M. A., and H. A. Margolis. 1989. Influence of short-term atmospheric CO_2 enrichment on growth, allocation patterns and biochemistry of black spruce seedlings at different stages of development. *Can. J. For. Res.* 19:773–782.

Carbon Dioxide Information Center. 1985. Growth and chemical responses to CO_2 enrichment—Virginia pine (*Pinus virginiana* Mill.). Document NDP-009. Information Resources Organization, Oak Ridge National Laboratory, Oak Ridge, Tenn.

Carlson, R. W., and F. A. Bazzaz. 1980. The effects of elevated CO_2 concentrations on growth, photosynthesis, transpiration, and water use efficiency of plants. In *Proceedings of the Symposium on Environmental and Climatic Impact of Coal Utilization,* ed. J. Singhjag and A. Deepek, pp. 609–612. Academic Press, New York.

Conroy, J. P., E. W. R. Barlow, and D. I. Bevege. 1986a. Response of *Pinus radiata* seedlings to carbon dioxide enrichment at different levels of water and phosphorus: Growth, morphology and anatomy. *Ann. Bot.* 57:165–177.

Conroy, J. P., R. M. Smillie, M. Kuppers, D. I. Bevege, and E. W. R. Barlow. 1986b. Chlorophyll a fluorescence and photosynthetic and growth responses of *Pinus radiata* to phosphorus deficiency, drought stress, and high CO_2. *Plant Physiol.* 81:423–429.

Conroy, J. P., M. Kuppers, B. Kuppers, J. Virgona, and E. W. R. Barlow. 1988. The influence of CO_2 enrichment, phosphorus deficiency and water stress on the growth, conductance and water use of *Pinus radiata* D. Don. *Plant Cell Environ.* 11:91–98.

Conroy, J. P., P. J. Milham, M. Mazur, and E. W. R. Barlow, 1990a. Growth, dry weight partitioning and wood properties of *Pinus radiata* D. Don after 2 years of CO_2 enrichment. *Plant Cell Environ.* 13:329–337.

Conroy, J. P., P. J. Milham, M. L. Reed, and E. W. R. Barlow. 1990b. Increases in phosphorus requirements for CO_2-enriched pine species. *Plant Physiol.* 92:977–982.

Conroy, J. P., P. J. Milham, and E. W. R. Barlow. 1992. Effect of nitrogen and phosphorus availability on the growth response of *Eucalyptus grandis* to high CO_2. *Plant Cell Environ.* 15:843–847.

Curtis, P. C., and J. A. Teeri. 1992. Seasonal response of leaf gas exchange to elevated carbon dioxide in *Populus grandidentata. Canad. J. For. Res.* 22:1320–1325.

Eamus, D., and P. G. Jarvis. 1989. The direct effects of increase in the global atmospheric CO_2 concentration on natural and commercial temperate trees and forests. *Adv. Ecol. Res.* 19:1–55.

El Kohen, A., H. Rouhier, and M. Mousseau. 1992. Changes in dry weight and nitrogen partitioning induced by elevated CO_2 depend on soil nutrient availability in sweet chestnut (*Castanea sativa* Mill). *Ann. Sci. For.* 49:83–90.

El Kohen, A., L. Venet, and M. Mousseau. 1993. Growth and photosynthesis of two deciduous forest tree species exposed to elevated carbon dioxide. *Func. Ecol.* 7:480–486.

Gates, D. M. 1985. Global biospheric response to increasing atmospheric carbon dioxide concentration. In *Direct Effects of Increasing Carbon Dioxide on Vegetation,* ed. B. R. Strain and J. D. Cure, pp. 171–184. U.S. Department of Energy, DOE/ER-0238, Carbon Dioxide Research Division, Washington, D.C.

Gifford, R. M. 1979a. Growth and yield of CO_2-enriched wheat under water limited conditions. *Aust. J. Plant Physiol.* 6:367–378.

Gifford, R. M. 1979b. Carbon dioxide and plant growth under water and light stress: Implications for balancing the global carbon budget. *Search* 10:316–318.

Gifford, R. M. 1980. Carbon storage by the biosphere. In *Carbon Dioxide and Climate,* ed. G. I. Pearman, pp. 167–181. Australian Academy of Science, Canberra.

Gifford, R. M. 1992. Implications of the globally increasing atmospheric CO_2 concentration and temperature for the Australian terrestrial carbon budget: Integration using a simple model. *Aust. J. Bot.* 40:527–543.

Gifford, R. M. 1993. Implications of CO_2 effects on vegetation for the global carbon budget. In *The Global Carbon Cycle*, Proceedings of the NATO Advanced Study Institute, ed. M. Heimann, pp.165–205. Il Ciocco, Italy, September 8–20, 1991. Springer-Verlag, Berlin.

Goudriaan, J., and P. Ketner. 1984. A simulation study for the global carbon cycle, including man's impact on the biosphere. *Climate Change* 6:167–192.

Graham, R. L., M. G. Turner, and V. H. Dale. 1990. How increasing CO_2 and climate change affect forests. *BioScience* 40:575–587.

Hardh, J. E. 1966. Trials with carbon dioxide, light and growth substances on forest tree plants. *Acta For. Fenn.* 81:1–10.

Higginbotham, K. O., J. M. Mayo, S. L'Hirondelle, and D. K. Krystofiak. 1985. Physiological ecology of lodgepole pine (*Pinus contorta*) in an enriched CO_2 atmosphere. *Can. J. For. Res.* 15:417–421.

Hollinger, D. Y. 1987. Gas exchange and dry matter allocation responses to elevation of atmospheric CO_2 concentration in seedlings of three tree species. *Tree Physiol.* 3:193–202.

Idso, S. B. 1991. Comment on "Modelling the seasonal contribution of a CO_2 fertilization effect of the terrestrial vegetation to the amplitude increase in atmospheric CO_2 at Mauna Loa observatory," by G. H. Kohlmaier et al. *Tellus* 43B:338–341.

Jenkinson, D. S. 1990. The turnover of organic carbon and nitrogen in soil. *Phil. Trans. R. Soc. London Ser. B* 329:361–369.

Jenkinson, D. S., D. E. Adams, and A. Wild. 1991. Global warming and soil organic matter. *Nature* 351:304–306.

Kaushal, P., J. M. Guehl, and G. Aussenac. 1989. Differential growth response to atmospheric carbon dioxide enrichment in seedlings of *Cedrus atlantica* and *Pinus nigra* spp Larico var. Corsicana. *Can. J. For. Res.* 19:1351–1358.

Keeling, C. D. 1973. The carbon dioxide cycle: Reservoir models to depict the exchange of atmospheric carbon dioxide with the oceans and land plants. In *Chemistry of the Lower Atmosphere,* ed. S. I. Rasool, pp. 251–329. Plenum Press, New York.

Kienast, F. 1991. Simulated effects of increasing atmospheric CO_2 and changing climate on the successional characteristics of Alpine forest ecosystems. *Landsc. Ecol.* 5:225–238.

Kienast, F., and R. J. Luxmoore. 1988. Tree-ring analysis and conifer growth responses to increased atmospheric CO_2 levels. *Oecologia* 76:487–495.

Kimball, B. A. 1983. Carbon dioxide and agricultural yield. An assemblage and analysis of 430 prior observations. *Agron. J.* 75:779–788.

Kohlmaier, G., H. Brohl, E. O. Sire, M. Plochl, and R. Revelle. 1987. Modelling stimulations of plants and ecosystem response to present levels of excess atmospheric CO_2. *Tellus* 39B: 155–170.

Kohlmaier, G., E. O. Sire, A. Janecek, C. D. Keeling, S. C. Piper, and R. Revelle. 1989. Modelling the seasonal contribution of a CO_2 fertilization effect of the terrestrial vegetation to the amplitude increase in atmospheric CO_2 at Mauna Loa observatory. *Tellus* 41B:487–510.

Korner, C. 1993. CO_2 Fertilization: The great uncertainty in future vegetation development. In *Vegetation Dynamics and Global Change,* eds. A. M. Soloman and H. H. Shugart, pp. 53–70. Chapman and Hall, New York.

LaMarche, V. C., D. A. Graybill, H. C. Fritts, and M. R. Rose. 1984. Increasing atmospheric carbon dioxide: Tree ring evidence for growth enhancement in natural vegetation. *Science* 225:1019–1021.

Lee, H., M. Murray, L. Evans, R. Pettersson, I. Leith, C. Barton, and P. Jarvis. 1992. Effects of elevated CO_2 on Sitka spruce seedlings. In Proceedings of the European Research Conference "Effects of elevated CO_2 levels, air pollutants and climate change on natural plant ecosystems—Impact on tree physiology." Capsis Beach Hotel, Aghia Pelaghia, Crete, Greece, April 3–7, 1992.

Miao, S. L., P. M. Wayne, and F. A. Bazzaz. 1992. Elevated CO_2 differentially alters the response of cooccurring birch and maple seedlings to a moisture gradient. *Oecologia* 90:300–304.

Mousseau, M. 1993. Effects of elevated CO_2 on growth, photosynthesis and respiration of sweet chestnut (*Castanea sativa* Mill.). *Vegetatio* 104/105:413–419.

Mousseau, M., and B. Saugier. 1992. The direct effect of increased CO_2 on gas exchange and growth of forest tree species. *J. Exp. Bot.* 43:1121–1130.

Nobel, R., K. F. Jensen, B. S. Ruff, and K. Loats. 1992. Response of *Acer saccharum* seedlings to elevated carbon dioxide and ozone. *Ohio J. Sci.* 92:60–62.

Norby, R. J. 1987. Nodulation and nitrogenase activity in nitrogen-fixing woody plants stimulated by CO_2 enrichment of the atmosphere. *Physiol. Plant.* 71:77–82.

Norby, R. J., E. G. O'Neill, W. G. Hood, and R. J. Luxmoore. 1987. Carbon allocation, root exudation and mycorrhizal colonization of *Pinus echinata* seedlings grown under CO_2 enrichment. *Tree Physiol.* 3:203–210.

Norby, R. J., and E. G. O'Neill. 1989. Growth dynamics and water use of seedlings of *Quercus alba* L. in CO_2-enriched atmospheres. *New Phytol.* 111:491–500.

Norby, R. J., and E. G. O'Neill. 1991. Leaf area compensation and nutrient interactions in CO_2-enriched seedlings of yellow-poplar (*Liriodendron tulipifera* L.). *New Phytol.* 117: 515–528.

Norby, R. J. 1987. Nodulation and nitrogenase activity in nitrogen-fixing woody plants stimulated by CO_2 enrichment of the atmosphere. *Physiol. Plant.* 71:77–82.

Oberbauer, S. F., B. R. Strain, and N. Fetcher. 1985. Effect of CO_2-enrichment on seedling physiology and growth of two tropical tree species. *Physiol. Plant.* 65:352–356.

Oechel, W. C., and B. R. Strain. 1985. Native species responses to increased atmospheric carbon dioxide concentration. In *Direct Effects of Increasing Carbon Dioxide on Vegetation,* ed. B. R. Strain and J. D. Cure, pp. 118–154. U.S. Department of Energy, DOE/ER-0238 Carbon Dioxide Research Division, Washington, D.C.

Oeschger, H., U. Siegenthaler, U. Schotterer, and A. Gugelmann. 1975. A box diffusion model to study the carbon dioxide exchange in nature. *Tellus* 27B:169–192.

O'Neill, E. G., R. J. Luxmoore, and R. J. Norby. 1987a. Elevated atmospheric CO_2 effects on seedling growth, nutrient uptake, and rhizosphere bacterial populations of *Liriodendron tulipifera* L. *Plant Soil* 104:3–11.

O'Neill, E. G., R. J. Luxmoore, and R. J. Norby. 1987b. Increases in mycorrhizal colonization and seedling growth in *Pinus echinata* and *Quercus alba* in an enriched CO_2 atmosphere. *Can. J. For. Res.* 17:878–883.

Pettersson, R., and A. J. S. McDonald. 1992. Effects of elevated carbon dioxide concentration on photosynthesis and growth of small birch plants (*Betula pendula* Roth.) at optimal nutrition. *Plant Cell Environ.* 15:911–919.

Polglase, P. J., and Y. P. Wang. 1992. Potential CO_2-enhanced carbon storage by the terrestrial biosphere. *Aust. J. Bot.* 40:641–656.

Post, W. M., T.-H. Peng, W. R. Emanuel, A. W. King, V. H. Dale, and D. L. DeAngelis. 1990. The global carbon cycle. *Am. Sci.* 78:310–326.

Radoglou, K. M., and P. G. Jarvis. 1990. Effects of CO_2 enrichment on four poplar clones. I. Growth and leaf anatomy. *Ann. Bot.* 65:617–626.

Rastetter, E. B., M. G. Ryan, G. R. Shaver, J. M. Melillo, K. L. Nadelhoffer, J. E. Hobbie, and J. D. Aber. 1991. A general biochemical model describing the responses of the C and N cycles in terrestrial ecosystems to changes in CO_2, climate and N deposition. *Tree Physiology 9:* 101–126.

Reekie, E. G., and F. A. Bazzaz. 1989. Competition and patterns of resource use among seedlings of five tropical trees grown at ambient and elevated CO_2. *Oecologia* 79:212–222.

Reynolds, J. F., D. W. Hilbert, J. Chen, P. C. Harley, P. R. Kemp, and P. W. Leadley. 1992. Modeling the response of plants and ecosystems to elevated CO_2 and climate change. U.S. Department of Energy, DOE/ER-60490T-H1 TRO19. Washington, D. C.

Rogers, H. H., J. F. Thomas, and G. E. Bingham. 1983. Response of agronomic and forest species to elevated atmospheric carbon dioxide. *Science* 220:428–429.

Samuelson, L. J., and J. R. Seiler. 1992. Fraser fir seedling gas exchange and growth in response to elevated CO_2. *Environ. Exp. Bot.* 32:351–356.

Samuelson, L. J., and J. R. Seiler. 1993. Interactive role of elevated CO_2, nutrient limitations and water stress in the growth responses of red spruce seedlings. *For. Sci.* 39:348–358.

Samuelson, L. J., and J. R. Seiler. 1994. Red spruce seedling gas exchange in response to elevated CO_2, water stress, and soil fertility treatments. *Canadian Journal of Forest Research* (in press).

Shipley, B., M. Lechowicz, S. Dumont, and W. H. Hendershot. 1992. Interacting effects of nutrients, pH-Al and elevated CO_2 on the growth of red spruce (*Picea rubens* Sarg.) seedlings. *Water Air Soil Pollut.* 64:585–600.

Silvola, J., and U. Ahlholm. 1992. Photosynthesis in willow (*Salix* × *dasyclados*) grown at different CO_2 concentrations and fertilization levels. *Oecologia* 91:208–213.

Sionit, N., B. R. Strain, H. Hellmers, G. H. Riechers, and C. H. Jaeger. 1985. Long-term atmospheric CO_2 enrichment affects the growth and development of *Liquidambar styraciflua* and *Pinus taeda* seedlings. *Can. J. For. Res.* 15:468–471.

Soloman, A. M. and W. Cramer. 1993. Biospheric implications of global environmental change. In *Vegetation Dynamics and Global Change,* eds. A. M. Soloman and H. H. Shugart, pp. 25–52. Chapman and Hall, New York.

Taylor, J. A., and J. Lloyd. 1992. Sources and sinks of atmospheric CO_2. *Aust. J. Bot.* 40:407–418.

Thomas, R. B., D. D. Richter, H. Ye, P. R. Heine, and B. R. Strain. 1991. Nitrogen dynamics and growth of seedlings of an N-fixing tree (*Gliricidia sepium* (Jacq.) Walp.) exposed to elevated atmospheric carbon dioxide. *Oecologia* 88:415–421.

Tinus, R. W. 1972. CO_2 enriched atmosphere speeds growth of ponderosa pine and blue spruce seedlings. *Tree Planters Notes* 23:12–15.

Tinus, R. W. 1976. Photoperiod and atmospheric CO_2 level interact to control black walnut (*Juglans nigra* L.) seedling growth. *Plant Physiol.* 57S:106.

Tolley, L. C., and B. R. Strain. 1984a. Effects of CO_2 enrichment on growth of *Liquidambar styraciflua* and *Pinus taeda* seedlings under different irradiance levels. *Can. J. For. Res.* 14:343–350.

Tolley, L. C., and B. R. Strain. 1984b. Effects of CO_2 enrichment and water stress on growth of *Liquidambar styraciflua* and *Pinus taeda* seedlings. *Can. J. Bot.* 62:2135–2139.

Tschaplinski, T. J., R. J. Norby, and S. D. Wullschleger. 1993. Responses of loblolly pine seedlings to elevated CO_2 and fluctuating water supply. *Tree Physiology 13:* 283–296.

Williams, W. E., K. Garbutt, F. A. Bazzaz, and P. M. Vitousek. 1986. The response of plants to elevated CO_2. IV. Two deciduous-forest tree communities. *Oecologia* 69:454–459.

Wong, S. C., P. E. Kriedemann, and G. D. Farquhar. 1992. CO_2 × nitrogen interaction on seedling growth of four species of Eucalypt. *Aust. J. Bot.* 40:457–472.

Yeatman, C. W. 1970. CO_2 enriched air increased growth of conifer seedlings. *Forestry Chronicle,* June, pp. 229–230.

Zak, D. R., K. S. Pregitzer, P. S. Curtis, J. A. Teeri, R. Fogel, and D. L. Randlett. 1993. Elevated atmospheric CO_2 and feedback between carbon and nitrogen cycles in forested ecosystems. *Plant Soil* 151:105–117.

Ziska, L. H., K. P. Hogan, A. P. Smith, and B. G. Drake. 1991. Growth and photosynthetic response of nine tropical species with long-term exposure to elevated carbon dioxide. *Oecologia* 86:383–389.

5

Indicators of Climatic
and Biospheric Change:
Evidence from Tree Rings

GORDON C. JACOBY AND ROSANNE D. D'ARRIGO

In the field of dendrochronology, annual growth ring measurements of old-aged trees are developed into high-resolution, precisely dated time series that reflect climatic and biospheric change over recent centuries to millennia. These records can help to assess whether recent temperature increases, coinciding with the period of enhanced levels of anthropogenic trace gases, exceed the natural variability of the climate system. This quantitative growth information can also be used to determine if natural vegetation is being influenced by direct CO_2 fertilization or nutrient change along with climate-induced (anthropogenic or natural) growth variation.

We find a general tendency for tree rings from severely temperature-stressed trees at many high-latitude and high-elevation sites in both hemispheres to show enhanced growth over the recent period of increasing trace gases, although there are, of course, exceptions for individual time series. In this chapter we describe some examples of recent growth trends at temperature-sensitive sites around the globe. It should be noted that this is by no means a comprehensive review.

NORTHERN HEMISPHERE

We have sampled, nondestructively, near the northern tree line because growth rates there are extremely limited by temperature-related effects. Recent northward and upward shifting of the tree line has been observed at this transition. Climate and vegetation models (Schlesinger and Mitchell 1985; Solomon 1986) suggest warming of a greater magnitude at high northern latitudes, a process that could cause shifts in the position of species distribution and nature of growth at the northern forest limit.

Northern North America

Many tree ring records from the northern boreal tree line of Alaska and Canada show increases in growth that coincide with large-scale instrumental temperature trends over

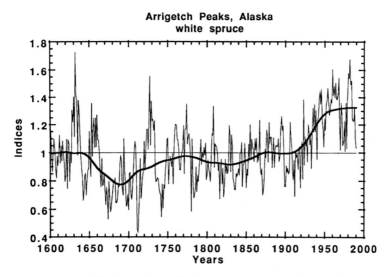

Figure 5.1. Tree ring width chronology of white spruce (*Picea glauca*) from the Arrigetch Peaks, Brooks Range, northern Alaska, from 1600 to 1990. Note the positive growth trend in the past century. Recent growth corresponds to recorded temperatures from nearby Bettles, Alaska.

the past century (e.g., Figure 5.1). In some cases (as in Figure 5.1) these trends are very nearly unprecedented with regard to the earlier (more unequivocally natural) climate variability inferred over the length of the tree ring record.

We have also documented that seedling growth is taking place north of the present latitudinal tree line and at higher elevations in Alaska. These observations indicate that tree line migration is occurring over the past century along the northern forest border, coinciding with the enhanced tree growth. MacDonald et al. (1993) determined that migration at the forest-tundra border has been very rapid in the past, based on pollen from central Canadian lakes. A change from tundra to closed-canopy black spruce (*Picea mariana*) was estimated to occur in only 150 years because of apparent warming between 5000 and 4000 years ago (MacDonald et al. 1993; cf. Payette and Morneau 1993). Rapid tree line responses were also found for Scotland by Gear and Huntley (1991). Such potentially rapid shifts during prior warm periods may serve as analogues for future trace-gas-induced vegetation changes and related feedbacks (Pastor and Post 1988; MacDonald et al. 1993).

Temperature-sensitive tree ring data from the northern tree line, or forest-tundra ecotone, of North America were combined to develop reconstructions of Arctic as well as Northern Hemisphere annual temperatures extending from 1671 to 1973 (Jacoby and D'Arrigo 1989). These reconstructions reflect high- and low-frequency variability, the latter including cooling during the latter part of the Little Ice Age, a recovery from this cold period, and a subsequent increase extending beyond recovery into a recent warming trend over the past century. An Arctic reconstruction (Figure 5.2) with additional coverage from Fennoscandia and Russia (1682–1968) shows similar long-term trends (D'Arrigo and Jacoby 1993a). A subset of these 20 chronologies that extends through

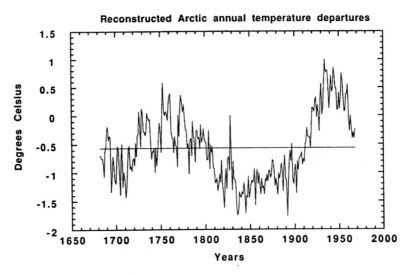

Figure 5.2. Reconstruction of Arctic annual temperature departures based on 20 chronologies from northern North America, Fennoscandia (provided by K. R. Briffa), and Russia (provided by D. A. Graybill and S. G. Shiyatov) from 1682 to 1968. Note that the warming of the past century exceeds the prior record back to the late 1600s. (R. D. D'Arrigo and G. C. Jacoby, 1993a.)

1982 shows recovery from the cooling of the 1960s–1970s (Kelly et al. 1982) and resumed warming.

There is no evidence of direct CO_2 fertilization in these trees for the period of record, as indicated by the absence of any significant trends after climate modeling (as shown in Jacoby and D'Arrigo 1989).

Although there have been laboratory and enclosed-tree experiments indicating a positive response to direct CO_2 fertilization in some types of plants, the issue of whether this phenomenon is currently occurring in natural vegetation is still unresolved and highly controversial (e.g., LaMarche et al. 1984; Graumlich 1991).

In a recent study, we evaluated tree ring width and maximum latewood density chronologies of white spruce (*Picea glauca*) from three temperature-sensitive tree line sites in northern North America: in the Brooks Range of Alaska, the Franklin Mountains of the Northwest Territories, and Churchill, Manitoba (D'Arrigo and Jacoby 1993a), When the ring width and density variations were estimated for each site using local temperature and precipitation data in principal components regression analysis, no substantial residual trends are detected that might require CO_2 or other nutrient fertilization as an additional explanation (Figure 5.3). Thus the growth estimates (Figure 5.3) appear valid for the recent decades for which we have instrumental climate data. Based in part on these studies and other observations, we conclude that CO_2 fertilization is not currently detectable as a significant factor influencing growth at the northern sites studied.

Although atmospheric CO_2 concentration has increased by as much as 30% since the mid-1800s (e.g., Neftel et al. 1985), a threshold level may need to be reached before a direct fertilization response can be detected. This level may not yet have

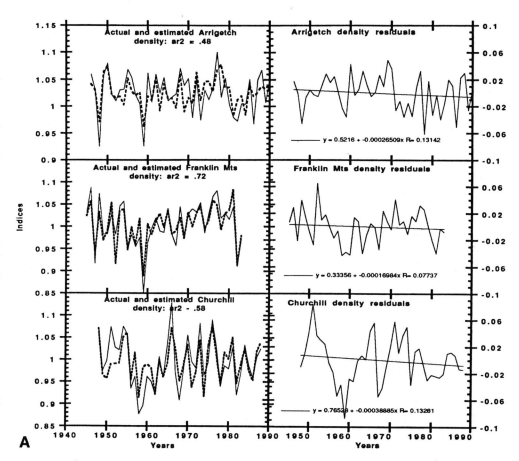

Figure 5.3A. Left: Plots of actual maximum latewood density (solid lines) and that estimated using climate variables in principal components regression analysis (dashed lines) for three sites. The variance explained or ar^2 (adjusted for degrees of freedom) for each climate growth model is shown for each site. Right: Residual estimates for regression models, indicating no significant trends that might suggest a response to CO_2 fertilization. (R. D. D'Arrigo and G. C. Jacoby, 1993b.)

beenreached for white spruce at the northern tree line (at least, as allocated to stem growth). Another likely explanation for the absence of a detectable fertilization effect is that other factors—including cold temperatures, a short growing season, and nitrogen deficiency (Pastor and Post 1988; McGuire et al. 1992)—are at present limiting to growth and are preventing a direct response to CO_2 in the northern boreal forests.

These results for the northern tree line are not necessarily applicable to the boreal forests as a whole, nor to even a sizable fraction of the world's northern forests, which might potentially serve as an added carbon sink (Tans et al. 1990; Sundquist 1993). In fact, preliminary analyses of a few sites within the interior boreal forests of Canada do not indicate unusual growth increases. However, the trees growing at their northern range limit could be monitors of rapid, greenhouse-induced climatic change. They are

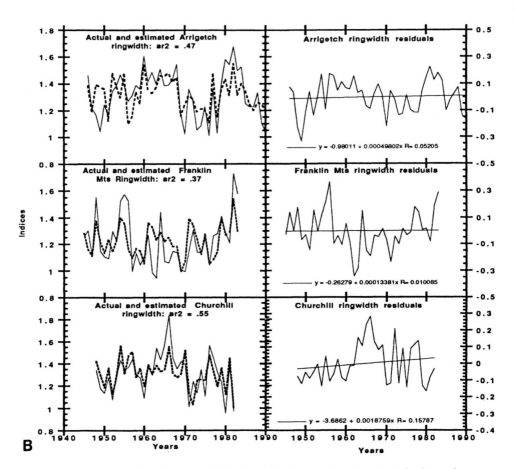

Figure 5.3B. As in A, but for ring width data. Again, no significant residual trends are evident following modeling with climate. (R. D. D'Arrigo and G. C. Jacoby, 1993b.)

thus an important component to consider in evaluating the response of different species and sites to the effects of increasing CO_2 concentration.

Central Canada and the Coterminous United States

In the Rocky Mountains of Canada, Luckman (1989) developed ring width chronologies whose recent growth reflects warming that is above any levels for the past 400 years. Brubaker and Graumlich (1989) found increases in net primary productivity over the past century in high-elevation conifers in the Cascade Mountains of Washington State. A reconstruction of mean annual temperature based on some of these data from Washington State show an increase in temperature of 1°C between the mid-19th and 20th centuries. No evidence of CO_2 fertilization or, conversely, forest decline owing to air pollution was found.

High-elevation bristlecone pine in California, Nevada, Colorado, and New Mexico show increased growth over the last century (LaMarche and Stockton 1974). In one very long ring width chronology from Sheep Mountain, California, the recent increase in growth is almost without precedent for the past 1500 years. Only at the beginning of the Medieval Warm period was the growth equivalent to that observed today. LaMarche et al. (1984) suggested that the enhanced growth in recent decades in high-elevation bristlecone and limber pine in California and Nevada might be attributable to CO_2 fertilization but did not present any quantitative modeling or discuss possible moisture effects in the wet years at the end of the chronologies. Conversely, Graumlich (1991) did not detect any evidence for CO_2 fertilization in high-elevation foxtail pine and other species in the Sierra Nevada. She speculated that the different conclusions of her work and that of LaMarche et al. (1984) could be explained by the fact that the trees in the latter study (unlike those in her study) were characterized by a strip-bark morphology that might enhance their sensitivity to CO_2 fertilization (as also suggested by Graybill and Idso 1993). Different nutrient and moisture requirements and efficiencies were also cited (Graumlich 1991) as possible explanations.

Europe

Temperature-sensitive coniferous tree ring width data from 69 sites covering wide geographical areas of Europe generally indicate increased growth or productivity over the past century (Briffa 1991). The magnitude and timing vary for different subregions and species but the increases tend to be in the late 1800s and early 1900s. There is then a renewed increase in the last few decades. The recent increases are generally unprecedented in terms of growth over the last 200 years. The central and southern spruce and fir trees indicate decreases in wood density during the times of increased ring width, with these trees showing some of the most dramatic width increases. In general, the results support a hypothesis of increasing growth during the last century with a possible recent acceleration indicated at the few sites with data for the 1980s. Both climatic and nonclimatic factors (the latter favored by Briffa 1991) may help explain the increases over the past hundred years. The increases may represent enhanced carbon sequestering by these particular trees (Briffa 1991). However, it is not known if these sites are representative of typical forest sites in Europe.

Russia

A reconstruction of summer temperature for the northern Urals has been developed based on near-tree line Siberian larch (*Larix sibirica*) tree ring data (Graybill and Shiyatov 1989). Their reconstruction exhibits a striking rise in values from near A.D. 1100 to the highs of the 1200s to 1300s, followed over the long term by an overall decline that nevertheless shows some increase since the lows of the 1600s. The low reconstructed values of the 1600s and those of the 1800s, which are followed by a sharp increase in the 20th century, are evident in other high-latitude and high-altitude tree ring chronologies (Graybill and Shiyatov 1989).

SOUTHERN HEMISPHERE

Compared to the instrumental and historical climate records of the Northern Hemisphere, the Southern Hemisphere has relatively few high-resolution climatic time series

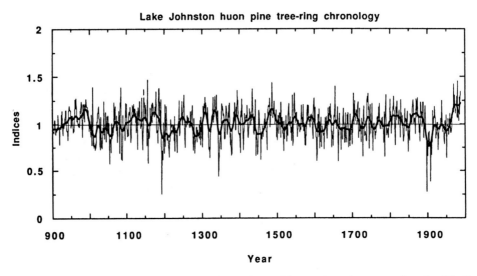

Figure 5.4. A 1089-year tree ring width chronology of huon pine (*Lagarostrobus franklinii*) from western Tasmania. Note the anomalous increase in growth since 1965, which corresponds to a regional temperature increase. (After Cook et al. 1991.)

that extend back more than 100 years. However, this situation has improved considerably in recent years with the development of long tree ring records from the southern middle- to higher-latitude subantarctic forests in Tasmania, New Zealand, and Argentina.

Tasmania

Although tree ring chronologies from Tasmania (and other Southern Hemisphere locales) were published in 1979 by LaMarche et al., the potential of Tasmanian forests to yield long dendrochronological records was not fully exploited at that time. More recently, a climatically sensitive huon pine (*Lagarostrobus franklinii*) tree ring chronology from western Tasmania (Figure 5.4) was developed. This record provides information about austral summer temperature change since A.D. 900 (Cook et al. 1991, 1992). The huon pine growth has been unusually rapid in recent decades. This growth increase is well-correlated with recent warming observed in Tasmania based on instrumental data and suggests that climatic change is now taking place in Tasmania. Although this temperature increase exceeds any inferred to have taken place over the past 1089 years of record, it is not yet clearly distinct from the natural variability of climate in this region (Cook et al. 1991). This time series can be used to help evaluate the Medieval Warm Epoch and Little Ice Age cold period for this part of the Southern Hemisphere. Warming conditions inferred for A.D. 940–1000 and 1100–1200, and colder conditions for the mid-15th and 17th centuries, may correspond to these two climatic intervals (Cook et al. 1991, 1992).

Analyses of the estimated and actual temperatures from this Tasmanian tree ring study do not indicate any trend in residual error that could be due to a monotonic forcing, such as an increase in CO_2. This tree ring record has now been extended to 300 B.C. (Cook et al. in press). Decadal- to century-scale oscillatory modes, based on singular spectrum analysis, have been found to be stable over the full length of this 2290-year record. These modes are thought to represent oscillations in the internal dynamics of the Southern Hemisphere climate system, rather than any external forcing (Cook et al. in press).

New Zealand

One of the few climatic reconstructions to be published for New Zealand was developed by Norton et al. (1989) for summer temperatures. Based on species of *Nothofagus*, it extends back to A.D. 1730. This reconstruction emphasizes variations of periods of less than 30 years (Norton et al. 1989). As noted by Norton et al. (1989), the reconstruction does not capture the cooling around 1900 or the warming around 1950 seen in the New Zealand instrumental record. This finding might be explained by the considerable spatial diversity of climate in New Zealand, so that these temperature changes may not have occurred in the areas represented by the chronologies (Norton et al. 1989). Additional tree ring research is underway in the Tasmania–New Zealand sector.

Southern South America

A network of 21 tree ring chronologies developed from Tierra del Fuego, Argentina, represents the southernmost forests in the world (Boninsegna et al. 1990). Some temperature-sensitive southern beech (*Nothofagus* spp.) from southernmost Tierra del Fuego show increases in growth in recent decades relative to the past few centuries (Boninsegna et al. 1990), but these trends are less anomalous than the huon pine from Tasmania described previously (Cook et al. 1991, 1992).

A 1000-year reconstruction of summer temperatures based on alerce (*Fitzroya cupressoides*) from farther north in Argentina (Villalba 1990) does not indicate any unusual recent trends. Tree growth at this site is positively related to precipitation and negatively related to temperature (Villalba 1990). A combined tree ring and glacial fluctuation study showed increased glacial retreat over the past century with a brief temporary resurgence in the 1970s (Villalba et al. 1990).

CONCLUSIONS

Analysis of tree ring records from severely stressed sites near the temperature-defined limits of growth suggests that we are now in a warm period relative to prior centuries. We note, however, that there are exceptions for individual time series at these locations. Trees from other, less stressed sites (e.g., those in interior forests) often do not show such trends. In northwestern Alaska, long tree ring records and observations of tree line movements indicate that the extent of the present warming may exceed that of comparable events for the last 1000 years. However, subarctic records from Fennoscandia and the northern Ural Mountains of Russia show equally warm precedents in the past 1000 years. High-elevation midlatitude records from the western United States show enhanced growth that is at least partly attributable to warming. One high-elevation millennium-long record from the southwestern United States shows only one positive growth period equal to that of the present, at the beginning of the Medieval Warm Epoch in the late 900s A.D.

For the Southern Hemisphere, a tree ring record from Tasmania indicates anomalous warming since 1965. Some tree ring series from southern Tierra del Fuego in Argentina, representing some of the southernmost forests of the world, show recent inferred warming, but their extent does not exceed that of previous warmer periods in the last several centuries. A 1000-year tree ring record and reconstruction from south-central Argentina does not indicate unusual warming, but these trees respond to both temperature and moisture.

The inference of relative warming over the past century in the Northern Hemisphere is supported by information from other proxy and recorded data, such as ice cores (e.g., Thompson 1992) and borehole temperature measurements (e.g., Mareschal and Beltrami 1992). The relatively sparse records from the few forested land areas at Southern Hemisphere middle to higher latitudes also indicate warming, particularly in recent decades. However, few other long, high-resolution paleoclimatic time series are available for comparative study. Recent growth increases may or may not have precedent during the past millennium, depending on the region or individual data set.

The tree growth increases described in this chapter were observed at unusual sites preselected to amplify a climatic signal. They are not necessarily representative of large components of the biosphere. Thus enhanced carbon sequestering in forests should not be inferred from these records. Also uncertain is whether natural vegetation is capable of responding consistently and persistently to CO_2 or nutrient fertilization effects. In the important boreal regions, changes in permafrost along with the varied moisture retention properties of soils could lead to negative forest responses to warmer conditions. Although some trees are presently increasing in growth, it would be misleading to over-interpret these changes until more extensive data are available from representative forest ecoregions. In addition, forests may have had insufficient time to adapt to the unprecedented rate and magnitude of projected anthropogenic climatic change, a situation that could lead to some negative responses. Improved understanding will require additional tree ring and other records from both hemispheres.

ACKNOWLEDGMENTS

This chapter is based on a report presented at the Senate Hearing on Global Change Research: Indicators of Global Warming and Solar Variability, held on February 27, 1992. Research reported herein was funded under grants ATM89–15353 and 87–16630 from the National Science Foundation's Climate Dynamics Program and Division of Polar Programs, grant EAR91–04922 from the National Science Foundation's Geologic Record of Global Change Program, grant INT84–19397 from the National Science Foundation, and CONICET-Argentina grant 3–92520385. This chapter is Lamont-Doherty Earth Observatory contribution No. 5051.

REFERENCES

Boninsegna, J., J. Keegan, G. C. Jacoby, R. D. D'Arrigo, and R. L. Holmes. 1990. Dendrochronological studies in Tierra del Fuego, Argentina. *Quat. S. Am. Ant. Penin.* 7:305–326.

Briffa, K. R. 1991. Final Report to the U.K. Forestry Commission: Detection of Any Widespread and Unprecedented Changes in the Growth of European Conifers.

Brubaker, L. B., and L. J. Graumlich. 1989. 100-year records of forest productivity at high elevations in western Washington, USA. In *Air Pollution Effects on Vegetation including Forest Ecosystems,* ed. R. D. Noble, J. L. Martin, and K. F. Jensen, pp. 9–56. United States Department of Agriculture Forest Service, Pennsylvania.

Cook, E. R., T. Bird, M. Peterson, M. Barbetti, B. Buckley, R. D'Arrigo, R. Francey, and P. Tans. 1991. Climatic change in Tasmania inferred from a 1089-year tree-ring chronology of huon pine. *Science* 253:1266–1268.

Cook, E. R., T. Bird, M. Peterson, M. Barbetti, B. Buckley, R. D'Arrigo, and R. Francey. 1992. Climatic change over the last millennium in Tasmania reconstructed from tree-rings. *Holocene* 2:205-217.

Cook, E. R., B. M. Buckley, and R. D. D'Arrigo. In press. Decadal-scale oscillatory modes in a millennia-long temperature reconstruction from Tasmania. In Proceedings, National Academy of Sciences Workshop on "The Natural Variability of the Climate System on 10–100 Year Time-Scales," Irvine, California, September 21–24, 1992.

D'Arrigo, R. D., and G. C. Jacoby. 1993a. Secular trends in high northern latitude temperature reconstructions based on tree rings. *Climatic Change* 25:163–177.

D'Arrigo, R. D., and G. C. Jacoby. 1993b. Tree growth-climate relationships at the northern Boreal forest treeline of North America: evaluation of potential response to increasing carbon dioxide. *Global Biogeochem. Cycles* 7:525–535.

Gear, A. J., and B. Huntley. 1991. Rapid changes in the range limits of Scots Pine 4000 years ago. *Science* 251:544–547.

Graumlich, L. J. 1991. Subalpine tree growth, climate, and increasing CO_2: An assessment of recent growth trends. *Ecology* 72:1–11.

Graybill, D. A., and S. B. Idso. 1993. Detecting the aerial fertilization effect of atmospheric CO_2 enrichment in tree-ring chronologies. *Global Biogeochem. Cycles* 7:81–95.

Graybill, D. A., and S. G. Shiyatov. 1989. A 1009 year tree-ring reconstruction of mean June-July temperature deviations in the Polar Urals. In *Air Pollution Effects on Vegetation Including Forest Ecosystems,* ed. R. D. Noble, J. L. Martin, and K. F. Jensen, pp. 37–42. United States Department of Agriculture Forest Service, Pennsylvania.

Jacoby, G. C., and R. D. D'Arrigo. 1989. Reconstructed Northern Hemisphere annual temperature since 1671 based on high-latitude tree-ring data from North America. *Climatic Change* 14:39–59.

Kelly, P. M., P. D. Jones, C. B. Sear, B. S. G. Cherry, and R. K. Tavakol. 1982. Variations in surface air temperatures: Part 2, Arctic regions, 1881–1980. *Mon. Weather Rev.* 110:71–83.

LaMarche, V. C., Jr., and C. W. Stockton. 1974. Chronologies from temperature-sensitive bristlecone pines at upper treeline, western U.S.A. *Tree-Ring Bull.* 44:21.

LaMarche, V. C., Jr., R. L. Holmes, P. W. Dunwiddie, and L. G. Drew. 1979. *Tree-Ring Chronologies of the Southern Hemisphere. 3. New Zealand, 4. Australia.* Chronology Series V, Laboratory of Tree-Ring Research, University of Arizona, Tucson.

LaMarche, V. C., Jr., D. A. Graybill, H. C. Fritts, and M. R. Rose. 1984. Increasing atmospheric carbon dioxide: Tree-ring evidence for growth enhancement in natural vegetation. *Science* 225:1019–1021.

Luckman, B. H. 1989. Global change and the record of the past. 1989. *GEOS* 18:1–8.

MacDonald, G. M., T. W. D. Edwards, K. A. Moser, R. Pienitz, and J. P. Smol. 1993. Rapid response of treeline vegetation and lakes to past climate warming. *Nature* 361:243–246.

McGuire, A. D., J. M. Melillo, L. A. Joyce, D. W. Kicklighter, A. L. Grace, B. Moore, III, and C. J. Vorosmarty. 1992. Interactions between carbon and nitrogen dynamics in estimating new primary productivity for potential vegetation in North America. *Global Biogeochem. Cycles* 6:101–124.

Mareschal, J. C., and H. Beltrami. 1992. Evidence for recent warming from perturbed geothermal gradients: Examples from eastern Canada. *Climate Dyn.* 6:135–143.

Neftel, A., E. Moor, H. Oeschger, and B. Stauffer. 1985. Evidence from polar ice cores for the increase in atmospheric CO_2 in the past two centuries. *Nature* 315:45–47.

Norton, D. A., K. R. Briffa, and M. J. Salinger. 1989. Reconstruction of New Zealand summer temperatures to 1730 AD using dendroclimatic techniques. *Int. J. Climatol.* 9:633–644.

Pastor, J., and W. M. Post. 1988. Response of northern forests to CO_2-induced climate change. *Nature* 334:55–58.

Payette, S., and C. Morneau. 1993. Holocene relict woodlands at the eastern Canadian treeline. *Quat. Res.* 39:84–89.

Schlesinger, M. E., and J. F. B. Mitchell. 1985. Model projections of the equilibrium climatic response to increased carbon dioxide. In *The Potential Climatic Effects of Increasing Carbon Dioxide,* ed. M. C. MacCracken and F. M. Luther, pp. 81–148. DOE/ER-0237. U.S. Department of Energy, Washington, D.C.

Solomon, A. 1986. Transient response of forests to CO_2-induced climatic change: Simulation modeling experiment in eastern North America. *Oecologia* 68:567–579.

Sundquist, E. T. 1993. The global carbon dioxide budget. *Science* 259:934–941.

Tans, P. P., I. Y. Fung, and T. Takahashi. 1990. Observational constraints on the global carbon dioxide budget. *Science* 247:1431–1438.

Thompson, L. G. 1992. Ice core evidence from Peru and China. In *Climate Since A.D. 1500,* ed. R. S. Bradley and P. D. Jones, pp. 517–548. Routledge, London.

Villalba, R. 1990. Climatic fluctuations in northern Patagonia during the last 1000 years as inferred from tree-ring records. *Quat. Res.* 34:346–360.

Villalba, R., J. C. Leiva, S. Rubulls, J. Suarez, and L. Lenzano. 1990. Climate, tree-ring, and glacial fluctuations in the Rio Frias Valley, Rio Negro, Argentina. *Arct. Alp. Res.* 22:215–232.

6

Global Climate Change: Disturbance Regimes and Biospheric Feedbacks of Temperate and Boreal Forests

WERNER A. KURZ, MICHAEL J. APPS,
BRIAN J. STOCKS, AND W. JAN A. VOLNEY

General circulation models that predict large-scale climatic changes at present account neither for the effects of these changes on terrestrial and aquatic biota nor for feedbacks associated with the effects. This chapter reviews some of the feedback mechanisms through which temperate and boreal forest ecosystems could affect future climatic changes. The effects of changes in forest distribution and changes in growth and decomposition rates are addressed by others in this volume. This chapter emphasizes the role of disturbance regimes and their effects on forest structure and function. We show that changes in climatic conditions have affected forest disturbance regimes in the past and that these changes in disturbance regimes must be considered when assessing biospheric feedbacks to future climate changes. The chapter first identifies indicators of forest ecosystems to be evaluated when assessing feedback potentials. The disturbance regimes in the last 60 years in Canadian forests are reviewed and the potential feedback to climatic changes is discussed. Mitigation options available through forest management and protection are addressed briefly.

FEEDBACK MECHANISMS

To address the issue of biospheric feedbacks to climate change through disturbance regimes, three questions must be answered: Do changes in disturbance regimes affect ecosystem structure and function? Do changes in climate result in changes in disturbance regimes? Will changes in ecosystem structure and function lead to climatic feedbacks?

Terrestrial ecosystems play an important role in the global and regional cycling of carbon, water, and nutrients and interact with climate through these cycles and through albedo, exchange of energy, and other mechanisms. All of these processes are affected by changes in the structure and function of terrestrial ecosystems. If the changes in ecosystem structure and function are brought about by changes in climate (or environmental change in general) and lead to further changes in climate, positive feedback is said to exist. Negative feedback results from changes in structure and function that reduce the climatic effects that initiated the ecosystem changes.

Forest ecosystems are major reservoirs of carbon and cover a large proportion of the land surface. In this chapter, we focus on feedback mechanisms of forest ecosystems through effects on the carbon cycle, but we emphasize that other feedback mechanisms also exist, as outlined by Bonan et al. (1992).

The effects of forest structure on the carbon cycle can be summarized through four indicators: forest area, forest age class distribution, species composition, and ecosystem carbon density. Changes in forest function (i.e., growth rate, decomposition rate, and other processes) also must considered. These processes affect and are affected by forest structure, and this discussion emphasizes forest structure.

Boreal and temperate forest ecosystems are adapted to frequent stand-replacing disturbances, such as wildfires and insect-induced stand mortality. The spatial heterogeneity and mix of age classes in forest ecosystems reflect past disturbance regimes. The effect of fire frequency on the forest age class structure is well established (Van Wagner 1978; Yarie 1981; Johnson and Larson 1991). Fire cycles (i.e., the average interval between fires at any given point; Van Wagner 1978) in boreal and temperate forests range from a few decades to a few centuries. Estimates of presuppression fire cycles in the boreal forest are 50–200 years and increase from the south to the north and from west to east (Rowe 1983; Bonan and Shugart 1989; Payette 1992). Fire cycles in cold and wet northern ecosystems can be 1000 years and longer (Payette et al. 1989).

Four aspects of disturbances and postdisturbance ecosystem recovery are important: the annual rate of disturbance, the type of disturbance, the rate of postdisturbance forest regeneration, and the rate of biomass and ecosystem carbon accumulation.

Rate of Disturbance

The annual rate of disturbance has a large effect on the forest age class structure. An increase in disturbance rates shifts the forest age class structure to the left, i.e., the proportion of young stands increases. Shifts in age class structures toward younger age classes result in reduced carbon storage because the average biomass decreases and because reductions in the average time between disturbances can affect coarse woody debris and detritus carbon pools (Harmon et al. 1990). A reduction in the length of the fire-free interval also affects species composition, favoring early successional and pioneer species.

Type of Disturbance

The different types of disturbance have significant effects on carbon storage and cycling. All disturbances transfer biomass carbon to soil and detritus carbon pools, where it decomposes in the years following the disturbance. The proportion of biomass carbon transferred to soil and detritus carbon pools is greatest for insect-induced stand mortality, intermediate for wildfires (some release to the atmosphere), and smallest for harvesting (transfer to the forest product sector). These differences in carbon transfer will affect future rates of carbon (and nutrient) cycling and can affect other ecosystem characteristics, such as soil temperatures and water-holding capacity.

Rate of Regeneration

The disturbances discussed here typically are stand replacing. The rate of regeneration and the associated regeneration delay are therefore of great importance to forest structure and function. Regeneration delay refers to a period of time after the disturbance during which the site is occupied by short-lived herbaceous or shrubby vegetation that is not accumulating significant amounts of carbon. The length of the regeneration delay depends on many factors, such as climatic conditions for seedling establishment, and on the presence of seed sources, seed banks, and bud banks. All of these could be significantly affected by climatic change. In some ecotones (e.g., the northern tree line) seed production and regeneration success are temperature limited, and increasing temperatures could reduce regeneration delays. Under present conditions, regeneration failure after (infrequent) major fires occurs in forest tundra ecotones in northern Quebec, resulting in gradual deforestation at the northern tree line (Payette and Gagnon 1985). In other ecotones, such as parts of the southern boreal forest, seedling establishment and survival are limited by soil moisture deficits, and greater evaporative stress associated with climatic changes will increase regeneration delays and favor the northward expansion of grassland (Zoltai et al. 1991).

Boreal forest species are adapted to frequent fires, and some produce serotinous cones that require high temperatures to release their seeds. Boreal tree species require several decades to start seed production (Nikolov and Helmisaari 1992). Regeneration failure could increase if an increased fire frequency results in a fire return interval that is shorter than the time required by boreal trees to reach reproductive age (Heinselman 1973). This would favor a shift in species composition toward species whose seeds are dispersed over longer distances by wind or animals or those species that regenerate vegetatively (Rowe 1983). Increased fire frequency could therefore also increase regeneration delay.

The length of the regeneration delay is of significance to biospheric feedbacks because it reduces the average amount of carbon stored in forest ecosystems in two ways. The area with delayed regeneration has less biomass carbon storage, thus reducing the regional carbon storage. Furthermore, the detritus and soil carbon pools could decrease during the regeneration period because decomposition proceeds while detritus inputs from biomass carbon pools are reduced.

Rate of Carbon Accumulation

Carbon is accumulating in two major pools: the living biomass pool and the soil and detritus carbon pool. In many boreal and temperate forests, biomass carbon pools increase in the first decades after disturbance and decrease in late stand developmental stages (Whynot and Penner 1990; Alban and Perala 1992). There is considerable uncertainty about the dynamics of the soil and detritus carbon pools, and it is not clear whether the reduction in biomass carbon pools in old age classes merely represents a transfer to detritus pools or whether it results in a decrease of total ecosystem carbon.

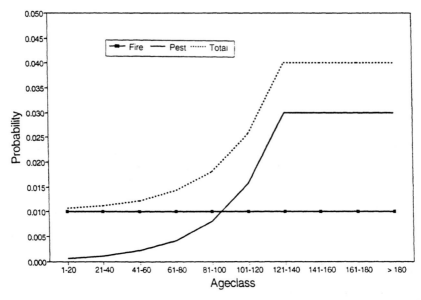

Figure 6.1. Probability of fire and pest disturbances as a function of forest age as used in the simulation model.

BIOSPHERIC FEEDBACKS THROUGH SHIFTS IN FOREST AGE CLASS DISTRIBUTION

The complex effects of disturbance regimes, age class structures, regeneration delays, and carbon accumulation rates on regional carbon storage and biospheric feedbacks are demonstrated through a simple model. The model represents 10,000 forest stands whose probability of disturbance is a function of stand age. It simulates the effects of two disturbance regimes and regeneration delays on forest age class structures. It assumes that the probability (p) of a stand's being affected by the first disturbance is constant with stand age (Figure 6.1, Fire). The probability of the second disturbance increases with stand age (Figure 6.1, Pest). In four sensitivity analysis runs, the fire disturbance rate and the length of the regeneration delay are modified (Table 6.1), whereas the probability of pest disturbance is not altered. After disturbance, the stands experience a regeneration delay (5 years in cases 1 and 3, 10 years in cases 2 and 4). Running the model to equilibrium with the parameters of case 1 generates the age class distribution shown in Figure 6.2 (open bars). Increasing the fire disturbance regime from $p = 0.01$ to $p = 0.02$ and the regeneration delay from 5 to 10 years (case 4) generates the second age class distribution (Figure 6.2, shaded bars). The increase in disturbance regimes shifts the age class structure to the left, i.e., it decreases the average age of the forest.

The forest, in this simple example, has a single biomass accumulation curve. Regional biomass carbon storage is the product of biomass carbon and the area in each age class. Table 6.1 summarizes the results of four sensitivity analysis runs that demonstrate the change in regional biomass carbon storage resulting from changes in disturbance regimes and regeneration delay. The combined increases in the probability of fire and the length of the regeneration delay reduce regional biomass carbon storage by 28%.

Table 6.1. Sensitivity Analysis of a Simple Forest Dynamics Model

Case	Fire probability	Regeneration delay (years)	Regional biomass carbon (percent of case 1)
1. Base case	0.01	5	100
2. More regeneration delay	0.01	10	94
3. More disturbances	0.02	5	78
4. No. 2 and No. 3	0.02	10	72
5. 10% more biomass carbon	0.01	5	110
6. No. 5 and No. 2	0.01	10	103
7. No. 5 and No. 3	0.02	5	86
8. No. 5 and No. 4	0.02	10	79

Note: Age class structure (cases 1 and 4) and biomass-over-age curve are as shown in Figure 6.2.

Increasing biomass storage in each age class by 10%, e.g., as a result of better growing conditions or CO_2 fertilization (Bazzaz 1990), decreases the reduction in regional biomass carbon storage (cases 5–8) but has less effect than the increase in disturbance regimes (cases 7 and 8). Given the age class structure of case 4, biomass carbon storage in each age class would have to increase by 39% to obtain the regional biomass carbon storage of case 1. The conclusions of this analysis are comparable with the effects on regional biomass carbon storage of reducing forest rotation length during the transition from natural to managed forests (Cooper 1983; Dewar 1991).

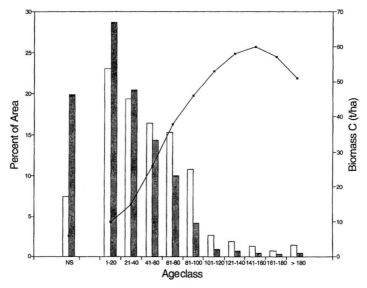

Figure 6.2. Percentage of total area (left axis) and biomass carbon density (C ha^{-1}, right axis) in 20-year age classes of a hypothetical forest. NS represents a nonstocked age class (regeneration delay). Open bars represent area for case 1 in Table 6.1, shaded bars represent case 4.

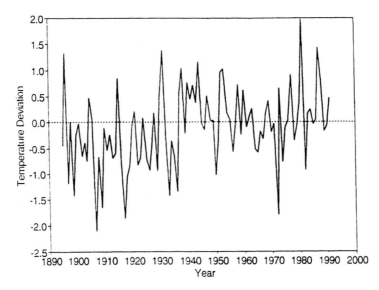

Figure 6.3. Canadian national temperature trend expressed as temperature departures (°C) from the average of the period 1951–1980. (Redrawn from Gullet and Skinner 1992.)

Our analysis ignores the contribution of soil and detritus carbon pools to regional carbon storage. The shift of forest age classes toward a younger age increases the proportion of the area in which decomposition rates are accelerated and detritus inputs reduced, thus decreasing soil and detritus carbon storage. The shift will also decrease the input of coarse woody debris during the stand breakup phases, reducing carbon accumulation in detritus carbon pools. Higher rates of disturbances (other than harvesting) accelerate the transfer of biomass carbon to soil and detritus carbon pools, thus offsetting some of the impacts of changes in age class structure on soil and detritus carbon storage (Kurz and Apps 1994).

HISTORIC EVIDENCE OF CLIMATE IMPACTS ON DISTURBANCE REGIMES

The long-term temperature record (1895–1991) in Canada shows three phases: a warming phase from the late 1890s to the 1940s, followed by a cooling period into the 1970s and a return to warming through the 1980s (Figure 6.3; Gullet and Skinner 1992). The overall warming trend of 1.1°C for the period is statistically significant.

The Canadian Forest Fire Weather Index (FWI) System (Van Wagner 1987), which is used throughout Canada, integrates the effects of temperature, precipitation, wind speed, and relative humidity on the moisture content of forest fuels to predict fire danger. The final component index of the FWI System, the Fire Weather Index itself, is correlated with the monthly area burned in Canadian provinces (Harrington et al. 1983). Long sequences of days without rain strongly influence the monthly provincial area burned in Canada (Flannigan and Harrington 1988). Correlations between the annual temperature deviation and the area annually burned in Canada are statistically significant only for individual regions but not for the entire country. Temperature data alone only account for a portion of the temporal variation, but detailed long-term precipitation records are not as readily available as temperature data.

Interpretation of the historic relationship between area burned and climatic conditions in Canada is confounded by the impact of organized forest fire protection, which began in the early 1920s. Use of forests for both industrial and recreational purposes has increased steadily over the past seven decades, resulting in significantly more ignition sources and fires. The effect has been somewhat offset by the development of increasingly sophisticated fire protection capability, but the relative impact of both developments on the total area burned is difficult to assess. In addition, the general forest fire management strategy in Canada is to provide intensive protection for high-value recreational and industrial areas, while applying a form of "modified suppression" in remote areas where fire is often allowed to burn naturally. This approach also serves to confound the statistics. Although it is impossible to establish with any accuracy the reduction in areas burned resulting from forest fire protection, it is clear that, at least in the southern parts of Canada, fire control efforts have reduced the area annually burned (Barney and Stocks 1983) and may have contributed to a shift in forest age class structure toward older ages (Blais 1983).

Forest fire protection can have a significant impact on forest age class structures, and many studies have demonstrated that the youngest age classes in regional forest inventories have much less area than older age classes (Heinselman 1973; Yarie 1981; Clark 1990). This deviation from the negative exponential decline in the age class structure (e.g., Figure 6.2) can be achieved through a reduction of the area annually disturbed or through a significant increase in regeneration failure that leads to a reclassification of previously forested land as nonstocked land. None of the studies cited previously addresses this latter possibility and all interpret the shift in age class structure as the result of successful fire protection measures.

Forest harvesting, in particular the clear-cut logging that is predominantly practiced in Canada, has introduced a new type of disturbance. Does it offset the reductions in areas annually disturbed that result from fire protection? Figure 6.4 summarizes the areas burned (10-year running average) and harvested for the period 1930–1989. National harvest area statistics are readily available only for the period after 1975, but harvest volume statistics have been recorded since the late 1920s. In Figure 6.4, the area harvested is approximated from the harvest volume to harvest area ratio established from the statistics for 1975—1990. This approach may lead to an error in the estimate of area harvested in the early part of this century because of changes in harvesting methods and efficiencies (underestimation of area) and the predominant harvesting of old-growth forests with high volumes (overestimation of area).

Figure 6.4 indicates that for the period 1930–1979 the area annually disturbed through fire and harvesting has averaged circa 1.63 million ha yr^{-1}, and that it has increased to circa 3.46 million ha yr^{-1} for the period 1980–1989, largely because of dramatic increases in the area annually burned. The figure also suggests that for the period 1930–1979 the reduction through climatic effects or fire suppression in areas annually burned offsets the increase in areas harvested, resulting in little change in the total area disturbed by fire and harvesting.

The impacts of insect-induced stand mortality on forest dynamics are more difficult to assess because of the nature of insect disturbances. Many forest insects cause growth

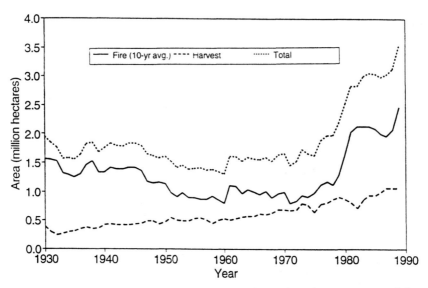

Figure 6.4. The 10-year average of area annually burned and an estimate of the area annually harvested in Canada for the period 1930–1989. Harvesting data are compiled from multiple sources; fire data are from Van Wagner (1988) and Canadian Forest Service.

reductions, selected mortality of host species, or stand-replacing mortality after repeated years of defoliation or bark beetle attacks. Although Canada has a long record of insect outbreak statistics, estimates of stand-replacing mortality are not available for the same period for which fire or harvesting statistics exist.

Regional climatic and forest conditions affect the relative importance of fire and insects as the predominant disturbance regimes. In the drier western boreal forest, fire is the dominant disturbance regime, whereas insects play a much more important role in the wetter eastern boreal forest (Apps et al. 1991; Kurz et al. 1992).

Climatic conditions have long been known to affect insect populations. The association of extreme weather with the occurrence of outbreaks in forest insect populations has been examined by correlating the outbreaks with unusual periods of weather (e.g., Volney 1988). Further indications of climatic controls on insect populations are the frequency of outbreaks and the geographic ranges of insects. Both have been documented for insects associated with northern coniferous forests or species that are likely to become important should climatic conditions permit an expansion of their range to the north. Table 6.2 is a listing of species that fall into the first category: boreal insect species whose outbreaks seem to be associated with extreme weather conditions. One pest species feeding on each major host species is represented in the table. Insect outbreaks are generally distributed in the southern portions of their respective host ranges. Should climatic warming occur, there is every reason to expect that the outbreak areas will expand. Table 6.3 lists insect species whose historical outbreak areas are not in the boreal forest. Yet the outbreak areas are likely to expand to the north if the distribution of the host species change in response to a warmer climate, or if the host range occurs in the boreal forests but the insect outbreaks are currently limited to the southern portion of that range.

Table 6.2. Pest Species of the Boreal Forest Whose Outbreaks Are Associated with Extreme Weather Conditions

Species	Hosts(s)	Factors favoring outbreaks	Reference
Choristoneura fumiferana (spruce budworm)	Spruce/fir	Warm springs	Ives (1974)
		Moderate winter temperatures	
		Dry years	
		Warm dry summers	Greenbank (1963)
Choristoneura pinus (jack pine budworm)	Jack pine	Drought and high temperatures	MacAloney (1944)
		Below average precipitation	Clancy et al. (1980)
		Warm springs	
		Warm, dry periods	Volney (1988)
Malacosoma disstria (forest tent caterpillar)	Aspen	Mild winters	Ives (1981)
		Warm springs	

Major insect outbreaks require a combination of prerequisites that include favorable climate and suitable host conditions. Many insects favor older stands (Volney 1988) and shifts in age class structure resulting from fire suppression can create conditions that are more favorable to insects (Heinselman 1973; Blais 1983; Ritchie 1987). However, the inverse is not necessarily true. Climatic stresses (warm and dry conditions) that lead to shifts to younger age classes because of increased fire activity can also increase host susceptibility to insects (witness insect problems in the southwestern United States during the recent drought conditions). Furthermore, stressed young forest stands may be susceptible to insect problems that are different from those affecting older stands.

FUTURE FEEDBACKS

In the past, disturbance regimes in boreal and temperate forests have been affected by climatic conditions. We must therefore assume that changes in climatic conditions will

Table 6.3. Pest Species of the Cordilleran Forest Whose Outbreaks May Expand to Boreal Regions under Climate Change

Species	Hosts	Factors favoring outbreaks	Reference
Choristoneura occidentalis (western spruce budworm)	Douglas-fir	Low rainfall	Kemp et al. (1985)
		Cool winters	
		Warm springs	
Orgyia pseudotsugata (Douglas-fir tussock moth)	Douglas-fir	Warm spring temperatures	Glendenen et al. (1978)
		Moderate temperatures	Shepherd et al. (1989)
		Dry conditions	
Dendroctonus ponderosae (Mountain pine beetle)	Lodgepole	High overwinter temperatures	Safranyik (1985)
		Low precipitation and warm periods	Thompson and Shrimpton (1984)

continue to result in changes in disturbance regimes that will further affect forest structure and function.

The effects of changes in temperature, precipitation, and relative humidity derived from general circulation models were used to explore the severity of the forest fire season in Canada (Flannigan and Van Wagner 1991). A 46% increase in the seasonal severity rating was predicted "with a possible similar increase in area burned," although much greater increases could not be ruled out. Global warming has also been predicted to lead to greater area burned and a larger number of escaped fires in Northern California (Torn and Fried 1992).

Other studies have predicted changes in forest area and distribution (Rizzo and Wiken 1992; Smith et al. 1992) on the assumption that vegetation and $2 \times CO_2$ climate conditions will reach a new equilibrium. The "transient" responses of vegetation to shifting climatic conditions could result in substantial release of carbon into the atmosphere (King and Neilson 1992).

At present, our ability to predict quantitatively the feedbacks of forest ecosystems to global climatic change is limited, and the analysis is complicated by the wide range of ecological conditions and the likely responses of ecosystem processes (Zoltai et al. 1991; Apps 1993). We are confident, however, in rejecting the assumption that the terrestrial biosphere will not provide feedback in response to changes in climatic conditions. Furthermore, we provide two observations and offer a suggestion about the nature of the feedback.

1. *Asymmetry of rate of change.* Forest ecosystems can remove carbon from the atmosphere or add carbon to it, thus providing negative or positive feedback to climatic warming. The speed at which carbon uptake and release occur can differ greatly because carbon uptake through net ecosystem productivity (the balance between carbon uptake and respiration) occurs much more slowly than the rapid carbon release associated with fires. Moreover, many of the processes that create more favorable conditions for carbon uptake also accelerate other biological processes, such as decomposition, making it harder to achieve large gains through net ecosystem production. On the other side, processes that create ecosystem stress and reduced growth rates, such as extreme temperatures and drought conditions, also favor such disturbances as fire and insect-induced tree mortality.

Changes in forest area are another example of asymmetrical rates of change. The rate at which forest area could expand where growing conditions become more favorable is much less than the rate at which forest area could be lost, e.g., by large-scale forest fires and regeneration failure.

2. *Asymmetry of risk.* To achieve carbon uptake at the regional scale, the sum of the products of carbon density in each age class and the area in each age class (the age class distribution) must increase. This increase can be accomplished through increases in the total forest area, through increases in carbon density in individual age classes, and through shifts in the age class distribution toward age classes with greater carbon density. In any case, though, forest trees must continue to take up carbon at a rate that is greater than ecosystem respiration release while maintaining an age class structure with large proportions of the total area in high-carbon-density age classes.

The factors that determine whether forest ecosystems remain carbon sinks, or at least maintain carbon storage, must all be favorable for many decades, whereas a single extreme year of environmental conditions can lead to tree mortality, insect attack, or loss from fires. As environmental conditions shift away from those to which present vegetation complexes are adapted, the risk of disturbance increases. The asymmetry of risk can be compared with the analogy that the weakest link in a chain leads to breakage. Any one factor that deviates too far from current conditions can induce ecosystem carbon losses, whereas maintenance or increase of carbon storage requires that all factors remain adequate. For example, provenance tests have repeatedly demonstrated that trees are adapted to the prevailing environmental conditions of their origin. Shifting environmental conditions away from that optimum often decreases forest productivity and increases tree stress and susceptibility to insects or other mortality factors (see Chapter 1, this volume).

Changes in forest area are also an example of asymmetrical risk. If environmental conditions shift the boundaries of the climatic zone suitable for boreal forest northward, all that is required to reduce forest area at the forest's warmer and drier limit is one extreme fire year followed by regeneration failure. To expand forest area in the north, however, many different factors (seed sources, soil conditions, disturbance regimes, growing conditions) must be appropriate for many years to ensure successful seedling establishment and survival. Furthermore, the forest area lost in the south will have had greater growth rates and higher carbon storage potential than the area gained in the north.

As a result of these two asymmetries, we suggest that it is more likely that the biospheric feedbacks from temperate and boreal forest ecosystems will be positive feedbacks that further enhance the carbon content of the global atmosphere. Although this discussion focuses on carbon cycling, the conclusion can be extended to all other feedback mechanisms that require that forest area, age class structures, and forest growth be maintained.

The potential to mitigate feedback mechanisms is limited. The success of fire protection measures in the past has contributed in part to calls for further increases in protection efforts to reduce the area annually burned, or at least to prevent future increases. Fire protection is costly and resources are limited. As fire conditions become more severe, a decreasing proportion of the emerging fires is acted upon and the area burned increases in response to climate forcing, albeit at a smaller rate than if no protection measures were in place.

Successful forest fire protection shifts forest age class distribution toward older ages. The structure and function of boreal forests are adapted to frequent natural disturbances, and reducing fire disturbances will increase the forests' susceptibility to insect attacks as the average forest age increases. Fire protection therefore can delay, but not prevent indefinitely, the time at which boreal forests are disturbed.

As already discussed, regeneration delay is one of the important determinants of the role of forest ecosystems in regional carbon cycling and storage. One option to mitigate feedbacks that result from changes in disturbance regimes is to reduce regeneration delays through seeding or planting efforts. Although an analysis of the quantitative implications for the carbon budget has not yet been conducted, seeding and planting areas disturbed through fire or insect-induced stand mortality might be a viable mitigation strategy in addition to forest protection efforts. For example, reducing the regeneration delay from 5 to 0 years in Table 6.1 increases the regional biomass storage of cases 1 and 4 by 6%.

CONCLUSIONS

Disturbances such as wildfire and insect-induced stand mortality affect forest structure and function because they influence forest age class distribution, forest area, species composition, and ecosystem carbon density. Disturbance regimes are affected by climatic conditions as drought and higher than normal temperatures increase fire frequency and insect outbreaks. Historic evidence suggests that fire regimes are closely coupled with climatic conditions, and recent temperature increases above long-term averages have resulted in very significant increases in the area burned annually in Canadian forests. These increases are large relative to the area burned annually in the period since 1918. A comparison with longer-term, natural fire cycles is complicated by lack of records and by multiple confounding factors.

Climate-induced changes in disturbance regimes affect forest structure and function, which in turn influence future climatic conditions through carbon, water, and nutrient cycling and other feedbacks. Assessing the direction of biotic feedback is a complex task because of the many nonlinear processes and internal feedbacks.

Assuming no change in forest structure and function under conditions of climatic change is not justifiable. The potential for feedbacks is significant. We conclude that— because (1) the potential changes in rates of carbon release from forest ecosystems are much higher than those of carbon uptake (asymmetry of rates) and (2) the probability of increased carbon release is much greater than that of increased carbon accumulation (asymmetry of risk)—positive feedback from temperate and boreal forest ecosystems to climate change is more likely than negative feedback. Mitigation options through forest management and protection are limited.

ACKNOWLEDGMENTS

Work for this project was funded in part by the Canadian Federal Panel on Energy Research and Development through ENFOR (Energy from the Forest) program of the Canadian Forest Service. We thank Sarah Beukema and Tamara Lekstrum for programming and research assistance.

REFERENCES

Alban, D. H., and D. A. Perala. 1992. Carbon storage in Lake States aspen ecosystems. *Can. J. For. Res.* 22:1107–1110.

Apps, M. J. 1993. NBIOME: A biome-level study of biospheric response and feedback to potential climate changes. *World Resource Rev.* 5:41–65.

Apps, M. J., W. A. Kurz, and D. T. Price. 1991. Estimating carbon budgets of Canadian forest ecosystems using a national scale model. In *Carbon Cycling in Boreal Forest and Subarctic*

Ecosystems: Biospheric Responses and Feedbacks to Global Climate Change, ed. T. Kolchugina and T. Vinson, pp. 241–250. Oregon State University, Corvallis.

Barney, R. J., and B. J. Stocks. 1983. Fire frequencies during the suppression period. In *The Role of Fire in Northern Circumpolar Ecosystems,* SCOPE 18, ed. R. W. Wein and D. A. MacLean, pp. 45–62, Wiley, Chichester, U.K.

Bazzaz, F. A. 1990. The response of natural ecosystems to the rising global CO_2 levels. *Annu. Rev. Ecol. Syst.* 21:167–196.

Blais, J. R. 1983. Trends in the frequency, extent, and severity of spruce budworm outbreaks in eastern Canada. *Can. J. For. Res.* 13:539–547.

Bonan, G. B., and H. H. Shugart. 1989. Environmental factors and ecological processes in boreal forests. *Annu. Rev. Ecol. Syst.* 20:1–28.

Bonan, G. B., D. Pollard, and S. L. Thompson. 1992. Effects of boreal forest vegetation on global climate. *Nature* 359:716–718.

Clancy, K. M., R. L. Giese, and D. M. Benjamin. 1980. Predicting jack pine budworm populations in northwestern Wisconsin. *Environ. Entomol.* 9:743–751.

Clark, S. J. 1990. Fire and climate change during the last 750 yr in Northwestern Minnesota. *Ecol. Monogr.* 60(2):135–159.

Cooper, C. F. 1983. Carbon storage in managed forests. *Can. J. For. Res.* 13:155–166.

Dewar, R. C. 1991. Analytical model of carbon storage in the trees, soils, and wood products of managed forests. *Tree Physiol.* 8:239–258.

Flannigan, M. D., and J. B. Harrington. 1988. A study of the relation of meteorology variables to monthly provincial area burned by wildfire in Canada 1953–1980. *J. Appl. Meteorol.* 27:441–452.

Flannigan, M. D., and C. E. Van Wagner. 1991. Climate change and wildfire in Canada. *Can. J. For. Res.* 21:66–72.

Glendenen, G., V. F. Gallucci, and R. I. Gara. 1978. On the spectral analysis of cyclical tussock moth epidemics and corresponding climatic indices, with a critical discussion of the underlying hypotheses. In *Time Series and Ecological Processes, Proceedings of the S.I.M.S.-S.I.A.M. Conference,* ed. H. H. Shugart, pp. 279–293, Society for Industrial and Applied Mathematics, Philadelphia.

Greenbank, D. O. 1963. Climate and the spruce budworm. In *The Dynamics of Epidemic Spruce Budworm Populations,* ed. R. F. Morris, *Mem. Entomol. Soc. Can.* 31:174–180.

Gullett, D. W., and W. R. Skinner. 1992. The state of Canada's climate: Temperature change in Canada 1895–1991. SOE Rep. No. 92-2. Environment Canada, Atmospheric Environment Service, Ottawa, Ontario.

Harmon, M. E., W. K. Ferrell, and J. F. Franklin. 1990. Effects on carbon storage of conversion of old-growth forests to young forests. *Science* 247:699–702.

Harrington, J. B., M. D. Flannigan, and C. E. Van Wagner. 1983. A study of the relation of components of the Fire Weather Index to monthly provincial area burned by wildfire in Canada 1953–80. Inf. Rep. PI-X-25. Canadian Forestry Service, Petawawa National Forestry Institute, Chalk River, Ontario.

Heinselman, M. L. 1973. Fire in the virgin forests of the Boundary Waters Canoe Area, Minnesota. *Quat. Res.* 3:329–382.

Ives, W. G. H. 1974. Weather and outbreaks of the spruce budworm, *Choristoneura fumiferana* (Lepidoptera: Tortricidae). Information Report NOR-X-118. Canadian Forestry Service, Northern Forestry Centre, Edmonton, Alberta.

Ives, W. G. H. 1981. Environmental factors affecting 21 forest insect defoliators in Manitoba and Saskatchewan. Information Report NOR-X-233. Canadian Forestry Service, Northern Forestry Centre, Edmonton, Alberta.

Johnson, E. A., and C. P. S. Larson. 1991. Climatically induced change in fire frequency in the southern Canadian Rockies. *Ecology* 72(1):194–201.

Kemp, W. P., D. O. Everson, and W. G. Wellington. 1985. Regional climatic patterns and western spruce budworm outbreaks. Technical Bulletin 1693. United States Department of Agriculture Forest Service, Washington, D.C.

King, G. A., and R. P. Neilson. 1992. The transient response of vegetation to climate change: A potential source of CO_2 to the atmosphere. *Water Air Soil Pollut.* 64:365–384.

Kurz, W. A., and M. J. Apps. 1994. The carbon budget of Canadian forests: A sensitivity analysis of changes in disturbance regimes, growth rates, and decomposition rates, *Environ. Pollut.* 83:55–61.

Kurz, W. A., M. J. Apps, T. M. Webb, and P. J. McNamee. 1992. The carbon budget of the Canadian Forest Sector: Phase I. Information Report NOR-X-326. Forestry Canada, Northwest Region, Northern Forestry Centre, Edmonton, Alberta.

MacAloney, H. J. 1944. Relation of root condition, weather, and insects to the management of jack pine. *J. For.* 42:124–129.

Nikolov, N., and H. Helmisaari. 1992. Silvics of the circumpolar boreal forest tree species. In *A Systems Analysis of the Global Boreal Forest,* ed. H. H. Shugart, R. Leemans, and G. B. Bonan, pp. 13–84. Cambridge University Press, Cambridge.

Payette, S. 1992. Fire as a controlling process in the North American boreal forest. In *A Systems Analysis of the Global Boreal Forest,* ed. H. H. Shugart, R. Leemans, and G. B. Bonan, pp. 144–169, Cambridge University Press, Cambridge.

Payette, S., and R. Gagnon. 1985. Late Holocene deforestation and tree regeneration in the forests-tundra of Quebec. *Nature* 313:570–572.

Payette, S., C. Morneau, L. Sirois, and M. Desponts. 1989. Recent fire history of the northern Quebec biomes. *Ecology* 70(3):656–673.

Ritchie, J. C. 1987. *Postglacial Vegetation of Canada.* Cambridge University Press, Cambridge.

Rizzo, B., and E. Wiken. 1992. Assessing the sensitivity of Canada's ecosystems to climatic change. *Climat. Change* 21:37–55.

Rowe, J. S. 1983. Concepts of fire effects on plant individuals and species. In *The Role of Fire in Northern Circumpolar Ecosystems,* SCOPE 18, ed. R. W. Wein and D. A. MacLean, pp. 135–154, Wiley, Chichester, U.K.

Safranyik, L. 1985. Effect of climatic factors on development, survival, and life cycle of the mountain pine beetle. In *Mountain Pine Beetle Symp. Proc.,* Pest Management Report #7, pp. 14–24, British Columbia Ministry of Forests, Smithers, British Columbia.

Shepherd, R. F., G. A. Van Sickle, and D. H. L. Clarke. 1989. Spatial relationships of Douglas-fir tussock moth defoliation within habitat and climatic zones. In *Proceedings, Lymantriidae: A Comparison of Features of New and Old World Tussock Moths,* ed. W. E. Wallner and K. A. McManus, pp. 381–400. General Technical Report NE-123. United States Department of Agriculture Forest Service Northeastern Forest Experiment Station, Broomall, Pa.

Smith, T. M., R. Leemans, and H. H. Shugart. 1992. Sensitivity of terrestrial carbon storage to CO_2-induced climate change: Comparison of four scenarios based on general circulation models. *Climat. Change* 21:367–384.

Thomson, A. J., and D. M. Shrimpton. 1984. Weather associated with the start of mountain pine beetle outbreaks. *Can. J. For. Res.* 14:255–258.

Torn, M. S., and J. S. Fried. 1992. Predicting the impacts of global warming on wildland fire. *Climat. Change* 21:257–274.

Van Wagner, C. E. 1978. Age-class distribution and the forest fire cycle. *Can. J. For. Res.* 8:220–227.

Van Wagner, C. E. 1987. The development and structure of the Canadian Forest Fire Weather Index System. Canadian Forestry Service Technical Report 35. Agriculture Canada, Canadian Forest Service, Ottawa, Ontario.

Van Wagner, C. E. 1988. The historical pattern of annual burned area in Canada. *Forestry Chronicle,* June, pp. 182–185.

Volney, W. J. A. 1988. Analysis of historic jack pine budworm outbreaks in the Prairie provinces of Canada. *Can. J. For. Res.* 18:1152–1158.

Whynot, T. W., and M. Penner. 1990. Growth and yield of black spruce ecosystems in the Ontario Clay Belt: Implications for forest management. Information Report PI-X-99. Forestry Canada, Petawawa National Forestry Institute, Chalk River, Ontario.

Yarie, J. 1981. Forest fire cycles and life tables: A case study from interior Alaska. *Can. J. For. Res.* 11:554–562.

Zoltai, S. C., T. Singh, and M. J. Apps. 1991. Aspen in a changing climate. In *Aspen Management for the 21st Century,* ed. S. Navratil and P. B. Chapman, pp. 143–152. Forestry Canada, Northwest Region, Northern Forestry Centre and Poplar Council of Canada, Edmonton, Alberta.

7

Permafrost and Vegetation Response to Global Warming in North Eurasia

A. A. VELICHKO, O. K. BORISOVA,
E. M. ZELIKSON, AND V. P. NECHAYEV

The largest variations in temperature and precipitation regimes from a global warming are predicted to occur in middle and high latitudes. In this chapter we have used both modeling and paleoclimatic analogues to examine the response of permafrost and vegetation in North Eurasia. The distribution of vegetation in this regime is dependent on permafrost amd will be affected by the warming.

CHOICE OF CLIMATIC SCENARIOS

There are three approaches to appraising climatic changes resulting from a global warming: general circulation models, study of trends in historical climatic data, and paleoclimatic reconstruction. In the latter approach, paleoclimatic data are used to reconstruct the environmental conditions of the past when global mean temperatures were higher than those of today and corresponded roughly to the temperatures now predicted. In the Laboratory of Evolutionary Geography at the Institute of Geography of the Russian Academy of Sciences, the mean annual temperatures of the northern hemisphere were obtained for various warm periods of the recently glaciated past that might reasonably be used as analogues for the current zwarming. It has been shown from this work that the Holocene climatic optimum of 5500 years B.P. was 0.8–1.0°C warmer than today. For the Mikulino (Eemian, Sangamon) interglacial stage of 125,000 B.P., northern hemispheric temperatures were 1.8–2.0°C above those of the present.

Quantitative paleoclimatic reconstructions for both the land and the sea have a much larger spatial resolution than general circulation model calculations. The reconstructions give a quasi-equilibrium picture of the spatial distribution of climate and are an important source of information on climatic conditions corresponding to the predicted rise in mean global temperature in the next century. Comparison of modeling results with paleoclimatic reconstructions also shows some similarities in the predicted climatic changes anticipated from a future enhanced greenhouse effect. In both approaches the

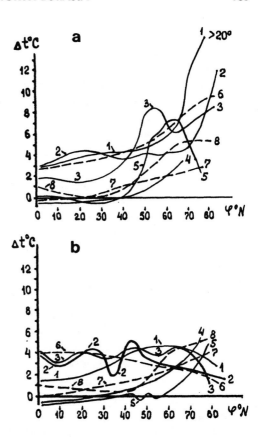

Figure 7.1. Mean temperature profiles for the Northern Hemisphere (deviations from present-day values). (a) Winter temperatures (December by general circulation models; January by paleoclimatic reconstructions). (b) Summer temperatures (June by general circulation models; July by paleoclimatic reconstructions). Profiles 1–5 show monthly mean temperatures along 30°E. Key: 1, GFDL (Geophysical Fluid Dynamics Laboratory 1990); 2, CCC (Canadian Climate Center 1989); 3, UKMO (United Kingdom Meteorological Organization 1990); 4, LEG (Laboratory of Evolutionary Geography 1990; mean global warming ±1°); 5, LEG (mean global warming ±2°C). Profiles 6–8 show monthly mean temperatures averaged for zonal belts. Key: 6, GISS (Goddard Institute for Space Studies 1990); 7, LEG (mean global warming ±1°C); 8, LEG (mean global warming ±2°C).

greatest rise in temperature is predicted for high latitudes and the least increase for low latitudes. There are, however, some differences in the results of the two approaches. In some of the model calculations, the summer temperatures at high latitudes do not increase; however, they are predicted to increase based on the paleoclimatic reconstructions. The model results predict a rise in temperature in low latitudes, but the paleoclimatic reconstructions suggest a cooling. Figure 7.1 is a comparison of temperature changes from model calculations and paleoclimatic data. Although there is a similarity in the prediction of temperature between these methods, the paleoclimatic reconstruction data generally give predicted temperatures lower than those obtained from the models.

Several methods (Velichko et al. 1985) make use of both paleobotanical and lithological information to reconstruct paleoclimate on land. In the work described here, paleoclimatic reconstructions have been based exclusively on paleobotanical data. The estimates of temperature and precipitation were obtained from quantitative correlations between the ancient pollen spectra of a geographic region and those observed under the present vegetation and climatic conditions of the region. This approach has been used in developing reconstructions of the Holocene climatic optimum. During that period, vegetation similar to that of today existed, although the areal distribution of the vegetation was somewhat different. A measure of the correlation between climatic variables and pollen composition or the significance of each component of the pollen was obtained with the help of statistical analysis (Klimanov 1976, 1984) of pollen spectra from recent vegetation zones.

The vegetation of the Mikulino interglacial was somewhat different from that of the present interglacial. To reconstruct the climate of this period, paleofloral methods were used based on plant species that inhabited Eurasia during the Mikulino interglacial. Two approaches were employed (Grichuk 1969, 1985). In the first, climatograms provided combinations of all the climatic variables considered that met the requirements of all the species of plants making up the ancient flora. In the second, arealograms defined the present region of the maximum concentration of plant species found in the ancient floral association. The climatic conditions of the region of interest are closest to those in space and time of the fossil floral associations. These approaches allow us to estimate the temperature within ±1°C and annual precipitation within ±50 mm.

Figure 7.2 shows the temperature and precipitation obtained for Eurasia for the Holocene climatic optimum from paleoclimatic reconstructions, given an increase in mean global temperature of the earth's surface of 1.8–2.0°C. The greater increase in temperature at the higher latitudes is evident. In both winter and summer, temperatures in the Arctic Ocean were 3–4°C higher than at present. In the subpolar regions, the deviations of July and January temperatures from their modern values were about +1–2°C and +2°C, respectively. In the southern part of Eurasia, no temperature change was found. The same conclusion holds true for the Mediterranean area in January and for Kazakhstan and Central Asia in July.

The pattern of precipitation change owing to a warming was complicated. At Arctic latitudes an increase of up to +100 mm/year was observed. In much of eastern Siberia, annual precipitation did not change. However, in parts of Siberia and in the center of the Russian Plain, annual precipitation decreased by 25–50 mm/year in comparison with recent values.

For a mean global temperature increase of the earth's surface of 1.8–2.0°C during the Mikulino interglacial (Figure 7.3), changes in temperature and precipitation regimes in Eurasia were characterized by a rather smooth change in temperature and precipitation across latitudinal zones. The greatest warming at high latitudes occurred from the west to the east: on the Yamal and Taimyr peninsulas, July temperatures deviated +6–8°C from their modern values, whereas in January temperatures in the Taimyr and eastward were +12°C greater than present-day values. On the Russian Plain, winter temperatures were found to increase by 8–10°C. However, in the Mediterranean region in January and in the eastern part of the Mediterranean and Central Asia in July, temperatures were below modern values.

Annual precipitation increased at high latitudes: on the Yamal and Taimyr peninsulas, the increase was about 200 mm/year. A similar increase was found for the northwestern portion of the Eurasian continent. At middle latitudes, precipitation significantly increased in western Europe. Even in the arid regions of southern Siberia, Kazakhstan, and Central Asia, precipitation increased by nearly 100 mm/year.

BASIC PRINCIPLES OF ESTIMATING LANDSCAPE RESPONSE TO CLIMATIC CHANGE

We can use the method of paleoanalogues to obtain an idea of the response of various landscape components to climatic change. To do so we must consider such properties of

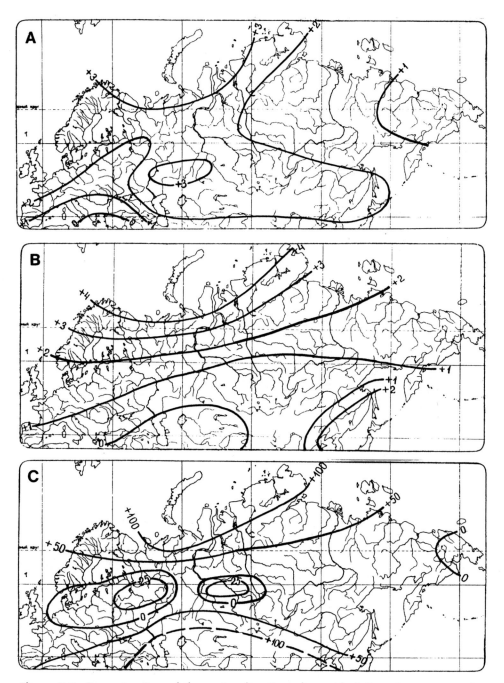

Figure 7.2. Reconstruction of the main climatic indexes (deviations from present-day values) of the Holocene climatic optimum (about 6000 to 5500 years B.P.): (A) mean January temperature (°C), (B) mean July temperature (°C), and (C) mean annual precipitation (mm).

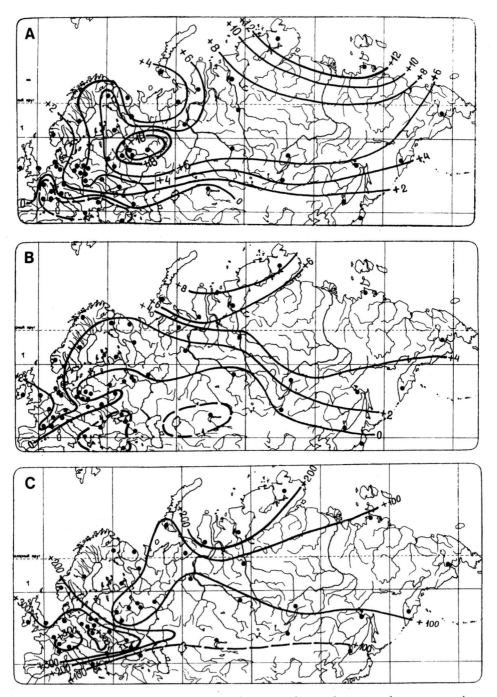

Figure 7.3. Reconstruction of the main climatic indexes (deviations from present-day values) of the last interglacial climatic optimum (125,000 years B.P.): (A) mean January temperature (˚C), (B) mean July temperature (˚C), and (C) mean annual precipitation (mm).

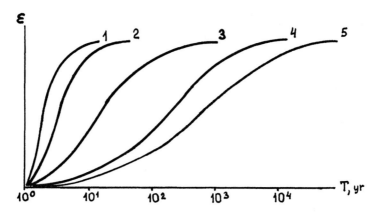

Figure 7.4. Characteristic times for the main elements of ecosystems to regain their equilibrium owing to changed climatic conditions. ε is equilibrium state. Key: 1, sea ice; 2, permafrost; 3, vegetation; 4, soil; 5, land forms and large continental ice sheets.

the system as inertia and the absence of simultaneity (Velichko 1991). Each component of a system has its own characteristic time of change and attainment of equilibrium with an established climatic regime. For example, changes in the distribution of sea ice brought about by a global warming will occur on a time scale of one to several years; changes in the thickness of the layer of seasonal thawing of the permafrost area will take decades; vegetation shifts will take a century or more to be noticeable; and changes in the distribution of soils and relief will take thousands of years (Figure 7.4). The combination of different characteristic times for various components of the system affects changes in the whole complex. Furthermore, within each component of the system, the separate parts in turn form a spectrum of characteristic times. For example, the reactions of different plant species and their communities, changes within communities and the interactions of the species constituting these communities, the rate of succession, and the displacement of vegetation boundaries are all characterized by different time scales. The overall process of change will continuously adapt to the changing climatic conditions of the next century caused by the enhanced greenhouse effect.

When analyzing the response to climatic change of a geosystem or its components, it is necessary to distinguish two states of the geosystem in relation to climate and environmental change. At equilibrium, or more precisely quasi-equilibrium, all components of the system are adapted to the combination of factors—such as seasonal variations in temperature, precipitation, and groundwater levels—that define the function of the system. The factors are relatively stable (in reality, their rate of change is slow, and the amplitude of any change is not great). Thus, all the components of a geosystem have time to adapt to change. In a nonequilibrium state, the rates of change of the geosystem or of external variables are relatively rapid. Under these conditions the system, because of its inertia and differences in the characteristic times of its individual components, is not able to catch up with changing climatic conditions. The system is thus not adapted to the various factors that define its state.

In the method of paleoanalogues that we will use to predict the response of the environment to anthropogenic warming, we will employ both equilibrium and nonequilibrium models. The equilibrium model will recreate the situation in which a system has reached a quasi-equilibrium with existing temperature and hydrologic regimes. It allows us to assess the maximum degree of change; that is, it enables us to define the overall state of the system when the system, temperature, and hydrologic regime have reached an equilibrium state. However, such a model is not appropriate on the decadal time scale of the expected warming during the next century.

Data on climate and vegetation of the past show that a quasi-equilibrium state was reached in almost 500 years during the Holocene climatic optimum when the mean global temperature was 1°C higher than its present-day value. During the Mikulino interglacial, for a warming of 2°C, it took about 1000 years to reach a quasi-equilibrium state. During these warm geologic periods, it is likely that most biotic and abiotic components of the global system might have achieved a quasi-equilibrium state. However, we must remember that climatic conditions were not strictly stable during such periods, and that paleoclimatic reconstructions reflect averaged prevailing climatic conditions over many years and therefore include an ensemble of climates.

Nonequilibrium models are designed to assess the state of the ecosystem by taking into account as realistically as possible the time scale of change in the temperature-precipitation regime of a global warming trend, that is, decades to several decades. Any such model should account for the full spectrum of characteristic rates of response of different ecosystem components, inertial mechanisms, and the dynamics of interactions between the individual components of the system.

The approaches we have described allow us on the one hand to develop an understanding of the true state of vegetation and permafrost during changing climatic conditions, and on the other hand to assess the prospects for further change, taking into account the inertial properties of the system. Such an understanding follows, in particular, from analysis of the two landscape components considered—permafrost and vegetation.

PERMAFROST PATTERNS

The permafrost area of Eurasia occupies about 15 million km². In addition to regions of permafrost in high mountain systems in inner Asia, the major area of permafrost is at high and middle latitudes, where it covers practically the whole area of Arctic desert, as well as tundra, forest-tundra, and a large part of the boreal forest (taiga). In the plains region north of 65–67°N, a subzone of continuous permafrost is found, with ground temperatures at the depth of zero annual variation (10–15 m) of −2°C and lower. The thickness of this permafrost varies from several hundred meters to greater than 1000 meters. The depth of the active layer of freezing and thawing usually does not exceed 1–1.4 m. The ice content of the near-surface horizons of frozen ground down to a depth of 20–30 m is occasionally as high as 50–90%. Subzones of discontinuous and sporadic permafrost are located farther south, where their areal extent and thickness are reduced substantially. The temperature of frozen ground is −2 to 0°C, and the thickness of the active layer is on the order of 1.5–2.0 m or more. The ice content of the near-surface frozen ground is less than that in regions farther to the north (with the exception of organic deposits).

Permafrost is highly sensitive to climatic change, and the prediction of changes in permafrost area and thickness resulting from a regional warming is of immediate scientific and practical interest. The Arctic and sub-Arctic zones will be subjected to the greatest warming, and the distribution of permafrost will be significantly affected. This conclusion is based on both paleoclimatic reconstructions and climate models, as well as on paleoreconstructions of the distribution of permafrost and its projected response to warming.

Reconstruction of Equilibrium and Nonequilibrium States of Permafrost

The reconstruction of the distribution of permafrost in the geologic past, as well as a number of its features, is possible using three major approaches. The first, the paleo-cryological approach, is based on the study of ancient structures in the landscape and deposits that are indicative of former frozen ground. The specific features studied include fossil wedgelike structures, involution and solifluction features, and relict cryogenic morphosculpture. It is also important to have some understanding of the depth of distribution of temperature in the permafrost layer and its trend with time, because the temperature of the frozen ground at depth is not necessarily in equilibrium with the modern climate.

The fact that the distribution of permafrost is governed primarily by climatic variables provides an excellent opportunity to apply paloeoclimatic methods, based on the use of data from climatic reconstructions, to obtain data on changes in the distribution of permafrost in the past. The intervals studied include past periods characterized by global warming of 1 and 2°C (Velichko and Nechayev 1988, 1992).

The time intervals under study in this analysis are on the order of several hundred to a thousand years. Thus, the reconstructions are of a situation in which permafrost properties were in a state of quasi-equilibrium with temperature and moisture regimes. By comparing maps of mean temperature for January and July, data on the amplitude of mean annual air temperature variations were obtained for the periods of the Holocene climatic optimum and the Mikulino interglacial for North Eurasia. We also used maps of mean annual air temperature as given by Frenzel et al. (1992). Because reconstructions of the thickness of snow cover have not as yet been made, snow cover thickness was approximated by assuming that it is directly proportional to the mean annual precipitation rate. It is also possible from the available data to calculate approximate ground temperatures at the depth of zero annual variation. Calculations were made using the equation

$$t = T + (A/2)K \tag{7.1}$$

where t is ground temperature, T is mean annual air temperature, A is annual air temperature amplitude, and K is a coefficient dependent on the thickness of snow cover (Shvetsov and Dostovalov 1959).

On the basis of the data obtained from the analysis of ground temperatures, we may derive the distribution of surface permafrost over large regions for various periods of time. Comparison of ground temperatures for past periods with recent values allows us to estimate approximately the changes in thickness of the layer of seasonal freezing and thawing (the active layer). An analysis of recent data shows that the thickness of the active layer will increase by about 0.15 m (and in subzones of discontinuous and sporadic permafrost by about 0.4 m) for an increase in ground temperature of 1°C. In general, these reconstructions of the depth of the active layer apply to sandy loam and loamy soils under typical hydrologic conditions. In depressions with thicker snow cover and also in course-grain deposits, the incremental increase in the thickness of the active layer could be larger.

Thus, equilibrium models are very important in the characterization of the distribution of permafrost during past periods of global climatic warming. However, for prediction purposes they may be used only as an indicator of potential change, because human-induced global warming will take place rapidly compared to geologic time. Thus, it is necessary to use both empirical and model data to develop nonequilibrium scenarios of change in the permafrost zone during the coming decade of potentially rapid global warming.

Analysis of observational and model-produced data for the northern part of West Siberia, where air and ground temperatures have been studied (Belopukhova et al. 1976), showed that during the period of warming in the 1930s and 1940s and subsequent cooling, variations in subsurface ground temperature amounted to about one-half of the variations in mean annual air temperature. A trend of increasing temperatures was shown to be correlated with an increase in the thickness of snow cover and vice versa. The length of the periods of warming and cooling was on the order of a dozen or so years, a period comparable to the interval of time appropriate to future climatic change.

Based on measured climatic changes during the historical period, it appears that if the climate warms in northern regions, a continuous rise in air and ground temperature and precipitation should occur, including an increase in snow cover. It may also be concluded from the nonequilibrium model that, with an increase in the thickness of snow cover, the increase in the temperature of frozen ground will be about 50% of the change in mean annual air temperature.

Now let us analyze the results of studies of the permafrost zone (cryolithozone) of Eurasia based on the approaches described.

Equilibrium Model of the State of Permafrost for a 1°C Temperature Increase

This model describes the state of the cryolithozone during the Holocene climatic optimum for North Eurasia. The area of permafrost was reduced in size in comparison to its current extent; its entire southern boundary retreated northward (Figure 7.5A). The most significant changes occurred on the Atlantic side of Eurasia and in the southern part of East Siberia. In the northern part of East Siberia and to the west of the lower reaches of the Pectoria River, permafrost completely disappeared from the surface. In the far northeastern part of Europe and in the northern part of West Siberia, the temperatures of frozen ground were elevated 3–4°C in comparison with modern values. The southern boundary of permafrost in western Siberia moved northward by 300 km; the area of continuous permafrost in this region was reduced by nearly 50%. The boundary of continuous permafrost in the northern part of Middle Siberia was displaced to the north by 100–150 km. The southern boundary of permafrost in the Angara River basin and in the southern Trans-Baikal region also moved northward by several hundred kilometers. In northeastern Asia permafrost deposits were preserved with frozen ground temperatures of –7°C.

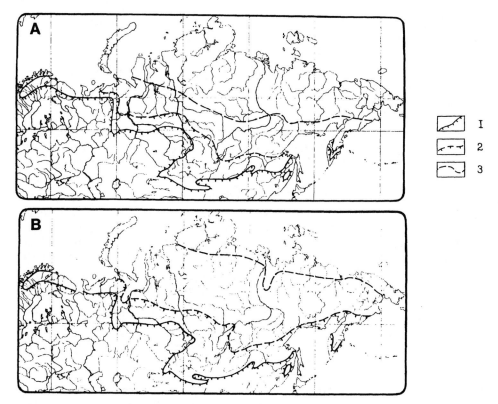

Figure 7.5. Permafrost response to global warming (equilibrium model). (A) mean global temperature increase up to +1°C. (B) mean global temperature increase up to +2°C. Key: 1, recent boundary of the permafrost zone; 2, permafrost boundary for the period of maximum warming (reconstruction); 3, continuous permafrost boundary for the period of maximum warming (reconstruction).

Equilibrium Model of the State of Permafrost for a 2°C Temperature Increase

This model describes the state of the cryolithozone at the climatic optimum of the last interglacial stage 125,000 years B.P. The increase in temperature and precipitation at high latitudes during this period of time caused a change in the distribution of permafrost over the whole region of northern Eurasia (Figure 7.5B). The southern boundary of an area of continuous permafrost with ground temperatures of –2 to –3°C retreated to the far northern rim of the continent. Continuous permafrost conditions existed only in the lower reaches of the Lena and Kolyma rivers and probably in the northern parts of the Taimyr and Chukotka regions. However, even here increases in frozen ground temperatures amounted to 5–7°C in comparison with modern values. A larger part of Middle and East Siberia than today was found within the zone of discontinuous and sporadic permafrost development. In northern West Siberia, frozen ground temperatures did not dip below –1 to –2°C. The southern boundary of permafrost in West Siberia was approximately at the latitude of the Arctic Circle. In the extreme northeastern part of Europe, only isolated regions of permafrost existed, with ground temperatures of –0.5°C.

In this region ice veins ceased to form, and thermokarst development was an active process. In areas farther to the south, the horizon of seasonal thawing was transformed into a zone of seasonal freezing.

Nonequilibrium Models of the State of Permafrost for a +1 to +2°C Temperature Change

These models describe the state of the cryolithozone given conditions that should have some relevance for the predicted global warming of the 21st century. For a warming of 1–2°C, in the southern part of the permafrost zone within a band several hundred kilometers wide, a gradual increase in ground temperatures to positive values and a reduction of permafrost area and thickness are predicted. Thus, the zone of sporadic permafrost development will shift northward to the modern region of discontinuous permafrost. However, the area of buried relict permafrost layers, for example in the Pechora and Ob river basins, will increase. The change in permafrost distribution in the more northern regions of Eurasia is especially important. For a global warming of 1°C, the change in frozen ground temperatures in the north, more noticeable in the western sector, will be on the order of +1–2°C. The increase in thickness of the soil layer undergoing seasonal thawing under typical hydrologic conditions and for sandy loam to loamy soils will be about 20–30 cm. With a global warming of +2°C, the increase in the temperature of frozen ground in the northern part of the continuous permafrost zone, including its eastern portion, may reach 3–4°C. The thickness of the active layer within this region of permafrost is predicted to increase by 40–50 cm.

These increases in active layer thickness and temperature in northern Eurasia will, at the turn of the 20th century and into the first third of the 21st century, initiate the processes of solifluction, thermokarst development, and thermoerosion. This situation may lead to deterioration of forests in the permafrost zone owing to swamping.

These models of permafrost zone dynamics should be considered preliminary. Further research must include development of models assuming greater increase in the CO_2 content of the atmosphere and a higher temperature, as expected in the second half of the 21st century. Nonequilibrium models of global warming of 1–2°C should also be constructed that take into account cryolithozone processes for the past 15–20 years of warming of the global climate.

VEGETATION PATTERNS

Equilibrium Models of Vegetation Conditions

Comparison of the vegetation map of the Mikulino interglacial optimum (Grichuk 1984, 1992) with that of the present (Lukicheva and Sochava 1964) shows the reconstruction of vegetation cover in North Eurasia to be radically different from that of today. The difference lies in the distribution of temperature and precipitation during the Mikulino interglacial, when global mean temperature was elevated by 2°C. The main changes in vegetation between the Mikulino interglacial and today involve

1. A shift in zonal boundaries (change in type of vegetation).
2. A shift in provincial boundaries (change in vegetation biomes).

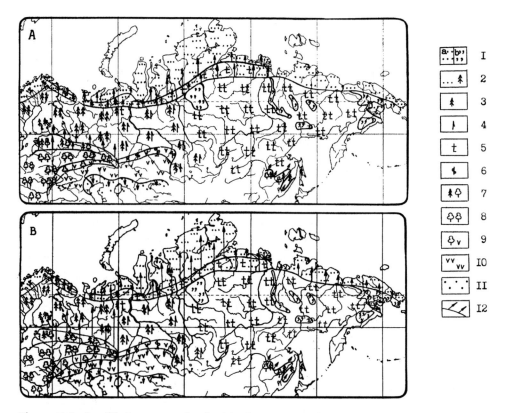

Figure 7.6. Equilibrium scenarios for Northern Hemisphere mean annual temperature deviations. (A) +1°C (Holocene optimum). (B) +2°C (last interglacial optimum). Key: 1, Tundra (a, on plains; b, on mountains). 2, forest-tundra; forest, main trees; 3, *Picea*; 4, *Abies*; 5, *Larix*; 6, *Betula*; 7, coniferous-broadleaf forest; 8, broadleaf forest; 9, forest-steppe; 10, steppe; 11, desert, semidesert; 12, shift of zonal boundaries compared to present-day configurations.

3. A change in vegetation within the framework of the existing biome (e.g., enrichment in floral composition, increased complexity of phytocoenose structure).

Because of these changes, in the European part of Russia, the forest boundary during the Mikulino interglacial was displaced 200–300 km northward of its present position. Forests extended as far north as the coast of the Arctic Ocean. In the Asian part of Russia, tundra was found at this time, but its area was greatly diminished in size: in the southern part of Asian Russia, tundra communities were replaced by forest-tundra and taiga forests. As a result, the northern boundary of forest was 500–600 km north of its present location (Figure 7.6B).

During the Mikulino interglacial, the area of broadleaf forests was greater than it is today; the northern boundary of the forest was found 600–700 km north of its present location. Climatic conditions were more favorable for arboreal species requiring higher temperatures and more moisture than are available in the region today and that are currently found in regions with an oceanic climate. Hornbeam, which now only occurs

west of the Dnieper River, occurred over a large part of the broadleaf forest zone to the
Volga River. In western Siberia the climate was suitable for the growth of oak, elm, and
lime trees. In the present-day larch taiga region of Middle Siberia, dark coniferous
species were abundant.

Warming during the Mikulino interglacial also led to an increase in the competitive
ability of plant species that required specific environmental conditions. As a result, even
where no immigrations of new arboreal species took place, the vegetation did not remain
unchanged: the role of thermophilous species in plant communities increased. In the
north of the coniferous forest area in European Russia and West Siberia, the proportion
of Siberian fir trees relative to larch increased. In coniferous-broadleaf forest, the
proportion of Siberian spruce decreased, and it actually disappeared from the forest
community in the southwestern part of the region. Broadleaf species were on the whole
more dominant than they are today. However, in several regions of North Eurasia, the
plant communities during the Mikulino interglacial were similar to those of today. For
example, the distribution of taiga in the northern half of West and Middle Siberia during
the interglacial was similar to that of the present day. Judging from the map of the
distribution of vegetation for the Holocene climatic optimum (Khotinsky 1984),
changes in the vegetation of North Eurasia were similar to those already described for
the Mikulino interglacial (Figure 7.6A).

These are the main structural changes in vegetation that one would expect for a
quasi-equilibrium state under warming climatic conditions. Before we discuss the results of the model of a nonequilibrium state of vegetation cover, it is necessary to
consider factors limiting the succession of plant communities.

Factors Limiting the Succession of Vegetation Communities

One of the most important factors limiting succession is the rate of migration of
arboreal species. The rate depends directly on the distance of spreading of seeds and
fruit and inversely on the age of the first fruit-bearing plants. These factors vary for
plants of the same species depending on habitat conditions (Table 7.1). Literature
data on the rate of dispersal of arboreal species are highly variable. Thus, according
to Udra (1988), the rate of dispersal of pioneer species with small, easily transported
seeds is about 100 m/year; the rate for broadleaf species is an order of magnitude
less. Other researchers, however, give higher rates; for example, for *Quercus robur*
in the Trans-Volga region, the rate of dispersal is 150 m/year at the boundary of its
area of growth (Denisov 1980). These quantitative dispersal estimates are valuable
because they offer initial conditions of the changes of vegetation cover. The estimates, however, are for a time when no drastic changes in climate, comparable to
those now anticipated, occurred. They characterize quasi-equilibrium states of vegetation with respect to climate.

Palynological data combined with radiocarbon measurements for the first half of the
Holocene may provide constraints on the rates of migration of arboreal species in a
nonequilibrium situation. A summary of a large volume of palynological data for
western Europe by Huntley and Birks (1983) showed that although migration rates of
the same species could vary, they still exceeded their rates for present conditions
(Table 7.1). Similar data were obtained for North America.

Table 7.1. Distance of Fruit and Seed Transportation, Beginning of the Age of Fruiting, and Rate of Dispersion of Arboreal Species

Arboreal species	Distance of transportation of fruit and seeds (km)		Age of beginning of fruiting (years)		Dispersion rate (m/year)	
	Observed (Udra 1988)	During storms (Muller 1955)	In the open	In the forest	Observed (Udra 1988)	Palynological data (Huntley and Birks 1983) with additions
Betula sect. *albae*	1–2	1.6	10	10–15*; 20	100	1000
Populus tremuloides	1.5–2	30	15	10–15*; 25	100 (133)	—
Pinus sylvestris	0.8	≤10†	2	5–8*; 15	15–25*; 30	60
Larix sibirica	0.25–0.6	—	20	15–20*; 30	30	65–80
Picea abies ssp. *excelsa*	0.35–0.5	—	15–25*	25–30*; 50	10–25	80–500
Quercus robur	0.5	—	20	50	10; 150‡	300
Tilia cordata	0.20–0.35	—	10–12*; 30	20–25*; 40	5–10	50–500
Fraxinus excelsior	0.2–0.4	0.5	—	40	5–10	200–500
Acer platanoides	0.17–0.25	4	—	10–15*; 30–50	5–8	500–1000[a]
Carpinus betulus	0.15	—	—	30	5	200–700

[a]Refers to genus *Acer* and not to *Acer platanoides*.
Note: Data are from Udra (1988) except for entries marked * (Bogdanov 1974), † (Pogrebnyak 1968), and ‡ (Denisov 1980).
*Estimated by authors.

Palynological data for a series of stratigraphic sections of Holocene deposits in the northwestern European part of Russia that include radiocarbon dates allow us to determine the rate of migration of the lime tree to the limit of its boundaries of areal extent at the Holocene climatic optimum. At the southern part of its extent, the rate of lime tree migration was about 210 m/year; farther northward the rate decreased gradually to 110 m/year. The migration values given are based on palynological data, and therefore they may not be representative of actual migration rates because of the possibility of long-distance atmospheric transport of pollen. To check this possibility, migration rates were obtained for larch by ^{14}C dating of wood, stubs, and trunks discovered in situ on the Taimyr Peninsula (Nikolskaya and Cherkasova 1982). It appears that even in this locality, representing the limit of existence of arboreal species, the migration rate of larch 6000–8000 years B.P. was 65–80 m/year. This value is close to that obtained from the palynological data.

The high rates of arboreal species migration in the Holocene compared with present rates are probably due to the fact that climatic conditions in the Holocene changed in a direction favorable for immigrants and decreased the competitive ability of the less thermophilic local species. The availability of large regions with relatively homogeneous ecological conditions suitable for these species contributed to the high migration rates. Furthermore, the homogeneous conditions led to the development of numerous animal populations capable of transporting fruit and seeds over long distances.

An important consideration in the discussion of vegetation reconstruction under global warming is the interrelationships of species of local flora under the new climatic conditions, as well as competition between the local flora and new arrivals. Several possible scenarios can lead to changes in vegetation, including the following:

1. One of the dominants of the plant community cannot exist under the new climatic conditions. Its place is taken by the more thermophilic members of the same community.
2. The single dominant plant of the community cannot exist under the new conditions. Its place is taken by a rapidly propagating pioneer species, such as a species of birch or aspen.
3. New climatic conditions are favorable for the local vegetation, but the dominant tree species are unable to compete with the immigrants. New plant communities will arise in such cases, but it will take a long time.

Nonequilibrium Models: Potential versus Expected Changes

The conclusions discussed above concerning the rates of migration of trees and their competitive interrelationships were used to develop nonequilibrium models as a means of assessing the potential versus the expectation for changes in vegetation caused by climatic change. Scenarios were developed for selected geographic regions in North Eurasia.

In the Yenisei region of West Siberia, larch forests are widely distributed today. North of 66°30′N the forest-tundra biome is widespread; its northern boundary is at 71°N. Closed forests in this region consist of larch growing at a mean July temperature not below +14°C, whereas larch forest-tundra is found in areas where the temperature is not lower than +8°C (Figure 7.7). With a mean global temperature increase of 1°C in northern West Siberia, mean January temperatures will exceed present-day values by 3°C, and mean July temperatures will exceed current values by 4°C. With an increase in the mean global temperature of 2°C, the positive deviations in January and July temperatures will be 10–12°C and 8°C, respectively. For an equilibrium model of vegetation change, we find that in the first case the northern boundary of continuous larch forest will move to 70°N, and the boundary of the forest-tundra will migrate northward to 73°N (Figure 7.7). Still greater vegetation shifts would be expected for an equilibrium model scenario with a global warming of 2°C. In the West Siberia region, the forest-tundra could reach the coast of the Arctic Ocean.

However, the rate of larch migration in this region in the Holocene did not exceed 80 m/year. If, under conditions of global warming, this species migrated at this rate for 10 years, the northern boundary of larch and the boundaries between forest-tundra and tundra would be displaced only 800 m. It would take 35 years for a displacement of 3 km to occur. It is clear that such shifts will not lead to major changes in the position of the boundaries of vegetation biomes.

South of 70°N in the Yenisei region of Siberia, in addition to larch, the spruce *Picea abies ssp. obovata* is also a forest-forming species. Judging by the mean rate of spruce migration during the first part of the Holocene, a shift in its area boundary could amount to 2.5 km in 10 years and 8.5 km in 35 years. If we consider the maximum rate of spruce migration in the Holocene, these estimates could be doubled. In any case we should not expect to find *P. abies ssp. obovata* in the northernmost forests of the Yenisei region of Siberia. In a region where both spruce and larch grow together, i.e., south of 70°N, changes in the relative proportions of these species in forest communities are possible. The larch *Larix sibirica* can grow in areas with a thin active layer of permafrost. *P. abies*

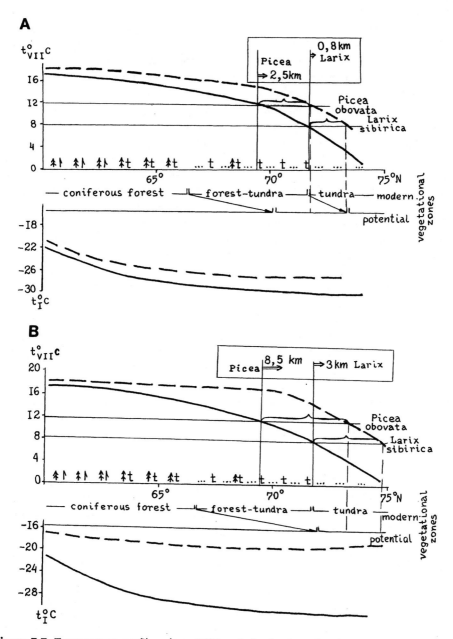

Figure 7.7. Temperature profiles along 85°E and distribution of main tree species in North Siberia (A) under global warming of +0.8°–+1°C and (B) under global warming of +1.8°–+2°C. Key: 1 (solid line), Present-day temperature; 2 (dash line), reconstructed temperature.

ssp. obovata grows in areas where the extent of seasonal ground thawing is deeper. Global warming will result in an increase in the thickness of the seasonally thawed layer and consequently in an increase in areas favorable for spruce growth.

In the forest-tundra biome, warming may lead to a change from open forests of larch to closed forests of this species because seeds of this species are abundant and warmer temperatures will lead to successful development of young plants. These processes will proceed gradually as the old trees die off. As a whole variations in the composition and migration of vegetation communities will proceed more slowly than the climatic changes causing them. Thus, the vegetation will be in a nonequilibrium state with respect to the rate of climatic change. This statement is consistent with the data on vegetation change in northeastern Europe (Figure 7.8). In this region coniferous-broadleaf forests are replaced by coniferous forests to the north, and even farther to the north they give way to forest-tundra and tundra. The northern boundary of coniferous-broadleaf forest is defined by the limit of *Quercus robur,* and the northern boundary of coniferous forest by the limit of *Larix sukaczewii. P. abies ssp. obovata,* a leading component of coniferous forests, forms open forests in the northern part of its area of distribution. In the northern part of coniferous forests and in forest-tundra, birch will become an important member of nondisturbed communities.

The temperature values that limit the existence of the trees in northeastern Europe are shown in Figure 7.8. In a quasi-equilibrium state under conditions of a global warming of 2°C, coniferous-broadleaf forests could potentially migrate approximately 550 km northward. Coniferous forests could migrate 220 km. Obviously, with a global warming of only 1°C, biome boundary displacements would not be as significant, but even in this case coniferous forests could reach the coast of the Arctic Ocean. However, if we consider the migration data for the above species in the Holocene, biome boundary shifts would be predicted to be no more than 10 km. Only birch, a pioneer species, could potentially migrate 35 km in 35 years.

In coniferous-broadleaf forests, global warming could lead to an increase in the competitive ability of oak and other broadleaf species, and their numbers could increase. Plants like spruce, which are adapted to cold, will be less able to combat diseases and pests. This situation will probably lead to a reduction of their role in plant communities. In the place of dead trees in open areas, birch and aspen will be dispersed in addition to broadleaf species, and sometimes even ahead of them. However, in the first several decades of a global warming, these processes will be reflected only in the vegetation composition of regrowth.

For the scenario of a nonequilibrium state of vegetation and climate, it can be concluded that, given global warming, the process of vegetational changes will differ measurably in different regions of North Eurasia. In some areas the changes will not lead to noticeable changes in existing vegetation; in other areas, however, these processes will cause a change in vegetation biomes (Figure 7.9).

For the main plant biomes, the expected changes owing to global warming are as follows:

1. *Tundra biome.* For a global warming and an increase in precipitation, the thickness of the seasonally thawed layer will increase. This change will lead to greater habitat differentiation based on the moisture content of the soil. In elevated regions desiccation would be possible because of higher summer temperatures and

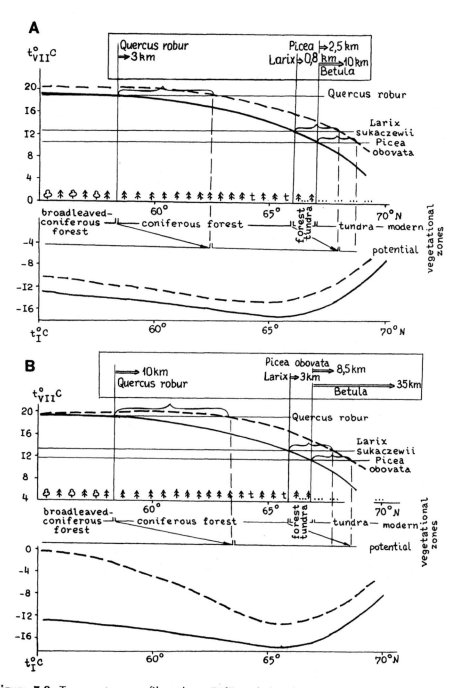

Figure 7.8. Temperature profiles along 50°E and distribution of main tree species in Northeast Europe (A) under global warming +0.8° – +1°C and (B) under global warming +1.8° – +2°C.

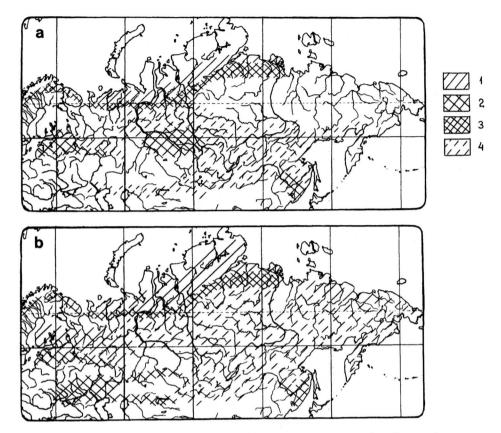

Figure 7.9. Vegetation will be in a nonequilibrium state owing to the climate change: a, mean global warming +1°C; b, mean global warming +2°C. 1, regions with slow changes showing no effects for several decades; 2, change of dominant abundance in zonal vegetation types; 3, tendency toward change of zonal vegetation type; 4, vegetation under a strong indirect influence of climatic changes (e.g., thawing of permafrost).

an increase in evaporation. Certain mesophytes and thermophilic species of plants (Gramineae, dicotyledonous perennial plants) may also increase in abundance. As a result it is likely that areas of moss and lichen will decrease, and areas of shrub and meadow communities will increase.

2. *Forest-tundra biome*. The variations in vegetation of the forest-tundra biome in the Asian part of Russia will be caused primarily by increases in the abundance of larch. Open forests containing this species will be replaced by closed forests. It is not likely that this replacement will occur quickly, because the development of young plants is possible only in areas free of the crowns and root systems of adult trees. The introduction of new individuals will take place only as old trees die off. Solifluction, an outcome of warming, will help to accelerate this process.

3. *Coniferous forest biome*. In the northern part of the forest biome of Siberia, the major factor causing changes in vegetation habitat conditions, aside from a temperature increase, will be an increase in the thickness of the seasonally thawed

layer. In combination with an increase in evaporation, this situation will lead to a reduction in the degree of swamping.

The composition of the dominant species in coniferous forests will change because of an increase in the numbers of thermophilic species; for example, spruce will replace larch in the northern part of the coniferous biome and *Abies sibirica* and *Pinus sibirica* will become widespread in other regions of Siberia. In areas disturbed as a result of solifluction, pioneer species will propagate.

In the middle latitudes of Middle Siberia, the temperature regime will be unfavorable for extensive growth of the larch *Larix dahurica*. This larch is widespread today in regions of extreme continental climate with mean January temperatures of –25° to –40°C and with mean July temperatures of 10–15°C (Bykov 1960). With a rise in the mean global temperature of 2°C, the mean January and July temperatures will be –24°C and 18°C, respectively, in this region. Under these conditions, we would anticipate a gradual disappearance of larch and its replacement by pine and birch because of the availability of local seed material. The more competitive dominants, *A. sibirica* and *P. abies ssp. obovata,* will not be dispersed into the region because of lack of time, although conditions for them could be favorable. In the central part of the Russian Plain and in the middle portion of the West Siberian lowland, for a global warming of 1°C, there will be a decrease in the amount of annual precipitation (Figure 7.2). It is likely that this change will affect the composition of the understory vegetation rather than the arboreal overstory of the forest communities. For a further rise in global temperature, it is likely that annual precipitation in these regions will increase.

4. *Coniferous-broadleaf forest biome.* Because of a rise in winter temperatures, the growth of European spruce will be slowed. In the southern part of these forests, for an increase in the mean global temperature of 2°C, mean January temperature will increase to 0°C. Because of this temperature change, the loss of spruce as a species within this biome is possible. These changes initially will probably affect only young trees, because our experience shows that the replacement of coniferous-broadleaf forest with principally broadleaf ones will take more than 50–100 years. Within this time period, recent seedlings and those trees that develop from their seeds will reach their maximum possible age.

In broadleaf forest, the role of late blossoming types of *Q. robur* in the forest will probably be diminished. It is likely that pioneer species and anthropophytes will increase in number.

5. *Steppe biome.* An increase in precipitation for a small increase in winter temperatures and summer temperatures as compared to those of today or lower in the southern steppe are predicted. This change will lead to an increase in the number of mesophilic components of the steppe flora because of the local availability of seeds. Afforestation of the steppe may increase owing to the presence of seeds of local origin. In spite of the relatively high rate of migration of herbaceous plants in comparison with that of arboreal species, we do not expect any noticeable shift in the areal boundaries of this biome.

6. *Desert and semidesert biome.* An increase in annual precipitation and a slight decrease in temperature are predicted for this biome. Because of these changes,

local mesophilic species of plants will increase in importance. It is likely that salinization will increase, and a longer and more continuous growing season of ephemeroids is possible.

DISCUSSION AND CONCLUSIONS

The results of these studies show that paleoclimatic reconstructions can provide a realistic assessment of the response of the landscape components of permafrost and vegetation to rapid climatic change. The predicted temperature changes derived from the paleoclimatic analysis are similar to those obtained using climate models. This similarity provides some support for the trends in vegetation and permafrost owing to a global warming presented here. However, it should be recognized that there are discrepancies between the climate change results obtained from paleoclimatic analysis and those derived from climate models.

The changes in landscape components obtained from paleoclimatic reconstructions apply to certain time intervals and are not synoptic analyses of climatic variables and vegetation change. The results are for an averaged condition. Thus, if in a certain region paleoclimatic data indicate a reduction in precipitation, this result should be interpreted as an increase in drought frequency in the region of interest. Such averaged climatic scenarios may also be obtained with the help of climate models. Because of the discrepancy between modeling and paleoclimatic results, and also to provide some control on results, the responses of landscape components to climatic change should in the future be investigated using both climate models and paleoclimatic reconstructions.

Paleoclimatic reconstructions provide important clues about the complex process of climatic change and the response of landscape components to that change. This is an invaluable feature of the approach. It allows us to foresee the state of the natural system in its initial stages (nonequilibrium models) of change and in the more distant future (equilibrium models). From our analysis of the two landscape components of permafrost and vegetation for North Eurasia, several conclusions can be drawn:

1. In spite of the fact that, given an enhanced greenhouse effect, the postulated climatic changes will be rapid during the coming decades, changes in the distribution of permafrost and vegetation will not be of a catastrophic nature. In the early stages, these changes will be gradual.
2. The present boundaries of permafrost will not change greatly owing to a global warming, but the depth of the layer of active freezing and thawing will increase. Especially important is the conclusion that the greatest increase in the thickness of the active layer will take place at high latitudes, where ecosystems will be extremely sensitive to climatic change. This result will lead to an increase in swamps of some regions and the enlargement of areas of unstable ground. In addition to having serious engineering and technical consequences, these changes will significantly influence the state of the vegetation and will probably lead to deterioration of wood quality in certain biotopes. These phenomena are of special interest to forestry specialists. With climatic change the tendency toward replacement of some dominants in the vegetation by others, or toward an increase in the abundance of one or more dominants, should be taken into account when afforestation plans are developed.

3. Finally, we showed that in some regions the new climatic conditions will be favorable for the existence of trees that are not present in these regions today. In particular, in the present tundra region the climatic changes predicted for the forthcoming decades favor the introduction and establishment of some arboreal plants.

These conclusions apply to possible climatic changes owing to an enhanced greenhouse effect for several decades into the future and the response of permafrost and vegetation to the global warming in the regions of eastern Europe and Siberia. Scenarios of this kind do not necessarily characterize actual situations. They should be interpreted rather as an indication of the tendency of change, that is, the direction and range in which climatic factors and landscape components may change within this region of study. Such scenarios are inherently probabilistic in nature.

REFERENCES

Belopukhova, Ye. B., I. P. Novikov, and A. L. Chekhovskiy. 1976. Short-period climatic oscillations and their influence upon ground temperatures in Northern West Siberia (in Russian). In *Inzhenerno-geologicheskiye i geokriologicheskiye issledovaniya Zapadnoi Sibiri*, Trudy PNIIIS, ed. G. I. Dubikov and M. M. Koreisha, pp. 32–47. Stroyizdat, Moscow.

Bogdanov, P. L. 1974. *Dendrologia* (in Russian). Nauka, Moscow.

Bykov, B. A. 1960. *Dominanty rastitelnogo pokrova Sovetskogo Soyuza* (in Russian), Vol. 1. Izdatelstvo Akademii Nauk Kazakhskoy SSR, Alma-Ata.

Denisov, A. K. 1980. Recent dynamics of the northern line of the area of English oak in the USSR (in Russian). *Lesovedeniye* 1:3–11.

Frenzel, B., M. Pecsi, and A. A. Velichko, eds. 1992. *Atlas of Paleoclimates and Paleoenvironments of Northern Hemisphere. Late Pleistocene–Holocene*. Geographical Research Institute of HAS, Budapest; Gustav Fischer Verlag, Stuttgart, Germany.

Grichuk, V. P. 1969. Experience of reconstruction of some climatic elements of Northern hemisphere in the Atlantic period of Holocene (in Russian). In *Golotsen*, ed. M. I. Neishtadt, pp. 41–57. Nauka, Moscow.

Grichuk, V. P. 1984. Late Pleistocene vegetation history. In *Late Quaternary Environments of the Soviet Union*, ed. A. A. Velichko, ed. of English-language edition H. E. Wright, Jr., and C. W. Barnosky, pp. 155–178. University of Minnesota Press, Minneapolis.

Grichuk, V. P. 1985. Reconstruction of the scalar climatic indices with the floristic data and assessment of its accuracy (in Russian). In *Metody rekonstruktsii paleoklimatov*, ed. A. A. Velichko, L. R. Serebryanny, and Ye. Ye. Gurtovaya, pp. 20–28. Nauka, Moscow.

Grichuk, V. P. 1992. Last Interglacial (about 120,000 yr B.P.). Vegetation (map). In *Paleoenvironments of the Northern Hemisphere. Late Pleistocene–Holocene*, ed. B. Frenzel, M. Pecsi, and A. A. Velichko, p. 11. Geographical Research Institute of HAS, Budapest; Gustav Fischer Verlag, Stuttgart, Germany.

Huntley, B., and H. I. B. Birks. 1983. *An Atlas of Past and Present Pollen Maps for Europe: 0–13,000 Years Ago*. Cambridge University Press, Cambridge.

Khotinsky, N. A. 1984. Holocene vegetation history. In *Late Quaternary Environments of the Soviet Union*, ed. A. A. Velichko, ed. of English-language edition H. E. Wright, Jr., and C. W. Barnosky, pp. 179–200. University of Minnesota Press, Minneapolis.

Klimanov, V. A. 1976. On methods of reconstruction of the past climatic indices (in Russian). *Vestn. Mosk. Univ. Geogr.* 2:92–98.

Klimanov, V. A. 1984. Paleoclimatic reconstructions based on the information statistical method. In *Late Quaternary Environments of the Soviet Union,* ed. A. A. Velichko, ed. of English-language edition H. E. Wright, Jr., and C. W. Barnosky, pp. 297–303. University of Minnesota Press, Minneapolis.

Lukicheva, A. N., and V. B. Sochava. 1964. Maps of the USSR. Vegetation (in Russian). In *Fiziko-geograficheskiy Atlas Mira,* ed. I. P. Gerasimov, pp. 240–241. AN SSSR i Glavnoye Upravleniye geodezii i kartografii GGK SSSR, Moscow.

Muller, P. 1955. Verbreitungsbiologie der Blutenpflanzen. *VerÖff. Geobot. Inst. Eidg. Tech. Hochsch. Stift. Ruebel Zuerich* 30.

Nikolskaya, M. V., and M. N. Cherkasova. 1982. Holocene floras dynamics at Taimyr (pale-ophytological and geochronological data) (in Russian). In *Razvitiye prirody territorii SSSR v pozdnem pleistotsene i golotsene,* ed. A. A. Velichko, I. I. Spasskaya, and N. A. Khotin-skiy, pp. 192–201. Nauka, Moscow.

Pogrebnyak, P. S. 1968. Obshcheye lesovodstvo (in Russian). Kolos, Moscow.

Shvetsov, P. R., and B. N. Dostovalov, eds. 1959. *Osnovy Geokriologii (Merzlotovedenia), Chast 1: Obshchaya Geokriologia* (in Russian). Izdatelstvo AN SSSR, Moscow.

Udra, I. F. 1988. *Rasseleniye rasteniy i voprosy paleobiologii* (in Russian). Naukova Dumka, Kiev.

Velichko, A. A. 1991. Global climate change and environment response (in Russian). *Izv. Akad. Nauk SSSR ser. geogr.* 5:5–22.

Velichko, A. A., and V. P. Nechayev. 1988. Empirical paleoclimatology and historical geocryology: An example of cryolithozone reconstruction for Mikulino (kazantsevo) Interglaciation (in Russian). In *Stratigrafia i correlatsia chetvertichnykh otlozheniy Azii i Tikhookeanskogo regiona.* Abstracts of symposium, Vol. 2, pp. 19–21. DVO AN SSSR, Vladivostok.

Velichko, A. A., and V. P. Nechayev. 1992. Estimation of permafrost zone dynamics in Northern Eurasia under the global warming (in Russian). *Dokl. Akad. Nauk* 324(3):667–671.

Velichko, A. A., L. R. Serebryanny, and Ye. Ye. Gurtovaya, eds. 1985. *Metody rekonstruktsii paleoklimatov* (in Russian). Nauka, Moscow.

B

Soils

Approximately three-quarters of the biotically controlled carbon stored on land is in the organic matter of soils. The pool of carbon in soils is about twice the amount in the atmosphere and is large enough that changes in it can affect the composition of the atmosphere significantly. The global distribution of carbon in soils is at once symmetrical and anomalous. Tropical soils in general contain little humus, although root systems may penetrate deeply and the total mass of roots may be high. Humus in soils and its C:N ratio increase as temperatures drop as we move into higher latitudes. Soils of the boreal forest are commonly podzolic, characterized by a deep surficial organic layer. Deep peat accumulations are common in tundra and boreal zones. There are, however, acid podzols in the tropics. The black, acid waters of the Rio Negro drain from the extensive tropical podzols of the northern Amazon Basin.

A rapid, continuous warming can be expected to extend the more common circumstance of tropical forests into higher latitudes, causing a poleward migration of the systems currently found at low latitudes. Such a migration can be expected to transfer carbon from soils into the atmosphere as carbon dioxide and methane as the organic matter of soils decays and the soils assume the characteristics of soils now found at lower latitudes. The rates are critical. If the transitions are measured in tenths of a degree centigrade per century, the effects on the atmosphere will probably be small. If the transitions are more rapid, on the order of tenths of a degree centigrade per decade, a substantial transfer of carbon to the atmosphere can be anticipated, and the feedback will be appreciable.

The chapters that follow show that the analysis is far more complex than the preceding statements suggest. Effects will in fact probably prove still more awkward, and perhaps surprising, than even these probing scholars anticipate. Substantial shifts in the ratios of carbon to major nutrient elements, especially nitrogen, can be anticipated with major changes in both the productivity of natural communities and the species they support. The potential exists for both positive and negative biotic feedbacks in these transitions, as shown so clearly by these authors, but the weight of evidence at present suggests that the anticipated rapid warming will favor additional releases of carbon from soils over large areas, especially in the higher latitudes.

8

Soil Respiration and Changes in Soil Carbon Stocks

WILLIAM H. SCHLESINGER

The pool of carbon in soils is one of the largest near-surface stores of carbon on earth. Although estimates vary widely, most workers now agree that between 1400 and 1500 petagrams (Pg) (= 10^{15} g)of carbon are contained in soil organic matter, which includes undecomposed litter on the soil surface and humic materials dispersed throughout the soil profile (Schlesinger 1977; Post et al. 1982). A large portion of this pool is held in the organic soils of tundra and boreal forest ecosystems, in which dramatic changes in climate are expected during the next century.

In addition to organic forms of carbon, soils contain about 800 Pg carbon in inorganic forms, largely as $CaCO_3$, in arid and semiarid regions, where it is deposited in pedogenic calcites, known variously as caliche or soil calcic horizons (Schlesinger 1982). Although the turnover of this pool is slow, accumulations of soil carbonate in noncarbonate terrain represent a net sink of CO_2 from the atmosphere (Schlesinger 1982).

Even small changes in such large pools of carbon would be expected to have dramatic feedbacks in the global climatic system. Losses of soil organic matter by oxidation could contribute to the atmospheric CO_2 increase and exacerbate global warming. However, increases in soil organic matter could slow the rise of atmospheric CO_2 and provide a negative feedback to global warming.

A significant portion of the carbon in soils is relatively labile—subject to return to the atmosphere as CO_2 as a result of decomposition and other soil processes. The annual flux of CO_2 to the atmosphere, known as soil respiration, amounts to about 68 Pg C globally (Raich and Schlesinger 1992). Adjusting for the contributions of roots, this flux of CO_2 from soils indicates a mean residence time (mass/flux) of 32 years for the global pool of carbon in soil organic matter. Most plant residues decompose rapidly at the soil surface, but a small fraction of the soil organic matter is very old; radiocarbon ages of soil humic materials often exceed 500 years (e.g., Campbell et al. 1967; Martel and Paul 1974).

The rates of CO_2 efflux vary as a function of soil temperature (Schleser 1982), and there is good reason to believe that the rates of soil respiration will increase globally with global warming (Figure 8.1). Even a 1% increase in the rate of CO_2 loss from soils globally (i.e., to 68.7 Pg carbon per year) would be equivalent to approximately 14% of the annual flux of CO_2 to the atmosphere from the burning of fossil fuels. Alternatively, if a slightly higher percentage of the annual net primary productivity on earth were to

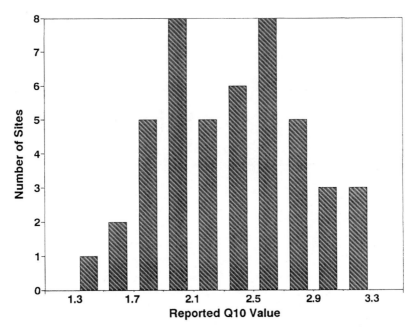

Figure 8.1. Range of Q_{10} values reported in the literature for the response of soil respiration to soil temperature. Q_{10} refers to the factor by which metabolic processes increase with a 10° increase in temperature. The median value for soils is 2.4. (From Raich and Schlesinger 1992.)

escape decomposition and accumulate in soils, a substantial net sink for fossil fuel CO_2 might be found in the terrestrial biosphere. Thus, an understanding of soils is essential for an improved understanding of the workings of the global carbon cycle and the potential biotic feedbacks to global climate change.

KNOWN CHANGES IN THE SOIL CARBON POOL

Ten thousand years ago, 29.5×10^6 km² of the earth's present land area was covered with ice and presumably contained little or no soil organic matter (Flint 1971; Bell and Laine 1985). Much of this area now supports tundra and boreal forest ecosystems, with substantial pools of soil organic matter. Adams et al. (1990) calculate that the global pool of soil organic matter increased by 490 Pg from the last glacial maximum (18,000 years B.P.) to today. Prentice and Fung (1990) estimate that the soil carbon pool has increased by 13–148 Pg during this interval. The average rate of accumulation on glaciated lands has been 0.04×10^{15} Pg carbon per year during the last 10,000 years (Schlesinger 1990), with current rates ranging from 0.075 to 0.18 Pg carbon per year (Harden et al. 1992). Accumulations in peatlands account for much of the total (Armentano and Menges 1986; Gorham 1991). Despite the large range in these estimates, there is no doubt that glaciated soils have served as a net sink for carbon as the climate has warmed from glacial conditions to the conditions seen today.

Over most of the rest of the earth, the geomorphic surfaces are much greater than 10,000 years old, and the accretion of soil organic matter is minimal. Schlesinger (1990) calculated an upper limit for the global formation of soil humic substances of 0.4 Pg carbon per year, or about 0.7% of terrestrial net primary production. Significantly, this value is about equivalent to the long-term riverine transport of organic carbon to the sea (Schlesinger and Melack 1981; Meybeck 1982), suggesting a steady-state condition in soil organic matter globally. Despite accumulations of organic matter in some glaciated soils, it is not likely that, globally, soils were a significant net source or sink for atmospheric CO_2 at the beginning of the industrial period.

The advent of mechanized agriculture in the late 1800s led to dramatic losses of organic matter from cultivated soils. Typically 20–40% of the native soil organic matter is lost when virgin lands are converted to agriculture (Detwiler 1986; Mann 1986; Schlesinger 1986). The losses are greatest during the first few years of land conversion and slow after about 20 years of cultivation. These percentage losses are consistent with independent estimates of the relative portion of organic matter that exists in labile versus refractory pools in the soil (Spycher et al. 1983). Schlesinger (1984) calculated that 36 Pg carbon were lost from soils between 1860 and 1960, with a current rate of loss of about 0.8 Pg carbon per year. Thus, the loss from soils is a significant component of the net biotic flux, which exceeded the flux from fossil fuels until about 1960 (Houghton et al. 1983). At present, this loss is mostly derived from the tropics, where the rates of new land conversion are greatest (Houghton et al. 1987). Nearly all of this loss is seen in greater rates of CO_2 emission from soils rather than greater transport of organic material in eroded soils.

POTENTIAL FUTURE CHANGES IN THE SOIL CARBON POOL

Vegetation is one of the major "state factors" determining soil development and the amount of carbon that a soil will contain (Jenny 1980). Thus, changes in the distribution of vegetation with changes in climate are likely to alter the input of plant residues to the soil and the accumulation of soil organic matter. From one of the first models predicting the future distribution of vegetation in response to climate (Emanuel et al. 1985), one may estimate that soils will lose 45 Pg carbon as the climate warms from today's conditions to those anticipated with a doubling of atmospheric CO_2. Smith et al. (1992) compare similar predictions derived from each of four newer climate models and find that potential changes in soil carbon range from a loss of 19.5 Pg to a gain of 57.3 Pg carbon under climatic shifts that include both temperature and precipitation (Table 8.1). In each scenario, the changes in soil carbon are dwarfed by changes in the pool of carbon in vegetation (see Chapter 17, this volume).

These models provide a picture of the carbon pool in the biosphere at two points in time; they do not include potentially transient states that may yield dramatic fluxes of CO_2 from the biosphere to the atmosphere. For example, although the climate may change rapidly, changes in the distribution of vegetation are likely to take 200–500 years (Overpeck et al. 1991). At the same time, soil microbial communities are likely to show an immediate response to higher soil temperature, increasing the rate of soil respiration. Billings et al. (1982) found that tundra soils became a significant net source of CO_2 at

Table 8.1. Changes in Carbon Storage in Aboveground Biomass and Soil for Four
Climate Change Models

Model	Aboveground biomass	Soil	Total	Change
Current	737	1158	1896	
Oregon State University	860	1216 (4.9)	2076	181 (9.5)
Geophysical Fluid Dynamics Laboratory (Princeton University)	782	1151 (–0.6)	1933	38 (2.0)
Goddard Institute of Space Studies (NASA)	830	1213 (4.7)	2043	147 (7.7)
United Kingdom Meterological Office	765	1139 (–1.7)	1904	9 (2.0)

Note: Values are in petagrams. Those in parentheses represent percentage change from the present.
Source: From Smith et al. (1992). Reprinted by permission of Kluwer Academic Publishers.

higher temperatures, and Jenkinson et al. (1991) suggest that during the next 60 years as much as 61 Pg carbon may be lost from the global pool of soil carbon and released to the atmosphere as CO_2. Schleser (1982) suggests that at least some of the current increase in atmospheric CO_2 may be due to this process.

All the models analyzed by Smith et al. (1992) show a decrease in the area of desert land globally. This prediction is consistent with anticipated increases in global rainfall on a warmer planet. However, Rind et al. (1990) suggest that the increase in rainfall will lag behind the changes in temperature as a result of the thermal buffering capacity of the oceans, where most of the water vapor for precipitation is generated. The land surface will warm faster than the ocean surface, resulting in an extended period when high rates of evapotranspiration will lower the soil moisture content in the central portions of continents. As deserts spread, the soils of the grasslands they replace are likely to lose a substantial portion of their native organic matter (Schlesinger et al. 1990). Again, the behavior of soils during the transient state differs considerably from their behavior under vegetation that is in equilibrium with the predicted future climate.

None of the climate models accounts for the effect of higher levels of atmospheric CO_2, which are likely to stimulate the rate of net primary production and the delivery of plant residues to the soil surface. Numerous authors have postulated that the size of the terrestrial biosphere has increased owing to CO_2 fertilization, providing an explanation for the "missing sink" of CO_2 that otherwise ought to be found in the atmosphere (Tans et al. 1990). There is ample evidence that the growth of crops in high-CO_2 atmospheres, with added water and fertilizer, is greater than that in ambient CO_2 (Strain and Cure 1985; Idso and Kimball 1992), and there is some evidence that the growth enhancement increases the content of organic matter in agricultural soils (Leavitt et al. 1994). If high levels of CO_2 stimulate plant production globally, then the pool of carbon in the terrestrial biosphere—vegetation and soils—may increase beyond the changes that might be expected because of global warming (Table 8.1). Alternatively, a growth stimulation could lead to greater rates of carbon turnover and little net sequestration (Korner and Arnone 1992).

Although there is much interest in a CO_2 fertilization effect, there is little or no empirical evidence that it exists in natural ecosystems. Neither Graumlich (1991) nor Jacoby and D'Arrigo (Chapter 5, this volume) found a statistically significant trend in tree rings that could be linked to a contemporary CO_2-induced stimulation of plant growth. Although much land has been removed from agriculture in the southeastern United States, thus increasing the storage of carbon in this region's forests (Delcourt and Harris 1980), the growth of individual trees shows no recent upward trend during the last several decades (Van Deusen 1992). Kauppi et al. (1992) suggest that the biomass of forests in Europe has increased by about 0.1 Pg carbon per year since 1971, mostly in stands that were in place in 1971. It is not clear, however, whether this biomass accumulation represents the effects of high levels of CO_2, high atmospheric deposition of nutrients, or normal growth trends during middle-age stand development. Houghton et al. (1987) suggested that the forests of Europe are increasing at a rate of 0.08 Pg carbon per year simply as a result of the regrowth of trees on abandoned lands and in areas previously harvested.

Field experiments in the tundra of Alaska showed that rapid acclimation of tundra plants to high concentrations of CO_2 reduced their potential for enhanced carbon acquisition (Tissue and Oechel 1987), and Grulke et al. (1990) found "little if any long-term stimulation of ecosystem carbon acquisition by increases in atmospheric CO_2." Billings et al. (1984) found that the storage of carbon in tundra soils increased at high CO_2 levels only if additional nitrogen was provided as a plant nutrient. Similarly, Norby et al. (1992) found that seedlings of *Liriodendron tulipifera* showed no net increase in size when grown under high concentrations of CO_2 with ambient levels of water and nutrients. In the latter experiment, however, the turnover of fine roots increased, potentially increasing the inputs of plant residues to the pool of soil organic matter.

The lack of response of natural vegetation to high levels of CO_2 is likely to result from the nutrient limitations on plant growth that are seen over much of the land surface. However, humans have inadvertently increased the atmospheric deposition of some plant nutrients, especially nitrogen, in regions downwind from major sources of air pollution, and the added deposition might be expected to alleviate partially nutrient deficiencies. Garrels et al. (1975) were among the first biogeochemists to note that the worldwide human mobilization of nitrogen and phosphorus approximately matched the stoichiometric requirements of these elements for increased plant growth that might sequester CO_2 from the atmosphere, and Peterson and Melillo (1985) calculate that the inadvertent fertilization of the biosphere might enhance carbon storage by 0.2 Pg carbon per year. Kauppi et al. (1992) suggest that at least a portion of the current increment in biomass in European forests (0.1 Pg carbon per year) is driven by the high nitrogen deposition in that region. Similar conditions might be expected in the northeastern United States, but I know of no data that demonstrate higher growth rates in that region over the last several decades (see Jacoby and D'Arrigo, Chapter 5, this volume). In any case, the maximum estimate of the carbon sink derived from greater atmospheric deposition is only a small portion of the "missing sink" that plagues current budgets for atmospheric CO_2.

Since only a small fraction of net primary production (NPP) escapes decomposition and accumulates in soils, any stimulation of the terrestrial biosphere by high CO_2 levels, high nutrient inputs, or their interaction is more likely to be seen in short-lived pools than in soils—i.e., vegetation > litter > soil humic matter. Despite dramatic increases in tree growth (97%), forests subject to long-term fertilization show much smaller increments in soil organic matter (21%, according to Neilsen et al. 1992). Thus, increases in the size of the soil carbon pool are not likely as an immediate negative feedback on higher atmospheric CO_2 concentrations and global warming.

One long-term feedback that may lead to greater storage of carbon in the terrestrial biosphere derives from the greater rates of decomposition that are expected as a result of soil warming. In a soil warming experiment in a black spruce forest in central Alaska, Van Cleve et al. (1990) measured higher nutrient concentrations in soil solutions and spruce foliage. Presumably, these nutrients were mineralized as a result of greater rates of decomposition. Thus, greater mineralization of nutrients in globally warmer soils may help alleviate the nutrient deficiencies experienced by vegetation. Since the C:N ratio of soil (circa 12) is much lower than the C:N ratio of wood (circa 160), a small amount of additional nitrogen mineralization could drive a large enhancement of vegetation production and net carbon sequestration, despite a higher rate of respiration of CO_2 from soils (Rastetter et al. 1991; Bonan and Van Cleve 1992; McGuire et al. 1992). We know little about whether such a sink is reasonable; it is also possible that the losses of nitrogen could increase with greater rates of mineralization (Davidson, Chapter 11, this volume).

ANAEROBIC RESPIRATION: METHANOGENESIS AND METHANOTROPHY

Flooded organic soils are sources of CH_4 as a result of anaerobic respiration (methanogenesis), and CH_4 emissions from natural wetlands appear to contribute about 20% to the global emissions of CH_4 to the atmosphere (Cicerone and Oremland 1988). Although early assessments suggested that emissions from boreal peatlands dominated this source, more recent workers have measured high rates of emission from tropical and subtropical wetlands, yielding a large tropical source from a relatively limited area (Fung et al. 1991; Bartlett and Harriss 1993). Other factors being equal, we can expect methanogenesis in wetland soils to increase with global warming, as a result of a greater rate of microbial activity and a longer growing season (Crill et al. 1988).

The emission of CH_4 from soils to the atmosphere is less than the gross production of CH_4 because of CH_4 oxidation (methanotrophy) by bacteria in the upper, aerobic layers of the soil (Oremland and Culbertson 1992). Yavitt et al. (1988, 1990) found that CH_4 oxidation consumed between 11% and 100% of the CH_4 production in some peatlands in West Virginia. Clearly, any climatic changes that might increase the rate of methanotrophy in wetland soils will result in a lower annual net flux to the atmosphere. For example, a lower water table in tundra ecosystems may decrease the emission of CH_4 to the atmosphere (Whalen and Reeburgh 1990). Changes in the relative role of methanogenesis and methanotrophy in wetland soils, where the diffusion of oxygen is limited, are consistent with observations that the ratio of CH_4 to CO_2 in soil respiration is strongly dependent on soil moisture content (Moore and Knowles 1989).

Upland soils can serve as a net CH_4 sink as a result of inward diffusion of atmospheric CH_4 and its consumption by soil bacteria. In desert soils where organic substrates are limited, methanotrophy by soil bacteria may consume up to 7 Tg/year of CH_4 (Striegl et al. 1992). Net CH_4 oxidation is also reported for moist, but not waterlogged, boreal forest soils (Whalen et al. 1991). The net CH_4 consumption by upland soils is limited by the rate of diffusion of atmospheric CH_4 into the soil pore space (Born et al. 1990; King and Adamsen 1992). Thus, the rate of methanotrophy should increase as the concentration of CH_4 in the atmosphere rises, and rates of CH_4 consumption will change as a result of climatic changes that alter the diffusion pathway in soils—e.g., changes in soil moisture. Methanotrophy shows relatively little temperature response (King and Adamsen 1992), so the consumption of CH_4 by soil bacteria is not likely to change in response to global warming alone.

CONCLUSIONS

The carbon content of soils has changed dramatically in response to past changes in climate, increasing as the earth's climate warmed at the end of the last glacial period. In the last century, however, humans have reduced the content of organic matter in agricultural soils throughout the world, and soils have been a significant contributor to the growth of atmospheric CO_2. We know little about the potential response of soils to future changes in climate, but there is no compelling reason to believe that soils will serve as a significant net sink for atmospheric CO_2 during the next century. Indeed, there is ample reason to believe that on a warmer planet soils will be a significant net source of atmospheric CO_2 and CH_4, especially in the face of continued growth of the human population and its management of the earth's land surface.

ACKNOWLEDGMENTS

I thank Lisa D. Schlesinger for her helpful review of the manuscript.

REFERENCES

Adams, J. M., H. Faure, L. Faure-Denard, J. M. McGlade, and F. I. Woodward. 1990. Increases in terrestrial carbon storage from the last glacial maximum to the present. *Nature* 348:711–714.

Armentano, T. V., and E. S. Menges. 1986. Patterns of change in the carbon balance of organic soils of the temperate zone. *J. Ecol.* 74:755–774.

Bartlett, K. B., and R. C. Harriss. 1993. Review and assessment of methane emissions from wetlands. *Chemosphere* 26:261–320.

Bell, M., and E. P. Laine. 1985. Erosion of the Laurentide region of North America by glacial and glaciofluvial processes. *Quat. Res.* 23:154–174.

Billings, W. D., J. O. Luken, D. A. Mortensen, and K. M. Peterson. 1982. Arctic tundra: A source or sink for atmospheric carbon dioxide. *Oecologia* 53:7–11.

Billings, W. D., K. M. Peterson, J. O. Luken, and D. A. Mortensen. 1984. Interaction of increasing atmospheric carbon dioxide and soil nitrogen on the carbon balance of tundra ecosystems. *Oecologia* 65:26–29.

Bonan, G. B., and K. Van Cleve. 1992. Soil temperature, nitrogen mineralization, and carbon source-sink relationships in boreal forests. *Can. J. For. Res.* 22:629–639.

Born, M., H. Dorr, and I. Levin. 1990. Methane consumption in aerated soils of the temperate zone. *Tellus* 42B:2–8.

Campbell, C. A., E. A. Paul, D. A. Rennie, and K. J. McCallum. 1967. Factors affecting the accuracy of the carbon-dating method in soil humus studies. *Soil Sci.* 104:81–85.

Cicerone, R. J., and R. S. Oremland. 1988. Biogeochemical aspects of atmospheric methane. *Glob. Biogeochem. Cycles* 2:299–327.

Crill, P. M., K. B. Bartlett, R. C. Harriss, E. Gorham, E. S. Verry, D. I. Sebacher, L. Madzar, and W. Sanner. 1988. Methane flux from Minnesota peatlands. *Glob. Biogeochem. Cycles* 2:371–384.

Delcourt, H. R., and W. F. Harris. 1980. Carbon budget of the southeastern U.S. biota: Analysis of historical change in trend from source to sink. *Science* 210:321–323.

Detwiler, R. P. 1986. Land use change and the global carbon cycle: The role of tropical soils. *Biogeochemistry* 2:67–93.

Emanuel, W. R., H. H. Shugart, and M. P. Stevenson. 1985. Climatic change and the broad-scale distribution of terrestrial ecosystem complexes. *Climat. Change* 7:29–43.

Flint, R. F. 1971. *Glacial and Quaternary Geology.* Wiley, New York.

Fung, I., J. John, J. Lerner, E. Matthews, M. Prather, L. P. Steele, and P. J. Fraser. 1991. Three-dimensional model synthesis of the global methane cycle. *J. Geophys. Res.* 96: 13033–13065.

Garrels, R. M., F. T. Mackenzie, and C. Hunt. 1975. *Chemical Cycles and the Global Environment.* W. Kaufmann, Los Altos, Calif.

Gorham, E. 1991. Northern peatlands: Role in the carbon cycle and probable responses to climatic warming. *Ecol. Appl.* 1:182–195.

Graumlich, L. J. 1991. Subalpine tree growth, climate, and increasing CO_2: An assessment of recent growth trends. *Ecology* 72:1–11.

Grulke, N. E., G. H. Riechers, W. C. Oechel, U. Hjelm, and C. Jaeger. 1990. Carbon balance in tussock tundra under ambient and elevated atmospheric CO_2. *Oecologia* 83:485–494.

Harden, J. W., E. T. Sundquist, R. F. Stallard, and R. K. Mark. 1992. Dynamics of soil carbon during deglaciation of the Laurentide ice sheet. *Science* 258:1921–1924.

Houghton, R. A., J. E. Hobbie, J. M. Melillo, B. Moore, B. J. Peterson, G. R. Shaver, and G. M. Woodwell. 1983. Changes in the carbon content of terrestrial biota and soils betweeen 1860 and 1980: A net release of CO_2 to the atmosphere. *Ecol. Monogr.* 53:235–262.

Houghton, R. A., R. D. Boone, J. R. Fruci, J. E. Hobbie, J. M. Melillo, C. A. Palm, B. J. Peterson, G. R. Shaver, G. M. Woodwell, B. Moore, D. L. Skole, and N. Myers. 1987. The flux of carbon from terrestrial ecosystems to the atmosphere in 1980 due to changes in land use: Geographic distribution of the global flux. *Tellus* 39B:122–139.

Idso, S. B., and B. A. Kimball. 1992. Seasonal fine-root biomass development of sour orange trees grown in atmospheres of ambient and elevated CO_2 concentration. *Plant Cell Environ.* 15:337–341.

Jenkinson, D. S., D. E. Adams, and A. Wild. 1991. Model estimates of CO_2 emissions from soil in response to global warming. *Nature* 351:304–306.

Jenny, H. 1980. *The Soil Resource.* Springer-Verlag, New York.

Kauppi, P. E., K. Mielikainen, and K. Kuusela. 1992. Biomass and carbon budget of European forests, 1971 to 1990. *Science* 256:70–74.

King, G. M., and A. P. S. Adamsen. 1992. Effects of temperature on methane consumption in a forest soil and in pure cultures of the methanotroph *Methylomonas rubra. Appl. Environ. Microbiol.* 58:2758–2763.

Korner, C., and J. A. Arnone. 1992. Responses to elevated carbon dioxide in artificial tropical ecosystems. *Science* 257:1672–1675.

Leavitt, S. W., E. A. Paul, B. A. Kimball, P. J. Pinter, G. F. Hendry, K. F. Lewin, J. Nagy, J. R. Mauney, R. Rauschkolb, H. Rogers, and H. B. Johnson. 1994. Carbon isotopes in soils indicate rapid input of carbon under free-air CO_2 enrichment. *Ag. Forest Metereol.* (in press).

McGuire, A. D., J. M. Melillo, L. A. Joyce, D. W. Kicklighter, A. L. Grace, B. Moore, and C. J. Vorosmarty. 1992. Interactions between carbon and nitrogen dynamics in estimating net primary productivity for potential vegetation in North America. *Glob. Biogeochem. Cycles* 6:101–124.

Mann, L. K. 1986. Changes in soil carbon storage after cultivation. *Soil Sci.* 142:279–288.

Martel, Y. A., and E. A. Paul. 1974. Effects of cultivation on the organic matter of grassland soils as determined by fractionation and radiocarbon dating. *Can. J. Soil Sci.* 54:419–426.

Meybeck, M. 1982. Carbon, nitrogen, and phosphorus transport by world rivers. *Am. J. Sci.* 282:401–450.

Moore, T. R., and R. Knowles. 1989. The influence of water table levels on methane and carbon dioxide emissions from peatland soils. *Can. J. Soil Sci.* 69:33–38.

Neilsen, W. A., W. Pataczek, T. Lynch, and R. Ryrke. 1992. Growth response of *Pinus radiata* to multiple applications of nitrogen fertilizer and evaluation of the quantity of added nitrogen remaining in the forest system. *Plant Soil* 144:207–217.

Norby, R. J., C. A. Gunderson, S. D. Wullschleger, E. G. O'Neill, and M. K. McCracken. 1992. Productivity and compensatory responses of yellow-poplar trees in elevated CO_2. *Nature* 357:322–324.

Oremland, R. S., and C. W. Culbertson. 1992. Importance of methane-oxidizing bacteria in the methane budget as revealed by the use of a specific inhibitor. *Nature* 356:421–423.

Overpeck, J. T., P. J. Bartlein, and T. Webb. 1991. Potential magnitude of future vegetation change in eastern North America: Comparisons with the past. *Science* 254:692–695.

Peterson, B. J., and J. M. Melillo. 1985. The potential storage of carbon caused by eutrophication of the biosphere. *Tellus* 37B:117–127.

Post, W. W., W. R. Emanuel, P. J. Zinke, and A. G. Stangenberger. 1982. Soil carbon pools and world life zones. *Nature* 298:156–159.

Prentice, K. C., and I. Y. Fung. 1990. The sensitivity of terrestrial carbon storage to climate change. *Nature* 346:48–51.

Raich, J. W., and W. H. Schlesinger. 1992. The global carbon dioxide flux in soil respiration and its relationship to vegetation and climate. *Tellus* 44B:81–99.

Rastetter, E. B., M. G. Ryan, G. R. Shaver, J. M. Melillo, K. J. Nadelhoffer, J. E. Hobbie, and J. D. Aber. 1991. A general biogeochemical model describing the responses of the C and N cycles in terrestrial ecosystems to changes in CO_2, climate and N deposition. *Tree Physiol.* 9:101–126.

Rind, D., R. Goldberg, J. Hansen, C. Rosenzweig, and R. Ruedy 1990. Potential evapotranspiration and the likelihood of future drought. *J. Geophys. Res.* 95:9983–10004.

Schleser, G. H. 1982. The response of CO_2 evolution from soils to global temperature changes. *Z. Naturforsch* 37a:287–291.

Schlesinger, W. H. 1977. Carbon balance in terrestrial detritus. *Ann. Rev. Ecol. Syst.* 8:51–81.

Schlesinger, W. H. 1982. Carbon storage in the caliche of arid soils: A case study from Arizona. *Soil Sci.* 133:247–255.

Schlesinger, W. H. 1984. Soil organic matter: A source of atmospheric CO_2. In *The Role of Terrestrial Vegetation in the Global Carbon Cycle: Measurement by Remote Sensing*, ed. G. M. Woodwell, pp. 111–127. Wiley, New York.

Schlesinger, W. H. 1986. Changes in soil carbon storage and associated properties with disturbance and recovery. In *The Changing Carbon Cycle: A Global Analysis*, ed. J. R. Trabalka and D. E. Reichle, pp. 194–220. Springer-Verlag, New York.

Schlesinger, W. H. 1990. Evidence from chronosequence studies for a low carbon-storage potential of soils. *Nature* 348:232–234.

Schlesinger, W. H., and J. M. Melack. 1981. Transport of organic carbon in the world's rivers. *Tellus* 33:172–187.

Schlesinger, W. H., J. F. Reynolds, G. L. Cunningham, L. F. Huenneke, W. M. Jarrell, R. A. Virginia, and W. G. Whitford. 1990. Biological feedbacks in global desertification. *Science* 247:1043–1048.

Smith, T. M., R. Leemans, and H. H. Shugart. 1992. Sensitivity of terrestrial carbon storage to CO_2-induced climatic change: Comparison of four scenarios based on general circulation models. *Climat. Change* 21:367–384.

Spycher, G., P. Sollins, and S. Rose. 1983. Carbon and nitrogen in the light fraction of a forest soil: Vertical distribution and seasonal patterns. *Soil Sci.* 135:79–87.

Strain, B. R., and J. Cure, eds. 1985. Direct effects of increasing carbon dioxide on vegetation. DOE/ER-0238. U.S. Department of Energy, Washington, D.C.

Striegl, R. G., T. A. McConnaughey, D. C. Thorstenson, E. P. Weeks, and J. C. Woodward. 1992. Consumption of atmospheric methane by desert soils. *Nature* 357:145–147.

Tans, P. P., I. Y. Fung, and T. Takahashi. 1990. Observational constraints on the global atmospheric CO_2 budget. *Science* 247:1431–1438.

Tissue, D. T., and W. C. Oechel. 1987. Response of *Eriophorum vaginatum* to elevated CO_2 and temperature in Alaskan tussock tundra. *Ecology* 68:401–410.

Van Cleve, K., W. C. Oechel, and J. L. Hom. 1990. Response of black spruce (*Picea mariana*) ecosystems to soil temperature modification in interior Alaska. *Can. J. For. Res.* 20:1530–1535.

Van Deusen, P. C. 1992. Growth trends and stand dynamics in natural loblolly pine in the southeastern United States. *Can. J. For. Res.* 22:660–666.

Whalen, S. C., and W. S. Reeburgh. 1990. Consumption of atmospheric methane by tundra soils. *Nature* 346:160–162.

Whalen, S. C., W. S. Reeburgh, and K. S. Kizer. 1991. Methane consumption and emission by taiga. *Glob. Biogeochem. Cycles* 5:261–273.

Yavitt, J. B., G. E. Lang, and D. M. Downey. 1988. Potential methane production and methane oxidation rates in peatland ecosystems of the Appalachian Mountains, United States. *Glob. Biogeochem. Cycles* 2:253–268.

Yavitt, J. B., D. M. Downey, E. Lancaster, and G. E. Lang. 1990. Methane consumption in decomposing *Sphagnum*-derived peat. *Soil Biol. Biochem.* 22:441–447.

9

The Biogeochemistry of Northern Peatlands and Its Possible Responses to Global Warming

EVILLE GORHAM

Plant remains accumulate as variably decomposed material, often in highly organic deposits above the mineral surface. These may occur as mor humus in the case of unsaturated forest and heath soils (Romell and Heiberg 1931), where the deposit is seldom more than 10–20 cm deep, or as peat (Clymo 1983) in waterlogged wetlands. Where the organic wetland deposit is shallow and mixed with a good deal of mineral matter, it is often designated muck. Where the wetland accumulates more than 30 cm of highly organic peat (40 cm in Canada), it is said to be a peatland. Peatlands, sometimes several meters in depth, are particularly common in the northern landscapes of Russia and the Baltic republics, Canada, the northern United States (especially Alaska), and Fennoscandia, where they cover 342×10^6 ha (Gorham 1991a) and account for about 9.7% of the total land surface. They are often—particularly in their later developmental stages—dominated by the unusual bog moss *Sphagnum* (Clymo and Hayward 1982), and the dry matter of bog peat is often well over 90% organic.

Peatlands, whose ecology has been studied for a long time (Gorham 1953), originate in two major ways: by the filling in of shallow water bodies and their invasion by semiaquatic peat-forming plants, or by the swamping and waterlogging (paludification) of unsaturated mineral soils in upland situations. Often the former process can serve as a focal point for the latter, which is by far the more important on an areal basis (Sjörs 1982). Large peatlands often possess extraordinary and beautiful landscape patterns as a result of the interactions of hydrology and vegetational succession (Sjörs 1961; Glaser et al. 1981; Glaser 1983, 1987, 1989; Siegel 1983; Siegel and Glaser 1987; Foster and Wright 1990). The development of peatlands can be extremely variable and follows a great variety of vegetation sequences (Kulczynski 1949; Walker 1970; Tallis 1983; Glaser et al. 1990). Most commonly, these lead from circumneutral fens, influenced by water from mineral soils, often dominated by sedges, and rich in species, to acid *Sphagnum* bogs (Gorham 1957; Gorham and Janssens 1992a) that receive their mineral supply solely from the atmosphere (see, however, Glaser et al. 1990). These bogs are usually open in oceanic climates and forested in continental climates (Glaser and Janssens 1986) but are generally poor in both vascular and bryophyte species, especially in midcontinental North America (Gorham 1990; Glaser 1992).

PEATLAND BIOGEOCHEMISTRY

Current Pools of Biophilic Elements

Northern peatlands currently store large amounts of carbon, nitrogen, and sulfur. The carbon stock is estimated (Gorham 1991a) at 455 petagrams (Pg) (= 10^{15} g) (133 kg m^{-2}): about 30% of the world pool of soil carbon excluding peat, about 64% of the atmospheric pool (Bolin 1983), and about 55% of total plant biomass (of which the plants in northern peatlands make up only a fraction of 1%).

Stocks of nitrogen and sulfur can be estimated by dividing the carbon stock by C:N (28.7) and C:S (350) ratios calculated (Gorham 1991b) from the very extensive, top-to-bottom peat core data of Riley (1987) and Riley and Michaud (1989) for a great variety of peatlands in northern Ontario. Such ratios suggest a nitrogen stock of 15.9 Pg (4.65 kg m^{-2}) and a sulfur stock of 1.30 Pg (0.38 kg m^{-2}). These compare well with global stocks in plant biomass, 11–14 Pg nitrogen and 0.76 Pg sulfur, but are much lower than global stocks in soil organic matter, 300 Pg nitrogen and 11 Pg sulfur (Freney et al. 1983; Rosswall 1983).

Production Versus Decomposition

Storage of organic carbon in peat obviously depends on an excess of net primary production over decomposition in waterlogged environments.

Production

Aselmann and Crutzen (1989) compiled productivity data from which they inferred a range of 300–700 g dry matter m^{-2} year^{-1} for boreal peatlands, contrasting with 100–300 and 400–800 g m^{-2} year^{-1} for polar and temperate peatlands, respectively. My overall estimate (Gorham 1990) for northern peatlands is 683 g m^{-2} year^{-1}, at the high end of their boreal range. Assuming a carbon content of 45% for living plants (Olson et al. 1983), the productivity would be 307 g carbon m^{-2} year^{-1}. Adding two taiga bog sites (Oechel and Billings 1992) would bring this value down to 296 g carbon m^{-2} year^{-1}. The productivity of pure *Sphagnum* bog communities is generally considerably lower (Wieder and Lang 1983; Grigal 1985). Tower measurements in the southern Hudson Bay Lowland suggest that carbon uptake there (with a 120-day growing season) is about 96 g carbon m^{-2} year^{-1} (King et al. 1991). It should be noted that the belowground component of the productivity of vascular plants in peatlands is very poorly known, so that overall harvest-based estimates may be subject to serious error (Bradbury and Grace 1983).

Peatland productivity is controlled by a variety of factors; especially important are temperature (Gorham 1974; Aselmann and Crutzen 1989) and carbon/nutrient interactions (Shaver et al. 1992).

Decomposition

Decomposition is a more complicated process than the photosynthetic fixation of atmospheric CO_2 by green plants insofar as it involves both aerobic and anaerobic processes, the latter resulting in emission to the atmosphere not only of CO_2 but also of CH_4 and volatile compounds of nitrogen and sulfur. Such emissions have been recognized qualitatively for a very long time (Gorham 1991c), whereas their significance has just begun to be assessed quantitatively.

Decomposition begins in the dead organic detritus at the surface of the peatland, often above the water table where the peat is relatively loose, conditions are aerobic, and hydraulic conductivity is high. This zone has been designated the "acrotelm" by Ingram (1978, 1983) in contrast to the "catotelm" beneath, where the peat is much more dense and therefore has a much lower hydraulic conductivity. Lower hydraulic conductivity leads in turn to waterlogging, slow oxygen transport, less activity by microbial and fungal decomposers, and anoxic conditions (Clymo 1987; Farrish and Grigal 1985). Under such waterlogged conditions most of the water reaching a peatland flows laterally through the uppermost peat and living moss layer, where the hydraulic conductivity is relatively high (Ivanov 1981; Clymo 1987; Gorham and Hofstetter 1971).

Decomposition is especially evident in the open pools and ponds that are often characteristic of large peatlands (Ivanov 1981). According to Hamilton et al. (1991), emissions of both CO_2 and CH_4 are relatively high, so that all of the 24 sites they sampled in the Hudson Bay Lowland are sites of net carbon loss.

Although decomposition is governed primarily by the availability of oxygen, moisture levels, temperature, and nutrients may all be important, as indicated in an excellent review by Clymo (1983).

Peat Accumulation

The overall (top to bottom) global rate of peat accumulation has been estimated at 29 g carbon m^{-2} year^{-1} over the past 4600 years (Gorham 1991a). Recent Finnish estimates, based on 19 peat cores, are considerably lower at 19 g carbon m^{-2} year^{-1} (Tolonen and Vasander 1992), chiefly because of lower bulk density. The Finnish data indicate higher rates in southern and western Finland than in the northern and eastern regions. They also indicate distinctly lower rates for 6 sedge fens (14 g carbon m^{-2} year^{-1}) than for 13 bogs (21 g carbon m^{-2} year^{-1}), whereas Canadian data (E. Gorham, J. A. Janssens, and S. C. Zoltai unpublished data) indicate that fen rates are about 23% higher than those in bogs. Paradoxically, unusually high overall rates of peat accumulation, about 37 g carbon m^{-2} year^{-1} (E. Gorham, unpublished data), characterize the Red Lake Peatland of northern Minnesota, which is very close to the prairie border in a region of warm, dry summer climate and relatively frequent droughts that can draw water tables down more than half a meter (Verry 1984).

Application in a very general fashion of Clymo's (1984) model of peat bog growth suggests that the current global rate of peat accumulation is 23 g carbon m^{-2} year^{-1}, or 79% of the overall rate (Gorham 1991a). By comparison, four Finnish peat cores (Tolonen and Vasander 1992) yield a current rate of about two-thirds of the overall long-term rate. The current global rate is 7.8% of plant productivity as estimated previously, a value that compares well with the estimate of 7–9% for Finnish peats (Tolonen and Vasander 1992). It is noteworthy that current rates of carbon accumulation in Finnish peats are similar to export rates of dissolved organic carbon in stream runoff, about 10 g carbon m^{-2} year^{-1}, from six Finnish catchments with about 50% average peatland cover (Sallantaus 1992). In these catchments most of the carbon undoubtedly comes from the peatlands (see Gorham et al. 1986).

Overall accumulation rates of nitrogen and sulfur, assuming a mean peatland age of 4600 years (Gorham 1991a, 1991b), are 1.0 g m^{-2} year^{-1} and 0.083 g m^{-2} year^{-1},

Table 9.1. Emissions of CH_4 Carbon from Northern Peatlands

Location	Latitude (°N)	Area represented (10 ha)	Annual flux (g carbon m^{-2})	Annual flux (Tg carbon)	Reference
Alaska	66–70	49[a]	4.1[b]	2.0	Whalen and Reeburgh (1990a)
Hudson Bay Lowland	50–58	27	1.9	0.5	Glooschenko and Roulet (1992)
Low boreal Ontario	45–50	9.2[c]	0.91	0.08	Roulet et al. (1992)
North central states, USA	47°32′	6.0	22[d]	1.3	Dise (1993a)
Overall	45–70	91.2	4.3	3.9	
All northern peatlands[e]		330	(4.3)	14	

[a]From Kivinen and Pakarinen (1981).
[b]Value for low brush-muskeg bog.
[c]Excluding marshes and beaver ponds.
[d]Average of bog and fen values.
[e]Excluding drained and mined areas (Gorham 1991a).

respectively. To estimate current rates of nitrogen and sulfur accumulation we may take the estimated current rate for carbon, 23 g m^{-2} year^{-1}, and divide it by the mean C:N ratio (45.0) and C:S ratio (417) for the top 10 cm of peat in northern Ontario (Riley 1987; Riley and Michaud 1989). By this means the rates are estimated to be 0.51 and 0.055 g m^{-2} year^{-1} for nitrogen and sulfur, respectively.

The question of lateral expansion (or contraction) of peatlands should also be addressed, although very few data are available (see Gorham 1991a). A recent compilation of 418 radiocarbon dates for basal peats in North America (Gorham and Janssens 1992b) suggests a slow beginning of peatland initiation after 14,000 years B.P., followed by a fairly steady spread between 10,000 and 4,000 years B.P. There appears to have been some slowing down since that time; it is likely, however, to be an artifact caused by undersampling of relatively shallow peats. However, this rate of lateral spread will not influence appreciably the rates of biogeochemical processes discussed in this chapter, amounting as it does to about 0.01% per year over 10,000 years.

Emissions of CH_4

Since my earlier estimate (Gorham 1991a) of CH_4 flux from northern peatlands, 46 teragrams (Tg) (= 10^{12} g) year^{-1}, several investigations have indicated distinctly lower areal fluxes than those in Minnesota (Crill et al. 1988) that provided the basis for my estimate. Several of these studies have calculated regional fluxes (Table 9.1), from which I have made a new and much lower estimate for northern peatlands of about 14 Tg carbon year^{-1}. Similar studies that show emissions lower than those in Minnesota, but do not allow calculation of regional fluxes, are reported by Moore and Knowles (1990), Moore et al. (1990), and Martikainen et al. (1992).

The anomalously high emissions in northern Minnesota are not presently explicable, particularly in view of much lower emissions in the nearby low boreal zone of southern Ontario (Roulet et al. 1992). Dise (1993a,b; Dise et al. 1993), in the most detailed studies yet made of the annual CH_4 cycle in northern peatlands, has shown that the Minnesota emissions respond very positively to higher water table and temperature, which account for 89% of their variance (on a logarithmic basis). Other studies, many of them dealing with plant communities closely similar to those of northern Minnesota, have not shown such a high degree of environmental control; in the low boreal wetlands of Ontario—with much lower emissions—only 56% of variance was ascribed to the same two variables (Roulet et al. 1992a; see also the Alaskan data of Whalen and Reeburgh 1990a).

Most of the CH_4 emitted annually is produced by decomposition of recently produced organic material in the uppermost part of the peat column, as shown by its strong seasonal cycle (Gorham 1991a). Nevertheless, radiocarbon dating suggests a minor contribution from the deeper parts of the catotelm (Wahlen et al. 1989; see also Clymo 1984).

The Carbon Cycle

Putting together the estimates given previously suggests (Figure 9.1) that current sequestration of carbon in peat amounts to nearly 8% of annual net primary production of carbon, whereas CH_4 emissions amount to a little over 1% and stream exports to nearly 7%.

On a global basis northern peatlands, according to my estimate (Gorham 1991a), have sequestered as peat over the past 4600 years about 96 Tg carbon year^{-1}, with the current estimate being 76 Tg carbon year^{-1}. About 8.5 Tg carbon year^{-1} are released as CO_2 by long-term drainage, and about 26 Tg carbon year^{-1} are released by combustion of fuel. CH_4 emissions (as revised in Table 9.1) amount to about 4.3 g carbon m^{-2} year^{-1}. If we take the global warming potential of CH_4 carbon (over a 100-year period) to be four times that of CO_2 carbon ($CH_4/CO_2 = 11$ on the basis of mass; Intergovernmental Panel on Climatic Change 1992), then the positive greenhouse effect of CH_4 emitted is equivalent to 17 g carbon m^{-2} year^{-1}, a little less than the present negative greenhouse effect of peat accumulation. All the numbers are, however, based on inadequate data that are in urgent need of amplification.

Emissions of N_2O

Data on fluxes of volatile nitrogen compounds are almost entirely lacking for northern peatlands. In northern Minnesota emissions of N_2O from a small kettle-hole peatland amounted to about 15 µg m^{-2} day^{-1}; in an experimentally acidified peatland in northwestern Ontario they amounted to <5 µg m^{-2} day^{-1} (Urban et al. 1988). A few measurements (Martikainen et al. 1992) of N_2O emissions by virgin Finnish peats indicate that they are similarly low, amounting to a few tens of micrograms per square meter per day. Some such peats exhibit uptake of similar amounts, perhaps owing to microbial reduction to molecular nitrogen under anaerobic conditions. Drainage led to strong emissions (mean 950 µg m^{-2} day^{-1}) from a mesotrophic sedge/pine peatland; oligotrophic sites did not respond similarly. A peatland drained and fertilized with nitrogen emitted even more strongly (mean 5200 µg m^{-2} day^{-1}).

POSSIBLE RESPONSES TO GLOBAL WARMING

Although it is possible to predict qualitatively a number of possible or even very likely responses of peatlands to global warming (Billings 1987; Gorham 1991a), quantitative

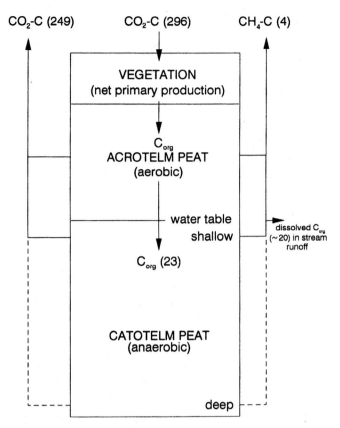

Figure 9.1. Diagram of the carbon cycle in northern peatlands. Numbers are in grams carbon m^{-2} $year^{-1}$ for 330×10^6 ha of undrained and unmined peatland (Gorham 1991a). The figure for stream runoff is from Sallantaus (1992), assuming that all the carbon comes from the 50% peatland cover in the catchments.

predictions are presently very hard to make (Oechel and Billings 1992; Shaver et al. 1992) because our understanding of peatland ecology and biogeochemistry is neither sufficiently detailed in relation to the considerable diversity of peatland processes nor sufficiently extensive in relation to the great variety of peatland ecosystems (Gore 1983). Moreover, we usually do not have sufficient data to infer past conditions (e.g., water table depth) against which to assess the influence of global change, although suitable techniques exist (Janssens et al. 1992; Gorham and Janssens 1992b).

Table 9.2 lists eight of the most significant ecological and biogeochemical processes to be considered and ten of the major environmental factors that influence them and whose effects are likely to be altered by global warming. Simple multiplication indicates the need to examine eighty interactions among them, without considering that the factors and processes interact among themselves, that each process and factor can be subdivided into a variety of components, and that responses may vary greatly among the diverse peatland ecosystems and according to their stages of development. Further complications may be induced by concurrent air pollution (Ferguson et al. 1978;

Table 9.2. Interrelated Processes and Factors Involved in
the Biogeochemistry of Peatlands and Likely to be Affected
by Global Warming

Ecological and biogeochemical processes	Environmental factors affecting such processes
Photosynthesis	Temperature
Aerobic respiration	Length of growing season
Anaerobic respiration	Precipitation
Storage of live organic matter	Water table
Storage of dead organic matter	Water flow
Evapotranspiration	Permafrost
Loss of species	CO_2
Gain of species	Nutrients
	Fire
	Erosion

Gorham et al. 1984, 1987; Aerts et al. 1992; but see Rochefort et al. 1990) and other kinds of human interference or exploitation (Gorham 1990). In the future these may be expected to include the influence of increased UV-B radiation caused by the loss of stratospheric ozone (Krupa and Kickert 1989); its effects have not yet been investigated in peatland ecosystems.

Broad Geographic Responses

With rising temperature the major zones of peat accumulation are likely to migrate northward wherever a flat relief allows, into or even beyond areas where peat formation has ceased owing to past climatic cooling and the development of permafrost (Zoltai and Tarnocai 1975; Zoltai and Pollett 1983). It seems probable, however (Gorham 1991a), that a greatly increased frequency of severe summer drought (Manabe and Wetherald 1986, 1987), despite overall increasing precipitation, will cause the degradation of southern peatlands much faster than northward migration into polar regions, particularly if they become subject to much more frequent droughts and fires. If the relative distribution of forested and open peatlands (Glaser and Janssens 1986) is altered by climatic warming, the consequent effects upon albedo are likely to produce a climatic feedback (Bonan et al. 1992).

Where temperatures rise and water tables fall in southern peatlands, emissions of CO_2 from the peat are likely to increase substantially as the deposits become sources rather than sinks for it (Armentano and Menges 1986; Silvola 1986). However, such emissions will be offset to some degree by increased carbon storage in live biomass owing to the improved tree growth (Hånell 1988; Glebov and Melentyeva 1992) caused by better root aeration and nutrient mineralization. According to Laine and Laiho (1992), after 28 years of drainage a Finnish peatland actually increased its store of carbon in the peat by 50 g m^{-2} year^{-1}, largely owing to accumulation of tree litter, while carbon in the aboveground forest biomass increased by 90 g m^{-2} year^{-1}. Laine et al. (1992) indicate that in minerotrophic peatlands increased tree litter, especially below ground, compensates—at least for a time—for peat oxidation following drainage, whereas in ombrotrophic sites it does not.

Another offset to this positive feedback on the greenhouse effect will be declining CH_4 emissions, or even CH_4 consumption, as water tables fall (Harriss et al. 1982; Whalen and Reeburgh 1990b; Dise 1993a; Martikainen et al. 1992; A. Braunschweig personal communication). Long-term drainage can also reduce the leaching of organic carbon from peatlands, despite an increase during and immediately after ditching (Kortelainen and Saukkonen 1992). Reverse phenomena will of course be occurring—probably more slowly—in northern areas newly invaded by peat-forming plants, and the balance of positive and negative feedbacks is impossible to forecast at this time.

It must be borne in mind that the phenomena described for artificially drained peatlands are transient, continuing until the ecosystem adjusts to the altered hydrologic regime. In a continually warming "greenhouse" climate, the water table is likely to continue to fall, with more profound consequences for the peatlands. Whether increasingly severe water table drawdowns will eventually cause a drought stress sufficient to reverse the initial improvement in tree growth remains to be seen.

An important focus for geographical study will be the overall carbon balance, including both CO_2 and CH_4, in peatlands occupying different climatic and phytogeographic regions (Tuhkanen 1984). Oechel and Billings (1992) indicate that Alaskan tussock tundra may now be losing carbon to the atmosphere at a rate of 180–360 g carbon m^{-2} $year^{-1}$ (but see Schell and Barnett 1994). Pools and ponds in the Hudson Bay Lowlands are also carbon sources for the atmosphere (Hamilton et al. 1991; see Kling et al. 1991). For that region (28×10^6 ha), however, the net storage of CO_2 carbon is estimated at about 30 Tg $year^{-1}$ (King et al. 1991), whereas the loss of CH_4 carbon is estimated at only 0.5 Tg $year^{-1}$ (Glooschenko and Roulet 1992). It would be of much interest to know the current status of more southerly peatlands along the United States/ Canada border, for instance the Red Lake Peatland of northern Minnesota (Glaser et al. 1981; Wright et al. 1992).

A possibility to consider is that the vast Hudson Bay Lowland of Canada (and perhaps similar regions in the West Siberian Plain) might, under warmer conditions and a substantially longer growing season, come to exhibit the unusually high annual rates of CH_4 emission (Crill et al. 1988; Dise 1993a) and peat accumulation presently characteristic of the much smaller Red Lake Peatland and kettle-hole peatlands of northern Minnesota, very near to the present prairie border.

Permafrost

Permafrost—ground that remains below 0°C and commonly contains large amounts of ice—is prevalent in polar regions, where its thickness amounts to 250 m, and extends in much shallower, discontinuous and sporadic form far to the south of the Arctic Circle, in some parts of Canada (Figure 9.2) to 50°N (Harris 1986) and in the vast peatland of the West Siberian Plain to 59°N (Baulin et al. 1984). Although permafrost is characteristic of very cold climates, distribution along its southern border is influenced by local relief, soil type, hydrology, vegetation, and fire frequency, factors that are themselves affected by climate. According to Harris (1987) the permafrost is stable where ground temperature is below –5°C; between –2 and –5°C it is metastable, and above –2°C it is unstable. A warming of 2°C would, therefore, shift its southern boundary to that of the metastable zone (Figure 9.3). A warming of 5°C would result eventually in melting the permafrost

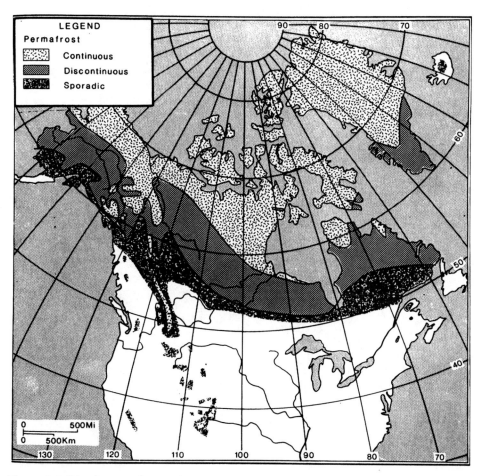

Figure 9.2. Distribution of permafrost in North America. (Reprinted, with permission of the author and of the Arctic Institute of North America, from Harris 1986.)

everywhere but in the far north. Climatic warming of even 2°C—very likely under most scenarios and particularly so in northern regions (Post 1990)—would shift most of the peatland regions of Canada, and notably the vast Hudson Bay Lowland, from the zone of discontinuous to the zone of sporadic permafrost (for Russia see Velichko 1989, and Velichko et al., this volume), with major consequences for their ecology and biogeochemistry. This is indicated by recent experiments (Van Cleve et al. 1990) on the responses of an Alaskan black spruce ecosystem on permafrost to soil heating. The results showed significant increases in thaw depth, rate of decomposition in the forest floor, nutrient release, and net seasonal photosynthesis.

Over the longer term permafrost melting is likely to have two opposing effects upon the trace gas chemistry of peatlands (Gorham 1991a). On the one hand it is likely to lower water tables in areas where runoff leads to thermokarst erosion and gully formation (Billings and Peterson 1990), leading to greatly increased emissions of CO_2 and the shutdown of CH_4 emissions. Likewise, where thaw lakes are formed they may initially

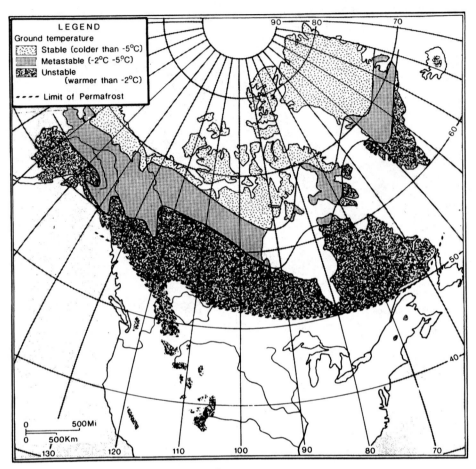

Figure 9.3. Stability of permafrost in North America. (Reprinted, with permission of the author and of the Arctic Institute of North America, from Harris 1986.)

be sources rather than sinks for CO_2. On the other hand, such lakes will eventually undergo plant succession (Drury 1956; Luken and Billings 1983) and develop into fens and bogs that again sequester CO_2 from the atmosphere as peat and emit CH_4 to it. In this connection, Zoltai and Wein (1990) have suggested that melting may cause black spruce–*Sphagnum*–lichen bogs on permafrost to revert to an earlier, wetter fen condition. The overall balance between landscape drainage and flooding as the permafrost melts cannot presently be assessed.

Fire

As climate warms, fire is likely to become a factor of greatly increased importance in many peatlands (Wein 1983; Maltby et al. 1990; Overpeck et al. 1990; see also Flannigan and Van Wagner 1991). Where and when climates are warmer, fires tend to be more frequent in northern forests (Wein et al. 1987; Payette et al. 1989; Clark 1990; Johnson and Larson 1991). It has long been known that wetlands are much less susceptible than uplands to forest fire (Smith 1835; see also Gorham 1955, Wein 1983), but fire may yet be important in some circumstances. For instance, Tolonen and Vasander (1992) suggest that in one Finnish bog peat accumulation was greatly reduced by repeated fires that burned the bog surface prior to the Subboreal Period. Little is known of forest fire frequency in peatland ecosystems. According to Payette et al. (1989), area-weighted fire frequency is less in shrub tundra (0.4 fire/year) than in closed forest (0.7 fire/year), and the average size of fires is 100-fold less: 80 versus 8000 ha.

One might expect that as the frequency of summer drought increases along the southern border of the boreal zone, peatland fires will become both more frequent (Clark 1989) and more severe, resulting in massive emissions of CO_2 to the atmosphere but also a substantial lessening of normal CH_4 emissions. Some CH_4 will, however, be emitted along with CO as products of incomplete combustion (Cofer et al. 1991). Smoke particles will also be emitted, weakening the greenhouse effect (Penner et al. 1992). Hogg et al. (1992) suggest that the deeper peats exposed by fire to decomposition may be relatively resistant even when they are both warm and aerobic, so that further losses of CO_2 after fire could be less than might be expected. They also note that mere warming of waterlogged peats is unlikely to increase emissions of CO_2 because the decay rate is limited by the rate of oxygen diffusion.

Other possibilities exist. For instance, Zoltai and Wein (1990) indicate that major fires have swept across northwestern Alberta at approximately 500-year intervals, in a region where localized permafrost has developed in peatlands that shifted from fens in the warm mid-Holocene to fire-prone black spruce–*Sphagnum*–lichen bogs when the climate cooled about 4000 years ago. In such a setting they suggest that permafrost melting is likely to cause a return to fen conditions, thus lessening the risk of fire. Paludification of upland forests is likely to have a similar result. According to Heinselman (1975) fires may retard paludification. However, Walter (1977) suggests that in certain circumstances fire may favor such a process by allowing *Sphagnum* to colonize burned upland surfaces.

Carbon and Nutrients

As pointed out by Shaver et al. (1992), the productivity of wet and moist tundra ecosystems—very similar to many northern peatlands though usually exhibiting less peat accumulation—is strongly limited by nutrients. These are most frequently nitrogen and phosphorus, and "virtually all of the limiting element supply to plants is provided by recycling from soil organic matter." Effects of global warming that increase the recycling of nutrients in this way, for instance (Van Cleve et al. 1990) warming and drawdown of water tables (and increasing thaw depths in permafrost), are likely on the one hand to increase carbon storage wherever long-lived tree and shrub biomass and litter is increased and on the other hand to decrease it by oxidation of the peat itself. Where global warming becomes severe the latter process seems likely to predominate, especially where fire frequency increases.

As regards direct effects of enrichment by CO_2 in a global-warming scenario, studies by Grulke et al. (1990) indicate a temporary increase in carbon acquisition by an Alaskan cotton-grass tussock community, but the authors believe that there will be little if any long-term stimulation of ecosystem carbon accumulation owing to such enrichment, particularly if global warming causes cloudiness to increase. Experiments by Billings and his colleagues, reviewed by Billings (1987), suggest that effects of increased temperature and of water table drawdown, both acting on nutrient recycling, are the truly important consequences of global warming for wet northern ecosystems. Changes in rates of waterflow are also likely to be important (Chapin et al. 1988).

Changes in Vegetation

One of the most important effects of climatic change is the alteration that takes place in plant communities (COHMAP members 1988; cf. Overpeck et al. 1991), biodiversity (Peters and Lovejoy 1992), and major floristic boundaries (cf. Wheeler et al. 1992). Indeed, many species may not be able to migrate sufficiently fast to adapt to the rapidity of climate change (Davis 1989; Davis and Zabinski 1992). Such alterations will have profound effects upon the nature of affected peatlands, their vegetational succession (Zoltai and Wein 1990), and consequently their biogeochemistry. Where water tables are drawn down severely, a vegetation cover may have great difficulty in re-establishing, at least in situations where nutrients are extremely limiting (Tamm 1951, 1965; Malmström 1952; Maltby et al. 1990). In such situations carbon loss from the peat may be expected to be severe, particularly where fire occurs.

CONCLUSION

In this chapter I have addressed the topic of peat accumulation and decomposition. Three questions were asked: (1) Are peatlands growing or shrinking? (2) At what rate? (3) How might this rate change in a warmer, drier continental climate? The evidence presented suggests that peatlands are still growing upward (Gorham 1991a; Tolonen and Vasander 1992) and that they have by no means reached their upper limit of vertical growth (Clymo 1984, 1987; Tolonen and Vasander 1992). The total area of pristine peatlands has, however, shrunk during the last century or two by about 3% owing to drainage and—to a lesser extent—peat mining (Gorham 1991a). In recent years the rate of drainage for forestry and agriculture together has been about 547,000 ha year^{-1} (Kivinen and Pakarinen 1981), or about 0.17% per annum. This rate is far faster than the approximate rate of spread of 0.01% per annum quoted earlier for the last 10,000 years.

Global warming may already be causing a net loss of carbon in some far northern wetlands (Oechel and Billings 1992; but see Schell and Barnett in press), and the question of whether this is now occurring extensively along the southern border of the boreal zone urgently deserves investigation (cf. Armentano and Menges 1986). I have speculated (Gorham 1991a) that such warming is likely to cause severe degradation of these southern boreal peatlands in the future, at a rate that will exceed the regeneration of far northern peatlands and their advance poleward into new territory wherever the terrain is sufficiently flat. At the present time, I see no reason to think otherwise.

ACKNOWLEDGMENTS

I thank George Woodwell for the stimulus to prepare this chapter, Jan Janssens for assistance in many ways, Nancy Dise for advice on CH$_4$, Leslie Beardsley for searching out numerous references, and the Andrew W. Mellon Foundation for continued financial support.

REFERENCES

Aerts, R., B. Wallén, and N. Malmer. 1992. Growth-limiting nutrients in *Sphagnum*-dominated bogs subject to low and high atmospheric nitrogen supply. *J. Ecol.* 80:131–140.

Armentano, T. V., and E. S. Menges. 1986. Patterns of change in the carbon balance of organic-soil wetlands of the temperate zone. *J. Ecol.* 74:755–774.

Aselmann, I., and P. J. Crutzen. 1989. Global distribution of natural freshwater wetlands and rice paddies, their net primary productivity, seasonality and possible methane emissions. *J. Atmo. Chem.* 8:307–358.

Baulin, V., Y. B. Belopukhova, and N. S. Danilova. 1984. Holocene permafrost in the USSR. In *Late Quaternary Environments of the Soviet Union*, ed. A. A. Velichko, pp. 87–91. University of Minnesota Press, Minneapolis.

Billings, W. D. 1987. Carbon balance of Alaskan tundra and taiga ecosystems: Past, present and future. *Quat. Sci. Rev.* 6:165–177.

Billings, W. D., and K. M. Peterson. 1990. Some possible effects of climatic warming on arctic tundra ecosystems of the Alaskan North Slope. In *Consequences of the Greenhouse Effect for Biological Diversity*, ed. R. L. Peters and T. Lovejoy, pp. 233–243. Yale University Press, New Haven, Conn.

Bolin, B. 1983. The carbon cycle. In *The Major Biogeochemical Cycles and Their Interactions*, ed. B. Bolin and R. B. Cook, pp. 41–45. SCOPE, Wiley, New York.

Bonan, G. B., D. Pollard, and S. L. Thompson. 1992. Effects of boreal forest vegetation on global climate. *Nature* 359:716–718.

Bradbury, I. K., and J. Grace. 1983. Primary production in wetlands. In *Ecosystems of the World. Mires: Swamp, Bog, Fen and Moor*, Vol. 4A: *General Studies*, ed. A. J. P. Gore, pp. 285–310. Elsevier, Amsterdam.

Chapin, F. S. III, N. Fetcher, K. Kielland, K. R. Everett and A. E. Linkins. 1988. Productivity and nutrient cycling of Alaskan tundra: Enhancement by flowing soil water. *Ecology* 69:693–702.

Clark, J. S. 1989. Effects of long-term water balances on fire regime, north-western Minnesota. *J. Ecol.* 77:989–1004.

Clark, J. S. 1990. Fire and climate change during the last 750 yr in northwestern Minnesota. *Ecol. Monogr.* 60:135–159.

Clymo, R. S. 1983. Peat. In *Ecosystems of the World. Mires: Swamp, Bog, Fen and Moor*, Vol. 4A: *General Studies*, ed. A. J. P. Gore, pp. 159–223. Elsevier, Amsterdam.

Clymo, R. S. 1984. The limits to peat bog growth. *Phil. Trans. R. Soc. London Ser. B* 303:605–654.

Clymo, R. S. 1987. The ecology of peatlands. *Sci. Prog. Oxford* 71:593–614.

Clymo, R. S., and P. M. Hayward. 1982. The ecology of *Sphagnum*. In *Bryophyte Ecology*, ed. A. J. E. Smith, pp. 229–289. Chapman and Hall, London.

Cofer, W. R. III, J. S. Levine, E. L. Winstead, and B. R. Stocks. 1991. Trace gas and particulate emissions from biomass burning in temperate ecosystems. In *Global Biomass Burning: Atmospheric, Climatic, and Biospheric Implications*, ed. J. S. Levine, pp. 203–208. Massachusetts Institute of Technology Press, Cambridge, Mass.

COHMAP members. 1988. Climatic changes of the last 18,000 years: Observations and model simulations. *Science* 241:1043–1052.

Crill, P. M., K. B. Bartlett, R. C. Harriss, E. Gorham, E. S. Verry, D. I. Sebacher, L. Mazdar, and W. Sanner. 1988. Methane flux from Minnesota peatlands. *Global Biogeochem. Cycles* 2:317–384.

Davis, M. B. 1989. Insights from paleoecology on global change. *Bull. Ecol. Soc. Am.* 70:222–228.

Davis, M. B., and C. Zabinski. 1992. Changes in geographical range resulting from greenhouse warming—Effects on biodiversity in forests. In *Consequences of CO₂-induced Climatic Warming on Biodiversity*, ed. R. L. Peters and T. Lovejoy, pp. 297–308. Yale University Press, New Haven, Conn.

Dise, N. B. 1993a. Methane emission from Minnesota peatlands: Spatial and seasonal variability. *Global Biogeochem. Cycles* 7:123–142.

Dise, N. B. 1993b. Winter fluxes of methane from Minnesota peatlands. *Biogeochemistry* 17:71–83.

Dise, N. B., E. Gorham, and E. S. Verry. 1993. Environmental factors controlling methane emissions from peatlands in northern Minnesota. *J. Geophys. Res. Atmos.* 98D:10,583–10,594.

Drury, W. H., Jr. 1956. Bog flats and physiographic processes in the Upper Kuskokwim River region, Alaska. *Contrib. Gray Herbarium Harvard Univ.* 178.

Farrish, K. W., and D. F. Grigal. 1985. Mass loss in a forested bog: Relation to hummock and hollow microrelief. *Can. J. Soil Sci.* 65:375–378.

Ferguson, P., J. A. Lee, and J. N. B. Bell. 1978. Effects of sulphur pollutants on the growth of *Sphagnum* species: *Environ. Pollut.* 16:151–162.

Flannigan, M. D., and C. E. Van Wagner. 1991. Climate change and wildfire in Canada. *Can. J. For. Res.* pp. 21:66–72.

Foster, D. R., and H. E. Wright Jr. 1990. Role of ecosystem development and climate change in bog formation in central Sweden. *Ecology* 71:450–463.

Freney, J. R., M. V. Ivanov, and H. Rodhe. 1983. The sulphur cycle. In *The Major Biogeochemical Cycles and Their Interactions*, ed. B. Bolin and R. B. Cook, pp. 56–65. SCOPE, Wiley, New York.

Glaser, P. H. 1983. Vegetation patterns in the North Black River Peatland, northern Minnesota. *Can. J. Bot.* 61:2085–2104.

Glaser, P. H. 1987. The ecology of patterned boreal peatlands of northern Minnesota: A community profile. *Biol. Rep.* 85 (7.14). U.S. Fish and Wildlife Service, U. S. Department of the Interior, Washington, D. C.

Glaser, P. H. 1989. Detecting biotic and hydrogeochemical processes in large peat basins with landsat and TM imagery. *Remote Sensing Environ.* 28:109–119.

Glaser, P. H. 1992. Raised bogs in eastern North America—Regional controls for species richness and floristic assemblages. *J. Ecol.* 80:535–554.

Glaser, P. H., and J. A. Janssens. 1986. Raised bogs in eastern North America: Transitions in landforms and gross stratigraphy. *Can. J. Bot.* 64:395–415.

Glaser, P. H., G. A. Wheeler, E. Gorham, and H. E. Wright Jr. 1981. The patterned mires of the Red Lake Peatland: Vegetation, water chemistry and landforms. *J. Ecol.* 69:575–599.

Glaser, P. H., J. A. Janssens, and D. I. Siegel. 1990. The response of vegetation to chemical and hydrological gradients in the Lost River Peatland, northern Minnesota. *J. Ecol.* 78:1021–1048.

Glebov, F. Z., and N. V. Melentyeva. 1992. Plant dynamics, balance and destruction of biomass in the soil of drained bogged birch forest. In *Proceedings of the 9th International Peat Congress,* ed. D. Fredriksson. pp. 457–469. Uppsala, Sweden.

Glooschenko, W. A., and N. T. Roulet. 1992. The northern wetlands study (NOWES): An assessment of trace gases from the Hudson Bay Lowland. In *Program and Abstracts, 4th Intecol International Wetlands Conference,* p. 83. Columbus, Ohio.

Gore, A. J. P., ed. 1983. *Ecosystems of the World. Mires: Swamp, Bog, Fen and Moor,* Vols. 4A: *General Studies,* and 4B: *Regional Studies.* Elsevier, Amsterdam.

Gorham, E. 1953. Some early ideas concerning the nature, origin and development of peatlands. *J. Ecol.* 41:257–274.

Gorham, E. 1955. Titus Smith (1768–1850), a pioneer of plant ecology in North America. *Ecology.* 36:116–123.

Gorham, E. 1957. The development of peatlands. *Q. Rev. Biol.* 32:146–166.

Gorham, E. 1974. The relationship between standing crop in sedge meadows and summer temperature. *J. Ecol.* 62:487–491.

Gorham, E. 1990. Biotic impoverishment in northern peatlands. In *The Earth in Transition: Patterns and Processes of Biotic Impoverishment,* ed. G. M. Woodwell, pp. 65–98. Cambridge University Press, Cambridge.

Gorham, E., 1991a. Northern peatlands: Role in the carbon cycle and probable responses to global warming. *Ecol. Appl.* 1:182–195.

Gorham, E. 1991b. Human influences on the health of northern peatlands. *Trans. R. Soc. Can.* Series VI, 2:199–208.

Gorham, E. 1991c. Biogeochemistry: Its origins and development. *Biogeochemistry* 13:199–239.

Gorham, E., and R. H. Hofstetter. 1971. The penetration of bog peats and lake sediments by tritium from atmospheric fallout. *Ecology* 52:898–902.

Gorham, E., and J. A. Janssens. 1992a. Concepts of fen and bog re-examined in relation to bryophyte cover and the acidity of surface waters. *Acta Soc. Bot. Pol.* 61:7–20.

Gorham, E., and J. A. Janssens. 1992b. The paleorecord of geochemistry and hydrology in northern peatlands and its relation to global change. *Suo* 43:9–19.

Gorham, E., S. E. Bayley, and D. W. Schindler. 1984. Ecological effects of acid deposition upon peatlands: A neglected field in "acid-rain" research. *Can. J. Fish. Aquat. Sci.* 41:1256–1268.

Gorham, E., J. K. Underwood, F. B. Martin, and J. G. Ogden III. 1986. Natural and anthropogenic causes of lake acidification in Nova Scotia. *Nature* 324:451–453.

Gorham, E., J. A. Janssens, G. A. Wheeler, and P. H. Glaser. 1987. The natural and anthropogenic acidification of peatlands. In *Effects of Atmospheric Pollutants on Forests, Wetlands and Agricultural Ecosystems,* ed. T. C. Hutchinson and K. Meema, pp. 493–512. Springer-Verlag, Berlin.

Grigal, D. F. 1985. *Sphagnum* production in forested bogs of northern Minnesota. *Can. J. Bot.* 63:1204–1207.

Grulke, N. E., G. H. Reichers, W. C. Oechel, U. Hjelm, and C. Jaeger. 1990. Carbon balance in tussock tundra under ambient and elevated atmospheric CO_2. *Oecologia* 83:485–494.

Hamilton, J. D., C. A. Kelly, and J. W. M. Rudd. 1991. Methane and carbon dioxide flux from ponds and lakes of the Hudsons Bay Lowland. *Eos* 72(17):84.

Hånell, B. 1988. Postdrainage forest productivity of peatlands in Sweden. *Can. J. For. Res.* 18:1443–1456.

Harris, S. A. 1986. Permafrost distribution, zonation, and stability along the Eastern Ranges of the Cordillera of North America. *Arctic* 39:29–38.

Harris, S. A. 1987. Effects of climatic change on northern permafrost. *North. Perspect.* 15(5):7–9.

Harriss, R. C., D. I. Sebacher, and F. P. Day, Jr. 1982. Methane flux in the Great Dismal Swamp. *Nature* 297:673–674.

Heinselman, M. L. 1975. Boreal peatlands in relation to environment. In *Coupling of Land and Water Systems,* ed. A. D. Hasler, pp. 93–103. Springer-Verlag, New York.

Hogg, E. H., V. J. Lieffers, and R. W. Wein. 1992. Potential carbon losses from peat profiles: Effects of temperature, drought cycles, and fire. *Ecol. Appl.* 2:298–306.

Ingram, H. A. P. 1978. Soil layers in mires: Function and terminology. *J. Soil Sci.* 29:224–227.

Ingram, H. A. P. 1983. Hydrology. In *Ecosystems of the World. Mires: Swamp, Bog, Fen and Moor*, Vol. 4A: *General Studies*, ed. A. J. P. Gore, pp. 67–158. Elsevier, Amsterdam.

Intergovernmental Panel on Climate Change, Working Group I. 1992. *Supplement to: Scientific Assessment of Climate Change*. World Meteorological Organization and United Nations Environmental Program.

Ivanov, K. E. 1981. *Water Movement in Mirelands*, trans. from the Russian by A. Thomson and H. A. P. Ingram. Academic Press, London.

Janssens, J. A., B. C. S. Hansen, P. H. Glaser, and C. Whitlock. 1992. Development of a raised-bog complex in Northern Minnesota. In *Patterned Peatlands of Northern Minnesota*, ed. H. E. Wright Jr., B. A. Coffin, and N. E. Aaseng, pp. 189–221. University of Minnesota Press, Minneapolis.

Johnson, E. A., and C. P. S. Larson. 1991. Climatically induced change in fire frequency in the southern Canadian Rockies. *Ecology* 72:194–201.

King, K. M., A. Chipanski, G. den Hartog, H. H. Neumann, C. Kelly, D. Hamilton, J. Rudd, I. MacPherson, R. L. Desjardins, P. H. Schnepps, and G. J. Whiting. 1991. CO_2 exchange between a northern wetland and the atmosphere by surface, tower and aircraft-based measurements. *Eos* 72(17):84.

Kivinen, E., and P. Pakarinen. 1981. Geographical distribution of peat resources and major peatland complex types in the world. *Ann. Acad. Sci. Fenn. Ser. A* 132.

Kling, G. W., G. W. Kipphut, and M. C. Miller. 1991. Arctic lakes and streams as gas conduits to the atmosphere: Implications for tundra carbon budgets. *Science* 251:298–301.

Kortelainen, P., and S. Saukkonen. 1992. The impact of ditching on the leaching of organic carbon from peatlands. In *The Finnish Research Programme on Climate Change: Progress Report*, ed. M. Kanninen and P. Anttila, pp. 233–236. VAPK, Helsinki, Finland.

Krupa, S. V., and R. N. Kickert. 1989. The "greenhouse effect": Impacts of ultraviolet B (UV-B) radiation, carbon dioxide (CO_2), and ozone (O_3) on vegetation. *Environ. Pollut.* 61:263–392.

Kulczynski, S. 1949. Peat bogs of Polesie. *Mem. Acad. Pol. Sci. Lett. Ser. B* 15.

Laine, J., and R. Laiho. 1992. Effect of forest drainage on the carbon balance and nutrient stores of peatland ecosystems. In *The Finnish Research Programme on Climatic Change: Progress Report*, ed. M. Kanninen and P. Anttila, pp. 205–210. VAPK, Helsinki, Finland.

Laine, J., H. Vasander, and A. Puhalainen. 1992. Effect of forest drainage on the carbon balance of mire ecosystems. In *Proceedings of the 9th International Peat Congress* 1:170–181. ed. D. Fredriksson. Uppsala, Sweden.

Luken, J. O., and W. D. Billings. 1983. Changes in bryophyte production associated with a thermokarst erosion cycle in a subarctic bog. *Lindbergia* 9:163–168.

Malmström, C. 1952. Svendka gödlingsförsök för belysande av de näringsekologiska villkoren för skogsväxt på torvmark. *Comm. Inst. For. Fenn.* 40(17).

Maltby, E., C. J. Legg, and M. C. F. Proctor. 1990. The ecology of severe moorland fire on the North York moors: Effects of the 1976 fires, and subsequent surface and vegetation development. *J. Ecol.* 78:490–518.

Manabe, S., and R. T. Wetherald. 1986. Reduction in summer soil wetness induced by an increase in atmospheric carbon dioxide. *Science* 232:626–628.

Manabe, S., and R. T. Wetherald. 1987. Large-scale changes of soil wetness induced by an increase in atmospheric carbon dioxide. *J. Atmo. Sci.* 44:1211–1235.

Martikainen, P. J., H. Nykänen, and J. Silvola. 1992. Emissions of methane and nitrous oxide from peat ecosystems. In *The Finnish Research Programme on Climate Change: Progress Report*, ed. M. Kanninen and P. Anttila, pp. 199–204. VAPK, Helsinki, Finland.

Moore, T., N. Roulet, and R. Knowles. 1990. Spatial and temporal variations of methane flux from subarctic/northern boreal fens. *Global Biogeochem. Cycles* 4:29–46.

Moore, T. R.,168 and R. Knowles. 1990. Methane emissions from fen, bog and swamp peatlands in Quebec. *Biogeochemistry* 11:45–61.

Oechel, W. C., and W. D. Billings. 1992. Effects of global change on the carbon balance of Arctic plants and ecosystems. In *Arctic Ecosystems in a Changing Climate,* ed. F. S. Chapin III, R. L. Jefferies, J. F. Reynolds, G. R. Shaver, and J. Svoboda, pp. 139–168. Academic Press, New York.

Olson, J. S., J. A. Watts, and L. J. Allison. 1983. Carbon in live vegetation of major world ecosystems. ORNL-5862. Environmental Sciences Division, Oak Ridge National Laboratory, Oak Ridge, Tenn.

Overpeck, J. T., D. Rind, and R. Goldberg. 1990. Climate-induced changes in forest disturbance and vegetation. *Nature* 343:51–53.

Overpeck, J. T., P. J. Bartlein, and T. Webb III. 1991. Potential magnitude of future vegetation change in eastern North America: Comparisons with the past. *Science* 254:692–695.

Payette, S., C. Moreau, L. Sirois, and M. Desponts. 1989. Recent fire history of the northern Quebec biomes. *Ecology* 70:656–673.

Penner, J. E., R. E. Dickinson, and C. A. O'Neill. 1992. Effects of aerosol from biomass burning on the global radiation budget. *Science* 256:1432–1433.

R. L. Peters and T. Lovejoy, eds. *Consequences of CO₂-induced Climatic Warming on Biodiversity.* Yale University Press, New Haven, Conn.

Post, W. M. 1990. Report of a workshop on climate feedbacks and the role of peatland, tundra, and boreal ecosystems in the global carbon cycle. ORNL/TM-11457. Oak Ridge National Laboratory, Oak Ridge, Tenn.

Riley, J. L. 1987. Peat and peatland resources of northeastern Ontario. Open-File Report, No. 5631. Ontario Geological Survey, Ministry of Northern Development and Mines, Ontario.

Riley, J. L., and L. Michaud. 1989. Peat and peatland resources of northwestern Ontario. Miscellaneous Paper No. 144. Ontario Geological Survey, Ministry of Northern Development and Mines, Ontario.

Rochefort, L., D. H. Vitt, and S. E. Bayley. 1990. Growth, production, and decomposition dynamics of *Sphagnum* under natural and experimentally acidified conditions. *Ecology* 71:1986–2000.

Romell, L.-G., and S. O. Heiberg. 1931. Types of humus layer in the forests of the northeastern United States. *Ecology* 12:567–608.

Rosswall, T. 1983. The nitrogen cycle. In *The Major Biogeochemical Cycles and Their Interactions,* ed. B. Bolin and R. B. Cook, pp. 46–50. SCOPE, Wiley, New York.

Roulet, N. T., R. Ash, and T. R. Moore. 1992a. Low boreal wetlands as a source of atmospheric methane. *J. Geophys. Res.* 97 (D4):3739–3749.

Roulet, N., T. Moore, J. Bubier, and P. Lafleur. 1992b. Northern Fens: Methane flux and climate change. *Tellus* 44B:100–105.

Sallantaus, T. 1992. The role of leaching in the material balance of peatlands. In *The Finnish Research Programme on Climate Change: Progress Report,* eds. M. Kanninen and P. Anttila, pp. 237–242. VAPK, Helsinki, Finland.

Schell, D. M., and B. Barnett. in press. Carbon dynamics in Arctic Alaskan tundra ecosystems. In *Landscape Function: Implications for Ecosystem Response to Disturbance,* ed. J. Reynolds and J. Tenhunen, Ecological Studies. Springer-Verlag, Berlin.

Shaver, G. R., W. D. Billings, F. S. Chapin III, A. E. Giblin, K. J. Nadelhoffer, W. C. Oechel, and E. B. Rastetter. 1992. Global change and the carbon balance of Arctic ecosystems. *BioScience* 42:433–441.

Siegel, D. I. 1983. Ground water and evolution of patterned mires, Glacial Lake Agassiz peatlands, northern Minnesota. *J. Ecol.* 71:913–922.

Siegel, D. I., and P. H. Glaser. 1987. Groundwater flow in a bog-fen complex, Lost River Peatland, northern Minnesota. *J. Ecol.* 75:743–754.

Silvola, U. 1986. Carbon dioxide dynamics in mires reclaimed for forestry in eastern Finland. *Ann. Bot. Fenn.* 23:59–67.

Sjörs, H. 1961. Surface patterns in boreal peatlands. *Endeavour* 20:217–224.

Sjörs, H. 1982. The zonation of northern peatlands and their importance for the carbon balance of the atmosphere. In *Wetlands: Ecology and Management. Proceedings of the 1st International Wetlands Conference,* ed. B. Gopal, R. E. Turner, R. G. Wetzel, and D. F. Whigham, pp. 11–14. New Delhi, India, 1980. National Institute of Ecology and International Scientific Publications, Jaipur, India.

Smith, T. 1835. Conclusions on the results on the vegetation of Nova Scotia, and on vegetation in general, and on man in general, of certain natural and artificial causes deemed to actuate and affect them. *Mag. Nat. Hist.* 8:641–662.

Tallis, J. H. 1983. Changes in wetland communities. In *Ecosystems of the World. Mires: Swamp, Bog, Fen and Moor,* Vol. 4A: *General Studies,* ed. A. J. P. Gore, pp. 311–347. Elsevier, Amsterdam.

Tamm, C. O. 1951. Chemical composition of birch leaves from drained mire, both fertilized with wood ash and unfertilized. *Sven. Bot. Tidskr.* 45:309–319.

Tamm, C. O. 1965. Some experiences from forest fertilization trials in Sweden. *Silva Fenn.* 117. Helsinki, Finland.

Tolonen, K., and H. Vasander. 1992. Rate of net accumulation of carbon in peat layers, with special emphasis on the past few hundred years. In *The Finnish Research Programme on Climate Change: Progress Report,* ed. M. Kanninen and P. Anttila, pp. 219–225. VAPK, Helsinki, Finland.

Tuhkanen, S. 1984. A circumboreal system of climatic-phytogeographical regions. *Acta Bot. Fenn.* 127:1–50.

Urban, N. R., S. J. Eisenreich, and S. E. Bayley. 1988. The relative importance of denitrification and nitrate assimilation in midcontinental bogs. *Limn. Oceanog.* 33:1611–1617.

Van Cleve, K., W. E. Oechel, and J. L. Hom. 1990. Response of black spruce (*Picea mariana*) ecosystems to soil temperature modification in interior Alaska. *Can. J. For. Res.* 20:1530–1535.

Velichko, A. A. 1989. Evolutionary analysis of the contemporary landscape sphere of the earth and prognosis. *Quat. Intl.* 2:35–42.

Verry, E. S. 1984. Microtopography and water table fluctuation in a *Sphagnum* mire. In *Proceedings of the 7th International Peat Congress, Dublin, Eire,* Vol. 2, pp. 11–31. International Peat Society, Helsinki, Finland.

Wahlen, M., N. Tanaka, R. Henry, B. Deck, J. Zeglen, J. W. Vogel, J. Southon, A. Shemesh, R. Fairbanks, and W. Broecker. 1989. Carbon-14 in methane sources and in atmospheric methane: The contribution from fossil carbon. *Science* 245:286–290.

Walker, D. 1970. Direction and rate in some British post-glacial hydroseres. In *Studies on the Vegetational History of the British Isles,* ed. D. Walker and R. G. West, pp. 117–139. Cambridge University Press, Cambridge.

Walter, H. 1977. The oligotrophic peatlands of Western Siberia—The largest peino-helobiome in the world. *Vegetatio* 34:167–178.

Wein, R. W. 1983. Fire behaviour and ecological effects in organic terrain. In *The Role of Fire in Northern Circumpolar Ecosystems,* ed. R. W. Wein and D. A. MacLean, pp. 81–95. SCOPE, Wiley, New York.

Wein, R. W., M. P. Burzynski, B. A. Sreenivasa, and K. Tolonen. 1987. Bog profile evidence of fire and vegetation dynamics since 3000 years B.P. in the Acadian forest. *Can. J. Bot.* 65:1180–1186.

Whalen, S. C., and W. S. Reeburgh. 1990a. A methane flux transect along the trans-Alaska pipeline haul road. *Tellus* 42B:237–249.

Whalen, S. C., and W. S. Reeburgh. 1990b. Consumption of atmospheric methane by tundra soils. *Nature* 346:160–162.

Wheeler, G. A., E. J. Cushing, E. Gorham, T. Morley, and G. B. Ownbey. 1992. A major floristic boundary in Minnesota: An analysis of 280 taxa occurring in the western and southern portions of the state. *Can. J. Bot.* 70:319–333.

Wieder, R. K., and G. E. Lang. 1983. Net primary production of the dominant bryophytes in a *Sphagnum*-dominated wetland in West Virginia. *Bryologist* 86:280–286.

Wright, H. E., Jr., B. A. Coffin, and N. E. Aaseng, eds. 1992. *The Patterned Peatlands of Minnesota.* University of Minnesota Press, Minneapolis.

Zoltai, S. C., and F. C. Pollett. 1983. Wetlands in Canada. In *Ecosystems of the World. Mires: Swamp, Bog, Fen and Moor,* Vol. 4B: *Regional Studies,* ed. A. J. P. Gore, pp. 245–268. Elsevier, Amsterdam.

Zoltai, S. C., and C. Tarnocai. 1975. Perennially frozen peatlands in the western Arctic and Subarctic of Canada. *Can. J. Earth Sci.* 12:28–43.

Zoltai, S. C., and R. W. Wein. 1990. Development of permafrost in peatlands of northwestern Alberta. In *Programme and Abstracts, Annual Meeting,* p. 195. Canadian Association of Geographers, Edmonton, Alberta.

10

Methane Output from Natural and Quasinatural Sources: A Review of the Potential for Change and for Biotic and Abiotic Feedbacks

E. G. NISBET AND B. INGHAM

Methane output from "natural" sources has changed rapidly in the recent geological past, is changing at present under human influence, and may change further as the earth warms. Unfortunately, the causes, feedback processes, and extent of geological changes are still only poorly understood, the relative strengths of modern sources of CH_4 remain controversial, and prediction is virtually impossible. One of the most difficult problems is in linking accurate but very imprecise, qualitative, and often anecdotal field biogeochemical observations with precise but not necessarily accurate quantitative synthetic models.

This discussion is confined to an analysis of those major "natural" sources and sinks that may have caused the postglacial fluctuations in atmospheric CH_4. The net effect of the latitudinal and seasonal distribution of sources, sinks, and atmospheric transport is shown in Figure 10.1, from Steele et al. (1992). This plot reveals the most important constraint on the global CH_4 budget; the other major constraint is the isotopic finding (e.g., Lowe et al. 1991) that roughly 20% of the atmospheric CH_4 is outut from fossil sources. The present budget is still not well understood (Watson et al. 1990; Tyler 1991), and prediction of future concentrations is difficult.

THE MAJOR SOURCES

Arctic and Sub-Arctic Hydrates

Very large stores of CH_4 exist in permafrost regions, held in soil and in sedimentary rock as gas hydrates (clathrates) (Kvenvolden 1988). Gas hydrates, composed of rigid cages of water molecules that trap molecules of gas (Cox 1983), are potentially stable where

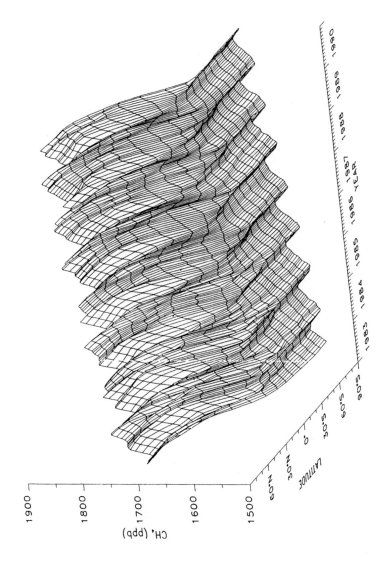

Figure 10.1. Plot of the CH₄ mixing ratio at the marine boundary layer, Pacific and Arctic. Note the seasonality in high latitudes, especially in the Northern Hemisphere. Southern seasonality in part reflects OH seasonality in the tropical upper troposphere, and in part may derive from northern seasonality, blown south over the equator in the midtroposphere by tropospheric circulation cells. (From U.S. NOAA data, courtesy of E. Dlugokencky and P. Tans. See also Steele et al. 1992.)

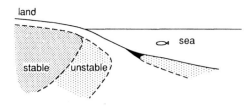

Figure 10.2. Arctic hydrate reservoirs of CH_4. Onshore, CH_4 hydrates are stable in permafrost, under appropriate pressure-temperature conditions, at depths that depend on surface temperatures, conductivity, density of cover, and ambient geothermal gradients. NaCl, if present, depresses the top of the hydrate stability zone; other impurities can raise it. Under the sea, the confining pressure is added to by the water column, and CH_4 hydrate is potentially stable at shallow levels in the seafloor sediments, at water depths in excess of about 250–300 m. In the Holocene and modern Arctic, between the two stable regions, unstable CH_4 hydrate may occur in shallow shelf settings that have been transgressed by the sea since deglaciation. Locally, such regions will at present be degassing CH_4 in seeps to the sea bottom. The black zone shown in the marine hydrate reservoir illustrates the hydrate that is at present stable but would be vulnerable if the Arctic water temperature at the sea bottom were to warm. CH_4 in this region would be rapidly released, with positive feedback. On land, the instability zone will increase with global warming, again with positive feedback, but more slowly and hence with weaker feedback. The size of the vulnerable regions is not known, so the strength of the feedback cannot at present be quantified, but the scale of anthropogenic "fossil" CH_4 emissions, and the isotopic limit on total "fossil" CH_4 output imply that the input from decaying hydrate is at present fairly small, perhaps a few teragrams annually. (From Nisbet 1989.)

pressure and temperature permit (Figure 10.2). In a region with a mean annual surface temperature of about $-10°C$, gas hydrate is stable in permafrost between depths of roughly 200 and 600 m, and in water at 600–1000 m or deeper (depending on local geothermal conditions). Although hydrate is widely stable, it is only locally important. Large parts of the Arctic and of Antarctica have an exposed or under-ice crystalline basement, which does not contain any significant amount of hydrate. However, in extensive areas of the Arctic onshore and in shallow seas, the surface rocks are sedimentary. During the Mesozoic and for part of the Tertiary a major interior sedimentary basin (the Western Interior Basin) extended from the Gulf of Mexico through Alberta and the Northwest Territories to the Arctic shelf of North America, crossing over to Western Siberia. Throughout this region, massive accumulations of hydrocarbons developed, especially in the Cretaceous. During the Tertiary, the Arctic Ocean opened, fragmenting the basin.

During the major glacial events, layers of permafrost developed over the top of these hydrocarbon basins. Within and under the permafrost, hydrate developed, and in some places huge pools of free gas formed. The sources of gas were varied and probably included decay of organic material in situ and the trapping of gas seeping upward from the slowly maturing hydrocarbon-rich deposits below. Gas seeping up from these layers would in nonglacial time reach the surface steadily and be oxidized atmospherically to CO_2. In contrast, in glacial time, upward-seeping gas would be trapped as hydrate in the developing permafrost and then in the groundwater below the permafrost, or eventually, when the groundwater reached saturation, in pools of free gas trapped beneath an overlying cap of saturated hydrate.

The total amount of gas stored in Arctic hydrate is clearly enormous. Kvenvolden (1988) estimated 540×10^{15} g carbon in CH_4 in the Arctic basin offshore from Alaska; the total hydrate or related store on land and offshore northern Russia including the giant

Western Siberian gas fields (Grace and Hart 1986) is probably much larger. Worldwide, estimates of total gas stores range from 10^{18} to 10^{21} g carbon in CH_4. The larger figures are probably overestimates, but a present-day Arctic hydrate-related gas store of the order of 10^{18}–10^{19} g carbon is nevertheless possible, contrasting with a total CH_4 inventory in the modern troposphere of about 4×10^{15} g.

At present, it is thought that clathrate sources are not major emitters of CH_4 to air, although evidence in support of this assumption is scant. Carbon-14 dating of atmospheric CH_4 has shown that some fraction (probably around one-fifth of the annual output, or roughly 100 teragrams (Tg) (= 10^{12} g) per year of atmospheric CH_4 is from fossil sources (including fossil fuel sources and gas hydrates). Wahlen et al. (1989) found that roughly 21% of atmospheric CH_4 was from fossil sources; Quay et al. (1988, 1991) estimated that about 16% was fossil, and Lowe et al. (1988, 1991) calculated that about 25% was from fossil sources. The errors are large, and all three measurements may be locally correct. There is a sense (although there is no definite proof) that this "fossil" fraction has been declining in recent years. Without detailed control by other isotopes, it is difficult to gauge how much of the "fossil" output is from decaying hydrate. A significant part of the "fossil" output in the 1980s may have been loss from the very leaky Russian gas fields, which in part tap hydrate reserves. The output of the Western Siberian gas fields is isotopically almost indistinguishable from much hydrate gas. During the 1980s, gas production in the former Soviet Union, much of it from Siberia but some from southern areas, rose to over 700 billion m^3 (circa 500 Tg CH_4) annually. The rapidly constructed extraction and transport system was notoriously leaky, according to anecdotal industry evidence: loss rates may have been as high as 10% or more. In the years since 1988, there is anecdotal evidence that gas losses from the former Soviet Union regions have been markedly reduced, a situation that could account for the recently observed slowing in the growth rate of global atmospheric CH_4 (Steele et al. 1992, Dlugokencky et al. 1994).

A second major source of "fossil" gas is coal mines (Beck 1993; Beck et al. 1993; Smith and Sloss 1992), especially in China. Many large-scale producers mine coal that is moderately well degassed (circa 1 m^3 gas per ton of coal). In contrast, deep mines can produce up to 20 m^3 per metric ton (gas content increases with depth in typical United Kingdom mines by 0.6 m^3 per 100 m depth increase [in A. Williams 1993]). Taking 15 m^3 per ton as typical, Chinese coal-derived gas loss from a coal production level of 10^6 metric tons may be as high as 10 Tg annually. These figures for Russian and Chinese releases are, unavoidably, rough estimates of poorly known quantities. Somewhat different figures for annual losses are given in World Resources Institute (1992): 13 Tg for former Soviet Union gas losses, 7.6 Tg for former Soviet Union coal losses, and 13 Tg for Chinese coal losses. Globally, CH_4 emitted as a by-product of the petroleum industry from oil extraction and refining probably totals 2 Tg annually, if detailed United Kingdom estimates (in A. Williams 1993) are scaled up.

Fossil gas releases from clathrate and other sources (Judd et al. 1993) that are not directly induced by human activity are difficult to assess. Large bursts are well documented in the geological record, for instance in submarine landslips (Paull et al. 1991). Less dramatic release from decaying hydrate by seepage is difficult to estimate, but C. S. Wong (personal communication 1991) has reported measurable CH_4 in Pacific water derived from the Arctic, implying that there is a significant steady release. In one

Table 10.1. Global CH_4 Budget

Source	Estimated flux (Tg)	Comment	Possible feedback (change after global warming)
"Quasi-natural"			
CH_4 hydrates	5	Variable	Significant danger
Wetlands			
Northern bogs, tundra	35	Too low ?	May increase or decrease
Swamps/alluvial	80	Too high ?	May increase
Biomass burning	55	Fluctuates	Substantial increase
Termites	20		May decrease
Oceans and freshwater	10		
	205		
Animals	80		May increase
Anthropogenic	285		
Rice	100		Will increase
Landfills	40	Too high ?	Decreasing?
Natural gas vents	10		Controllable
Natural gas leaks	30		Controllable
Coal mining	35	Poorly known	Controllable
	215		
Total	500		

Source: Estimates of flux from Fung et al. (1991): scenario 7.

well-studied area, albeit a small one (in global terms) without permafrost, the North Sea seepage losses to the atmosphere have been thought to be small, on the order of a few kilotons (Judd in A. Williams 1993). Globally, however, the flux from shallow marine sediment has been estimated as being as large as 8–65 Tg annually (Hovland et al 1993).

The isotopic data imply that between 80 and 125 Tg of fossil CH_4 are released annually (Table 10.1). Fung et al. (1991) took the lower figure and allocated 35 Tg to loss from coal mining, 40 Tg to loss from the natural gas industry, and 5 Tg to loss from CH_4 hydrates. However, the 5-Tg figure, which is ultimately derived from the estimate of Cicerone and Oremland (1988), is essentially a "placeholder," to use Cicerone and Oremland's term. The true figure may be rather different and is very poorly constrained. The 40-Tg figure for gas industry losses may be a snapshot of a moving figure, roughly correct in the 1970s and too low for the 1980s, but perhaps an attainable target for the 1990s as losses from the huge Russian natural gas industry are reduced. Table 10.2 is a rough estimate of the "fossil" CH_4 burden of the atmosphere.

Subtracting fossil fuel losses from the isotopically derived total of 80–125 Tg of fossil CH_4 emitted annually gives, by difference, the hydrate loss. Tables 10.1 and 10.2 make the assumption that hydrate gas emission at present is roughly 5 Tg annually. This estimate is highly approximate, and within the isotopic constraints it is possible that the hydrate output is either virtually nil or perhaps as high as 10 Tg. The only way to improve the estimate of the hydrate contribution is through steady isotopic monitoring; even then, since the Russian gas is derived partly from hydrate, it may be impossible to quantify hydrate losses until detailed knowledge of Russian gas industry losses is available. Nevertheless, the hope is that hydrate losses are at present fairly small.

Table 10.2. Model of the "Fossil" CH_4 Content of the Atmosphere: Natural Gas and Coal Industry CH_4 Production and Losses, and Contribution from the Oil Industry and Hydrates: A Simple Model to Calculate Atmospheric Burden of "Fossil" Methane

	Natural gas[a]		Coal[b]		Total[c]	
	Production $10^9 m^3$	Loss (Tg)	Production (10^9 metric ton)	Loss (Tg)	Annual (Tg)	Cumulative (Tg)
1981	1503	49	3814	27	61	770
1982	1484	49	3930	28	63	776
1983	1490	51	3951	28	66	783
1984	1626	52	4122	29	66	790
1985	1686	55	4345	31	71	803
1986	1738	57	4518	32	74	817
1987	1830	60	4630	33	77	834
1988	1906	63	4730	34	80	853
1989	1975	61	4816	34	77	868
1990	2028	57	4736	34	72	877
1991	2059	52	4566	33	66	881
1992	2066	47	4484	32	60	878

[a]The loss is calculated on the assumption that 95% of natural gas is CH_4 and that in the industry (excluding the territory of the former Soviet Union) the average loss rate is 3%. For the former Soviet Union, a loss rate of 9% is assumed until 1983, 8% from 1984 to 1988, 7% in 1989, 6% in 1990, and 5% from 1991 to 1992. These assumptions are arbitrary but within the range of anecdotal information.

[b]The loss for the global coal industry excluding China is calculated on the basis of 10 m^3 gas per metric ton, using a conversion factor of 714 g/m^3 from United Kingdom sources. Production figures include brown coal with low losses of CH_4, and hard coal with losses on the order of 15 m^3/metric ton. The assumption of a global loss rate of 10 m^3/metric ton falls between estimates of Smith and Sloss (1992) and Beck et al. (1993).

[c]The total figure includes the loss of CH_4 as a by-product of the oil industry, approximately 2 Tg/year, scaled up from United Kingdom industry data to a global figure. It also includes 5 Tg/year from hydrate release. This figure is simply a "placeholder" (Cicerone and Oremland 1988): the true figure is unknown. The cumulative total assumes a pre-1981 content of 820 Tg of fossil-derived CH_4 in the atmosphere and an annual exponential decay of one-tenth of the previous year's value. The cumulative total should be compared with the total CH_4 content of the troposphere from all sources, which is roughly 4000 Tg. This model is very approximate, but it illustrates the scale of "fossil" emissions of CH_4.

Sources: *Encyclopedia Britannica Yearbooks*, 1982–1992; *BP Statistical Review of World Energy, 1993*, and previous years; and *BP Review of World Gas 1993*, British Petroleum Company, London.

Vulnerability of Hydrate to Climatic Change

Any changes that increase temperature or reduce pressure may liberate CH_4 from hydrate (Figure 10.3A). Specific changes that can occur include heating from a rise in atmospheric temperature, heating from a change in surface albedo, and heating because of marine transgression. Pressure release can occur as a result of sea-level drop, either as a global effect or (a more important cause at present) from local uplift as the lithosphere recovers after glacial loading. Pressure release also occurs in slumping. Perhaps the major cause for concern is the risk of a sudden massive release of CH_4 either from a marine slump or from the rupturing of a major pool of Arctic gas. Such massive release would not be directly attributable to modern warming in the past few decades; rather it would be a stochastically timed part of a longer-term process. The largest source would be a major marine slump (Figure 10.3B). Marine slumps are more likely at times of lowered sea-level (i.e., just before the end of the last major glaciation, around 13–15 kaBP), but can occur at any time and release enormous quantities of CH_4 (Paull et al.

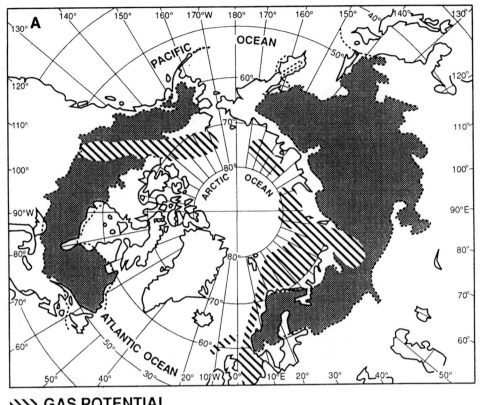

\\\\\ GAS POTENTIAL

Figure 10.3A. Areas of potential for gas release in the high Northern Hemisphere. Important regions include Western Siberia and the nearby Arctic shelf, and the Arctic shelf of Alaska and Canada. Shading indicates extent of boreal forest: the northern boundary of the forest very roughly coincides with the southern boundary of extensive permafrost hydrates. Thus the areas of major potential hydrate output are in the Siberian shelf and the McKenzie Delta region. The major region of CH_4 emission from northern wetlands coincides broadly with the area of boreal forest, as shown, although some CH_4 is also emitted from wetlands north of this zone.

1991; Nisbet 1992). Temperature effects are simplest to model, as permafrost normally transfers heat by conduction, and deep convective heat transfer appears to be rare. Any surface warming (Figure 10.3C) will enter the ground penetrating downward at a rate depending on the square root of the time since perturbation. Initial penetration is rapid; deeper penetration is slower. The geotherm, which is the temperature profile in the ground with depth, thus keeps a record of former surface temperature, each variation in surface temperature penetrating slowly downward in sequence through the permafrost (Nisbet 1989, 1990). In parts of Russia, double layers of permafrost may reflect the climatic fluctuations over the past 15,000 years (Baulin and Danilova 1984; Velichko 1984).

Many models of global change in the near future, in an atmosphere richer in greenhouse gases, have concluded that warming will be most marked in the Arctic (e.g., Houghton et al. 1990, Figure 5.4; Houghton et al. 1992, Figures B4, B5, B11). This

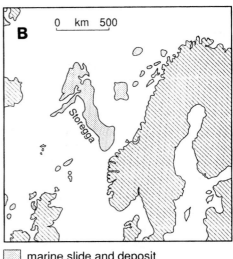

Figure 10.3B. Region affected by the Storegga slips, off Norway. These enormous marine slumps occurred during the last glaciation and in the early Holocene. They may have been hydrate-lubricated and may have emitted much CH_4. Worldwide, there appears to have been a major episode of slumping off the continental shelves and slopes at the time of late glacial climate change. (Diagram from Nisbet 1992, using sources cited therein.)

☐ marine slide and deposit

inference is strongly supported by geothermal measurements (e.g., Lachenbruch and Marshall 1986; Lachenbruch 1988), which measure real, not proxy, heat. Retrodictive models suggest that the surface temperatures in the Prudhoe Bay area have fluctuated by about 3°C in the last millennium, but with a net 4°C warming in the last 120 years (Kakuta 1993). Meteorological results are more ambiguous (Kahl et al. 1993): it has not been proven that the Arctic as a whole is warming. A thermal wave sent into the permafrost may, after a time interval of decades to centuries (depending on local temperature), destabilize hydrate. Marine transgression caused by rising sea level can, in

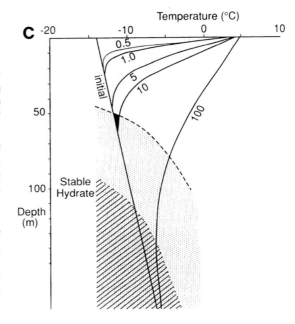

Figure 10.3C. Potential for gas release from CH_4 hydrate. The diagram shows the feedback effect on hydrate of a significant warming, from −14°C to 5°C, as might be expected, for example, in a marine transgression. The diagram shows the number of years taken by a thermal pulse to reach the hydrate zone and destabilize it. The lightly shaded area shows the approximate stability field (Cox 1983) of hydrate under lithostatic pressure; under hydrostatic pressure, the top of the stability field lies deeper (dark shading), and a warming pulse takes longer to reach the hydrate. (From Nisbet 1990.)

the high Arctic, change the mean annual surface temperature from a subaerial temperature of –10°C or even –20°C to the bottom temperature of the Arctic Ocean, slightly
below 0°C. In hydrate-rich areas, this change could rapidly (10–100 years) cause release
of CH_4. In many areas, hydrate is already unstable (Figure 10.2) as a result of sea-level
rise over the past 10,000 years, and in these areas the release of CH_4 may occur in bursts
that reflect the penetration of warming events that may be millennia old.

Curiously, although the long-term effect of marine transgression is CH_4 release, the
short-term effect is stabilization of hydrate, because pressure increases. However, in
large areas of the Arctic, especially Hudson Bay and Scandinavia, sea level is falling as
the land rises in response to the removal of the glacial ice dome loads. In these areas,
hydrate is further destabilized by pressure release as the load of sea water above
decreases. Locally, the development of thermokarst, by reducing load pressure on the
underlying hydrate, may also help destabilize hydrate.

Kvenvolden and Lorenson (1993) have used measurements of CH_4 concentrations in
shallow cores from permafrost to construct a speculative model of emissions under $2 \times CO_2$
conditions. They concluded from typical GCM model results that warming would release
25–30 Tg of CH_4 a year, although it would take at least 30 years for emission to start.

In summary, many Arctic hydrates are currently unstable (Kvenvolden 1988) and
likely to become more so if global warming occurs. There is a clear danger of positive
feedback (MacDonald 1990). However, the potential output is very difficult to quantify
and may be low in the steady state. Much of the unstable hydrate that may currently be
producing CH_4 is a long-term consequence of Holocene warming. This long-term
release may be intensified by new thermal pulses from global warming, as a consequence either of direct heating of the land surface, of albedo changes, or of marine
transgression. Most processes except slumping are slow, but once established they are
long-lived and almost impossible to counteract.

Wetland Sources of CH_4: Controls on Production

CH_4 is ubiquitous in terrestrial wetland systems, especially in shallow ponds and bogs
and in rice fields. To emit significant amounts of CH_4, a system needs a supply of
organic matter, the correct conditions for methanogenesis, and an emission pathway by
which the CH_4 can reach the atmosphere without being oxidized to CO_2 en route
(Figure 10.4). Conditions favorable to methanogenesis (Boone 1993) are common in
northern bogs and ponds and in rice fields, but although production is widespread
emission to the air is more restricted.

CH_4 production within a particular ecosystem is controlled by a variety of factors:

1. *Presence of anoxic and reduced conditions.* Methanogens can metabolize only in
   the absence of oxygen and require a redox potential (E_h) of less than –200 mV
   (Conrad 1989), although some CH_4 production can occur at between –200 and
   0 mV. Such oxygen-deficient zones develop where oxygen has been consumed by
   respiration and resupply by diffusion from the atmosphere is limited. Consequently, methanogenic ecosystems are typically highly stratified (Moore et al.
   1990; Whalen and Reeburgh 1992). In a system such as a small pond, the top of
   the system is aerobic, the second layer is more reduced, below this is a zone of
   sulfate reduction, and last is found a zone of CH_4 generation. This stratification

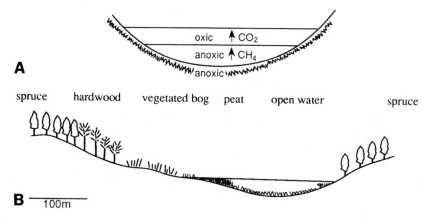

Figure 10.4. (A) Oxic and anoxic zones in a shallow pond. (B) Environment of a typical forest bog.

coincides with the redox potential of the different zones such that as depth increases, E_h decreases (Bouwman 1990). Other important factors include the variation in animal activity and carbon supply to the wetland and changes in physical factors such as free-up or mud overturn.

2. *Soil pH.* Most methanogens grow optimally over a narrow range around pH 7.0, with a few bacteria having optimum conditions around pH 8–10 (R. T. Williams and Crawford 1984; Conrad 1989). Methanogenic microbial communities in acidic environments such as peatlands (pH 3.0–4.0) thus operate under suboptimal conditions. R. T. Williams and Crawford (1984) and Yavitt et al. (1987) found that adjustment of pH to near 6.0 stimulated CH_4 production. Flooding increases the pH of acid soil, probably through reduction of Fe^{3+} to Fe^{2+}, and decreases the pH of alkaline soil, probably through accumulation of CO_2 (Bouwman 1990). Thus rainfall changes or large-scale hydroelectric schemes forming shallow lakes may change CH_4 output by altering pH and, probably more important, E_h.

3. *Substrate and nutrient availability.* CH_4 production depends on the availability of organic matter, via fermentative production of CH_4 precursors. In many cases, however, methanogenesis appears to be substrate-limited (Conrad 1989). Yavitt and Lang (1990) noted different rates of CH_4 production depending on the type of peat that dominated an ecosystem. Shrub-dominated sites, composed of mostly acid-insoluble organic matter, appeared resistant to microbial decomposition and yielded 3 µmol CH_4 liter^{-1}day^{-1}, whereas peats from moss- and sedge-dominated sites, composed of acid-soluble organic matter, yielded up to 216 µmol CH_4 liter^{-1}day^{-1}. Thus it is the quality rather than the quantity of substrate that is important. Certain animals, especially beaver, muskrat, and moose, may be very important in controlling site quality (e.g., McInnes et al. 1992); in tropical regions, the presence of vegetation such as papyrus and the activities of animals such as hippopotamuses may be very significant. When considering the impact of climatic change on biotic CH_4 production, these factors may be crucial.

The primary substrates for bacterial production of CH_4 are acetate and CO_2. Acetate dissimilation predominates in soils and peatlands, accounting for 50–80% of the CH_4 produced (Burke and Sackett 1986; Thebrath et al. 1992). CO_2 predominates in freshwater lake sediments and marine environments, accounting for up to 90% of production (Burke and Sackett 1986; Whiticar et al. 1986). Other compounds such as $(CH_3)_3N$ and CH_3OH may also serve as substrates. Yavitt et al. (1987) suggested that stratification may occur, with CH_4 production in deep layers in peat occurring primarily through CO_2 reduction, and acetate being the primary substrate in upper layers.

Compounds such as nitrate and sulfate inhibit CH_4 production by competing with bicarbonate as electron acceptors (Conrad 1989). Yavitt et al. (1987) suggested that the effect of sulfate concentration on CH_4 production varied within the strata. CH_4 production generally decreased with increasing sulfate concentration in upper peat layers, but in deeper layers production was unaffected or enhanced. Nitrate and sulfate also serve as substrates for competitive nonmethanogenic bacteria. Such factors mean that future CH_4 production under an altered climate may be partly dependent on nutrient supply, for instance, nutrients washed down after grassland fires in the tropics or after forest fires or cutting in the boreal realm.

4. *Temperature*. Methanogenesis shows a typical Arrhenius relationship with temperature (Westermann and Ahring 1987). Most methanogens grow optimally at 30–40°C, although some acetate-utilizing bacteria may prefer cooler conditions (Conrad 1989). R. T. Williams and Crawford (1984) noted increasing CH_4 production with increasing temperature in the range 4–30°C for slurries of Minnesota peat. Temperatures above 30°C are not uncommon in the tropics but may also occur on hot days in the boreal realm. Furthermore, local temperatures on peat hummocks may be well above ambient conditions, and in some small ponds in the boreal forest temperatures near 30°C may be sustained for long periods.

However, temperature may not be the critical factor controlling production, let alone emission, of CH_4. Yavitt et al. (1987) noted that in surface samples of winter peat, CH_4 production was controlled by factors other than temperature, most probably substrate availability. Westermann (1993) showed that the apparent activation energy and hence temperature sensitivity of methanogenesis decreased with decreasing substrate concentrations.

Controls on CH_4 Emission from Wetland

Wetland emissions of CH_4 are massive but not yet accurately quantified (Bartlett and Harriss, 1993). CH_4 produced in a soil, peat, or shallow marine setting may be trapped in pore water, oxidized, or emitted to the air. Only a small proportion of the production, however, successfully reaches the atmosphere (Galchenko et al. 1989).

Much CH_4 is trapped in pore water in mud and in peat (R. T. Williams and Crawford 1984; Brown et al. 1989). Dinel et al. (1988) and Brown et al. (1989) found substantial storage of CH_4 in peat, especially at moderate depth (e.g., 125–145 cm, according to Dinel et al.). Similarly, Moore et al. (1990) found lower concentrations in pore water near the surface than at depth, and Whalen and Reeburgh (1992), studying Arctic tundra, noted a low concentration of CH_4 on the surface (<10 μM), increasing to 200–500 μM in the deep subsurface. This store of trapped CH_4 at depth in peat bogs may reflect stored summer production; it is possible that some of it is liberated on winter freeze-up.

CH_4 oxidation is also critically important (Rosswall et al. 1989), as much of the ascending CH_4 in a pond or bog is oxidized. Oxidation is carried out primarily by oxygen-requiring methanotrophic bacteria. Where sulfate concentration is high, sulfate-reducing bacteria may oxidize CH_4 (Rudd and Taylor 1979; Galchenko et al. 1989). This process may occur in marine settings where relatively high sulfate concentrations (on the order of 100 times higher than those in freshwater sediments) exist.

The main controls on CH_4 oxidation are the availability of oxidizers and the physical structure of the local system, particularly the length of the transport pathway by which CH_4 reaches the air. Where emission rates are low or diffusion paths long, there is a high probability that most of the CH_4 produced on, for example, a pond bottom will be oxidized prior to reaching the atmosphere. However, where emission occurs in large bubbles, say after animal burrowing (e.g., worms) or after the passage of the flat-footed moose family, or if emission is channeled via vascular plants, a substantial part of the CH_4 that is produced will be emitted. Thus boreal ponds and bogs may have low steady-state emissions in summer, but very high catastrophic emissions on the days on which moose browse. Similarly, the summer CH_4 store in peat and mud may not be emitted until freeze-up, when a sudden burst may occur (Rudd and Hamilton 1978). Methanotrophic bacteria and thus CH_4 oxidation are generally concentrated in the oxic surface layers of water bodies, sediment, peat, or soil (Figure 10.4). Fechner and Hemond (1992), studying a northern *Sphagnum* bog, noted that most of the CH_4 oxidation occurred in a zone 6–15 cm below the surface. Aquatic plant life is also directly important in oxidation (apart from its effect in attracting moose or hippopotamuses). Vascular transport, which allows transport of oxygen diffusing into roots and eventually into sediment, also facilitates egress of CH_4 into the atmosphere. The amount of CH_4 oxidation that occurs depends on the root microenvironment, but the presence of vegetation may substantially increase CH_4 flux into the air (Chanton et al. 1992). Schutz et al. (1989) found that as rice grew in Italian paddy fields, the percentage of CH_4 emitted via plant-mediated transport changed. Early in the season, when the rice plants were not developed, 100% of the CH_4 emission took place by ebullition. By the end of the season, only 4–9% of the CH_4 that was emitted to air was released by ebullition, and plant-mediated transport carried over 90%. However, during the same period CH_4 oxidation increased, so that by the end of the season only 6% of the CH_4 produced was emitted. It is possible that changes in cultivation practice and in rice varieties may substantially alter levels of CH_4 emission. The importance of vascular processes is not limited to rice, but extends to a wide variety of plants, from tundra sedges to tropical grasses.

NORTHERN WETLAND SOURCES OF CH_4: OUTPUT AND VULNERABILITY TO CHANGE

In late Holocene time, shortly before major human disruption, wetland sources were the most important contributors to global atmospheric CH_4 quantities. Throughout the high latitudes of the Northern Hemisphere wetland is abundant because, despite the low level of precipitation, evaporation levels are even lower through much of the year. Northern wetland, including tundra and northern forest (Figure 10.3A), remains a major source of natural global atmospheric CH_4. However, there is a strong controversy over the exact amount of CH_4 due to this source and its relative importance when contrasted with tropical and subtropical wetland.

Assessment of the total output of northern wetland depends both on site information and on detailed time sequence studies of atmospheric CH_4 concentration (Figure 10.1). Bartlett and Harriss (1993), using an extensive data base of emission measurements together with the global wetland inventory of Matthews and Fung (1987), concluded that total global wetland emissions are about 109 Tg annually. Sixty-six Tg of those emissions are from broadly tropical regions (20°N–30°S), 5 Tg are from subtropical and temperate wetland, and 38 Tg are from northern wetland and tundra (north of 40°N). To evaluate CH_4 output, Fung et al. (1991) derived a series of models. In their preferred scenario, constrained also by isotopic data, they concluded that the northern wetland poleward of 50°N (i.e., north of the Canadian border in North America, and the equivalent in Asia) in general contributed about 35 Tg CH_4 to the air annually, less than one-tenth of the global total. In contrast, they concluded that the tropical swamps contributed about 80 Tg annually. Aselmann and Crutzen (1989) came to broadly similar conclusions, stressing the importance of rice paddies. A third model is that of J. A. Taylor et al. (1991), who depended in part on the Matthews and Fung (1987) data base for wetland area, as also discussed by Fung et al. (1991). Taylor et al. (their Figure 17) calculated a model of annual total flux of CH_4 from 2.5° grid squares over the globe and inferred several major wetland source areas: Western Siberia, Canada (especially on the flanks of Hudson's Bay and in the Slave region), and the Parana swamps of South America, as well as in Southeast Asia and Uganda-Rwanda-Burundi.

Models such as these are elegant integrations of global data bases, but they also illustrate the more general problem of linking essentially site-specific field information that is accurate but not precise with quantitative data bases that are precise but not accurate. Russian output is very poorly known. In Canada, for example, our personal, unquantified field impression is that the maximum CH_4 output is probably from point sources in the warm, carbon-rich ponds of the southern fringe of the boreal forest, not the more extensive but less productive tundra wetlands farther north. Similarly, in Africa, the output from small areas of papyrus swamps of the Sudd, Niger, Chad, and Okavango may compare in significance with that of the Uganda region. This impression is anecdotal, derived from personal observation in the areas and not quantitative data. The task of ground-truthing global models is very difficult; it is a measure of the excellence of the work of the modelers that they do appear to have succeeded in approximating reality.

There is some debate about the details of these conclusions, however. Perhaps the most interesting puzzle lies in understanding the controls on the seasonality and latitudinal distribution of global CH_4 sources (e.g., as shown in Figure 10.1 and by Steele et al. 1992). Northern wetland sources are probably strongly seasonal (e.g., Rudd and Taylor 1979; Dise 1993), although data are surprisingly scarce, especially from boreal forest bogs. The global north-south gradient in atmospheric CH_4 concentration reflects the preponderance of northern terrestrial sources. It has been assumed that northern wetlands are a major cause of the seasonality in the north, and also, after transport via the Hadley cells (Tie et al. 1991), a major cause of the smoothed Southern Hemisphere seasonality (Fraser et al. 1987), which may lag by nearly a year. However, the OH sink of CH_4 is also seasonal by latitude and contributes strongly to global seasonality.

In their preferred scenario, Fung et al. (1991) infer a relatively small natural northern source and ascribe much northern production to other, possibly less markedly seasonal, sources, such as landfills. It is, however, notable that the least successful part of the otherwise very impressive Fung et al. model is its fit to seasonality in the high Northern Hemisphere. Tie et al. (1991), however, assume in their model that the north, especially the Canadian forest, is an intense source of CH_4 (their Figure 3). The fit of their calculated CH_4 mixing ratios to actual observed ratios is moderately good on an annualized basis, though the calculated model is slightly low, especially in the southern tropics. J. A. Taylor et al. (1991) also achieve a fairly successful fit. The Taylor et al. model, like that of Fung et al., does not fully reproduce the northern seasonality. In the Taylor et al. model, either the CH_4 production in summer at high northern latitudes is overestimated, there is some other problem in the data set, or the OH model is inadequate.

There is much controversy about the size of the northern contribution to the global CH_4 budget. In the tundra, Whalen and Reeburgh (1992) estimated that Arctic wet meadow and tussock-shrub tundra emit about 40 Tg CH_4 annually (nearly 10% of the global budget), but with wide error margins. A small counterbalance to this flux is consumption by soil, at up to 0.8 Tg/year (Whalen et al. 1991). Moore et al. (1990) estimated 14 Tg annually for the northern fen output. In the boreal forest, widely varying measurements and estimates have been made of CH_4 output (see emissions data in Bartlett and Harriss 1993). Roulet et al. (1992), in an excellent study, estimated a comparatively very small output of 0.15 Tg/year from the low boreal region of Canada, a figure that contrasts strongly with, for example, the model fits to atmospheric data of Tie et al. (1991) and J. A. Taylor et al. (1991), which assume that the region is a major global source. The work of Dise (1993) suggests that the total northern wetland flux may be well in excess of 50 Tg. By integrating daily flux values from a variety of site types, she estimated annual emissions around 20 g CH_4 per m^2 from bogs and 39 g CH_4 per m^2 from fens. If so, annual emission from Minnesota, Wisconsin, and Michigan peatlands (approximately 6 million ha) would be 1.2–2.3 Tg. Applying similar emission estimates to the peatlands of the Soviet Union (245 million ha according to Neustadt [1984]) would give 49–96 Tg. The caveat is that the data are from Minnesota, in the very productive southern fringe of the boreal forest, and so may be unrepresentative of average productivity. Nevertheless, Dise's data would suggest a global boreal peatland source on the order of 50–150 Tg.

In particular, the impact of emission from small ponds and bogs in the boreal realm, especially as point source bursts in the autumn, is very difficult to quantify. There is strong evidence that much of the output from boreal forest is from very localized sources, such as beaver ponds (Roulet et al. 1992, although farther south Yavitt et al. [1990] find rather different results). In much of the Canadian forest, conifer swamps probably produce little CH_4, but thicket swamps and especially beaver ponds are more productive. The great importance of beaver ponds has been stressed by Ford and Naiman (1988), Nisbet (1989), Roulet et al. (1992), and Yavitt et al. (1990), although in the more southerly temperate forests studied by Yavitt et al. running open water was even more significant. One example of the impact of beaver is the change in wetland in northern

Minnesota in recent years (Naiman et al. 1991). By A.D. 1900, beaver were nearly extinct in the area. More recently, the population has recovered sharply. As this has happened, the open water area of beaver ponds in one study area increased 86-fold, from 0.16 km² to 13.7 km² in a total area of 294 km². Generalizing from this finding, Naiman et al. (1991) concluded that about 1% of the recent increase in global atmospheric CH₄ was due to pond creation by beavers in North America. Beaver ponds are fed with carbon from hardwoods from wide surrounding areas, and the shallow ponds, over-loaded with organic debris, are, in comparison with other parts of the system, efficient producers of CH₄. The output of such point sources may have been underestimated, both because it is difficult to quantify such factors as animal disturbance in summer and autumn and because it is very difficult to quantify the total productive area of small point-source ponds. Informal personal observations on the margins of running open water where beavers were active suggested that episodic ebullition of CH₄ (e.g., in late afternoon, possibly increased as a result of temperature and worm burrowing) was a major factor. Ebullition was from small "volcanoes" in sand over mud, which were probably difficult to sample.

In the preferred scenario of Fung et al., the landfill output may be too high at 40 Tg/year. The carbon in newspapers and diapers comes, ultimately, from the forest. The purely subjective and qualitative impression gained by walking over the debris and through new deciduous growth, beaver ponds, and bogs of old clearcuts is that more CH₄ is produced from carbon remaining in anaerobic decay in the former forest and early hardwood regrowth in the region that has supplied the pulp for the landfill than is emitted in the landfill itself from slow decay of that pulp converted to paper and packaging and now under soil cover. A study of United Kingdom landfill emission concluded that about 1 Tg annually was released from landfills (see discussion in A. Williams 1993). If it is proportionate to United King-dom emissions, global total emission from rich waste would be circa 22 Tg annually (using data in World Resources Institute 1990). Global estimates of landfill emis-sions of CH₄ vary widely, from 9 to 70 Tg/year (Bogner and Spokas 1993), and such estimates are poorly constrained (Peer et al. 1993). Thornloe et al. (1993) estimated global landfill emissions in the range of 11–33 Tg. Emissions from wastewater treatment are also large. Reviewing the literature, Thornloe et al. (1993) estimated 12–38 Tg annually.

Possibly, the Fung et al. model would better fit the seasonality data in the Northern Hemisphere if it used a somewhat higher estimate of boreal realm wetland output, partly emitted in the autumn (Rudd and Hamilton 1978), and a correspond-ingly lower estimate of the less seasonal or aseasonal landfill output. Similarly, the J. A. Taylor et al. model would more closely approximate observations if boreal output were skewed toward autumn. However, there are few observational data to prove this assumption of skewed emission. The work of Dise (1993) shows that output is strongly pulsed, and at a maximum in later summer, with some pulses of emission in autumn. CH₄ produced in July-September may find its way to the marine boundary layer in August-October, yielding (Nisbet 1989) the sharp autumn rise in output recorded by the global sampling network (Figure 10.1), especially as OH levels are falling at this time.

VULNERABILITY OF NORTHERN WETLAND SOURCES TO CLIMATIC CHANGE

The northern wetlands are today far from their late Holocene "natural" state. In the productive southern fringe of the forest, very extensive exploitation has drastically changed the vegetation patterns, both in Canada and in Russia (Barr and Braden 1988). Farther north, radical changes in animal populations, induced by humanity, have probably had a marked impact on the wetland and the carbon supply to that wetland. Consequently, the present CH_4 output may be very different from the output in the late Holocene. Further changes, however, are likely to take place in the near future, as a consequence of climatic change, changes in animal impact, and changes in direct and indirect human impact (Chapter 9, this volume). These changes may be massive in scope (Overpeck et al. 1991) but constrained by carbon-nutrient interactions (Shaver et al. 1992).

The most obvious variable is climatic change. Most models of future climate infer markedly warmer temperatures in the boreal realm. There is less agreement, however, about the consequences for surface soil moisture. In some models, precipitation and wetness generally increase over much of the boreal realm, while melting of permafrost contributes water. In others, the increase in precipitation is counterbalanced by increased evaporation. In yet others, significant areas may become substantially drier (e.g., in central Canada and the Russian interior). Although interesting studies have been carried out on the impact of this change on forest productivity (e.g., Wheaton et al. 1987), it is clearly premature to assess the impact on CH_4 production in quantitative terms. It is possible, however, to draw general inferences. Most models (e.g., Canadian Climate model, Geophysical Fluid Dynamics model, and United Kingdom Hadley Institute model, in Houghton et al. 1990) suggest generally higher soil moisture and precipitation in the winter months in the boreal realm. This situation would imply generally wetter conditions at spring melt, and thus broadly more favorable conditions for spring CH_4 generation. However, the same models imply substantially warmer summers in the boreal realm, possibly with less precipitation, giving much lower soil moisture in the summer months. If correct, these circumstances would imply reduced wetland in the late summer months—those months that may be the most important for CH_4 generation. Furthermore, low moisture levels in July-August-September, when pond and bog floor CH_4 inventories are high, would imply that much of the CH_4 would be oxidized on ascent and might not reach the atmosphere. To some extent, these effects would be countered by the impact of generally higher temperatures on methanogenesis, increasing output, but such impact would be rapidly limited by nutrient and substrate availability. Thus, in the most general terms, the direct impact of physical changes in the climatic system would probably be to reduce CH_4 output—a negative biotic feedback to the system.

However, a variety of more subtle second-order effects may prove to be very important. The most obvious is that there may be substantial shifts in the ecotones in the boreal realm, with general northward retreat of the southern forest boundary (most probably occurring after fires, when forest would be succeeded by parkland) and a simultaneous northward advance of the northern forest boundary. The advance of the northern boundary of the forest would be substantially slower than the changes in the south, however, possibly taking millennia, and in the interim extensive productive wetland would develop along the southern fringe of the tundra. These shifts would change the areas of CH_4-producing systems and hence change output. Brown et al. (1989) have drawn attention to the very large storage of CH_4 in peat. Globally, they suggested that as much as 144 Tg of CH_4 may be held as gas trapped in peat bogs, much of it in the northern realm. Sharp changes in climate, exposing this peat, could have the effect of rapidly liberating this store, with a significant short-term global impact on atmospheric CH_4. Gorham (Chapter 9) discusses the potential response of peatlands to global change at greater length. More generally, the total carbon store in some peat-rich areas is huge. This carbon has been accumulated through the Holocene and may be at risk of conversion to CO_2 or CH_4 if climate changes.

The northern forest depends heavily on fire for renewal. Over the past half century, there has been widespread fire suppression, and forest in protected areas may be overmature; in other areas, cutting practice may have disrupted the preindustrial coniferous-deciduous balance and hence changed the supply of carbon to bogs and small ponds. Furthermore, many fires are now being set by humans during hot periods, rather than by lightning during storms: consequently the impact of the fires may be changing. If climate changes to reduce summer soil moisture, then fire incidence may increase, increasing the flux of CH_4, CO, and NO_x (with impact on OH) and the supply of nutrients to wetland, and in the early stages of regrowth may increase the supply of food for animals such as beavers that have a major impact on CH_4 production.

The extent of the impact of animals such as beaver, moose, and muskrat may be disputed, but animal population cycles, affected both directly and indirectly by climatic change and changing human behavior, can have a measurable impact on global atmospheric processes. Some past changes have had massive effect. Prior to 1950, trapping and disease had driven beavers almost to extinction in both North America and Russia. The subsequent recovery has caused a sharp increase in wetland area in Canada and the northern United States, and most likely will have a parallel effect in Russia in the near future. Moose population cycles can also affect CH_4 output. Moose browsing favors the growth of spruce forest (McInnes et al. 1992), which would eventually reduce the supply of carbon to wetland. It is possible that the recent introduction of moose into the island of Newfoundland may have locally reduced total CH_4 output, although increasing bubble-release by physical disruption of bogs and ponds.

Another example is the population cycling of spruce budworm. Waves of budworm infestation can spread across forest regions, massively destroying mature spruce forest so that it is replaced by immature new growth. Budworm has been long established in eastern Canada, where it is a major natural control on forest generational stands, but it has only recently become recognizably significant in the forests west of Hudson Bay. In this latter region, it has in very recent years caused major changes in forest management. One plausible though unproven explanation of the appearance of budworm in the west is that its increase may be related to the decline of birds such as the Tennessee warbler that have lost winter habitat in central America. Should there be major loss of mature spruce forest in the western boreal realm, there will be an increase in the area under younger successional stages, which is rich in deciduous hardwood, providing organic matter for ponds and sustaining a much larger population of beavers and other wetland fauna. The increase in CH_4 output from new beaver ponds fed by the new organic matter may be significant. All these impacts are very difficult to quantify, or even in some cases to identify, but it is clear that simplistic physically derived conclusions about the change in CH_4 output under an altered climate are unreliable.

Direct human impact can also be important. An obvious example is flooding for hydroelectric power. J. W. M. Rudd (personal communication 1991) warned of the danger of degeneration of flooded peat bogs with conversion of carbon to CH_4. Projects such as the James Bay hydroelectric scheme, by creating a large area of shallow water, supplied with ample stores of carbon in flooded peat and forest, may become globally significant CH_4 producers. Another example of human intervention is acid rain, which affects areas of boreal forest in eastern Canada, Scandinavia, and Russia. The impact of sulfate on ecosystems has already been discussed.

To summarize, the initial impression one gains from physical models of climate is that the boreal realm may become substantially drier in summer and that accordingly CH_4 output will decline, giving a negative feedback. However, there are a wide variety of other effects—including shifts in vegetation, changes in the population dynamics of animals such as beaver and moose, insect outbreaks, and direct human intervention— that collectively may overwhelm the direct impact of changing climate (for further discussion, see Gorham, Chapter 9, this volume). The overall nature of the changes is not yet predictable and will depend strongly on carbon-nutrient balances. Perhaps large parts of the northern biomes will become unstable. They may continue to exist simply because they already exist, but with little resilience to challenges such as drought or major fire, as they will be essentially relict communities out of their optimum environmental setting. It is impossible to estimate in which way these factors will collectively change CH_4 output, but intuition suggests that CH_4 output from the boreal zone may increase overall as climate changes.

LOW-LATITUDE WETLAND SOURCES OF CH_4

Low-latitude wetlands include some of the most important global sources of CH_4 (Aselmann and Crutzen 1989; Bartlett and Harriss 1993). In the preferred scenario of Fung et al. (1991), of the total global output of circa 500 Tg annually (see also Crutzen [1991], who gives an annual emission flux of 505 ± 105 Tg), 100 Tg were thought to come from rice fields and 80 Tg from swamps and alluvial sources in low latitudes (Table 10.1). Fung et al. considered that much of the rice field output came from the region between 40°N and 10°N; in contrast, the swamp output was mostly from the region between the equator and 30°N. In general, tropical wetland sources of CH_4, like the northern wetland sources, are difficult to measure. Areas of relatively low output with oxic conditions may be widespread, whereas intense output may come from point sources of anaerobic decay. Most probably, the most productive areas are regions of papyrus swamp, such as the Sudd. Much depends on local vegetation type, on animal and insect activity, and on the vagaries of seasonal rainfall in zones where interannual fluctuation in runoff may be considerable.

In northern tropical Africa the major sites are in the Sahel margin, the Niger inland delta, the Lake Chad area, and the Sudd. The Tigris-Euphrates system of Iraq is similar. It is likely that these regions were much more extensive in the early Holocene (Street-

Perrott 1992), and they may have played a critical role in initiating the postglacial climate. In west equatorial Africa, the major swamps and wetland sources are in Cameroon, the Congo (Brazzaville), the Central African Republic, and Zaire. Some of these include seasonally inundated forest, comparable to the flooded forests of Amazonia. In the Congo basin, high emissions come from 10^5 km of flooded forest, which alone may produce 1.6–3.2 Tg of CH_4 annually (Tathy et al. 1992). East equatorial Africa has extensive swamp in Rwanda and Uganda. In Southern Africa, Zambia has widespread swamp, as has the inland delta of the Okavango in Botswana. South America has the celebrated Amazonian wetland (Bartlett et al. 1990) and the very extensive but much less studied Parana Basin and Gran Chaco swamps of Paraguay and northern Argentina. Elsewhere in the tropics, natural wetland exists but has been much modified into rice field in Bangladesh, Burma, Thailand, Cambodia, Vietnam, China, and Indonesia. Seasonal inundation also occurs in Queensland and in the Murray-Darling Basin of Australia.

Bartlett et al. (1990), studying the Amazonian wetland, contrasted the output of open water areas (circa 75 mg CH_4 per m^2d), floating grass mats (circa 200 mg CH_4 per m^2d), and flooded forest (circa 125 mg CH_4 per m^2d). Wassmann et al. (1992) obtained slightly different results but also reached the general conclusion that the region is a major CH_4 source. Ebullition was a significant component of CH_4 emissions. Bartlett et al. (1990) and Bartlett and Harriss (1993) used these data to generalize, using the Matthews and Fung (1987) model of wetland distribution, to obtain the global output from tropical wetland. They concluded (1993) that tropical sources produced perhaps 60% of the total wetland output globally, or 66 Tg annually. This conclusion was broadly accepted by the IPCC assessment (Houghton et al. 1990) and in the Fung et al. (1991) model of global output, which allocated 80 Tg annually to tropical swamps and alluvial sources in the preferred scenario 7.

If this conclusion is correct, tropical sources are the most significant natural component of the CH_4 budget. However, the estimate of tropical wetland areas in the Matthews and Fung (1987) compilation, used by Fung et al., J. A. Taylor et al. (1991), and Bartlett and Harriss (1993), may be locally in error, and it may be unsafe to generalize from Amazonian data to the Parana Basin and East Africa. For instance, in East Africa some basins are seasonally or rarely inundated, or are salt lakes and salt flats, producing little CH_4. Thus it is possible Bartlett et al. (1990), Fung et al. (1991), and Bartlett and Harriss (1993), as well as the IPCC assessment, may have overestimated the relative significance of tropical wetland sources. Nevertheless, in this review the Fung et al. model is accepted as the best one currently available. Aselmann and Crutzen (1989), using a different wetland data base, come to broadly comparable conclusions.

HISTORICAL RECORD AND VULNERABILITY TO CHANGE

The tropical wetland sources of CH_4 are probably most intense in the flooded forest and floating grass mats of Amazonia and the papyrus swamps of Africa. These sources have the interesting capacity of being able to respond extremely quickly to climatic change. Because of this property, they may be among the most important components of the natural response to global climatic change.

The areal extent of African and Amazonian swamps is closely dependent on rainfall, especially in the north African monsoonal system. The dependence is nonlinear, since runoff is greatly disproportionate to rainfall and is vegetation-dependent. In the drought of the early 1980s, for instance, the area of Lake Chad shrank dramatically, to the extent that it appeared that the lake might be lost entirely. Conversely, an increase in rainfall can, over a few seasons, massively increase the area of a swamp. However, this areal increase in flooded area would have little impact if organic carbon were not available in the correct environment for conversion to CH_4. Here lies the significance of such vegetation as papyrus. This vegetation very rapidly captures CO_2 from air; after annual growth, the organic matter rots, creating anaerobic conditions ideal for the efficient production of CH_4. This system may be contrasted with the boreal forest, where annual CO_2 capture (by the leaves of deciduous hardwoods) is significant, but where much of the longer-term carbon store and supply to wetland is in wood; consequently, the response time of the tropical system to short-term climatic change is much faster than that of the boreal system.

The rapid response of the tropical CH_4 output may have been a major cause of the sharp rises in atmospheric CH_4 at the end of the last glaciation (Figure 10.5). The geological record of the early Holocene (Petit-Maire et al. 1991; Street-Perrott 1992) shows that vast swamps developed in the Sahel region of Africa; this was the time of the "wild Nile," when major CH_4-producing swamps were probably widespread in the Nile and Niger basins. Historically, in the geological record, there is evidence that sudden and major changes in CH_4 output probably took place. The most important of these changes (Figure 10.6) took place at the end of the last glaciation and at the start of the Holocene, at Termination 1A (about 13,500 years ago) and Termination 1B (slightly over 10,000 years ago). (These dates may be revised in light of the results of Johnsen et al. [1992] and K. C. Taylor et al. [1993], which suggest that Termination 1A was at 14,600 years B.P. and Termination 1B at 11,600 years B.P.) In both these events, the CH_4 mixing levels in the atmosphere increased. It is possible that they jumped suddenly, over a period that may have been as short as a few decades or even a few years (e.g., see conductivity results in K. C. Taylor et al. 1993), although this interpretation is a matter of dispute and the published records are ambiguous. At Termination 1A, the levels increased from roughly 350 ppb to roughly 650 ppb, a near doubling (see the summary of the evidence in Raynaud et al. [1988] and Chappellaz et al. [1990]). Chappellaz et al. (1990, 1993) suggested that increased output from tropical sources in the strengthened monsoon system caused the rise in mixing ratio. Nisbet (1992) suggested that a thermal runaway was induced by bursts of CH_4 from Arctic hydrates, which then warmed the global climate briefly, increasing rainfall in the Sahel and creating major swamps in Africa. It is possible that sudden flooding caused the rapid formation of large swamps in Amazonia and Zaire, then substantial grassland, but the evidence for this conclusion is fragmentary. The CH_4 that these swamps would rapidly have produced would then have reinforced the Arctic warming, setting off more hydrate release. It is possible that a resonance developed between formation of tropical CH_4 swamps and release of CH_4 from Arctic hydrates, each reinforcing the other to cause the sharp climatic warming in the early Holocene.

These speculations are as yet not widely accepted. However, there is excellent geological evidence that the area of the African swamps has been markedly larger in

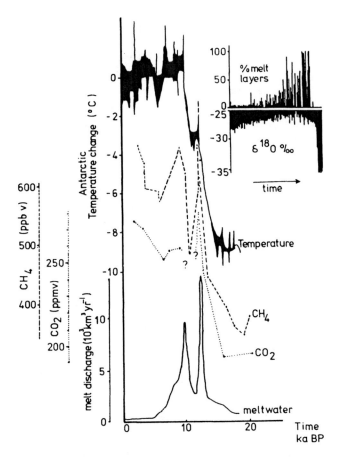

Figure 10.5. Summary of changes at the end of the last glaciation. Note that CH_4 levels and possibly also CO_2 levels may have declined in the Younger Dryas, around 11,000 years B.P., although the evidence for CO_2 decline is ambiguous. The figure shows, from the bottom, meltwater discharge, CO_2 level, CH_4 level, and temperature in Antarctica, as measured from the Vostok Core. The inset, on a different time scale, shows the incidence of melt layers and oxygen isotope variation in a Canadian ice core to illustrate the frequency of warm events in the early Holocene. (From Nisbet 1992, using sources cited therein.)

times of past climatic optima in the early Holocene, and that global warming would be likely to recreate these conditions. Global climatic models are mixed in their predictions about the African and South American swamp areas. Some models predict increased soil moisture over northern tropical Africa; however, these models also predict lower soil moisture in Amazonia. Yet soil moisture is not necessarily crucial in swamp formation; what is much more important is runoff to river systems, especially after brief very wet periods or after intense tropical storms or hurricanes. A single cyclone that penetrates far inland can produce more runoff to interior swamp basins than a decade of normal rainfall. If the frequency of tropical storms in north central Africa increases, there may be a very sharp increase in CH_4 output from the papyrus swamps to the Nile, Chad, and Niger basins.

Figure 10.6. Vegetation changes in tropical Africa in the past 20,000 years. This region is a major source of CH_4, produced by swamps and other wetlands in the equatorial region, by biomass burning in the savanna regions, and by seasonal swamps and fluctuating lakes in the northern and southern tropical margins. In equatorial areas in Africa and Amazonia, major swamps may have developed very quickly in the earliest Holocene after sudden massive flooding. Forest growth reestablished the interglacial pattern of vegetation more slowly. The figure shows the scale of the changes in vegetation that can occur within a time interval that, by geological standards, is a very short one. By implication, major future climatic shifts would have an impact on a similar scale. However, future changes, although they may be equally large, are most unlikely to replicate this pattern: the climatic shifts will be different, and the most important shaper of vegetation distribution today is humanity. (From Nisbet 1992, using sources cited therein.)

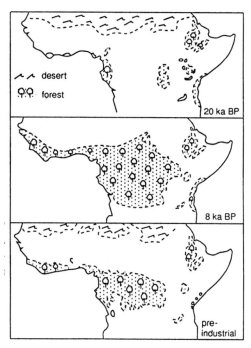

The impact of rice production is not discussed here, as it falls outside the scope of a review dealing with natural and quasi-natural systems. However, it should be noted that the conversion by water impoundment of tropical and subtropical habitat to rice paddy fields is continuing; this process has the effect of considerably increasing CH_4 output. There may be large feedbacks to climate change as rice production changes. Furthermore, CH_4 output may be altered by changes in crop management. For a general summary, see Neue and Roger (1993).

SOILS AND TERMITES

Soils can act both as emitters and as consumers of CH_4, depending on local conditions, which may vary daily and seasonally. Fung et al. (1991), in their preferred scenario, estimated that the net global absorption of CH_4 by soil (i.e., the net sink, mostly for eventual release as CO_2) is 10 Tg/year, but this estimate is poorly constrained. Dorr et al. (1993), for example, find a range of 9.0–55.9 Tg of annual uptake with a best estimate of 28.7 Tg/year.

Much of the soil absorption occurs in the drier parts of forests that in other local areas are net emitters. In boreal forest, CH_4 consumption is linked to the maturity of the sites (Whalen et al. 1991), with mature stands of spruce being net consumers of CH_4. At the sites measured, CH_4 consumption increased with temperature and at some sites decreased with increasing soil moisture. In general, it is likely that well-drained boreal forest soils are a net CH_4 sink. Whalen et al. estimated the global size of the upland and floodplain taiga sink at 0–0.8 Tg annually, although other estimates have been higher. In temperate soils, CH_4 fluxes can vary considerably, depending on moisture and temperature (Yavitt et al. 1990), with soils at times acting as sinks, but also occasionally acting as net sources. Yavitt et al. estimated that the net effect of temperate forest soils was production of roughly 2 Tg annually but placed wide error limits (roughly 3 Tg) on this estimate. Ojima et al. (1993), in contrast, inferred a temperate soil sink of approximately 20 Tg/year. More generally, Crill (1991) has pointed out how poorly constrained is our knowledge of soil behavior and the seasonality of CH_4 uptake and CO_2 release.

In tropical systems, the role of soils is also poorly constrained. In general, forest soils probably have higher consumption than soils converted to agricultural use (Keller et al. 1990). The effect of this conversion alone may have been to reduce the global sink by 2 Tg annually. In scrub grass savanna in the northern Guyana Shield, Venezuela, Scharfe et al. (1990) observed emission of CH_4 from "dry" savanna soils but its consumption by forest. A significant dispersed source, probably from termites and small tracts of flooded soils (termed "vleis" in southern Africa), is present in the tropical savanna region worldwide. Termites probably emit on the order of 20 Tg annually. Scharfe et al. extrapolated an annual global emission of 30–60 Tg from this type of source, including in this figure the very substantial contribution from termites (e.g., Tyler et al. 1988).

BIOMASS BURNING

One of the striking impressions gained from a night flight southward down the length of Africa during the later months of the year is the intensity and widespread distribution of grass fires in the savanna margins of west and central Africa; six months later, in the period after the northern spring equinox, the same is true of east central and southern Africa. Biomass burning is a massive CH_4 source (Crutzen and Andreae 1990). Virtually all of these fires are lit by humans. The spread of grassland, partly induced by climatic change but mostly due to direct human intervention, will greatly increase the level of this burning, and thus increase the net CH_4 flux from the tropical savannas at the expense of forest and woodland output.

Delmas et al. (1991) studied African savanna. In Congolese savanna, there was significant CH_4 uptake by soil, at the same rate on both natural and cultivated land, but on the Niger River floodplain in Mali only very small amounts of uptake occurred. In contrast, the amount of CH_4 liberated by biomass burning was considerable, much greater than that absorbed by soil. Delmas et al. (1991) estimated that, after allowing for the effect of fire, emission exceeds absorption in African savanna soils by about an order of magnitude, and that total CH_4 emission from combustion of organic material in tropical Africa is on the order of 9 Tg, about half of which is from fire in the savanna zones and a quarter of which is from charcoal production.

Estimates of the total CH_4 output from biomass burning vary greatly, from about 28 to 80 Tg/year (Hao and Ward 1993; Levine et al 1993), or 5–15% of total global emissions. Hot dry flaming fires produce mostly CO_2, but smoldering biomass produces CO, CH_4, and other hydrocarbons. In forest fires, the smoldering phase may last a day or more, and in savanna grassland, after a brief flaming phase, smoldering may continue for an hour.

Anecdotal evidence implies that a significant increase in biomass burning has occurred between 1850 and 1990. Certainly, in my own personal observations during frequent flights across Africa in the 1968–1992 period, there has been a marked increase since the early 1980s. Savanna burning is very dependent on the stand of grass and hence on wet season rainfall. Toward the end of the dry season, widespread burning occurs to clear land for cultivation and to bring on green shoots from perennial grasses for cattle to eat. Even after a single drought year, biomass burning can be intense since well-rooted perennial grasses remain abundant. However, in multiyear drought perennial grasses die and little material is available to burn. This may have occurred between 1992 and 1993 in Africa (Andreae et al. 1993) and may partly explain a marked C isotopic shift detected in the Southern Hemisphere by Lowe et al. (1994). They concluded that reduction in burning was a partial cause of the cessation of growth in Southern Hemisphere atmospheric CH_4 between 1992 and 1993.

Any substantial future change in rainfall in the savanna grassland areas, especially in Africa, will thus have significant impact on CH_4 output from burning. It should be stressed, however, that dry season burning is almost entirely anthropogenic and the product of poor agricultural practice. Reduction in burning can be achieved by agricultural education.

ANIMAL-DERIVED CH_4

Enteric fermentation in animals is a major CH_4 source, but most such animals today are domestic and thus out of the scope of this review of natural and quasi-natural sources. The emissions by domestic animals reflect the priorities of humanity, not natural controls. However, a brief mention of this large source is justified for the sake of completeness. The topic is reviewed by Johnson et al. (1993).

Total emission from enteric fermentation in animals is around 80 Tg annually (Johnson et al. 1993) at present (see also Table 10.1), mostly from cattle (bovine eructation), sheep, and water buffalo, which account for over 90% of emissions from domestic animals. Emission from this source is obviously dependent on global climatic change, and episodes of famine will cause major reductions in emission. However, if it is assumed that major famine will be avoided, Anastasi and Simpson showed that it is possible to use demographic projections of human populations to estimate future populations of domestic animals. If this approach is valid, it implies that emissions will increase by about 1 Tg annually, to 119 (± 12) Tg by A.D. 2025.

RECENT CHANGES TO THE GLOBAL CH_4 BUDGET

In the early 1990s, radical changes have begun to take place in the behavior of global atmospheric CH_4. Dlugokencky et al. (1994) report a dramatic decrease in the growth rate. The average Northern Hemisphere trend in the period 1983–1991 was a growth of 11.6 ± 0.2 ppbv per year; in 1992, in contrast, there was negligible growth of 1.8 ± 1.6 ppbv, with some stations, especially in high latitudes, showing falls. In the Southern Hemisphere in 1992 the growth rate, which in 1983–1991 had been similar to that in the north, fell to 7.7 ± 1.0 ppbv. Since the end of 1992, growth in the Southern Hemisphere has slowed yet more. Lowe et al. (1994) consider that the behavior of the 1992–1993 Southern Hemisphere is now similar to that of the 1992 Northern Hemisphere, with evidence for isotopic changes in addition to the change in concentration growth.

There are probably several explanations for this change in the behavior of atmospheric CH_4. These include changes in fossil CH_4 output, in biomass burning flux, and in atmospheric chemistry (Lowe et al. 1994). The change seems to have commenced in the north and reached the south a year later by atmospheric circulation. One factor is the change in destruction of CH_4 by OH partly as a result of the stratospheric impact of the eruption of Mt. Pinatubo in June 1991. This effect may have reduced global CH_4 levels, but it does not explain the strong northern bias to the change. Most likely, the dominant cause of the change is a decline in losses of "fossil" CH_4 from the formerly centrally planned economies of Russia and Eastern Europe. Here there have been reductions in leakage from gas pipelines and wellheads, together with sharp reductions in coal mining. Simultaneously, drought in the Southern Hemisphere, especially Africa, sharply reduced biomass burning there.

The suddenness of the change in the behavior of CH_4 demonstrates how easily CH_4 systematics can alter. Much of the change may be the result of direct human intervention (fixing gas leaks, closing coal mines), but the reduction in biomass burning is a response to a climatic event. Changes in other species, such as CO, tropospheric ozone, and nitrogen oxides, are also involved, with their important role in CH_4 chemistry (Law and Pyle 1991). Strong feedback factors are probably at work in the atmospheric chemistry controlling CH_4. Feedbacks that in the future could be important include climatic shifts in tropical grassland areas, changing the amount of burning; changes in forest flooding; changes in moisture levels, vegetation patterns, and animal ecology in the boreal realm; and Arctic warming (Lachenbruch 1988; Hinzman and Kane 1992). A strong case can be made that reduction in natural gas, coal, and landfill emissions is both feasible and necessary.

There is a significant danger of a massive increase in output from several of the natural and quasi-natural sources, although the more conservative view is that dramatic shifts, though possible, are unlikely. The major potential dangers include massive emission from Arctic hydrate, especially in Western Siberia, or release from a major marine slump. In the boreal realm, decomposition of carbon held in bogs may occur, with resultant CH_4 production. Biomass burning, either on a vast scale in forests or of bog, can also accompany climatic change. In the tropics, climatic change may produce large-scale flooding and sudden development of huge swamps, which would massively increase CH_4 output. It should be stressed, however, that all these scenarios are "nightmares": most are unlikely to happen. They are mentioned here only because the geological record at Terminations 1A and 1B shows that such events are indeed possible; although there is no present evidence to show that we are on the threshold of such a catastrophe, the possibility should not be forgotten.

In conclusion, global climatic change may have major impact on the CH_4 output of natural abiotic and biotic systems. Nevertheless, at present our understanding of the CH_4 budget remains poor, and the response of human societies to global change, particularly as that response affects vegetation, is unpredictable. Thus it is impossible to predict the likely strength and direction of feedbacks. The only certainty is that change will occur; there is an urgent need for better monitoring of that change as it takes place.

ACKNOWLEDGMENTS

We thank George Woodwell for asking us to prepare this review, Dominique Raynaud for his very helpful criticism and thoughtful arguments against some of the hypotheses outlined here, and Eric Davidson for a most helpful critical review. I thank Dave Lowe, Keith Lassey, Martin Manning, and Gordon Brailsford for much erudition, and Ed Dlugokencky, Pieter Tans, and Meg Fung for their kindness and help to a newcomer to the field. I would also like to thank a very patient copyeditor. This work was supported in part by an NSERC Canada operating grant to E. G. Nisbet and by the University of Saskatchewan and the University of London.

REFERENCES

Andreae, M. O., T. W. Andreae, W. Ebert, G. W. Harris, F. G. Weingold, T. Zenker, H. Annegarn, F. Beer, H. Cachier, W. Maenhut, I. Salma, and R. Swap. 1993. Airborne studies of aerosol emissions from savannah fires in Southern Africa. *EOS 74* 43:128.

Aselmann, I., and P. J. Crutzen. 1989. Global distribution of natural freshwater wetlands and rice paddies, their net primary productivity, seasonality and possible CH_4 emissions. *J. Atmos. Chem.* 8:307–358.

Barr, B. M., and K. E. Braden. 1988. *The Disappearing Russian Forest.* Hutchinson, London.

Bartlett, K. B., P. M. Crill, J. A. Bonassi, J. E. Richey, and R. C. Harriss. 1990. Methane flux from the Amazon river floodplain: Emissions during rising water. *J. Geophys. Res.* 95:16773–16788.

Bartlett, K. B., and R. C. Harriss. 1993. Review and assessment of methane emissions from wetlands. *Chemosphere* 26:261–320.

Baulin, V. V., and N. S. Danilova. 1984. Dynamics of late Quaternary permafrost in Siberia. In *Late Quaternary Environments of the Soviet Union,* ed. A. A. Velichko, English language eds. H. E. Wright and C. W. Barnosky, pp. 69–78. University of Minnesota Press, Minneapolis.

Beck, L. L. 1993. A global methane emissions program for landfills, coal mines and natural gas systems. *Chemosphere* 26:447–452.

Beck, L. L., S. D. Piccot, and D. A. Kirchgessner. 1993. Industrial sources. In *Atmospheric Methane: Sources, Sinks and Role in Global Change,* ed. M. A. K. Khalil, pp. 399–431. Springer-Verlag, Berlin.

Bogner, J., and K. Spokas. 1993. Landfill CH_4: Rates, fates and role in global carbon cycle. *Chemosphere* 26:369–386.

Boone, D. R. 1993. Biological formation and consumption of methane. In *Atmospheric Methane: Sources, Sinks and Role in Global Change,* ed. M. A. K. Khalil, pp. 102–127. Springer-Verlag, Berlin.

Bouwman, A. F., ed. 1990. *Soils and the Greenhouse Effect.* Wiley, Toronto.

Brown, A., S. P. Mathur, and D. J. Kushner. 1989. An ombrotrophic bog as a CH_4 reservoir. *Glob. Biogeochem. Cycles* 3:205–213.

Burke, R. A., and W. M. Sackett. 1986. Stable hydrogen and carbon isotopic compositions of biogenic CH_4 from several shallow aquatic environments. In *Organic Marine Chemistry,* ed. M. L. Sohn, pp. 297–313. American Chemical Society, Washington, D.C.

Chanton, J. P., G. J. Whiting, W. J. Showers, and P. M. Crill. 1992. Methane flux from *Pelttandra virginica:* Stable isotope tracing and chamber effects. *Glob. Biogeochem. Cycles* 6:15–31.

Chappellaz, J., J. M. Barnola, D. Raynaud, Y. S. Korotkevich, and C. Lorius. 1990. Ice core record of atmospheric CH_4 over the past 160,000 years. *Nature* 345:127–131.

Chappellaz, J. A., I. Y. Fung, and A. M. Thompson. 1994. The atmospheric CH_4 increase since the last glacial maximum: 1. Source estimates. *Tellus* 45B:228–241.

Cicerone, R. J., and R. S. Oremland. 1988. Biogeochemical aspects of atmospheric methane. *Glob. Biogeochem. Cycles* 2:299–327.

Conrad, R. 1989. Control of CH_4 production in terrestrial ecosystems. In *Exchange of Trace Gases between Terrestrial Ecosystems and the Atmosphere,* ed. M. O. Andreae and D. S. Schimel, pp. 39–58. Wiley, Chichester, U.K.

Cox, L. J. 1983. ed. *Natural Gas Hydrates: Properties, Occurrence and Recovery.* Butterworth, Woburn, Mass.

Crill, P. M. 1991. Seasonal patterns of CH_4 uptake and carbon dioxide release by a temperate woodland soil. *Glob. Biogeochem. Cycles* 5:319–334.

Crutzen, P. J. 1991. Methane's sources and sinks. *Nature* 350:380–381.

Delmas, R. A., A. Marenco, J. P. Tathy, B. Cros, and J. G. R. Baudet. 1991. Sources and sinks of CH_4 in the African savanna. CH_4 emissions from biomass burning. *J. Geophys. Res.* 96: 7282–7299.

Dinel, R. D., S. P. Mathur, A. Brown, and M. Levesque. 1988. A field study of the effect of depth on CH_4 production in peatland waters: Equipment and preliminary results. *J. Ecol.* 76: 1083–1091.

Dise, N. B. 1993. Methane emission from Minnesota peatlands: Spatial and seasonal variability. *Global Biogeochem. Cycles* 7:123–142.

Dlugokencky, E. J., K. A. Masarie, P. M. Lang, P. P. Tans, L. P. Steele, and E. G. Nisbet. 1994. A dramatic decrease in the growth rate of atmospheric CH_4 in the Northern Hemisphere during 1992. *Geophys. Res. Lett.* 21:45–48.

Dorr, H., L. Katruff, and I. Levin. 1993. Soil texture parameterization of the methane uptake in aerated soils. *Chemosphere* 26:697–713.

Fechner, E. J., and H. F. Hemond. 1992. Methane transport and oxidation in the unsaturated zone of a *Sphagnum* peatland. *Glob. Biogeochem. Cycles* 6:33–44.

Ford, T. E., and R. J. Naiman. 1988. Alteration of carbon cycling by beaver: Methane evasion rates from boreal forest streams and rivers. *Can. J. Zool.* 66:529–533.

Fraser, P. J., S. Coram, and N. Derek. 1987. Atmospheric CH_4, carbon monoxide and carbon dioxide by gas chromatography, 1978–85. In *Baseline Atmospheric Program (Australia), 1985,* ed. B. W. Forgan and P. J. Fraser, pp. 48–50. Department of Science, CSIRO, Canberra, Australia.

Fung, I., J. John, J. Lerner, E. Matthews, M. Prather, L. P. Steele, and P. J. Fraser. 1991. Three-dimensional model synthesis of the global CH_4 cycle. *J. Geophys. Res.* 96:13033–13065.

Galchenko, V. F., A. Lein, and M. Ivanov. 1989. Biological sinks of methane. In *Exchange of Trace Gases between Terrestrial Ecosystems and the Atmosphere,* ed. M. O. Andreae and D. S. Schimel, pp. 59–71. Wiley, Chichester, U.K.

Grace, J. D., and G. F. Hart. 1986. Giant gas fields of northern west Siberia. *Am. Assoc. Petrol. Geol. Bull.* 70:830–852.

Hao, W. M., and D. E. Ward. 1993. Methane production from global biomass burning. *J. Geophys. Res.* 98:20,657–20,661.

Hinzman, L. D., and D. L. Kane. 1992. Potential response of an Arctic watershed during a period of global warming. *J. Geophys. Res.* 97:2811–2820.

Houghton, J. T., G. J. Jenkins, and J. J. Ephraums. 1990. *Climate Change: The IPCC Scientific Assessment.* Cambridge University Press, Cambridge.

Houghton, J. T., B. A. Callander, and S. K. Varney. 1992. *Climate Change 1992: The Supplementary Report to the IPCC Scientific Assessment.* Cambridge University Press, Cambridge.

Hovland, M., A. G. Judd, and R. A. Burke. 1993. The global flux of methane from shallow submarine sediments. *Chemosphere* 26:559–578.

Johnsen, S. J., H. B. Clausen, W. Dansgaard, K. Fuhrer, N. Gundestrup, C. U. Hammer, P. Iversen, J. Jouzel, B. Stauffer, and J. P. Steffensen. 1992. Irregular glacial interstadials recorded in a new Greenland ice core. *Nature* 359:311–313.

Johnson, D. E., T. M. Hill, G. M. Ward, K. A. Johnson, M. E. Branine, B. R. Carmean, and D. W. Lodman. 1993. Ruminants and other animals. In *Atmospheric Methane: Sources, Sinks and Role in Global Change,* ed. M. A. K. Khalil, pp. 199–229. Springer-Verlag, Berlin.

Judd, A. G., R. H. Charlier, A. Lacroix, G. Lambert, and C. Rouland. 1993. Minor sources of methane. In *Atmospheric Methane: Sources, Sinks and Role in Global Change,* ed. M. A. K. Khalil, pp. 432–456. Springer-Verlag, Berlin.

Kahl, J. D., D. J. Charlevoix, N. A. Zaltseva, R. C. Schnell, and M. C. Serreze. 1993. Absence of evidence for greenhouse warming over the Arctic ocean in the past 40 years. *Nature* 361:335–337.

Keller, M., M. E. Mitre, and R. F. Stallard. 1990. Consumption of atmospheric CH_4 in soils of central Panama; Effects of agricultural development. *Glob. Biogeochem. Cycles* 4:21–27.

Kvenvolden, K. A. 1988. Methane hydrates and global climate. *Glob. Biogeochem. Cycles* 2: 221–229.

Kvenvolden, K. A., and T. D. Lorenson. 1993. Methane in permafrost–Preliminary results from coring at Fairbanks, Alaska. *Chemosphere* 26:609–616.

Lachenbruch, A. H. 1988. Permafrost temperature and the changing climate. *Eos* 69:1043.

Lachenbruch, A. H., and B. V. Marshall. 1986. Changing climate: Geothermal evidence from permafrost in the Alaskan Arctic. *Science* 234:689–696.

Law, K. S., and J. A. Pyle. 1991. Modelling the response of tropospheric trace species to changing source gas concentrations. *Atmo. Environ.* 25:1863–1871.

Levine, J. S., W. R. Cofer, and J. P. Pinto. 1993. Biomass burning. In *Atmospheric Methane: Sources, Sinks and Role in Global Change,* ed. M. A. K. Khalil, pp. 299–313. Springer-Verlag, Berlin.

Lowe, D. C., C. A. M. Brenninkmeijer, M. R. Manning, R. Sparks, and G. Wallace. 1988. Radiocarbon determination of atmospheric CH_4 at Baring Head, New Zealand. *Nature* 332:522–525.

Lowe, D. C., C. A. M. Brenninkmeijer, S. C. Tyler, and E. J. Dlugokencky. 1991. Determination of the isotopic composition of atmospheric CH_4 and its application in the Antarctic. *J. Geophys. Res.* 96:15455–15467.

Lowe, D. C., C. A. M. Brenninmeijer, G. W. Brailsford, K. R. Lassey, A. J. Gomez, and E. G. Nisbet. 1994. Concentration and ^{13}C records of atmospheric methane in New Zealand and Antarctica: Evidence for changes in methane sources. *J. Geophys. Res.* 99: in press.

MacDonald, G. J. 1990. Role of methane clathrates in past and future climate. *Climat, Change* 16:247–282.

Matthews, E., and I. Fung. 1987. Methane emission from natural wetlands: Global distribution, area, and environmental characteristics of sources. *Glob. Biogeochem. Cycles* 1:61–86.

McInnes, P. F., R. J. Naiman, J. Pastor, and Y. Cohen. 1992. Effects of moose browsing on vegetation and litter of the boreal forest, Isle Royale, Michigan USA. *Ecology* 73:2059–2075.

Moore, T., N. Roulet, and R. Knowles. 1990. Spatial and temporal variations of methane flux from subarctic/northern boreal fens. *Glob. Biogeochem. Cycles* 4:29–46.

Naiman, R. J., T. Manning, and C. A. Johnston. 1991. Beaver population fluctuations and tropospheric methane emissions in boreal wetlands. *Biogeochemistry* 12:1–15.

Neue, H.-U., and P. A. Roger. 1993. Rice agriculture: Factors controlling emissions. In *Atmospheric Methane: Sources, Sinks and Role in Global Change,* ed. M. A. K. Khalil, pp. 254–298. Springer-Verlag, Berlin.

Nisbet, E. G. 1989. Some northern sources of atmospheric methane: Production, history, and future implications. *Can. J. Earth Sci.* 26:1603–1611.

Nisbet, E. G. 1990. The end of the ice age. *Can. J. Earth Sci.* 27:148–157.

Nisbet, E. G. 1992. Sources of atmospheric CH_4 in early post-glacial time. *J. Geophys. Res.* 97:12859–12867.

Ojima, D. S., D. W. Valentine, A. R. Mosier, W. J. Parton, and D. S. Schimel. 1993. Effect of land use change on methane oxidation in temperate forest and grassland soils. *Chemosphere* 26:675–685.

Overpeck, J. T., P. J. Bartlein, and T. Webb. 1991. Potential magnitude of future vegetation change in eastern North America: Comparisons with the past. *Science* 254:692–695.

Paull, C. K., W. Ussler, and W. P. Dillon. 1991. Is the extent of glaciation limited by marine gas-hydrates? *Geophys. Res. Lett.* 18:432–434.

Peer, R. L., S. A. Thornloe, and D. L. Epperson. 1993. A comparison of methods for estimating global methane emissions from landfills. *Chemosphere* 26:387–400.

Petit-Maire, N., M. Fontugne, and C. Rouland. 1991. Atmospheric methane ratio and environmental changes in the Sahara and Sahel during the last 130 yrs. *Palaeogeog. Palaeoclimatol. Palaeoecol.* 86:197–204.

Quay, P. D., S. L. King, J. M. Landsdowne, J. M., and D. O. Wilbur. 1988. Isotopic composition of methane released from wetlands: Implications for the increase in atmospheric methane. *Glob. Biogeochem. Cycles* 2:385–397.

Quay, P. D., S. L. King, J. Stutsman, D. O. Wilbur, L. P. Steele, I. Fung, R. H. Gammon, T. A. Brown, G. W. Farwell, P. M. Grootes, and F. H. Schmidt. 1991. Carbon isotopic composition of atmospheric CH_4: Fossil and biomass burning source strengths. *Glob. Biogeochem. Cycles* 5:25–47.

Raynaud, D., J. Chappellaz, J. M. Barnola, Y. S. Korotkevich, and C. Lorius. 1988. Climatic and CH_4 change in the Vostok ice core. *Nature* 333:655–657.

Rosswall, T., F. Bak, D. Baldocchi, R. J. Cicerone, D. H. Ehhalt, M. K. Firestone, I. E. Galbally, V. F. Galchenko, P. M. Groffman, H. Papen, W. S. Reeburgh, and E. Sanhueza. 1989. Group report: What regulates production and consumption of trace gases in ecosystems: Biology or physiochemistry? In *Exchange of Trace Gases between Terrestrial Ecosystems and the Atmosphere,* ed. M. O. Andreae and D. S. Schimel, pp. 73–95. Wiley, Toronto.

Roulet, N. T., R. Ash, and T. R. Moore. 1992. Low boreal wetlands as a source of atmospheric CH_4. *J. Geophys. Res.* 97:3739–3749.

Rudd, J. W. M., and R. D. Hamilton. 1978. Methane cycling in a eutrophic shield lake and its effects on whole lake metabolism. *Limnol. Ocean.* 23:337–348.

Rudd, J. W. M., and C. D. Taylor. 1979. Methane cycling in aquatic environments. *Adv. Aquat. Microbiol.* 2:77–150.

Scharfe, D., W. M. Hao, L. Donoso, P. J. Crutzen, and E. Sanhueza. 1990. Soil fluxes and atmospheric concentration of CO and CH_4 in the northern part of the Guayana Shield, Venezuela. *J. Geophys. Res.* 95:22475–22480.

Schutz, H., A. Holzapfel-Pschorn, R. Conrad, H. Rennenberg, and W. Seiler. 1989. A 3-year continuous record on the influence of daytime, season, and fertilizer treatment on CH_4 emission rates from an Italian rice paddy. *J. Geophys. Res.* 94:16405–16416.

Shaver, G. R., W. D. Billings, F. S. Chapin, A. E. Giblin, K. J. Nadelhoffer, W. C. Oechel, and E. B. Rastetter. 1992. Global change and the carbon balance of Arctic ecosystems. *Bioscience* 42:434–441.

Smith, I. M., and L. L. Sloss. 1992. *Methane Emissions from Coal.* International Energy Agency Coal Research, Report IEAPER/04, London.

Steele, L. P., E. J. Dlugokencky, P. M. Lang, P. P. Tans, R. C. Martin, and K. A. Masarie. 1992. Slowing down of the global accumulation of atmospheric methane during the 1980s. *Nature* 358:313–316.

Street-Perrott, F. A. 1992. Atmospheric methane: Tropical wetland sources. *Nature* 355:23–24.

Tathy, J. P., B. Cros, R. A. Delmas, A. Marenco, J. Servant, and M. Labat. 1992. Methane emission from flooded forest in central Africa. *J. Geophys. Res.* 97:6159–6168.

Taylor, J. A., G. P. Brasseur, P. R. Zimmerman, and R. J. Cicerone. 1991. A study of the sources and sinks of CH_4 and methyl chloroform using a global three-dimensional Lagrangian tropospheric tracer transport model. *J. Geophys. Res.* 96:3013–3044.

Taylor, K. C., G. W. Lamorey, G. A. Doyle, R. B. Alley, P. M. Grootes, P. A. Mayewski, J. W. C. White, and L. K. Barlow. 1993. The "flickering switch" of late Pleistocene climate change. *Nature* 361:432–436.

Thebrath, B., H-P. Mayer, and R. Conrad, 1992. Bicarbonate-dependent production and methanogenic consumption of acetate in anoxic paddy soils. *FEMS Microb. Ecol.* 86:295–302.

Thornloe, S. A., M. A. Barlaz, R. Peer, L. C. Huff, L. Davis, and J. Mangino. 1993. Waste management. In *Atmospheric Methane: Sources, Sinks and Role in Global Change,* ed. M. A. K. Khalil, pp. 362–398. Springer-Verlag, Berlin.

Tie, X.-X., F. N. Alyea, D. M. Cunnold, and C.-Y. J. Kao. 1991. Atmospheric methane: A global three-dimensional model study. *J. Geophys. Res.* 96:17339–17348.

Tyler, S. C. 1991. The global methane budget. In *Microbial Production and Consumption of Greenhouse Gases: Methane, Nitrogen Oxides and Halomethanes,* ed. J. E. Rogers and W. B. Whitman, pp. 7–38. American Society for Microbiology, Washington, D.C.

Tyler, S. C., P. R. Zimmerman, C. Cumberbatch, J. P. Greenberg, J. P. Westberg, and J. P. E. C. Darlington. 1988. Measurements and interpretation of $\delta^{13}C$ of methane from termites, rice paddies, and wetlands in Kenya. *Glob. Biogeochem. Cycles* 2:341–355.

Velichko, A. A. 1984. *Late Quaternary Environments of the Soviet Union,* English eds. H. E. Wright and C. W. Barnosky. University of Minnesota Press, Minneapolis.

Wahlen, M., N. Tanaka, R. Henry, B. Deck, J. Zeglen, J. S. Vogel, J. S. Southon, A. Shemesh, R. Fairbanks, and W. S. Broecker. 1989. Carbon-14 in methane sources and in atmospheric methane: The contribution from fossil carbon. *Science* 254:286–290.

Wassmann, R., U. G. Thein, M. J. Whiticar, H. Rennenberg, W. Seiler, and W. J. Junk. 1992. Methane emissions from the Amazon floodplain: Characterization of production and transport. *Glob. Biogeochem. Cycles* 6:3–13.

Watson, R. T., H. Rodhe, H. Oeschger, and U. Siegenthaler. 1990. Greenhouse gases and aerosols. In *Climate Change: The IPCC Scientific Assessment,* ed. J. T. Houghton, G. J. Jenkins, and J. J. Ephraums, pp. 1–40. Cambridge University Press, Cambridge.

Westermann, P. 1993. Temperature regulation of methanogenesis in wetlands. *Chemosphere* 26:321–328.

Westermann, P., and B. K. Ahring. 1987. Dynamics of methane production, sulfate reduction and denitrification in a permanently waterlogged alder swamp. *Appl. Environ. Microbiol.* 53: 2554–2559.

Whalen, S. C., and W. S. Reeburgh. 1992. Interannual variations in tundra methane emission: A 4-year time series at fixed sites. *Glob. Biogeochem. Cycles* 6:139–159.

Whalen, S. C., W. S. Reeburgh, and K. S. Kizer. 1991. Methane consumption and emission by taiga. *Glob. Biogeochem. Cycles* 5:261–273.

Wheaton, E. E., T. Singh, R. Dempster, K. D., Higgenbotham, J. P. Thorpe, G. C. Kooten, and J. S. Taylor. 1987. An exploration and assessment of the implications of climatic change for the boreal forest and forestry economics of the Prairie Provinces and Northwest Territories, 282 pp. Tech. Rep. No. 211. Saskatchewan Research Council, Innovation Place, Saskatoon.

Whiticar, M. J., E. Faber, and M. Schoell. 1986. Biogenic methane formation in marine and freshwater environments: CO_2 reduction vs. acetate fermentation—Isotope evidence. *Geochem. Cosmochem. Acta* 50:1266–1271.

Williams, A. 1993. Methane emissions. Paper presented to the 29th Consultative Conference of the Watt Committee on Energy.

Williams, R. T., and R. L. Crawford. 1984. Methane production in Minnesota peatlands. *Appl. Environ. Microbiol.* 47:1266–1271.

World Resources Institute. 1990. *World Resources 1990–91.* Oxford University Press, New York.

World Resources Institute. 1992. *World Resources 1992–93.* Oxford University Press, New York.

Yavitt, J. B., and G. E. Lang. 1990. Methane production in contrasting wetland sites: Response to organic-chemical components of peat and to sulfate reduction. *Geomicrobiol. J.* 8:27–46.

Yavitt, J. B., G. E. Lang, and R. K. Wieder. 1987. Control of carbon mineralization to CH_4 and CO_2 in anaerobic, *Sphagnum*-derived peat from Big Run Bog, West Virginia. *Biogeochemistry* 4:141–157.

Yavitt, J. B., G. E. Lang, and A. J. Sexstone. 1990. Methane fluxes in wetland and forest soils, beaver ponds, and low-order streams of a temperate forest ecosystem. *J. Geophys. Res.* 95:22463–22474.

11

Linkages between Carbon and Nitrogen Cycling and Their Implications for Storage of Carbon in Terrestrial Ecosystems

ERIC A. DAVIDSON

Nitrogen is an essential element for life, but the transformation of elemental nitrogen into organic forms that can be used by living organisms requires large inputs of energy. Organic carbon is one of the currencies by which energy is transferred among cells, among organisms, and among communities of people, so it is not surprising that the biogeochemical cycles of carbon and nitrogen are linked. One such linkage occurs in the soil, where the presence of organic carbon and organic nitrogen, and the ratio of the two, affect soil fertility.

When humans first began agricultural activities, they probably observed a relationship between the blackness of soil rich in organic matter and the productivity of their crops. Identification of nitrogen as one of the components in soil organic matter (SOM) that contributes to plant productivity awaited the development of soil science as an academic discipline in the 19th century. Using a new analytical technique developed by Kjeldahl, E. W. Hilgard noted that "humus with much less than 2.5 per cent of nitrogen was to be suspected of nitrogen hungriness" (Jenny 1961). Hilgard also wrote in 1893 that "on the average the humus of the arid soils contains three times as much nitrogen as that of the humid." Hans Jenny (1941) later established climate as a "state factor" that affects many soil properties, and he clarified Hilgard's observations by describing the narrowing of the C:N ratio of soils with increasing mean annual temperature (Figure 11.1). The importance of nitrogen as a factor limiting agricultural and silvicultural yields and limiting plant productivity in many natural ecosystems has since received wide attention (Vitousek and Howarth 1991 and references therein).

As humans now consider the effects of a possible rapid change in climate resulting largely from increasing concentrations of CO_2 in the atmosphere, the C:N ratios of plant tissues and SOM may play an important role in feedback mechanisms. Plant responses to increased atmospheric concentrations of CO_2 levels may be constrained by nitrogen availability. On the other hand, mineralization of carbon and nitrogen resulting from enhanced microbial respiration in the soils of a warmer world could increase nitrogen

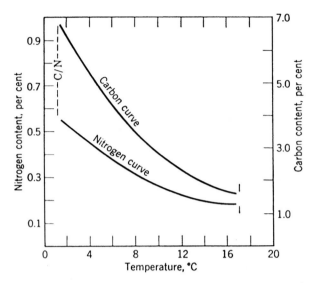

Figure 11.1. Schematic representation of the narrowing C:N ratio of soil humus with increasing temperature. (Data from Jenny in Waksman 1952. Redrawn based on figure in Waksman 1952.)

availability to plants. Current and past research on biogeochemical cycles of carbon and nitrogen provides an imperfect guide to how ecosystems will respond to higher CO_2 concentrations, higher temperatures, and perhaps a drier or wetter climate in various regions of the world. Reasonable predictions can be made about how some individual processes may respond to specific changes in climate and availability of resources, but integrated ecosystem responses to simultaneous changes in climate and resources are more difficult to predict. In this chapter, I do not propose to give the right answer or the best prediction. Rather, I identify the mechanisms by which linkages of carbon and nitrogen cycles may affect carbon storage in terrestrial ecosystems and discuss our degree of understanding of the individual mechanisms and how they are integrated.

ENHANCED RATES OF DECOMPOSITION

The probable response of one component of ecosystem function in a warmer world is not controversial: soil respiration and microbial decomposition of SOM are strongly temperature-dependent. Raich and Schlesinger (1992) report that the Q_{10} for soil respiration (which includes microbial and root respiration) varies from 1.3 to 3.3 with a median of 2.4 across a broad range of ecosystems. Dyer et al. (1990) and Meentemeyer (1978) have shown a strong correlation on regional and global scales between rates of litter decomposition and actual evapotranspiration, which can be viewed as a measure of energy and moisture available to the ecosystem. There is little doubt that increases in soil temperatures will result, at least initially, in higher rates of microbial respiration and decomposition of SOM. Melting of permafrost and drainage of northern wetlands would also expose considerable stores of SOM to aerobic conditions that would promote decomposition (Billings 1987).

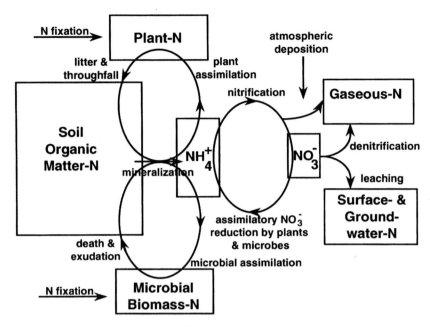

Figure 11.2. Schematic representation of some of the important inputs, transfers, losses, and pools of the nitrogen cycle. Arrows indicate inputs, transfers, or losses; boxes indicate pools.

When SOM is mineralized during decomposition, organic carbon is transformed into CO_2, which is lost to the atmosphere. Similarly, organic nitrogen is transformed into NH_4^+, but several fates of that NH_4^+ are possible. The dominant fates of NH_4^+ are microbial assimilation (immobilization), plant uptake, and nitrification to NO_3^- (Figure 11.2). Where NO_3^- is produced, its dominant fates are microbial assimilation, plant uptake, loss from the ecosystem via leaching and hydrologic export, and loss from the ecosystem via denitrification to gaseous nitrogen that diffuses into the atmosphere. Knowing the fate of mineralized nitrogen is critical to predicting the total ecosystem response to higher rates of decomposition; three plausible scenarios are considered.

THREE SCENARIOS FOR THE FATE OF MINERALIZED NITROGEN

Nitrogen Is Retained by Plants

In the first scenario, which is gaining considerable attention, the mineralized nitrogen is taken up by plants. If plants are not stressed by other factors, such as drought, soil acidification, or ozone pollution, then the plants may be able to use additional nitrogen to assimilate more carbon. Because plant productivity in many temperate ecosystems responds to nitrogen fertilization (Vitousek and Howarth 1991), it is reasonable to postulate that plants could also utilize the nitrogen mineralized when rates of decomposition increase. Greater plant demand for nitrogen might also be facilitated by higher concentrations of atmospheric CO_2 that enhance rates of photosynthesis and that improve water use efficiency (Gifford 1992). Because the C:N ratio of plant tissue (espe-

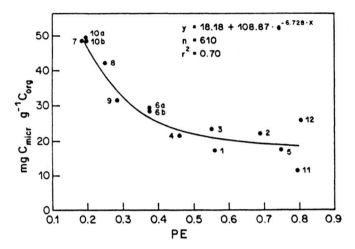

Figure 11.3. Amount of microbial biomass carbon per unit soil organic carbon (milligrams microbial carbon per gram soil organic carbon) plotted against the quotient (PE) between mean annual precipitation (millimeters) and annual pan evaporation (millimeters). The left side of the x axis represents hot and dry conditions; the right side, cold and/or wet conditions. The points represent means for a variety of sites reviewed by Insam et al. (1989). Redrawn based on figure in Insam et al. (1989).

cially woody tissue) is higher (usually ranging from 40 to 400) than the C:N ratio of the soil (usually ranging from 10 to 25), the additional carbon assimilation by plants resulting from the transfer of nitrogen from the soil to the plants would more than compensate for the carbon lost from the soil. The net effect of this transfer of nitrogen and the stoichiometric difference in the C:N ratios of soils and plant tissues would be greater ecosystem storage of carbon (Rastetter et al. 1992; Shaver et al. 1992). If correct, this scenario describes an important negative feedback mechanism of global warming.

Nitrogen Is Retained by Soil Microorganisms

A second scenario, in which nitrogen mineralized from SOM is largely retained by soil microorganisms, is also plausible. If the mineralized nitrogen is immobilized by soil microbial biomass, thus allowing the microorganisms to utilize more of the soil carbon to produce more microbial biomass, then the ratio of microbial biomass to total soil carbon would increase. Microbial biomass has a narrower C:N ratio (usually 5–10) than SOM (usually 10–25), so an increase in the microbial biomass relative to the SOM pool would result in a narrower average C:N ratio of the soil. In this case, nitrogen is retained within the soil, carbon is lost from the soil, and little or no additional carbon assimilation by plants compensates for the carbon lost from the soil.

This scenario is supported by a strong relationship between climate and the ratio of microbial biomass to organic carbon in soil (Insam et al. 1989). The microbial biomass constitutes a larger fraction of the total soil organic carbon in warmer and drier climates than in cooler and wetter climates (Figure 11.3). A larger fraction of microbial biomass and microbial products (extracellular enzymes, dead biomass) helps account for the narrowing of the C:N ratio of soils with increasing mean annual temperature described by Hilgard and Jenny (Figure 11.1).

This scenario is further supported by evidence from forest fertilization studies using ^{15}N as a tracer. Most of the added nitrogen remains within the soil and only 10–25% is taken up by the trees (Binkley 1986). Rates of gross immobilization of NH_4^+ and NO_3^- by soil microorganisms have also been shown to exceed rates of plant uptake in an annual grassland (Jackson et al. 1989; Schimel et al. 1989).

The microbial biomass is very dynamic; the mean residence time of nitrogen in the soil microbial biomass was only 1–2 months in temperate conifer forest soils (Davidson et al. 1992). Some fraction of the nitrogen immobilized in the microbial biomass may gradually become available to plants during cycles of mineralization and reimmobilization. Hence the first two scenarios are not mutually exclusive, and an intermediate scenario is possible, but the fraction of nitrogen that eventually becomes available to plants is unknown.

Nitrogen Is Lost from the Ecosystem

A third scenario involves loss of nitrogen from the ecosystem via leaching and hydrologic export and/or via gaseous loss of nitrogen during nitrification and denitrification (Firestone and Davidson 1989). Although most aggrading temperate forest ecosystems export less nitrogen in stream water than they receive from atmospheric inputs (Swank and Waide 1980), the ratio of output to input of nitrogen could change as rates of decomposition and nitrogen mineralization change. Loss of nitrogen occurs when the processes of microbial mineralization and plant uptake are decoupled in either space or time. Several examples of this decoupling in disturbed and undisturbed ecosystems are instructive.

Disturbance, of both human and insect origin, has yielded rich insight into ecosystem function. In the classic clearcut and herbicide experiment at Hubbard Brook (Bormann and Likens 1979), the plant sink for nitrogen and other nutrients was decoupled from the microbial processes of decomposition and mineralization. Mineralization continued during the herbicide treatment, but the plant sink for inorganic nitrogen was eliminated. The result was substantial loss of inorganic nitrogen via the stream water until the herbicide treatment was stopped and plants were allowed to regenerate. Similarly, severe insect defoliation in a watershed at the Coweeta Hydrologic Laboratory resulted in nitrogen loss in stream water (Swank et al. 1981). These two examples of disturbance demonstrate extreme decoupling of mineralization and plant uptake, resulting in significant nitrogen loss from the ecosystem. Climate change may be a less severe and less rapid disturbance, but nitrogen loss might also be expected if plants and microorganisms respond differently to changes in climate.

Even undisturbed, nitrogen-limited, forested ecosystems lose some nitrogen because of the asynchrony of plant and microbial activity and atmospheric inputs of nitrogen. Nearly all of the 4 kg nitrogen ha^{-1} yr^{-1} lost from the Hubbard Brook control watershed is exported by the stream during the fall, winter, and early spring (Figure 11.4). Significant mineralization of nitrogen and possibly nitrification occur early in the spring. Atmospheric inputs of nitrogen also accumulate in the soil and the snowpack during the autumn and winter, resulting in a large pulse of nitrogen in the stream during snowmelt.

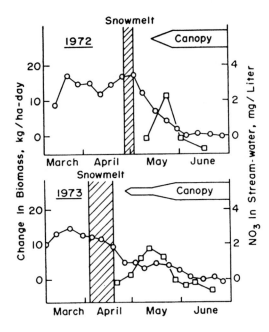

Figure 11.4. Stream water nitrate concentrations (○) and biomass increase (□) in *Erythronium americanum* in 1972 and 1973 at Hubbard Brook (From Bormann and Likens 1979. Copyright © 1979 by Springer-Verlag New York, Inc. Reprinted by permission.)

Understory plants respond more quickly than overstory plants to spring thaw and assimilate some of the inorganic nitrogen that is available in the spring (Muller and Bormann 1976). Immobilization of nitrogen by growing microbial biomass may be a more important sink than understory plants for conserving nitrogen within the ecosystem early in the spring (Zak et al. 1990). Although these "vernal dam" mechanisms help mitigate the amount of nitrogen lost, they are not perfect, as is evidenced by stream export of nitrogen in the winter and early spring (Figure 11.4). Ecosystem losses of gaseous nitrogen via denitrification have also been shown to peak during the spring and fall in temperate forests, when soil temperature is high enough to permit denitrification but when plant uptake of nitrogen is low (Goodroad and Keeney 1984; Groffman and Tiedje 1989; Davidson and Swank 1990).

Reasonable arguments could be made that autumn, winter, and spring losses of nitrogen could either increase or decrease in a warmer climate. On the one hand, a longer growing season could permit more plant uptake of nitrogen and a narrower window for nitrogen loss when plants are dormant. On the other hand, a generally warmer climate could still have late winter storms and early autumn frosts, so the length of the growing season may not change substantially. In fact, early bud break followed by a late winter storm could damage plants, and nitrogen losses could increase. Furthermore, casual observations of present interannual variation in weather suggest that the extant plant genotypes may be able to accelerate spring bud break and delay autumn anthesis by a week or two, but no one knows if the growing season could be extended significantly without migration of new plant genotypes. In contrast, microbial responses to milder winters and earlier springs are not limited by photoperiod or any other adaptation that plants have evolved for responding to seasonal trends in weather. Microbial mineralization of nitrogen, nitrification, and denitrification can occur whenever the soil is warm enough to support microbial activity. Finally, climate change may involve seasonal

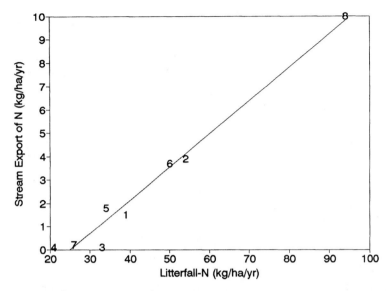

Figure 11.5. Annual stream export of nitrogen plotted against nitrogen in annual litterfall of forested ecosystems for which both values have been reported. Data for sites 1–4 are from Swank and Waide (1980): (1) Walker Branch, Tennessee—mixed hardwood; (2) Hubbard Brook, New Hampshire—mixed hardwood; (3) Coweeta, North Carolina—mixed hardwood; (4) H. J. Andrews, Oregon—mixed conifer. Data for sites 5 and 6 are from Kimmons et al. (1985): (5) New Zealand—nothofagus; (6) Australia—eucalyptus. Site 7 is pine at Coweeta, North Carolina (Swank in Cole and Rapp 1981). Site 8 is mixed hardwood in the Netherlands (Tietema and Verstraten 1991). The least squares linear regression is $y = 0.14x–3.66$; $R^2 = 0.97$.

redistribution of precipitation. Summer drought, which would limit plant uptake of nitrogen during the growing season, is predicted from many general circulation models. Greater precipitation during the dormant season could result in greater flushing of soil nitrogen into the streams and greater loss of nitrogen via denitrification.

There are some indications, albeit imperfect ones, that higher rates of nitrogen loss should be expected. If plants are "fertilized" by higher rates of nitrogen mineralization of SOM, then rates of nitrogen transfer within the ecosystem should be enhanced. Rates of nitrogen uptake by plants, deduced from measures of rates of net nitrogen mineralization, have been correlated with rates of nitrogen transfer in litterfall (Nadelhoffer et al. 1983), and with rates of nitrogen allocation to roots (Nadelhoffer et al. 1985). As with interconnected gears, it appears that when one part of the internal nitrogen cycle spins faster, other parts of the nitrogen cycle also spin faster (Figure 11.2). Using litterfall nitrogen as an index of the rate of nitrogen cycling within the ecosystem, losses of nitrogen from the ecosystem can be related to cycling rates. In those forested watersheds where both have been measured, nitrogen loss in streams is correlated with litterfall nitrogen (Figure 11.5). This relationship suggests that the amount of nitrogen lost is a function of the rate at which nitrogen is cycled within the ecosystem. In other words, nitrogen losses increase with site fertility. If ecosystems become more "fertile" because nitrogen previously bound in SOM is mineralized in a warmer climate, some of that mineralized nitrogen will probably be lost from the ecosystem.

Another area of uncertainty is the role of dissolved organic nitrogen (DON). Evidence is growing that leaching of DON is the dominant output of nitrogen in some ecosystems (Binkley et al. 1992), but the microbial processes involved in the formation of DON are very poorly understood. I have left DON out of Figure 11.2 partly to avoid confusing overlapping arrows, but also partly because it is not clear whether DON is formed from microbial products or leached from plant and soil residues. Nor is it known to what extent DON participates in mineralization and immobilization cycles. Without this fundamental understanding of what may be one of the most important nitrogen loss mechanisms, it would be premature to predict how nitrogen losses and related carbon storage will be affected by climate change.

The process of "nitrogen fertilization" caused by increased mineralization of SOM in a warmer climate is also confounded with increasing anthropogenic inputs of nitrogen into many forested ecosystems of North America and Europe. Increased losses of nitrogen have been predicted if anthropogenic inputs of nitrogen via atmospheric deposition continue to increase or remain at current high levels, causing many forests to become "nitrogen-saturated" (Aber et al. 1989). A system that is already N-saturated because of anthropogenic inputs of nitrogen would not be able to retain the nitrogen mineralized from SOM in a warmer climate.

One mechanism by which enhanced nitrogen loss could be avoided is the production of lower-quality litter, i.e., litter with higher C:N and lignin:N ratios. Poorer-quality litter decomposes more slowly (Melillo et al. 1982), and this process would slow all ecosystem transfers of nitrogen, including nitrogen loss. Plants exposed to higher concentrations of atmospheric CO_2 may be able to assimilate more carbon per unit nitrogen in leaf tissue (Gifford 1992). When yellow poplar and white oak saplings were exposed to elevated levels of CO_2 (+300 ppm) during an entire growing season, early morning starch content of the leaves at the end of the growing season was nearly double the starch content of control plants (Wullschleger et al. 1992). Starch is relatively easily decomposed, however, and no data are available for the effects of elevated CO_2 treatments on less labile structural carbon, either in leaves or in other tissues. Nor do we know the degree of plasticity of extant plant genotypes in altering structural components of their litter.

NITROGEN FIXATION

A warmer climate and higher concentrations of atmospheric CO_2 might also affect rates of nitrogen input to ecosystems from biological nitrogen fixation. In several studies reviewed by Sinclair (1992), nodulation and accumulation of plant biomass nitrogen increased when soybeans were grown with long-term exposure to elevated CO_2. With a few notable exceptions, however, legumes and other plant hosts of nitrogen-fixing symbionts are usually not important components of late successional temperate and boreal ecosystems (Vitousek and Howarth 1991). Epiphytic lichens contribute modest amounts of nitrogen fixed by cyanobacteria symbionts (blue-green algae), but their activity is also limited by levels of light and water. It is unlikely that symbiotic nitrogen fixation would increase in most late successional temperate and boreal forests as a result of increasing atmospheric CO_2. Symbiotic nitrogen fixation may be stimulated in tropical forests, where legumes often make up a significant part of the canopy, but it is unclear whether plant productivity would be stimulated because nitrogen is often an abundant nutrient relative to phosphorus where tropical forests occur on highly weathered soils (Vitousek and Sanford 1986).

Cyanobacteria and some heterotrophic bacteria can also fix nitrogen without symbiotic associations, and this process is known as "free-living nitrogen fixation." The cyanobacteria are important in many aquatic ecosystems and in exposed soil crusts in a few terrestrial ecosystems, but light limits their importance in most soils.

Free-living heterotrophic nitrogen fixers depend on abundant supplies of readily available organic carbon in soil for energy to support nitrogen fixation. Oxygen must also be eliminated from the immediate environs of these bacteria because oxygen is a potent inhibitor of the nitrogen-fixing enzyme, nitrogenase. Increased allocation of carbon to roots by plants that experience elevated concentrations of atmospheric CO_2 could stimulate free-living nitrogen fixation by these bacteria in two ways. First, increases in root exudates might be used directly as energy sources to fuel nitrogen fixation. Second, increased root respiration would help consume oxygen that would otherwise inhibit nitrogen fixation. The importance of free-living nitrogen fixation has not been demonstrated in any ecosystem because it is inherently difficult to measure (but see the review of existing evidence in Bormann et al. [1993]). Some intriguing results from mass balance calculations in temperate forests suggest that free-living nitrogen fixation cannot be discounted, although the current rates of free-living nitrogen fixation are so poorly known that guesses as to the magnitude of the potential increase in nitrogen fixation resulting from enhanced inputs of carbon are futile at this point.

Increased nitrogen fixation, if it occurs, might mean greater availability of nitrogen to plants and enhanced plant growth, but not necessarily. Additional carbon inputs to the soil would also increase microbial biomass, which would strengthen the microbial sink competing for plant-available N. Another possible mitigating factor is that the same microaerophilic conditions needed for nitrogen fixation also favor denitrification, although significant losses of nitrogen via denitrification would be likely only if nitrogen availability increased substantially. In summary, although there is some basis for expecting increases in free-living nitrogen fixation if elevated atmospheric CO_2 results in increased allocation of carbon to roots, uncertainty as to the magnitude of this flux, both now and in the future, and the uncertain fate of newly fixed nitrogen cast doubt on the probability of a significant net increase in the nitrogen balance.

LESSONS FROM ECOSYSTEM MODELS

Because experiments that impose higher temperatures and elevated concentrations of atmospheric CO_2 are difficult to conduct in forested ecosystems, most of the projections of ecosystem response are obtained from models. These models are less useful as absolute predictors of expected responses than as heuristic tools that reveal the sensitivity of those predictions to underlying mechanisms and assumptions.

A good example of what can be learned from models was presented by Scott et al. (1992). They compared output of the General Ecosystem Model (GEM; Rastetter et al. 1991) and the CENTURY model (Parton et al. 1987). In simulations with elevated atmospheric CO_2 concentrations and higher mean annual temperatures, GEM predicted greater total carbon storage by both grassland and forest ecosystems than did CENTURY. One plausible explanation for this difference is that the two models have very different approaches for predicting nitrogen loss. In GEM, loss of nitrogen is proportional to the concentration of inorganic nitrogen in the soil, whereas in CENTURY nitrogen loss is a function of the rate of nitrogen mineralized from plant litter. Concentrations of inorganic nitrogen in soil are not related to rates of nitrogen mineralization (Davidson et al. 1992); in other words, the rates of transfer and loss (arrows in Figure 11.2) are not necessarily related to the size of the soil inorganic nitrogen pools (NH_4^+ and NO_3^- boxes in Figure 11.2). Using an inorganic nitrogen concentration rather than a rate of nitrogen transfer to model nitrogen losses resulted in lower predictions of nitrogen loss in the GEM model. The GEM model also partitions decomposition evenly across all months, whereas CENTURY allows mineralization, and hence nitrogen loss, to vary seasonally and asynchronously with plant uptake. These mechanisms that allow greater nitrogen loss in the CENTURY model result in lower projected storage of carbon in the ecosystem than are predicted by GEM. The predictions of neither model can be verified, but they show that nitrogen loss is a critical mechanism that affects model forecasts of carbon storage. The functions for nitrogen loss used in these predictive models need closer scrutiny. Until the seasonality of microbial and plant responses to warming is better characterized, it would be premature to assume that nitrogen mineralized from SOM will be taken up by plants, thereby increasing ecosystem storage of carbon. A recent prediction of increased potential net primary productivity in temperate and boreal ecosystems in response to global warming (Melillo et al. 1993) is based on a model that implicitly favors plant uptake of mineralized nitrogen, which may not be correct.

SUMMARY

Much is known about individual processes of decomposition, plant uptake, nitrogen fixation, nitrification, denitrification, and leaching, but we cannot predict with confidence the integrated ecosystem response to changing climate. Nevertheless, it is clear that projections of carbon storage in a climate with higher temperatures and higher atmospheric concentrations of CO_2 depend on how these mechanisms of nitrogen transfer and loss are comprehended and modeled.

Assuming that drought, soil acidification, ozone pollution, or some other stress does not become more limiting than nitrogen in temperate and boreal ecosystems, linkages between carbon and nitrogen cycling will affect the amount of carbon stored within the soils and vegetation of a warmer world. Carbon will probably be lost from soils, but we do not know whether the nitrogen that is mineralized will be taken up by plants, retained within the soil, or lost from the ecosystem. Plausible arguments can be made for each scenario. However, several lines of evidence suggest that at least some of the nitrogen will either be retained within the soil or lost from the ecosystem. Microbial biomass may retain a larger fraction of soil nitrogen under warmer and drier conditions. Hydrologic loss of nitrogen appears to increase with increasing rates of nitrogen cycling, as would occur with enhanced rates of decomposition and nitrogen mineralization. Finally, nitrogen loss occurs when plant and microbial processes are decoupled, either in space (as occurs when a site is disturbed) or in time (as occurs across seasons in undisturbed systems). If plants and soil microorganisms respond differently to climate change, further decoupling of plant and microbial processes and further losses of nitrogen would be expected.

REFERENCES

Aber, J. D., K. J. Nadelhoffer, P. Steudler, and J. M. Melillo. 1989. Excess nitrogen from fossil fuel combustion may stress the biosphere. *Bioscience* 39:378–386.

Billings, W. D. 1987. Carbon balance of Alaskan tundra and taiga ecosystems: Past, present, and future. *Quat. Sci. Rev.* 6:165–177.

Binkley, D. 1986. *Forest Nutrition Management.* Wiley, New York.

Binkley, D. P. Sollins, R. Bell, D. Sachs, and D. Myrold. 1992. Biogeochemistry of adjacent conifer and alder-conifer stands. *Ecology* 73:2022–2033.

Bormann, B. T., F. H. Bormann, W. B. Bowden, R. S. Pierce, S. P. Hamburg, D. Wang, M. C. Snyder, C. Y. Li, and R. C. Ingersoll. 1993. Rapid N$_2$ fixation in pines, alder, and locust: Evidence from the sandbox ecosystem study. *Ecology* 74:583–598.

Bormann, F. H., and G. E. Likens. 1979. *Pattern and Process in a Forested Ecosystem.* Springer-Verlag, New York.

Cole, D. W., and M. Rapp. 1981. Elemental cycling in forest ecosystems. In *Dynamic Properties of Forest Ecosystems,* ed. D. E. Reichle, pp. 341–409. Cambridge University Press, Cambridge.

Davidson, E. A., and W. T. Swank 1990. Nitrous oxide dissolved in soil solution: An insignificant pathway of nitrogen loss from a southeastern hardwood forest. *Water Resources Res.* 26:1687–1690.

Davidson, E. A., S. C. Hart, and M. K. Firestone. 1992. Internal cycling of nitrate in soils of a mature coniferous forest. *Ecology* 73:1148–1156.

Dyer, M. L., F. Meentemeyer, and B. Berg. 1990. Apparent controls of mass loss rate of leaf litter on a regional scale. *Scand. J. For. Res.* 5:1–13.

Firestone, M. K., and E. A. Davidson. 1989. Microbiological basis of NO and N$_2$O production and consumption in soil. In *Exchange of Trace Gases between Terrestrial Ecosystems and the Atmosphere,* ed. M. O. Andreae and D. S. Schimel, pp. 7–21. Wiley, New York.

Gifford, R. M. 1992. Interaction of carbon dioxide with growth-limiting environmental factors in vegetation productivity: Implications for the global carbon cycle. *Adv. Bioclimatol.* 1:24–58.

Goodroad, L. L., and D. R. Keeney. 1984. Nitrous oxide emissions from forest, marsh, and prairie ecosystems. *J. Environ. Qual.* 13:448–452.

Groffman, P. M., and J. M. Tiedje. 1989. Denitrification in north temperate forest soils: Spatial and temporal patterns at the landscape and seasonal scales. *Soil Biol. Biochem.* 21:613–620.

Insam, H., D. Parkinson, and K. H. Domsch. 1989. Influence of macroclimate on soil microbial biomass. *Soil Biol. Biochem.* 21:211–221.

Jackson, L. E., J. P. Schimel, and M. K. Firestone. 1989. Short-term partitioning of ammonium and nitrate between plants and microbes in an annual grassland. *Soil Biol. Biochem.* 21:409–415.

Jenny, H. 1941. *Factors of Soil Formation.* McGraw-Hill, New York.

Jenny, H. 1961. *E. W. Hilgard and the Birth of Modern Soil Science.* Collana Della Rivista "Agrochimica," Industrie Grafiche V. Lischi & Figli, Pisa, Italy.

Kimmons, J. P., D. Binkley, L. Chatarpaul, and J. de Catanzaro. 1985. Biogeochemistry of temperate forest ecosystems: Literature on inventories and dynamics of biomass and nutrients. Information Report PI-X-47E/F. Canadian Forestry Service, Ottawa, Ontario.

Meentemeyer, V. 1978. Macroclimate and lignin control of litter decomposition rates. *Ecology* 59:465–472.

Melillo, J. M., J. D. Aber, and J. F. Muratore. 1982. Nitrogen and lignin control of hardwood leaf litter decomposition dynamics. *Ecology* 63:621–626.

Melillo, J. M., A. D. McGuire, D. W. Kicklighter, B. Moore III, C. J. Vorosmarty, and A. L. Schloss. 1993. Global climate change and terrestrial net primary production. *Nature* 363:234–239.

Muller, R. N., and F. H. Bormann. 1976. Role of *Erythronium americanum* Ker. in energy flow and nutrient dynamics of a northern hardwood forest ecosystem. *Science* 193:1126–1128.

Nadelhoffer, K. J., J. D. Aber, and J. M. Melillo. 1983. Leaf-litter production and soil organic matter dynamics along a nitrogen-availability gradient in Southern Wisconsin (U.S.A.). *Can. J. For. Res.* 13:12–21.

Nadelhoffer, K. J., J. D. Aber, and J. M. Melillo. 1985. Fine roots, net primary production, and soil nitrogen availability: A new hypothesis. *Ecology* 66:1377–1390.

Parton, W. J., D. S. Schimel, C. V. Cole, and D. S. Ojima. 1987. Analysis of factors controlling soil organic matter levels in Great Plains grasslands. *Soil Sci. Soc. Am. J.* 51:1173–1179.

Raich, J. W., and W. H. Schlesinger. 1992. The global carbon dioxide flux in soil respiration and its relationship to climate. *Tellus* 44B:81–99.

Rastetter, E. B., M. G. Ryan, G. R. Shaver, J. M. Melillo, k. J. Nadelhoffer, J. E. Hobbie, and J. D. Aber. 1991. A general biogeochemical model describing the responses of the carbon and nitrogen cycles in terrestrial ecosystems to changes in CO_2, climate, and nitrogen deposition. *Tree Physiology* 9:101–126.

Rastetter, E. B., R. B. McKane, G. R. Shaver, and J. M. Melillo. 1992. Changes in carbon storage by terrestrial ecosystem: How C-N interactions restrict responses to CO_2 and temperature. *Water Air Soil Pollut.* 64:327–344.

Schimel, J. P., L. E. Jackson, and M. K. Firestone. 1989. Spatial and temporal effects on plant-microbial competition for inorganic nitrogen in a California annual grassland. *Soil Biol. Biochem.* 21: 1059–1066.

Scott, N. A., D. S. Ojima, W. J. Parton, R. B. McKane, and E. B. Rastetter. 1992. Carbon storage in terrestrial ecosystems: A comparison of the CENTURY and GEM ecosystem simulation models. *Bull. Ecological Soc. America* 73(2):340.

Shaver, G. R., W. D. Billings, F. S. Chapin III, A. E. Giblin, K. J. Nadelhoffer, W. C. Oechel, and E. B. Rastetter. 1992. Global change and the carbon balance of arctic ecosystems. *Bioscience* 42:433–441.

Sinclair, T. R. 1992. Mineral nutrition and plant growth response to climate change. *J. Exp. Bot.* 43:1141–1146.

Swank, W. T., and J. B. Waide. 1980. Interpretation of nutrient cycling research in a management context: Evaluating potential effects of alternative management strategies on site productivity. In *Forests: Fresh Perspectives from Ecosystem Analysis,* ed. R. H. Waring, pp. 137–158. Oregon State University Press, Corvallis.

Swank, W. T., J. B. Waide, D. A. Crossley Jr., and R. L. Todd. 1981. Insect defoliation enhances nitrate export from forest ecosystems. *Oecologia* 51:297–299.

Tietema, A., and J. M. Verstraten. 1991. Nitrogen cycling in an acid forest ecosystem in the Netherlands under increased atmospheric nitrogen input. *Biogeochemistry.* 15:21–46.

Vitousek, P. M., and R. W. Howarth. 1991. Nitrogen limitation on land and in the sea: How can it occur? *Biogeochemistry* 13:87–115.

Vitousek, P. M., and R. L. Sanford. 1986. Nutrient cycling in moist tropical forests. *Annu. Rev. Ecol. Syst.* 17:137–167.

Waksman, S. A. 1952. *Soil Microbiology.* Wiley, New York.

Wullschleger, S. D., R. J. Norby, and D. L. Hendrix. 1992. Carbon exchange rates, chlorophyll content, and carbohydrate status of two forest tree species exposed to carbon dioxide enrichment. *Tree Physiol.* 10:21–31.

Zak, D. R., P. M. Groffman, K. S. Pregitzer, S. Christensen, and J. M. Tiedje. 1990. The vernal dam: Plant-microbe competition for nitrogen in northern hardwood forests. *Ecology* 71:651–656.

C

Oceans and Estuaries

The oceans are absorbing carbon dioxide at the rate of 2 ± 1 petagrams of carbon per year. Experience and theory based on experience suggest that the warming projected on the basis of meteorological considerations will reduce the oceanic absorption of CO_2. The reduction will be the result of the warming of the surface waters and the reduction of their capacity for holding dissolved CO_2, and of a further reduction in absorptive capacity from the gradual titration of the carbonate ion of oceanic water by CO_2 from the atmosphere. Both processes speed the accumulation of CO_2 in the atmosphere and augment the warming. Although this process is a positive feedback, it is not a biotic feedback, the emphasis of this analysis. Nevertheless, there are biotic processes in the oceans that have the potential for being influenced by, and influencing, the warming of the earth. Some of these are addressed in the chapters that follow.

One of the most important questions is whether the warming will affect the net ecosystem production of the oceans in the context of the metabolism of ecosystems discussed by Woodwell (Chapter 1). The question is unanswerable if the potential of a warming for effecting profound changes in the density, and therefore in the circulation, of the oceans is included. The best we can do is fall back on the observation that changes in the upwelling of nutrients do not affect the net metabolism of the ocean significantly because upwelling waters deliver dissolved carbon, nitrogen, and phosphorus in the Redfield ratio of production. The carbon is fixed again immediately, and storage is probably not enhanced.

Over the period of years to decades that is of interest in these discussions, changes in marine net ecosystem production are probably most closely coupled to additions of inorganic nutrients and organic carbon from the land, as discussed in Chapter 2 by Mackenzie, and in this section in greater detail by Rowe and Baldauf and by Stephen Smith.

Negative feedbacks affecting the warming of the next several decades appear to be few. A potentially significant negative feedback is the CLAW hypothesis. The suggestion is that complex connections exist among the phytoplankton community, zooplankton, bacterial decomposition, and the flux of dimethyl sulfide (DMS) from the sea surface. The DMS becomes a sulfate aerosol that has a net cooling effect on the earth. The magnitude of the process as a feedback is uncertain and controversial. The issue is examined here by Charlson, long a careful student of the biogeochemistry of sulfur. The hypothesis is important and provocative. At the very least it illuminates a fascinating aspect of global biogeochemistry.

The oceanic production of trace gases other than DMS (OCS, N_2O, CO, NH_3) that are involved in climatic forcing may also be proportional to net ecosystem production. In a warmer world, it is likely, as with DMS, that the production of these gases will be enhanced, creating the potential for further feedbacks. The magnitude of the feedbacks on the time scale of the warming has been addressed by Mackenzie and appears to be small.

There are, however, several other potentially significant changes in the oceans that may influence climates globally and are of interest in these discussions. Increases in UV-B flux owing to stratospheric ozone depletion may affect net oceanic metabolism, particularly in the Antarctic, as outlined by Raymond Smith. The composition and distribution of phytoplankton appear to be especially sensitive to the abundance of iron, molybdenum, and other trace metals. Decomposition of a fraction of the large reservoir of dissolved organic carbon in the ocean may occur under certain circumstances. The biotic processes associated with such changes would, of course, alter the partial pressure of CO_2 in surface waters and the strength of the oceanic sink. However, there is little experiential or theoretical information to substantiate the argument that these potential effects are large enough to be significant in determining the composition of the atmosphere on the decadal time scale.

The evidence available at present seems to support the conclusion that physical and chemical processes in the oceans under a systematic warming will reduce the capacity of the oceans for removing carbon from the atmosphere. In the case of the oceans, on the time scale of the warming of the next several decades, the warming appears to feed the warming.

12

Biofeedback in the Ocean in Response to Climate Change

GILBERT T. ROWE AND JACK G. BALDAUF

This chapter considers the likelihood that biological processes in the ocean, in response to changes in climate, may produce feedback mechanisms that accentuate climate change. The oceans cover 71% (361×10^6 km²) of the earth's surface and sustain a significant fraction of global photosynthesis. The rate of net primary production in the ocean is estimated to be about 29 petagrams (Pg) (=10^{15} g) carbon per year (J. Walsh 1988), although estimates of more than 40 Pg carbon per year (S. Smith and Hollibaugh 1993) have been made. This total represents 25–50% of the annual global photosynthetic production. A significant fraction of the photosynthesis in the ocean is confined to microscopic single-celled phytoplankton rather than large attached plants. It is reasonable to assume that this primary production influences global biogeochemical cycles and the potentially profound exchanges between the oceanic and atmospheric carbon reservoirs. For example, the ocean carbon reservoir is about 60 times larger than that of the atmosphere. Although actual exchanges of trace gases between the ocean and the atmosphere are a function of the direction and magnitude of chemical gradients, it is photosynthesis and the ocean's "biological pump" that establish and maintain the gradients between the oceanic and atmospheric reservoirs.

In this chapter we present the hypothesis that gradual warming will diminish the intensity of the interoceanic circulation. The result, we predict, will be to diminish the surface primary productivity, which is a function of upwelling and equatorial divergence. With a more homogeneous, diffuse productivity over the entire global ocean, export flux out of the surface ocean will diminish because the heterotrophic and autotrophic processes in the surface layers will be more tightly coupled on ecological as well as geological time scales. As a result, less fixed carbon will be exported to the deep sea. Although these ideas are speculative, atmospheric CO_2 levels, for example, were reduced during glacial periods (e.g., Oeschger et al. 1984). We present paleontological evidence from deep-sea sediments that an intensification in the production of deep water and concomitant upwelling resulted in mass accumulation of siliceous sediments. We conclude that warming the global ocean will cause it to retain less fixed carbon, that is, that the biological feedback will be positive, thus exacerbating the greenhouse effect.

The continental margins, however, deserve special consideration. These coastal or "green ocean" regions constitute only about 15% of the world ocean's surface area

(Sverdrup et al. 1942) but support rates of primary production that are about four times greater than those in the open ocean (in g carbon m^{-2} $year^{-1}$). If the coastal oceans are predominantly heterotrophic (produce more CO_2 than they consume), the warming and sea level rise will also allow them to add to the trace gas problem. However, the burial rate of particulate organic carbon (POC) in the coastal zone is about 100 times greater than that in the "blue ocean" (Berger et al. 1989). The net transfers of carbon in the coastal ocean are currently a point of contention, as is their relative role in the global ocean.

THE GEOLOGICAL RECORD

A propitious initial approach for predicting future climate change is to consider the geological record. Most sediment data on global change are based on proxy indicators, such as stable and radioactive isotopes, organic carbon, nutrient proxies, species indexes, and biogenic opal (Herbert et al. 1989). Numerous hypotheses address the impact of variations in biological productivity on atmospheric CO_2 (e.g., see Barnola et al. 1987; Mix 1989). It is generally agreed that atmospheric CO_2 fluctuations, on short time scales (10^2–10^3 years), are largely controlled by the physical exchange rates of CO_2 between the atmosphere and the oceanic reservoir. This exchange is only indirectly influenced by variations in primary production, being principally dependent on such physical parameters as "skin" temperature and sea state. However, dissolved compounds such as CO_2 are removed from the surface layer of the ocean during photosynthesis. The particulate matter thus formed is then transferred by sinking or vertical migration to deeper water where it can be remineralized by heterotrophic metabolism back to inorganic compounds such as CO_2. It is this biological pump that establishes and maintains the vertical gradient in CO_2. If the transfers are confined to the surface mixed layers, then exchanges with the atmosphere can continue; only if the particulate matter is transferred to below the mixed layer can it be considered removed from the potential for short-term exchanges with the atmosphere. How the pump might be affected by climate change is discussed in the next section.

Baldauf and Barron (1990) and Barron and Baldauf (1989) documented the spatial and temporal distribution of siliceous microfossils in sediments since the Eocene (circa 40–50×10^6 years B.P. In doing so they determined that biosiliceous sedimentation, and hence surface water productivity, evolved in a stepwise fashion in response to changing oceanographic, tectonic, and climatic conditions. Significant periods of geologic reorganization occurred throughout the Tertiary period (2–65×10^6 years B.P.). The general trend consisted of a very broad and diffuse distribution of siliceous sediments (and hence diffuse surface productivity) during the Eocene through a stepwise development to geographically restricted regions having increased concentrations of siliceous sediments. The latter was interpreted as representing elevated surface productivity during the Pliocene up to the present. This evolution to higher levels of primary production was thought to be in response to a steepening of both vertical and latitudinal thermal gradients during the last 40×10^6 years.

Abelmann et al. (1990) refined the work of Baldauf and Barron (1990) by focusing on the more recent Pliocene-Pleistocene interval. In doing so they identified climatic optima in the Atlantic sector of the southern ocean during the Pliocene. They suggested that gradients were less well developed compared to the present and that export productivity was high owing to extensive upwelling. Latitudinal temperature gradients that were steep during the late early Pliocene led to the development of the oceanic frontal system, which acted as a biogeographic barrier. Development of northern hemispheric glaciation, deep water formation, and the development of the Weddell Gyre led to localized productivity associated with the oceanic frontal system and a decrease in the regional export productivity at the poles.

Within this broad and generalized pattern were localized events. For example, Kemp and Baldauf (1993) have documented the occurrence of distinct laminae for specific intervals of the sedimentary record of the eastern equatorial Pacific for the time interval from about 15 to 4.4×10^6 years B.P. These siliceous laminae are thought to represent extensive diatom blooms preserved in the sedimentary record. These mats imply massive productivity and a greater export flux than that observed today.

More quantitative information (Shackleton and Pisias 1985; Mix 1989) was obtained through the use of isotopes and foraminiferal indexes to examine variation in paleoproductivity during the Pleistocene (from 2×10^6 years B.P. to the present). Shackleton and Pisias (1985) used $\delta^{13}C$ differences between benthic and planktonic foraminifera to determine the effect of the biological pump on CO_2 variations for the last 340,000 years. In doing so they documented cyclic changes coinciding with the common Milankovitch orbital periods and determined that changes in atmospheric CO_2 occurred prior to changes in ice volume (i.e., CO_2 is in part forcing ice volume changes; Pisias and Shackleton 1984). Broecker (1982) suggested that an increase in atmospheric CO_2 of about 100 ppm occurred at the end of the last glacial stage, which in turn contributed to ice sheet decay. Comparisons of the $\delta^{13}C$ records of a marine core (Shackleton and Pisias 1985) and the Vostok Core (Barnola et al. 1987) appear at first glance to yield different values ($r = 0.54$, $n = 63$; Mix 1989). However, this finding may reflect differences in the time scales and chronological resolution of the records.

Of note is the more recent suggestion, based on Greenland ice cores (Dansgaard et al. 1993), that the ocean-atmosphere system may be more variable than previously thought, and that this system may be able to reorganize rapidly, perhaps within a few decades. This possibility implies that the stability observed in the Holocene may be the exception rather than the norm for earth history.

Mix (1989) reconstructed paleoproductivity for the last glacial maximum based on foraminifera and showed that the largest increase was along the equator. This conclusion is supported by the organic carbon data of, for example, Sarnthein et al. (1987). They estimated that primary production in the North Atlantic was 18% higher than that presently observed. Applying the empirical relationship between primary production and new production assembled by Eppley and Peterson (1978), Sarnthein et al. went on to infer that this primary production would represent a 38% increase in "new" production during the last glacial period. This estimate is of course speculative. (The "new" production of Eppley and Peterson can be equated roughly for our purposes to the export production or flux of the biological pump.)

THE OCEANIC BIOLOGICAL PUMP

Ecological Efficiency of the Biological Pump

The geological record of what have been inferred to be the remnants of export production clearly varies dramatically; and, as indicated previously, investigators have tried to infer how the preserved signals relate to climate and paleoproductivity. Just as the past may be the key to the future, an understanding of processes in the present ocean may be a key to understanding the past.

Whereas the relatively thin mixed layer at the ocean's surface remains in equilibrium with the atmosphere, the deep ocean is oversaturated with dissolved CO_2. This CO_2 gradient is maintained throughout most of the blue ocean by the biological pump. The pump is the long-recognized rain of small debris consisting of phytoplankton cells and zooplankton fecal pellets and molts that sink out of the surface waters of the ocean (Agassiz 1888). These sinking particles are important because they remove POC from surface mixed layers into stratified, relatively deep layers where, on a millennial time scale, it is no longer susceptible to exchange with the atmosphere. When particulate matter is removed in this manner, it is termed "export flux" (Berger et al. 1989).

The relative importance of biological transfers compared to physical exchanges is not well understood. For example, Longhurst (1991) states, "The biologically mediated flux that maintains the geochemical disequilibrium is small compared with the transfer of CO_2 across the ocean surface driven simply by diffusion and solubility." If this were true on all scales, however, the pump would not maintain the gradient. Only if carbon is transferred across the permanent thermocline (circa 500 m depth) or buried in sediments by the pump can it be considered sequestered or removed from surface ocean–atmosphere exchanges, which are purely physical. The rate of this export flux, to use the terminology of Berger et al. (1989) for the rain of detritus, has only since the mid-1970s been quantified using sediment particle traps (Deuser and Ross 1980; Martin et al. 1987).

Several empirical expressions have been developed from sediment trap data to describe POC fluxes (Lorenzen et al. 1983). These expressions allow us to estimate these fluxes over a wide range of conditions. Fluxes at 1000- to 4000-m depth in the oceanic central gyres amount to 1–2% of surface primary production (Knauer et al. 1990), confirming the work of Menzel (1970) that most remineralization of organic matter occurs within several hundred meters of the surface. Over a large depth interval in the Atlantic, the POC flux appears to control both total sediment community biomass and total sediment community respiration (Rowe et al. 1991).

A widely cited model for estimating export flux is the Suess (1980) equation:

$$C_{flux(z)} = C_{prod}/0.0238z + 0.212 \qquad (12.1)$$

where $z = >50$ m depth. C_{prod} is the primary production rate at the surface and C_{flux} is the organic carbon flux at depth measured with a variety of different kinds of sediment traps. The biological pump equates "new" production, which is roughly equivalent to export flux, or $C_{flux(z)}$ in the above equation, with the production driven by new nitrate (Eppley and Peterson 1978). It has been pointed out by Longhurst (1991) that this simple though elegant relationship does not apply if nitrification or nitrogen fixation is significant in the euphotic zone. It also does not recognize the possible importance of vertical diurnal migration as an enhancement to transfers of organic matter down and inorganic nutrients up in the water column (Longhurst and Harrison 1989).

It is clear from surface water analysis, sediment trap data, and the distribution of surface sediments that biosiliceous components (such as diatoms and silicoflagellates) are a major contributor to particle flux in regions of high surface water productivity.

Heath (1974) suggested that these phytoplankton may represent as much as 50–70% of ocean primary production. The distribution of biogenic opal in surface sediments corresponds well with regions of high surface water productivity, such as the equatorial divergence, the polar fronts, and regions of coastal upwelling (Lisitsin 1972; Leinen et al. 1986). Diatom abundance in the sedimentary record appears to be a useful indicator of export production. The difficulty lies in relating the quantity preserved in the sediments to that produced in the photic zone. Generally less than 5% of that produced in the surface waters is incorporated into the sedimentary record.

The ecological efficiency of oceanic ecosystems is inversely related to the quantitative importance of the biological pump, and this varies around the world. The depth of the mixed layer is thought to be inversely related to the flux of material through it (Hargrave 1975). At low latitudes the ecological efficiency of the mixed layer is higher than that at high latitudes, and little escapes the mixed layer. That is, at low latitudes autotrophic production and heterotrophic consumption are more tightly coupled, owing to temperature and stability (measured as annual variation in temperature). At high latitudes the surface water ecosystems are less tightly coupled and more material is transferred to, and therefore sequestered in, the deep sea. That is, the pump is more important. Direct evidence of this finding is the higher biomass of zooplankton (Vinogradov 1968) and benthos (Rowe 1983) supported in deep water at high latitudes. The North Pacific differs from the North Atlantic as well, with the apparent absence of a spring bloom in the Pacific being attributed to tighter coupling between primary production and grazing. Poor coupling in the North Atlantic results in a pronounced spring bloom.

Efficiency of Vertical Transfer and Decay Rates

The simplest description of the vertical distribution of organic particles is a first-order decay equation (Craig 1971):

$$d(C)/dz = -k_z (C_o) \qquad (12.2)$$

where C is the concentration of organic carbon, z is depth, k_z is a coefficient for the rate at which carbon disappears with depth, and C_o is organic carbon concentration in the surface water. The solution of this equation is

$$C_z = (C_o) e^{-k(z)} \qquad (12.3)$$

Given a single average rate of net vertical transfer (sinking and vertical migration) for all the particles, the equation can be rewritten with the concentration of carbon varying with time instead of depth. In this form the decay constant can be considered the average remineralization rate (k_t) for all particulate carbon at any depth. In an ecologically efficient system, the decay rate would be high, similar to that found at low latitudes, but at high latitudes and in those systems in which coupling between particle production and consumption is poor, the efficiency would be low. This would be reflected in a smaller (less negative) k_t value. By varying either the k_z or the k_t value, one can calculate changes in concentration with depth that would result from alterations in ecosystem gross remineralization efficiency.

A qualitative example of poor recycling efficiency in surface water on seasonal time scales is the occasional photographic documentation of extensive flocculent layers covering the sea floor (Billet et al. 1983; Lampitt 1985). An intriguing and important question concerns the relationship between such patchy, serendipitous observations and the diatom mats in the sedimentary record of the eastern equatorial Pacific (Kemp and Baldauf 1993).

If global climate changes increase the decoupling between production and consumption in surface waters, then greater transfer of POC and particulate inorganic carbon through the pycnocline will occur, and the deep ocean will sequester more carbon. This increase in the vertical flux, i.e., the importance of the biological pump, can be estimated grossly by altering vertical decay constants and summing resultant increases in POC stocks at depth for given areas or for the entire global ocean.

THE COASTAL OCEAN

Productivity in Response to Nutrient Loading

J. Walsh et al. (1981) and Mackenzie (1981) suggested that coastal ocean productivity is increasing on a global scale because of nutrient loading attributable to such factors as agricultural runoff, sewage effluents, and rain. Although evidence for gradual increases in inorganic nitrogen loading is persuasive, the evidence that coastal productivity is increasing is equivocal.

At this point, it is difficult to distinguish between open, POC-exporting (autotrophic) and closed, POC-recycling (heterotrophic) continental margins. For example, Rowe et al. (1986) revised the Walsh et al. (1981) budget for the mid-Atlantic bight and found that the same area had a greater respiratory organic carbon demand than was being produced. They suggested that the original positive imbalance resulted from a failure to include bacteria and a failure to recognize that the system might be able to "store" excess carbon briefly (on a time scale of weeks to months) following bloom periods. The early estimates of shelf export (circa 50% of total shelf primary production, e.g., 90% of the spring bloom) have been lowered to 5–15% of annual production (Falkowski et al. 1988; Christensen 1989).

S. Smith and Mackenzie (1987) and S. Smith and Hollibaugh (1993) suggest that most shelves are actually heterotrophic. The added organic matter from which the excess CO_2 is produced is terrestrial and delivered by rivers. However, there is nothing to prevent continental shelves from both exporting fixed carbon to the adjacent deep basin and producing a net excess of CO_2, if the excess organic matter provided by the rivers and associated wetlands is sufficient. They make the case that the preindustrial ocean was heterotrophic, with the fluvial load driving this flux of carbon, but that the ocean has become autotrophic since then in response to increases in atmospheric CO_2 concentration (Mackenzie et al. 1991).

Lateral Input or Export from Continental Margins

Several lines of evidence indicate that continental margins are important sources of organic matter in the deep ocean. Benthic ecologists have long recognized that seafloor biomass is higher near the continent (Rowe 1971; Carey 1981). J. Walsh et al. (1981) suggested that excess POC is exported from dynamic shelf environments to low-energy, quiescent "depocenters" on the adjacent slopes that are below the permanent pycnocline. Premuzic et al. (1982) have mapped the worldwide distribution of surface sediment organic matter and agree that most continental slopes act as transient depocenters. Gardner and Richardson (1992) have summarized information on the apparent increase in particles in the lower several hundred meters above the seafloor at continental rise and abyssal plain depths in the western North Atlantic. This has long been assumed to be resuspended material, but whether it is local resuspension or margin export remains to be determined. The near-bottom POC in this "nepheloid" layer appears to be more refractory than primary or export POC flux at midwater depths (J. Walsh 1992). Recent studies of sediment traps on moorings along the continental slope (Biscaye et al. 1988; Biscaye and Anderson 1994), radionuclide budgets in a North Atlantic continental slope depocenter (Anderson et al. 1994), and transects of sediment oxygen demand in the eastern North Pacific (Jahnke and Jackson 1987, 1992; Archer and Devol 1992) and the western North Atlantic (Rowe et al. 1994) confirm that shelf export is a good possibility. The absolute fluxes are up to 70 mg carbon m^{-2} day^{-1} in the few measurements made to date. The input off shelves and down slopes into the ocean's interior has not included fluxes down submarine canyons, however.

Coastal Sediment Anaerobic Metabolism

Greater runoff of fresh water would increase stratification as well as nutrient loading. The two results combined will increase intermittent hypoxia and anoxia in east coast environments (e.g., the New York bight, the Mississippi River plume area, and Chesapeake Bay) and broaden permanent low-oxygen zones below intense upwelling systems along the west coasts of continents (Rowe et al. 1994). The general shift to anaerobic processes might tend to increase carbon burial rate (Canfield 1989; Jorgensen et al. 1990), a result that would prevent carbon from potentially returning to the atmosphere as CO_2. Berner (1982) noted that although the coastal ocean, including estuaries and deltaic sediments, covers only 10% of the ocean area, 90% of the organic carbon deposition occurs there. To a large degree this percentage depends on the recycling efficiency, which is the ratio of the total deposited organic matter to that which is buried. A crucial point of disagreement lies in whether sediment organic concentrations are a function of surface water productivity or low oxygen concentrations (Henrichs and Reeburgh 1987; Jorgensen et al. 1990). The observation that sapropels or black shales, with high carbon loading, do not necessarily correlate with low oxygen (Pedersen and Calvert 1990) needs further explanation.

A shift to anaerobic processes may also increase the CH_4 or volatile fatty acid gas flux to the atmosphere from shallow organic-rich environments. Estimates of CH_4 fluxes to the atmosphere from natural ecosystems, including aquatic environments, range from about 5 to 10×10^7 tons carbon per year (J. Walsh 1988). The phenomenon of increasing the area of hypoxic and anoxic coastal zone sediments might remove greater quantities of fixed organic matter by burial in the coastal zone, but there may also be a shift to radiatively important trace gases (RITGs) other than CO_2, which accompanies such a shift in metabolic pathways. A widely cited coastal zone burial rate is circa 0.1 Pg carbon per year (Berner 1982), but a number of values are on the order of twice that (Mackenzie [1981] gives circa 0.2 Pg carbon per year, for example). Berner's value is a long-term average; Mackenzie's estimate includes perturbations.

Benthic Primary Production and Coralline Algae

Primary production by large, attached marine plants is fairly well documented at high rates of 500 to 2,000 g carbon m^{-2} (S. Smith 1981; Mann 1982). Primary production by microscopic bottom-dwelling algae is well documented and thought to be less important on a global scale, but this is perhaps because the measurements have been made principally in coastal embayments (Revsbech et al. 1981, 1986). It has now been discovered that single-celled algae are abundant on the sediments of many continental shelves, and that they are active primary producers (Cahoon et al. 1992; G. T. Rowe and others, unpublished data). Coralline algae fix organic matter and calcium carbonate (Hillis-Colinvaux 1974; S. Smith 1978). The role of such carbonate banks on a global scale is not clear, although on a local scale (G. Rowe and G. Boland, preliminary unpublished data for the Gulf of Mexico), they can double the present estimates of primary production for the water column. One would expect their importance to be confined to clear, shallow-water environments at low latitudes, which are not characterized by upwelling or fluvial runoff.

Estimates of total benthic primary production are as high as 2.9 Pg carbon per year (Charpy-Roubaud and Sournia 1990), or about 10% of the global rate. There is the possibility of extensive growth in biomass and CaCO$_3$ deposition in coastal waters with either global warming or sea level rise.

BIOTIC RESPONSE TO GLOBAL WARMING

Numerous models address the ocean's response to CO$_2$-induced climate warming. Most focus on high-latitude regions, because these are the regions where deep water is formed and the sea surface is undersaturated with CO$_2$ relative to the atmosphere. This situation differs from CO$_2$ saturation at low latitudes. For example, the equatorial divergences and coastal upwelling zones are by contrast characterized by oversaturated levels of CO$_2$. In addition, CO$_2$ is less soluble in warmer water. Induced global warming could result in a decreased temperature gradient between the high and low latitudes, as evidenced by proxy indicators in the geological record. This situation in turn would result in the partial collapse of thermohaline circulation (e.g., Bryan and Spelman 1985) and in more diffuse and less intense upwelling and surface water productivity compared with that observed today. Nutrient availability also should be reduced (Broecker 1982), with increased dependence on nutrient recycling to support photosynthesis. Siegenthaler and Wenk (1984) suggest that such changes in nutrient availability and in the concentration of atmospheric CO$_2$ can occur quickly enough to change the atmospheric CO$_2$ level by as much as 50–100 ppm on a time scale of 100 years.

Today, productivity in the blue ocean is divided distinctly between siliceous primary production dominating regions of upwelling and the high latitudes and calcareous primary production dominating the low latitudes. With global warming, this partitioning would be less well defined, with increased precipitation of shallow-water carbonates (Berger et al. 1989).

Sarmiento and Toggweiler (1984) suggest that the biological pump at high latitudes might be embellished by a more sluggish circulation. This would be logical during glacial recession because there is less ice cover and more open water. Granted, if the coupling of the autotrophic and heterotrophic metabolism in the surface ocean decreases owing to global warming, then the biological pump will transfer more material to depth. A surface-water warming and a decrease in deep-water mass formation, however, could decrease the amount of CO_2 taken up at high latitudes by physical and chemical processes. With global warming the polar ice caps recede, giving this newly exposed water greater light in an ocean already having plentiful nitrate. One might expect that productivity at the poles would increase somewhat, but the fact that open waters are not presently producing up to their NO_3 carrying capacity suggests that this increase would have only marginal effects, the iron limitation hypothesis (Berger and Wefer 1991; Peng and Broecker 1991) notwithstanding. In our view the opposite will occur. Productivity on a global scale is increased by global cooling and polar ice accretion because upwelling, including equatorial divergence, is embellished.

Abrupt alteration of climate could promote the growth of opportunistic phytoplankton without accompanying longer-lived populations of grazers. This situation could result in blooms of opportunistic species. If the production of uneaten or unpalatable species (e.g., *Phaeocystis* spp.) increases (W. Smith et al. 1994), and these organisms are not eaten and do not sink out of the surface layers, then most of the carbon will be returned via microbial heterotrophy to the atmosphere as CO_2. If the processes increase respiration in the mixed layer, then more RITGs will be transferred back into the atmosphere. Alterations of weather patterns termed El Niño Southern Oscillations have had well-documented, catastrophic effects on fisheries in oceanic ecosystems for well over a century off the west coast of South America. There is no reason not to expect that these large-scale phenomena will have effects on trace gas transfers as well, but again the coupling between such patterns and biological responses to them is poorly understood.

Serious attention should be directed to the coastal regions when considering the impacts of climate warming. Global warming will melt the ice caps and cause sea level to rise, in turn increasing the width of continental shelves. In addition to flooding cities, this incursion of the ocean, when combined with sluggish circulation, decreasing oxygen levels, and higher nutrient loading, could enlarge areas of local hypoxia or anoxia over much of our continental shelves. Overall global ocean productivity will increase because productivity is higher on shelves than in the open ocean. If the coastal zone is a net producer of RITGs, that is, if it is heterotrophic, then the warming will ultimately accelerate trace gas production. The greenhouse warming will be exacerbated by biological feedback. Feedback in the opposite direction might occur if the RITGs produced are potential cloud condensation nuclei (CCN), such as dimethyl sulfide. The importance of the coastal ocean in the production of CCN is unknown at present. Increasing cloud cover in the coastal zone will decrease insolation and primary productivity there.

REFERENCES

Abelmann, A., R. Gersonde, and V. Spiess. 1990. Pliocene-Pleistocene paleoceanography in the Weddell Sea—siliceous microfossil evidence. In *Geological History of the Polar Oceans: Arctic versus Antarctic*, ed. U. Bleil and J. Thiede, pp. 729–760. Kluwer, Dordrecht, The Netherlands.

Agassiz, A. 1888. *Three Cruises of the Blake*. Museum of Comparative Zoology, Harvard University, Cambridge, Mass.

Anderson, R., G. Rowe, P. Kemp, S. Trumbore, and P. Biscaye. 1994. Carbon budget for the mid-slope depocenter of the middle Atlantic bight. *Cont. Shelf Res.* In press.

Archer, D., and A. Devol. 1992. Benthic oxygen fluxes on the Washington shelf and slope: A comparison of in situ microelectrode and chamber flux measurements. *Limnol. Oceanogr.* 37:614–629.

Baldauf, J. G., and J. A. Barron. 1990. Evolution of biosiliceous sedimentation patterns—Eocene through Quaternary: Paleoceanographic response to polar cooling. In *Geological History of the Polar Oceans: Arctic versus Antarctic,* ed. U. Bleil and J. Thiede, pp. 575–608. Kluwer, Dordrecht, The Netherlands.

Barnola, J., D. Raynaud, Y. Korotkevich, and C. Lorius. 1987. Vostok ice core provides 160,000-year record of atmosphere CO_2. *Nature* 329:408–414.

Barron, J. A., and J. G. Baldauf. 1989. Tertiary cooling steps and paleoproductivity as reflected by diatoms and biosiliceous sediments. In *Productivity of the Ocean: Present and Past,* ed. W. H. Berger, V. S. Smetacek, and G. Wefer, pp. 341–354. Wiley, New York.

Berger, W., and G. Wefer. 1991. Productivity of the glacial ocean: Discussion of the iron hypothesis. *Limnol. Oceanogr.* 36:1899–1918.

Berger, W. H., V. S. Smetacek, and G. Wefer. 1989. Ocean productivity and paleoproductivity—An overview. In *Productivity of the Ocean: Present and Past,* ed. W. H. Berger, V. S. Smetacek, and G. Wefer, pp. 1–34. Wiley, New York.

Berner, R. A. 1982. Burial of organic carbon and pyrite sulfur in the modern ocean: Its geochemical significance. *Am. J. Sci.* 282:451–473.

Billet, D., R. Lampitt, A. Rice, and R. Mantoura. 1983. Seasonal sedimentation of phytoplankton to the deep-sea benthos. *Nature* 302:520–522.

Biscaye, P., and R. Anderson. 1994. Particle fluxes on the slope of the southern Mid-Atlantic Bight: SEEP II. *Cont. Shelf Res.* In press.

Biscaye, P., R. Anderson, and B. Deck. 1988. Fluxes of particles and constituents to the eastern United States continental slope and rise: SEEP-I. *Cont. Shelf Res.* 8:888–904.

Broecker, W. S. 1982. Ocean chemistry during glacial time. *Geochem. Cosmochim. Acta* 46:1689–1705.

Bryan, K., and M. Spelman. 1985. The ocean's response to a CO_2-induced warming. *J. Geophys. Res.* 90:11,679–11,688.

Cahoon, L., and J. Cooke. 1992. Benthic microalgal production in Onslow Bay, North Carolina, USA. *Mar. Ecol. Prog. Ser.* 84:185–196.

Canfield, D. 1989. Sulfate reduction and oxic respiration in marine sediments: Implications for organic carbon preservation in euxinic environments. *Deep-Sea Res.* 36:121–138.

Carey, A. 1981. A comparison of benthic infaunal abundance on two abyssal plains in the northeast Pacific Ocean with comments on deep-sea food sources. *Deep-Sea Res.* 28:467–479.

Charpy-Roubaud, C., and A. Sournia. 1990. The comparative estimation of phytoplanktonic, microphytobenthic and macrophytobenthic primary production in the oceans. *Mar. Microb. Food Webs* 4:31–57.

Christensen, J. 1989. Sulphate reduction and carbon oxidation rates in continental shelf sediments, an examination of offshelf carbon transport. *Cont. Shelf Res.* 9:223–246.

Craig, H. 1971. The deep metabolism: Oxygen consumption in abyssal ocean water. *J. Geophys. Res.* 76:5078–5086.

Dansgaard, W., S. J. Johnsen, H. B. Clausen, D. Dahl-Jensen, N. S. Gundestrup, C. U. Hammer, C. S. Hvidberg, J. P. Steffensen, A. E. Sveinbjornsdottir, J. Jouzel, and G. Bond. 1993. Evidence for general instability of past climate from a 250-kyr. ice-core core. *Nature* 364:218–219.

Deuser, W., and E. Ross. 1980. Seasonal change in the flux of organic carbon to the deep Sargasso Sea. *Nature* 283:364–365.

Eppley, R., and B. Peterson. 1978. Particulate organic matter flux and planktonic new production in the deep ocean. *Nature* 282:677–680.

Falkowski, P., C. Flagg, G. Rowe, S. Smith, T. Whitledge, and C. Wirick. 1988. The fate of a spring phytoplankton bloom: Export or oxidation? *Cont. Shelf Res.* 8:457–484.

Gardner, W., and M. Richardson. 1992. Particle export and resuspension fluxes in the western North Atlantic. In *Deep-Sea Food Chains and the Global Carbon Cycle,* ed. G. Rowe and V. Pariente, pp. 339–364. NATO ASI Series. Kluwer, Dordrecht, The Netherlands.

Hargrave, B. 1975. The importance of total and mixed-layer depth in the supply of organic material to bottom communities. *Symp. Biol. Hung.* 15:157–165.

Heath, G. R. 1974. Dissoved silica and deep-sea sediments. In *Studies in Paleo-oceanography,* ed. W. W. Hay, pp. 77–93. Society of Economic Paleontologists and Mineralogists Special Publication 20. Tulsa, Okla.

Henrichs, S., and W. Reeburgh. 1987. Anaerobic mineralization of marine sediment organic matter: Rates and the role of anaerobic processes in the oceanic carbon economy. *Geomicrobiol. J.* 5:191–237.

Herbert, T. D., W. B. Curry, J. A. Barron, L. A. Codispoti, R. Gersonde, R. S. Kier, A. C. Mix, B. Mycke, H. Schrader, R. Stein, and H. R. Thierstein. 1989. Geological reconstructions of marine productivity. In *Productivity of the Ocean: Present and Past,* ed. W. H. Berger, V. S. Smetacek, and G. Wefer, pp. 409–428. Wiley, New York.

Hillis-Colinvaux, L. 1974. Productivity of the coral reef alga *Halimeda* (order Siphonales), In *Proceedings of the Second International Coral Reef Symposium,* ed. A. Cameron, pp. 35–42. The Great Barrier Reef Committee, Brisbane, Australia.

Jahnke, R., and G. Jackson. 1987. Role of the sea floor organisms in oxygen consumption in the deep North Pacific Ocean. *Nature* 329:621–623.

Jahnke, R., and G. Jackson. 1992. The spatial distribution of sea floor oxygen consumption in the Atlantic and Pacific oceans. In *Deep-Sea Food Chains and the Global Carbon Cycle,* ed. G. Rowe and V. Pariente, pp. 295–307. NATO ASI Series. Kluwer, Dordrecht, The Netherlands.

Jorgensen, B., M. Bang, and T. Henry Blackburn. 1990. Anaerobic mineralization in marine sediments from the Baltic Sea–North Sea transition. *Mar. Ecol. Progr. Ser.* 59:39–54.

Kemp, A., and J. G. Baldauf. 1993. Vast Neogene laminated diatom mat deposits from the eastern equatorial Pacific Ocean. *Nature* 362:141–143.

Knauer, G., D. Redalje, W. Harrison, and D. Karl. 1990. New production at the VERTEX time-series site. *Deep-Sea Res.* 37:1121–1134.

Lampitt, R. 1985. Evidence for the seasonal deposition of detritus to the deep-sea floor and its subsequent resuspension. *Deep-Sea Res.* 22:885–897.

Leinen, M., D. Cwienk, R. Heath, P. Biscaye, V. Kolla, J. Thiede, and P. Dauphin. 1986. Distribution of biogenic silica and quartz in recent deep-sea sediments. *Geology* 14:199–203.

Lisitsin, A. P. 1972. *Sedimentation in the world ocean.* Society of Economic Paleontologists and Mineralogists Special Publication 17. SEPM, Tulsa.

Longhurst, A. 1991. Role of the marine biosphere in the global carbon cycle. *Limnol. Oceanogr.* 36:1507–1526.

Longhurst, A., and G. Harrison. 1989. The biological pump: Profiles of plankton production and consumption in the upper ocean. *Prog. Oceanogr.* 22:47–123.

Lorenzen, C., N. Welschmeyer, and A. Copping. 1983. Particulate organic carbon flux in the subarctic Pacific. *Deep-Sea Res.* 30:639–643.

Mackenzie, F. 1981. Global carbon cycle: Some minor sinks for CO_2. In *Carbon Dioxide Effects Research and Assessment Program Conference 016: Flux of Organic Carbon by Rivers to the Ocean,* pp. 359–384. U.S. Department of Energy, Washington, D.C.

Mackenzie, F., J. Bewers, R. Charlson, E. Hofmann, G. Knauer, J. Kraft, E. E. Nöthig, B. Quack, J. Walsh, M. Whitfield, and R. Wollast. 1991. Group report: What is the importance of oceanic margin processes in global change? In *Ocean Margin Processes in Global Change,* ed. R. Mantoura, J. Martin, and R. Wollast, pp. 433–454. Wiley, New York.

Mann, K. 1982. *The Ecology of Coastal Waters. A Systems Approach.* Blackwell, Oxford.

Martin, J., G. Knauer, D. Karl, and W. Broenkow. 1987. VERTEX: Carbon cycling in the northwest Pacific. *Deep-Sea Res.* 34:267–285.

Menzel, D. W. 1970. The role of *in situ* decomposition of organic matter on the concentration of non-conservative properties in the sea. *Deep-Sea Res.* 17:751–764.

Mix, A. 1989. Pleistocene paleoproductivity: Evidence from organic carbon and foraminiferal species. In *Productivity of the Ocean: Present and Past,* ed. W. H. Berger, V. S. Smetacek, and G. Wefer, pp. 313–340. Wiley, New York.

Oeschger, H. J. Beer, U. Siegenthaler, B. Stauffer, W. Dansgaard, and C. Zangway. 1984. Late glacial climate history from ice cores. In Climate Processes and Climate Sensitivity, ed. J. Hansen and T. Takahashi, 29:299–306. Geophysical Monograph 29. American Geophysical Union, Washington, D. C.

Pedersen, T., and S. Calvert. 1990. Anoxia vs. productivity: What controls the formation of organic-carbon-rich sediments and sedimentary rocks? *Am. Assoc. Petrol. Geol. Bull.* 74:454–466.

Peng, T.-H., and W. Broecker. 1991. Factors limiting the reduction of atmospheric CO_2 by iron fertilization. *Limnol. Oceanogr.* 36:1919–1927.

Pisias, N., and N. Shackleton. 1984. Modelling the global climate response to orbital forcing and atmosphere carbon dioxide changes. *Nature* 310:757–759.

Premuzic, E., C. Benkovitz, J. Gaffney, and J. Walsh. 1982. The nature and distribution of organic matter in the surface sediments of world oceans and seas. *Org. Geochem.* 4:63–77.

Revsbech, N., B. Jorgensen, and O. Brix. 1981. Primary production of microalgae in sediments measured by oxygen microprofile, $H^{14}CO_3$ fixation and oxygen exchange methods. *Limnol. Oceanogr.* 26:717–730.

Revsbech, N., B. Madsen, and B. Jorgensen. 1986. Oxygen production and consumption in sediments determined at high spatial resolution by computer simulation of oxygen microelectrode data. *Limnol. Oceanogr.* 31:293–304.

Rowe, G. 1971. Benthic biomass and surface productivity. In *Fertility of the Sea, Vol. 2,* ed. J. Costlow Jr., pp. 441–454. Gordon & Breach, New York.

Rowe, G. 1983. Biomass and production in the deep-sea macrobenthos. In *The Sea,* Vol. 8: *Deep-Sea Biology,* ed. G. Rowe, pp. 97–121. Wiley, New York.

Rowe, G., S. Smith, P. Falkowski, T. Whitledge, R. Theroux, W. Phoel, and H. Ducklow. 1986. Do continental shelves export organic matter? *Nature* 324:559–561.

Rowe, G., M. Sibuet, J. Deming, A. Khripounoff, J. Tietjen, S. Macko, and R. Theroux. 1991. "Total" sediment biomass and preliminary estimates of organic carbon residence time in deep-sea benthos. *Mar. Ecol. Prog. Ser.* 79:99–114.

Rowe, G., G. Boland, W. Phoel, R. Anderson, and P. Biscaye. 1994. Deep sea-floor respiration as an indication of lateral input of biogenic detritus from the continental margins. *Cont. Shelf Res.* In press.

Sarmiento, J., and J. Toggweiler. 1984. A new model for the role of the ocean in determining atmospheric pCO_2. *Nature* 308:621–624.

Sarnthein, M., K. Winn, and R. Zahn. 1987. Paleoproductivity of oceanic upwelling and the effect on atmosphere CO_2 and climatic change during deglaciation times. In *Abrupt Climatic Change,* ed. W. H. Berger and L. D. Labeyrie, pp. 311–337. Reidel, Dordrecht, The Netherlands.

Shackleton, N. J., and N. G. Pisias. 1985. Atmospheric carbon dioxide, orbital forcing and climate. In *The Carbon Cycle and Atmospheric CO_2: Natural Variations Archean to Present,* ed. E. T. Sundquist and W. S. Broecker, pp. 303–317. Geophysical Monograph 32. American Geophysical Union, Washington, D.C.

Siegenthaler, U., and T. Wenk. 1984. Rapid atmospheric CO_2 variations and ocean circulation. *Nature* 308:624–626.

Smith, S. 1978. Coral reef area and the contributions of reefs to processes and resources of the world's oceans. *Nature* 273:225–226.

Smith, S. 1981. Marine macrophytes as a global carbon sink. *Science* 211:838–840.

Smith, and F.Mackenzie. 1987. The ocean as a net heterotrophic system:Implications from the carbon biogeochemical cycle. *Global Biogeochem. Cycles* 1:87–198.

Smith, S., and J. Hollibaugh. 1993. Role of coastal ocean organic metabolism in the oceanic organic carbon balance. *Rev. Geophys.* 31:75–89.

Smith, W., L. Codispoti, D. Nelson, T. Manley, E. Buskey, J. Niebauer, and G. Cota. 1994. Importance of *Phaeocystis* blooms in the high-latitude ocean carbon cycle. *Nature* 352: 514–516.

Suess, E. 1980. Particulate organic carbon flux in the oceans—Surface productivity and oxygen utilization. *Nature* 288:260–263.

Sverdrup, H., M. Johnson, and R. Fleming. 1942. *The Oceans*. Prentice Hall, New York.

Vinogradov, M. 1968. *Vertical Distribution of Oceanic Zooplankton*. Academica Nauk, U.S.S.R., Institute of Oceanology, Moscow. English transl. Israel Program for Scientific Translations. National Technical Information Service, U.S. Department of Commerce, Washington, D.C.

Walsh, J. 1992. Large aggregate flux and fate at the seafloor: Diagenesis during the rebound process. In *Deep-Sea Food Chains and the Global Carbon Cycle*, ed. G. Rowe and V. Pariente; pp. 365–373. NATO ASI Series. Kluwer, Dordrecht, The Netherlands.

Walsh, J. 1988. *On the Nature of Continental Shelves*. Academic Press, London.

Walsh, J., G. Rowe, R. Iverson, and P. McRoy. 1981. Biological export of shelf carbon is a sink of the global CO_2 cycle. *Nature* 291:196–201.

13

Net Carbon Metabolism of Oceanic Margins and Estuaries

STEPHEN V. SMITH

COASTAL OCEAN NET METABOLISM AT STEADY STATE

Primary production in the oceans fixes approximately 4×10^{15} mol CO_2 into organic matter annually. This rate is about 80% that of primary production on land and about 10 times the anthropogenic emissions of CO_2 into the atmosphere via burning of fossil fuel. The coastal ocean, which is the focus of this summary, apparently accounts for about 10% of the total primary production in the global ocean.

Net oceanic metabolism (net ecosystem production for the oceans), that is, the difference between gross primary production and total respiration, is more closely associated with CO_2 transfer as a gas across the air-sea interface than is either gross primary production or total respiration taken individually. Three lines of evidence suggest that total respiration of organic matter in the ocean slightly exceeds gross primary production:

1. Estimated long-term burial of organic matter in marine sediments (about 0.13 petagrams [Pg] [=10^{15} g] carbon or 11×10^{12} mol carbon per yr) apparently accounts for only about one-third of estimated organic input from land (~0.41 Pg carbon or 34×10^{12} mol carbon per yr).

2. Considerations of the chemical reactivity of land-derived organic matter suggest that about one-third of this material is readily oxidized.

   If both of these observations are correct, the remainder of the terrigenous organic matter, about one-half of the total, oxidizes more slowly. The most plausible alternative explanation to this suggestion is that there are separable rapidly and slowly reactive terrigenous carbon reservoirs reaching the ocean. The discrepancy between budgetary and biochemical estimates of terrigenous carbon reactivity may simply reflect statistical uncertainty in some combination of the organic loading rate, the burial rate, and the biochemical reactivity.

3. Independently of the first two sets of arguments, community metabolism data for about 20 estuarine and shallow shelf sites suggest that total respiration tends to exceed gross primary production in this region of the ocean. We estimate the net respiration of the coastal ocean to be at least 0.08 Pg carbon or 7×10^{12} mol carbon per yr.

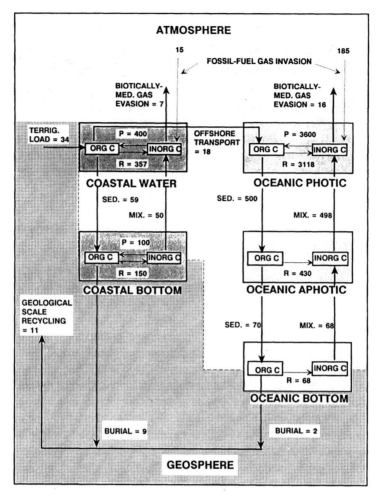

Figure 13.1. Characteristics of the organic carbon cycle in the coastal ocean and the open ocean. The data shown are estimates of steady-state fluxes (in 10^{12} mol carbon per year). Also shown for comparison, as dashed-line arrows, are today's approximate rates of non-steady-state penetration of fossil fuel CO_2 into surface waters of the coastal ocean and the open ocean. (From Smith and Hollibaugh 1994.)

From these observations, we estimate that the ocean oxidizes approximately 0.28 Pg carbon or 23×10^{12} mol carbon per yr more organic matter than it produces (Figure 13.1). At least 0.08 Pg carbon or 7×10^{12} mol carbon per yr net oxidation is likely to occur in the coastal ocean, with the remainder probably occurring farther offshore. The coastal ocean takes on particular importance in this budgetary analysis, because the surface area involved is only about 8% of the total oceanic area. Thus, the coastal ocean net metabolic "signal" per unit area or per unit water volume, in the form of elevated CO_2 partial pressure, is likely to exceed the air-to-ocean partial pressure gradient generated by the human-induced rise of atmospheric CO_2 partial pressure.

Much of the coastal ocean net organic oxidation apparently occurs in bays and estuaries, rather than on the open shelf. This further geographic restriction will make the local signal of net organic oxidation even larger. By contrast, the net oxidation that I calculate for the open ocean generates a small PCO_2 signal.

The explicit suggestion that the ocean at steady state oxidizes slightly more organic material than it produces has been recognized at least since Garrels and Mackenzie (1972) called attention to the topic. Berner (1982) reiterated the suggestion, and Smith and Mackenzie (1987) introduced the term "net heterotrophy" to describe the net oxidation of organic matter by the ocean. Other recent statements of this conclusion are given by Holligan and Reiners (1992) and Sarmiento and Sundquist (1992).

ANTHROPOGENIC CHANGES IN COASTAL OCEANIC METABOLISM

A major question is the degree to which human activities may already have altered this net heterotrophy and may further alter it in the future (Wollast and Mackenzie 1989). The actual amount of net heterotrophy in the ocean is only poorly known, so it should not be surprising that the degree of disturbance is essentially unknown. Nevertheless, some attempts can be made to guess at the changes.

Smith and Mackenzie (1987) concluded from data in Meybeck (1982) that only about 3% of land-derived organic matter entering the ocean would have to be oxidized for oxidation to exceed the potential new production from inorganic nutrient loading from land. Meybeck (1982) did not have data for pollutant transport of organic matter by rivers, but his data for pollutant transport of inorganic nitrogen and phosphorus suggested that these inputs have increased to about two to three times the natural rates. From the carbon/nutrient ratio of marine organic matter, it can be calculated that the increased autotrophic activity that might be associated with increased nitrogen and phosphorus loading would be equivalent to about 0.01–0.04 Pg carbon or $1–3 \times 10^{12}$ mol carbon per year.

Long-term change in organic transport to the ocean is more difficult to estimate. Judson and Ritter (1964) estimated that the long-term sediment delivery to the ocean has been about 9 Pg/year. Organic delivery from land to the ocean is divided approximately evenly between particulate and dissolved organic matter. So, from a long-term delivery of about 0.02 Pg or 17×10^{12} mol/year of particulate organic carbon, we calculate that the suspended material delivered to the ocean has averaged about 2.2% organic carbon. Milliman and Syvitski (1992) estimated the present sediment delivery to be about 20 Pg/year. If the percent organic carbon in this suspended load has remained constant, we can calculate that present particulate organic carbon delivery is about 0.44 Pg carbon or 37×10^{12} mol carbon per year—an increase of 0.24 Pg carbon or 20×10^{12} mol/year over preanthropogenic delivery. If one-third to two-thirds of land-derived organic matter is reactive, then increased particulate organic delivery has been sufficient to increase organic oxidation (i.e., net heterotrophy) in the ocean by 0.08–0.17 Pg carbon or $7–14 \times 10^{12}$ mol carbon per year. There may also be an increase in organic oxidation associated with elevated delivery of dissolved organic material.

It therefore seems likely that change in the rate of terrigenous organic loading has exceeded the pollutant changes in inorganic nutrient delivery. There are undoubtedly many locations where local increases of inorganic nutrient discharge have exceeded local increases of labile organic matter discharge, but the total changes appear to favor increased oceanic heterotrophy as a result of human activity.

Finally, consider the specific role of future global warming in altering net metabolism in the coastal ocean. If the arguments offered previously are generally reasonable, then the net metabolism of the coastal ocean is probably most directly controlled by the delivery of organic matter rather than by temperature. Perhaps at polar latitudes a significant amount of potentially labile organic matter is buried without oxidation; elevating water temperature would probably result in increased oxidation of this material. Nevertheless it can be assumed that, to a first approximation, temperature elevation might increase the rate of organic cycling in the ocean (hence possibly elevate both primary production and respiration) without having much effect on the absolute difference between production and respiration.

A second approximation can be offered. Enhanced oxidation of terrigenous detrital organic matter may occur on land and/or in freshwater aqueous systems in response to elevated temperature. In this case, the amount of organic matter reaching the ocean as a proportion of total suspended load may decrease. One can therefore speculate that elevated temperature might well lead to some decrease in the net respiratory potential of organic matter reaching the coastal ocean.

Balancing these considerations of increased organic matter supply and more rapid oxidation of that organic matter before it reaches the ocean, it seems likely that the net effect of changing global climate on oceanic carbon storage will be an increase in the amount of carbon oxidized in the coastal zone. The major environmental controls on oceanic carbon storage would seem to be related to altered land use and population density, rather than to elevated temperature, although elevated temperature could accelerate the rate at which newly delivered organic matter is oxidized. Much of the changing oceanic metabolism and most of the oceanic signal from elevated respiration can be expected to occur very near shore: in estuaries and near river mouths.

ACKNOWLEDGMENTS

I have relied heavily in this chapter on Smith and Mackenzie (1987), as recently modified and amplified by Smith and Hollibaugh (1993).

REFERENCES

Berner, R. A. 1982. Burial of organic carbon and pyrite sulfur in the modern ocean: Its geochemical significance. *Am. J. Sci.* 282:451–473.

Garrels, R. M., and F. T. Mackenzie. 1972. A quantitative model of the sedimentary rock cycle. *Mar. Chem.* 1:22–41.

Holligan, P. M., and W. A. Reiners. 1992. Predicting the responses of the coastal zone to global change. *Adv. Ecol. Res.* 22:211–255.

Judson, S., and D. F. Ritter. 1964. Rates of regional denudation in the United States. *J. Geophys. Res.* 69:3395–3401.

Meybeck, M. 1982. Carbon, nitrogen, and phosphorus transport by world rivers. *Am. J. Sci.* 282:401–450.

Milliman, J. D., and J. P. M. Syvitski. 1992. Geomorphic/tectonic control of sediment discharge to the ocean: The importance of small mountainous rivers. *J. Geol.* 100:525–544.

Sarmiento, J. L., and E. T. Sundquist. 1992. Revised budget for the oceanic uptake of anthropogenic carbon dioxide. *Nature* 356:589–593.

Smith, S. V., and J. T. Hollibaugh. 1993. Coastal metabolism and the oceanic organic carbon balance. *Rev. Geophys.* 31:75–89.

Smith, S. V., and F. T. Mackenzie. 1987. The ocean as a net heterotrophic system: Implications from the carbon biogeochemical cycle. *Global Biogeochem. Cycles* 1:187–198.

Wollast, R., and F. T. Mackenzie. 1989. Global biogeochemical cycles and climate. In *Climate and Geo-Sciences,* ed. A. Berger, S. Schneider, and J.-Cl. Duplessy, pp. 453–510. Kluwer, Dordrecht, The Netherlands.

14

The Vanishing Climatic Role of Dimethyl Sulfide

R. J. CHARLSON

Among the factors that influence the heat balance of the globe, the amount, location, and microphysical properties of aerosol particles and water clouds pose some of the largest uncertainties for forecasting future climate change. Of the key microphysical properties of aerosols, the scattering of shortwave (solar) radiation and the concentration of cloud condensation nuclei (CCN) are of particular interest to chemists because most scattering particles and CCN are produced in the atmosphere through chemical reactions of reactive gases. Of these gases, dimethyl sulfide (DMS) produced by marine phytoplankton is of singular importance because it appears to dominate the natural CCN production over the oceans, which cover about two-thirds of the area of the globe. The possibility that CCN effects on climate could influence DMS production by phytoplankton led Charlson, Lovelock, Andreae, and Warren to propose a hypothetical feedback loop (the CLAW hypothesis; Figure 14.1). Although many substantial questions remain regarding the measurement of individual steps in this loop, and even the sign of the putative feedback has not been defined (see "0/+/–" in Figure 14.1), many features of this system have been demonstrated. The purpose of this chapter is to review the topic, define the major gaps in our knowledge, and discuss the relative roles of biogenic DMS and anthropogenic SO_2 in global climatic change.

EVIDENCE FOR THE EXISTENCE OF A FEEDBACK

In the course of photosynthesis and the formation of structural protein, marine phytoplankton reduces inorganic aqueous SO_4^{2-} in successive steps to the two essential sulfur-bearing amino acids, cysteine and methionine. Figure 14.2 depicts the biochemical pathway by which these are formed as well as one normal degradation pathway for methionine via the zwitterion dimethyl sulfonium propionate (DMSP). Curiously, land plants exhibit a different set of processes without DMSP, a biochemical fact with profound significance for the global sulfur cycle. DMSP, which may serve as an osmolyte in oceanic biota, cleaves enzymatically to form DMS and acrylic acid. If land plants produced abundant amounts of DMS, the atmospheric sulfur cycle could be much more active than at present, depending ultimately on the availability of SO_4^{2-} in soils. The

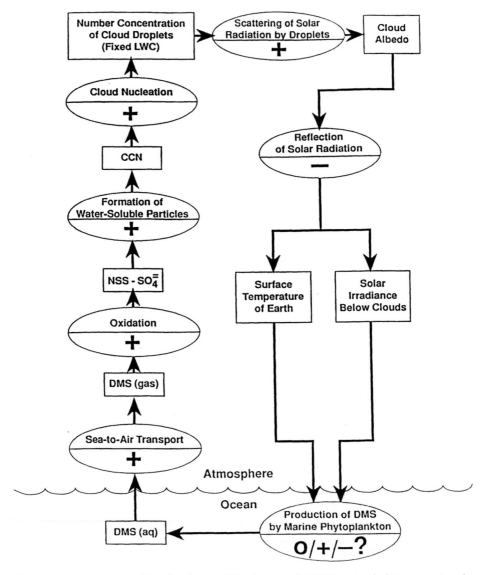

Figure 14.1. Hypothetical feedback loop (Charlson et al. 1987) amended to recognize the possibility of zero feedback. The sign in the oval indicates the sign of the response of the succeeding box to a change in the preceding one.

degrees to which this cleavage occurs in living phytoplankton—during senescence, during grazing by zooplankton, via their symbiotic bacteria, and/or by bacterial consumption of DMSP dissolved in seawater—are all questions of current concern. There is, however, no doubt left that DMS in seawater may be either emitted to the atmosphere or consumed chemically or biologically in the water column. Estimates of the fraction of DMS in seawater that enters the atmosphere range from 2 to 10% (Wolfe 1992). This flux into the atmosphere must be sensitive not only to the amount and type of algal

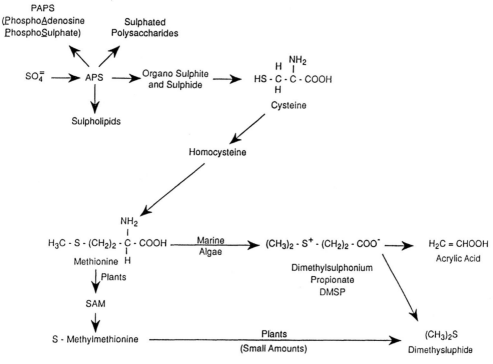

Figure 14.2. Main pathways for biochemical transformation of inorganic SO_4^{2-} into DMS. (Adapted from Andreae 1986.)

photosynthesis and metabolism but also to other microbiological as well as chemical processes in the photic zone.

The escape of DMS into the atmosphere can be demonstrated empirically by stripping seawater with a carrier gas in laboratory experiments and by finding it in the atmosphere above the sea surface. The flux of DMS to air has not been measured directly, and it is still calculated by a semiempirical model. Fluxes estimated in this way range from 1.4 to 6.6 μmol sulfur m^{-2} day^{-1} for different parts of the world ocean and for different seasons. The estimates result in a global marine DMS-sulfur flux into the atmosphere of 0.5±0.3 Tmol year^{-1} (16±10 Tg year^{-1}) (Bates et al. 1991). The flux leads to concentrations of DMS in marine air of circa 4–30 pmol liter^{-1} (Calhoun 1991).

It also has been demonstrated that variations in biological primary production and in phytoplankton speciation result in changes in the seawater DMS concentration. Seasonal studies (Leck 1989; Gras 1990; Berresheim et al. 1991) in mid- to high-latitude regions clearly show a seasonality due to a spring/summer maximum and a winter minimum of photosynthetic activity. Increased insolation, increased temperature, and/or increased nutrient levels produce increased DMS. However, this seasonality also clearly shows that the DMS production rate is not just proportional to primary productivity but is also dependent on more subtle details of the microbial ecology, especially the dominant species of phytoplankton (Andreae 1986; Leck 1989). Studies of the geographical

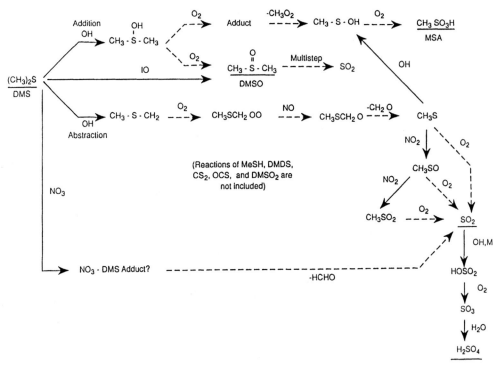

Figure 14.3. Oxidation pathways for DMS. Solid lines are pathways for which rate constants have been measured. The broken lines are suggested by product analysis. (After Plane 1989 and Tyndall and Ravishankara 1991.)

dependence of DMS production also show variability, reflecting, for instance, differences in zones of coastal upwelling in contrast to the open ocean (Bates et al. 1991).

The lifetime of DMS in the atmosphere is governed by its reactions with OH (and perhaps NO_3) radicals and is generally given as ~1 day, a value that must vary with oxidant concentrations. The reaction products include SO_2, H_2SO_4, and CH_3SO_3H (methanesulfonic acid, MSA) (Figure 14.3), as shown by laboratory studies and atmospheric measurements.

As a result of the oxidation of DMS, two condensible species (H_2SO_4 and MSA) form and produce submicrometer aerosol mass (Kreidenweiss et al. 1989). This mass either forms new particles by heteromolecular nucleation with H_2O vapor (possibly NH_3 as well) or it condenses on the surfaces of preexisting aerosol particles, causing them to grow. The ubiquitous presence in remote marine air masses of submicrometer SO_4^{2-} and MSA in aerosol particles (Ayers and Gras 1991; Berresheim 1991), along with the range of concentration, is given as evidence that the oxidation of DMS produces aerosol particles. The range of SO_4^{2-} concentration (1–7 nmol m^{-3}) is about what is expected from the source strength from DMS oxidation (a few nanomoles per cubic meter per day) and aerosol particle lifetimes on the order of days in marine air. This mass concentration is too small to be of importance to aerosol optical depth—at least under current conditions. Hence, the DMS-derived SO_4^{2-} aerosol probably has a currently inconsequential *direct* effect on solar radiation.

The measured size range encompassing the main number concentration (0.05–0.5 μm diameter) and the molecular composition of the sulfur-containing aerosol particles allow them to act as CCN at the low supersaturations (S < 1%) found in the atmosphere, particularly in climatically dominant, marine stratus clouds. The necessary chemical feature of water solubility along with ionic dissociation yields a large amount of vapor pressure depression for a given amount of solute, making H_2SO_4, MSA, and their reaction products with NH_3 into nearly ideal CCN. The largest chemical unknown that remains regarding the CNN is the possible role of organic species, also acting as CCN, coating the droplets or acting as surfactants, and thereby influencing the cloud nucleating characteristics.

There have been few studies of aerosol microphysical and chemical composition that were connected to studies of the clouds that actually formed. Most observations have instead been carried out with thermal diffusion cloud chambers. The DMS-CCN hypothesis has not yet been fully tested, e.g., to show that changes in DMS concentration cause a corresponding change of measurable magnitude in clouds. Such experiments are difficult because large amounts of data are required to produce stable average values of the key parameters. Boers et al. (1994) have shown coherence between seasonal variations in satellite-derived cloud optical depth and CNN and DMS levels at Cape Grim, Tasmania. Such satellite data appear to offer the only way to detect subtle changes in CCN levels.

The strongest proof to date that connects DMS to CCN necessarily involves the use of cloud chambers as surrogates for real clouds. Ayers and Gras (1991) have shown from observations over 20 months at Cape Grim that MSA and DMS concentrations are highly and linearly correlated, an observation that appears to justify the use of MSA as a surrogate for DMS. The correlation has enabled reanalysis of an eight-year record of MSA mass concentration with CCN number concentrations at two supersaturations. Figure 14.4 illustrates the possible nonlinear relationship of monthly average CCN level (active at 0.23% and 1.2% supersaturation) as a function of MSA (and hence DMS) level. This result, along with the biological and chemical studies mentioned earlier, provides a complete, semiquantitative link of production of DMS by phytoplankton to the CCN number concentration, as was proposed by Charlson et al. (1987). The coincident summertime maxima of all the chemical variables and CCN suggest that the sign in Figure 14.1 should be positive. That is, the biota produce more DMS and CCN under the influence of enhanced solar radiation in summer.

The most tenuous physical aspect of the loop in Figure 14.1 (not including the question of biological feedback) remains the identification and measurement of the effect of CCN on cloud albedo. Twomey (1971, 1977) proposed the theory that cloud albedo is a strong function of the CCN concentration, as shown in Figure 14.5 (Charlson et al. 1991). Coakley et al. (1987) provided substantial evidence as well as a useful approach by studying systematically the change in reflectivity of marine stratus clouds at visible wavelengths as they were modified by CCN from ships passing below the clouds. Figure 14.6 shows cloud reflectivity at 0.63 μm wavelength versus 3.7 μm cloud radiance. The data compare parts of a stratus cloud layer that was influenced by ship-produced aerosol particles to unaffected parts. Coakley et al. (1987) suggested on the basis of theory that the enhancement of radiance at 3.7 μm in the ship tracks is an indication of smaller drop sizes; the simultaneous lack of any corresponding change in

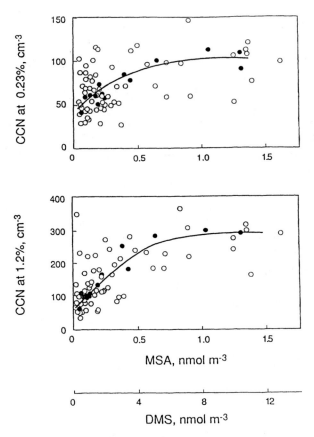

Figure 14.4. Empirical dependence of CCN number concentration at 0.23 and 1.2% supersaturation on MSA as a surrogate for DMS. Open data points are individual observations; filled points are monthly means. (Adapted from Ayers and Gras 1991.)

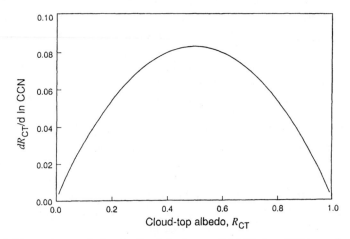

Figure 14.5. Dependence of cloud reflectivity for sunlight (R_{CT}) on CCN number concentration (CNN). The derivative $dR_{CT}/d\ln$ CCN is given as a function of R_{CT}, showing a maximum at $R_{CT} \approx 0.5$. At the R_{CT}, a 12% increase in CCN yields a change of circa 1% in cloud albedo.

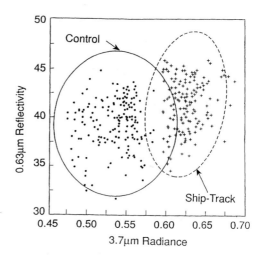

Figure 14.6. Observed change in stratus cloud reflectivity at 0.63-μm wavelength versus 3.7-μm radiance due to ships passing below the clouds. (Adapted from Coakley et al. 1987.)

emission at 11 μm indicated that there was no change in liquid water content. Thus, the coincident increase of 0.63-μm reflectivity ($1.6 \pm 2.7\%$), though small, was evidence for the dependence of cloud albedo on the CCN. Subsequent in situ observations by Radke et al. (1989) confirmed this finding. Again, the results of Boers et al. (1994) were consistent with the hypothesis; that is, they failed to disprove it.

However suggestive and encouraging, these analyses do not prove that the correlation between CCN concentrations and variation in MSA (or DMS) levels in Figure 14.4 is sufficiently high to be significant to cloud albedo. Because of the very large natural variability of all of the key parameters, large amounts of data are required if the connections are to be statistically meaningful. One way around gathering large data sets for statistical analysis might be to perform intensive experiments on a volume of air that can be tracked in a Lagrangian sense, with simultaneous measurements of the necessary set of chemical variables, physical aerosol properties, and cloud micro- and macrophysical (optical) properties. Another approach would be to add deliberately a reactive, CCN-producing gas (SO_2, DMS, or even SO_3) in a known amount and in a temporally and spatially known, modulated fashion that would allow the temporal-spatial variability in the atmosphere to be used to extract quantitative information about the deliberate perturbation from the natural, large variability.

Thus, studies of the individual steps in Figure 14.1 have confirmed qualitatively (and quantitatively in some instances) that DMS from oceanic biota does act as the chemical source for most CCN in the unpolluted marine atmosphere, that seasonal variations of DMS level coincide with seasonal variations of CCN concentration, and that increases in DMS concentration should therefore cause increases in cloud albedo and vice versa. However, two key features of Figure 14.1 remain to be explored. The first of these, which Figure 14.4 seems to indicate as feasible, is to establish the response of CCN concentration at low supersaturation to regional scale changes in DMS source strength. Better still would be to describe empirically the relationship between CCN concentration and some appropriate measures of the photosynthetic activity of key species of phytoplankton in seawater. The study would include factors that influence the oceanic biota.

The remaining problem is to develop an understanding of the response of phyto-plankton and DMS production to the specific climatic factors that are influenced by CCN. These factors must include the solar irradiance below clouds and/or the temperature of the water in the photic zone. Other factors may also exist, such as the influence of cloud properties on the wind profile in the atmospheric boundary layer and hence on the mixing rates of sea-surface nutrients. The essential question regarding Figure 14.1 is to establish the sign of the response of the biota. If increased temperature or solar irradiance causes enhanced DMS production (i.e., a plus in the "Production of DMS" oval in Figure 14.1), the sense of the feedback would be to produce more CCN and thereby counteract the initial increase. A minus in Fig-ure 14.1 would amplify the initial change. Of course, it is possible that DMS production is not sensitive at all to the CCN-induced changes in temperature or solar irradiance, or that it is sometimes a plus and sometimes a minus (hence the "0/+/–?" notation in Figure 14.1). Regardless of the sign, a way to evaluate and understand the gain of the system also is needed. It will be useful to know such partial derivatives as: ∂DMS-sea/∂CCN, ∂DMS-sea/∂T, and ∂DMS-sea/∂Irradiance. This combined chemical-physical-biological problem obviously poses serious challenges for a diverse community of scientists. Until results are available from observations of the response of biological production of DMS to climatic factors, we may be able to obtain clues regarding the nature of the feedback (if any) from the contemporary data mentioned earlier and from the paleoclimatic record in glacial ice.

PALEOCLIMATIC RECORD OF ATMOSPHERIC SULFUR SPECIES

Two ionic sulfur species have been analyzed in deep ice cores taken from Vostok in East Antarctica (Figure 14.7) (Legrand et al. 1991). Non–sea-salt sulfate (the excess of SO_4^{2-} after subtracting sea-salt SO_4^{2-} inferred from Na^+) and MSA both show *elevated* levels during the low-temperature periods of the most recent (Wisconsin) ice age. This increase may be due to enhanced source strength of SO_4^{2-} and MSA (i.e., DMS), but it could also be due in part to changes in the efficiency of the scavenging process and/or in the annual accumulation of snow.

Non–sea-salt sulfate also has been studied over recent centuries in snow and ice cores from both Greenland and Alaska. Acidic ingredients contributing H^+ (as inferred from electrical conductivity), presumably often dominated by H_2SO_4, have been followed over several time scales in numerous ice cores. Samples both from the most recent century and over the current millennium show natural variability on annual, decadal, and century time scales (Hammer et al. 1980). Volcanic episodes are evident from spikes in SO_4^{2-} or H^+ concentration that lasted a few years. Another component of natural variation occurs annually with $\Delta[SO_4^{2-}]/[SO_4^{2-}] \approx 0.3$, where $\Delta[X]$ denotes variability around the mean of concentration $[X]$. The annual maxima occur in summer, with more sunlight and warmer water, presumably because of the previously mentioned summer maximum of DMS emission. But again some $\Delta[SO_4^{2-}]$ could be due to *either* variations in the source strength of reduced sulfur gases (e.g., DMS) or variations in the efficiency of the atmospheric scavenging process. However, the record of the past two millennia when compared to the temperature inferred from oxygen isotope analysis is that higher

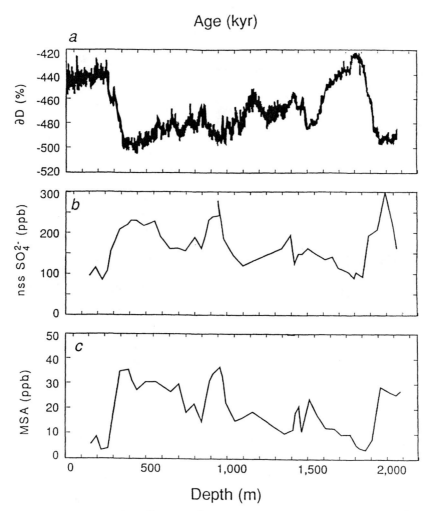

Figure 14.7. Non–sea-salt SO_4^{2-} (nss SO_4^{2-}) and MSA versus depth in the Vostok Core, backward in time through the entire previous ice age. Variation in temperature is inferred from the variation in the ratio of deuterium to hydrogen represented by the geochemical index ∂D. Both nss SO_4^{2-} and MSA were elevated in concentration during the last ice age (13–130,000 years B.P.).

conductivity (higher SO_4^{2-}?) occurred preferentially during periods of *lower* temperature, a finding that is just the opposite of the annual variation but in agreement with the deep ice cores.

Samples over the most recent century from Greenland also show a secular increase of $[SO_4^{2-}]$ from circa 0.2 to 0.6 μM, which is generally believed to be due to anthropogenic enhancement of the flux of SO_4^{2-} (Mayewski et al. 1990). Interestingly, high-altitude ($z = 3000$ m) snow samples from Alaska do not show this secular increase, a finding attesting to the geographical inhomogeneity of the concentration and flux of anthropogenic SO_4^{2-} in the atmosphere.

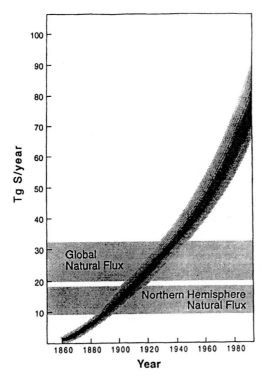

Figure 14.8. Growth of the flux of anthropogenic SO_2 sulfur into the atmosphere since A.D. 1850. (After Charlson 1991, Bates et al. 1991, and Penner et al. 1994.) The range of estimates of the global and Northern Hemisphere natural fluxes is given as the stippled range of values. Because anthropogenic sulfur is not efficiently mixed into the Southern Hemisphere, and because 90% of the anthropogenic emission is in the Northern Hemisphere, the relevant comparison is between the anthropogenic and Northern Hemisphere natural fluxes.

Thus, the longer ice core records suggest (but do not prove) the possibility that on ice age time scales the DMS feedback is an amplifier of climate change rather than a thermostat. That is, the sign in Figure 14.1 may be a minus on the century to millennium time scale. On the other hand, the annual cycle at high latitudes with a summertime maximum of DMS, SO_4^{2-}, MSA, and CCN suggests a positive sign.

CONCLUSION

Whether or not the feedback exists or the sign of it is positive or negative, we may reach several conclusions. First, most of the natural CCN of the globe currently appears to be derived from marine phytoplankton. If the world's oceanic biota changed dramatically, we would expect the properties of unpolluted marine clouds also to change. Second, because global heat balance is calculated to be very sensitive to CCN, even small changes in the amount or type of phytoplankton might have large effects on climate. Third, because SO_4^{2-} in seawater never limits the production of DMS, control of the DMS-CCN loop must be due to some still unknown combination of nutrients, temperature, solar radiation, and marine microbial ecology. It seems likely that the natural marine DMS production could have been either much smaller or much greater (or both) than it is at present, with attendant climatic influences. Fourth, and last, because the atmospheric sulfur cycle is now dominated by anthropogenic SO_4^{2-}, control of the direct and indirect (cloud) effects of aerosol on the heat balance of the Northern Hemisphere has been taken away from the natural system. Therefore, if the world's climate depended in some way on this biological feedback, that feedback has been superseded. Figure 14.8

(based on Bates et al. 1991, Charlson 1991, and Penner et al. 1994) shows the historical development of anthropogenic SO_2-S, illustrating this shift away from the natural system.

This overwhelming mass flux of anthropogenic SO_2-S may constitute an unplanned, inadvertent experiment, as outlined by Schwartz (1988). Inasmuch as the temperature of the globe has warmed slightly since circa 1850 (presumably due to the greenhouse effect), this aerosol forcing, *direct* plus *indirect*, has not totally reversed the course of the warming. The *direct* effect of the anthropogenic SO_4^{2-} is calculated to have approximately balanced the greenhouse gas forcing over a large part of the Northern Hemisphere (Charlson et al. 1990). Thus, this anthropogenic flux of sulfur may have had a relatively small *indirect* effect, and (as Schwartz suggested) by extension DMS variations may have a similar effect. If the indirect forcing were a linear function of the SO_2 source strength, the severalfold anthropogenic increase of aerosol sulfur would have resulted in a negative forcing of 10 or more W m^{-2}, which clearly has not occurred. Does this mean that there should be a zero somewhere in Figure 14.1? If the CCN number concentration were a linear function of SO_4^{2-} mass concentration, that might be so. However, there are many nonlinear processes that determine the droplet population and hence the albedo of clouds. Indeed, the CCN number concentration might even have two stable states, corresponding to the chemical reaction source being balanced by two different removal mechanisms (Baker and Charlson 1990). This leaves us with a number of still-open questions and a continuing challenge to understand this coupled biological-chemical-meteorological system.

REFERENCES

Andreae, M. O. 1986. The ocean as a source of atmospheric sulfur compounds. In *The Role of Air-Sea Exchange in Geochemical Cycling,* ed. P. Buat-Menard, pp. 331–362. D. Reidel, Dordrecht, The Netherlands.

Ayers, G. P., and J. L. Gras. 1991. Mean seasonal relationship between CCN and aerosol MSA in marine air at Cape Grim. *Nature* 353:834–835.

Baker, M. B., and R. J. Charlson. 1990. Bistability of CCN concentrations and thermodynamics in the cloud-topped boundary layer. *Nature* 345:142–145.

Bates, T. S., B. K. Lamb, A. Guenther, J. Dignon, and R. E. Stoiber. 1992. Sulfur emissions to the atmosphere from natural sources. *J. Atmos. Chem.* 14:315–337.

Berresheim, H., M. O. Andreae, R. L. Iverson, and S. M. Li. 1991. Seasonal variations of dimethylsulfide emissions and atmospheric sulfur and nitrogen species over the western Atlantic Ocean. *Tellus* 43B:353–372.

Boers, R., G. P. Agers, and J. L. Gros. 1994. Coherence between seasonal variation in satellite-derived cloud optical depth and boundary layer CCN concentrations at a mid-latitude Southern Hemisphere site. *Tellus.* In press.

Charlson, R. J. 1991. Anthropogenic sulfur from phytoplankton versus sulfur from industry: Which dominates cloud condensation nuclei? In *Scientists on Gaia,* ed. S. Schneider. MIT Press, Cambridge, Mass.

Charlson, R. J., J. E. Lovelock, M. O. Andreae, and S. G. Warren. 1987. Oceanic phytoplankton, atmospheric sulfur, cloud albedo and climate. *Nature* 326:655–661.

Charlson, R. J., J. Langner, and H. Rodhe. 1990. Sulphate aerosol and climate. *Nature* 348:22.

Coakley, J. A., Jr., R. L. Berstein, and P. A. Durkee. 1987. Effect of ship-stack effluents on cloud reflectivity. *Science* 237:1020–1022.

Gras, J. L. 1990. Cloud condensation nuclei over the Southern Ocean. *Geophys. Res. Lett.* 17: 1565–1567.

Hammer, C. U., H. B. Clausen, and W. Dansgaard. 1980. Greenland ice sheet evidence of post-glacial volcanism and its climatic impact. *Nature* 288:230–235.

Kreidenweiss, S. M., R. C. Flagan, J. H. Seinfeld, and K. Okuyama. 1989. Binary nucleation of methanesulfonic acid and water. *J. Aerosol Sci.* 20:585–607.

Legrand, M., C. Fenist-Saigne, E. S. Saltzman, C. Germain, N. I. Barkov, and V. N. Petrov. 1991. Ice core record of oceanic emissions of dimethylsulfide during the last climate cycle. *Nature* 350:144–146.

Mayewski, P. A., W. B. Lyons, M. J. Spencer, M. S. Twickler, C. F. Buck, and S. Whitlow. 1990. An ice core record of atmospheric response to anthropogenic sulfate and nitrate. *Nature* 346:554–556.

Penner, J. E., R. J. Charlson, J. M. Haleg, N. S. Laulainen, R. Leifer, T. Novakov, J. Ogren, L. F. Radke, S. E. Schwartz, and L. Travis. 1994. Qualifying and minimizing uncertainties of climate forcing by anthropogenic aerosol. *Bull. Am. Meteorol. Soc.* 75:375–400.

Plane, J. M. 1989. Gas phase atmospheric oxidation of biogenic sulfur compounds. In *Biogenic Sulfur in the Environment*. eds. E. S. Saltzman and W. M. Cooper, American Chemical Society, Washington, D.C.

Radke, L. F., J. A. Coakley Jr., and M. D. King. 1989. Direct and remote sensing observations on the effects of ships on clouds. *Science* 246:1146–1149.

Schwartz, S. E. 1988. Are global cloud albedo and climate controlled by marine phytoplankton? *Nature* 336:441–445.

Twomey, S. 1971. Figure 8.9, *Inadvertent Climate Modification*. C. L. Wilson and W. H. Mottben (directors). MIT Press, Cambridge, Mass.

Twomey, S. 1977. Figure 12.6, *Atmospheric Aerosols*. Elsevier, Amsterdam.

Tyndall, G. S., and A. R. Ravishankara. 1991. Atmospheric oxidation of reduced sulfur species. *Int. J. Chem. Kinet.* 23:483–527.

Wolfe, G. V. 1992. The cycling of climatically active dimethyl sulfide (DMS) in the marine euphotic zone. Ph.D. dissertation, University of Washington, Seattle.

15

Implications of Increased Solar UV-B For Aquatic Ecosystems

RAYMOND C. SMITH

[Editors' Note: The increase in ultraviolet radiation at the surface of the earth is one of a host of toxic effects of intensified human activities, not a feedback system intrinsic to the processes now thought to be affecting temperature and climates globally. It is, however, a global effect that is now recognized as having potential for reducing the rate of absorption of carbon as CO_2 into the oceans. This discussion has been included because of the importance of the mechanism and the fact that, if it is as important as indicated here, it simply emphasizes the general trend toward an increased rate of accumulation of CO_2 in the atmosphere during the period when the feedback mechanisms discussed elsewhere will be accentuating the warming.

Smith and his colleagues have been leaders in sorting through the difficult studies of the biotic effects of ultraviolet radiation, long a topic of great uncertainty and confusion. In this chapter he outlines the various studies that confirm the general observation that the effect of enhanced UV-A and -B at the surface is a contribution to the impoverishment of marine systems with effects on primary productivity and the composition of the communities as well.]

There is considerable and increasing evidence that higher levels of UV-B radiation (280–320 nm) at the surface of the earth, the result of a reduction of the ozone layer, may be detrimental to various forms of marine life in the upper layers of the sea. It is now widely documented that reduced ozone concentration will result in increased levels of UV-B incident at the surface of the earth (NAS 1979, 1982, 1984; Watson 1988; Anderson et al. 1991; Frederick and Alberts 1991; Schoeberl and Hartmann 1991). With respect to aquatic ecosystems, we also know that this biologically damaging mid-ultraviolet radiation can penetrate to ecologically significant depths in the ocean (Jerlov 1950; Lenoble 1956; R. C. Smith and Baker 1979, 1980, 1981). The United Nations Environmental Programme (UNEP) has provided recent summaries of the effects of ozone depletion on aquatic ecosystems (Häder, Worrest, and Kumar in UNEP 1989, 1991), and the Scientific Committee on Problems of the Environment (SCOPE) has provided (SCOPE 1992) a summary of the effects of increased ultraviolet radiation on biological systems. SCOPE has also produced (SCOPE 1992) a report on the effects of increased ultraviolet radiation on the biosphere. Earlier reviews include National Academy of Science and National Research Council (1984), Caldwell et al. (1986), Worrest (1986), National Oceanic and Atmospheric Administration (1987), R. C. Smith (1989), R. C. Smith and Baker (1989), and Voytek (1990). As Häder et al. have summarized

(UNEP 1989, 1991), "UVB radiation in aquatic systems: 1) affects adaptive strategies (e.g., motility, orientation); 2) impairs important physiological functions (e.g., photosynthesis and enzymatic reactions); and 3) threatens marine organisms during their developmental stages (e.g., the young of finfish, shrimp larvae, crab larvae)." Possible consequences to aquatic systems include reduced biomass production, changes in species composition and bio-diversity, and altered biogeochemical cycles and aquatic ecosystems associated with the foregoing changes. This chapter outlines research subsequent to these recent reviews and emphasizes studies concerned with phytoplankton.

ENHANCED UVB AND AQUATIC ORGANISMS

Atmospheric and Biooptical Models

Relatively high-resolution full-spectral models are required to examine quantitatively the biological effects of UV-B radiation on aquatic ecosystems. R. C. Smith (1989) has shown that these models allow the quantitative calculation of spectral irradiance at the surface, $E_d(0^+,\lambda)$, and as a function of depth, $E_d(z,\lambda)$, within the water column. Given the spectral irradiance and an appropriate biological weighting function, $\varepsilon(\lambda)$, the biologically effective fluence rate,

$$E_B(Z, \theta) \ [\text{W m}^{-2}]_{\varepsilon(\lambda)} = \int E(Z, \theta, \lambda) \ [\text{W m}^{-2} \text{ nm}^{-1}] \cdot \varepsilon(\lambda) \cdot d\lambda [\text{nm}] \qquad (15.1)$$

as a function of depth can be computed. The units notation $[\text{W m}^{-2}]_{\varepsilon(\lambda)}$ indicates a physically measured (or measurable) absolute irradiance weighted by a biological effi-ciency. In other words, the biologically active ultraviolet radiation incident at the surface of the ocean and as a function of depth for various times of day and season and under conditions of various ozone layer thickness can be estimated.

Discovery of the Antarctic ozone hole has motivated the development of a number of improved atmospheric models suitable for low-sun-elevation (long-pathlength) condi-tions at high latitudes. This work includes studies by Lubin et al. (Lubin and Frederick 1990, 1991; Lubin et al. 1992), Stamnes et al. (Stamnes et al. 1988, 1990; Tsay and Stamnes, 1992), and R. C. Smith et al. (R. C. Smith et al. 1992a). These models permit accurate space-time extrapolation of satellite or ground measurements, quantitative estimates of the effects of ozone depletion on spectral solar ultraviolet and visible irradiance at the surface of the earth, and relatively accurate estimation of column ozone concentrations above high-resolution spectral ground stations. They are also suitable for polar regions where high solar zenith angles dominate.

Biological Weighting Function

The last few years have seen significant progress in determining a biological weighting function for the inhibition of photosynthesis by phytoplankton. The biological weighting function takes account of the wavelength dependency of biological action. It is a critical parameter in the assessment of the potential biological effects of ozone-related enhanced ultraviolet radiation (National Academy of Science and National Research Council 1979, 1982, 1984). A number of authors (Caldwell 1968; Rundel 1983; Caldwell et al.

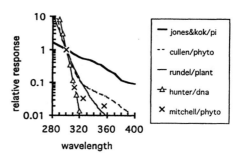

Figure 15.1. Relative response of action spectra normalized at 300 nm versus wavelength.

1986) have shown that an accurate knowledge of ε(λ) is essential to estimate quantitatively biological fluence rate. Biological weighting functions have tradition- ally been determined by evaluating biological responsiveness to monochromatic radiation with the objective of identifying potential chromophore targets and photo- biological mechanisms. Caldwell et al. (1986) reviewed evidence suggesting that weighting functions determined using polychromatic radiation and intact organisms may have more ecological relevance with respect to assessing the ozone reduction problem. Work during the mid-1980s (Rundel 1983; Caldwell et al. 1986) provided biological weighting functions for plant damage by ultraviolet radiation that showed relatively strong dependence on UV-B but also showed a significant contribution from the UV-A (320–400 nm) region.

Cullen et al. (1992) used a novel application of principal components analysis to estimate a biological weighting function that provides both polychromatic radiation to the cells and a relatively high-resolution, smoothly-varying spectral response. Their work provides an e(l) for the inhibition of phytoplankton photosynthesis by ultraviolet light. Other recent estimates of ε(λ) of natural populations, using relatively coarse broadband spectroscopy, include the work of Mitchell (1990) and Helbling et al. (1992). Figure 15.1 compares these published action spectra, all normalized to 1.0 at 300 nm. There is now broad agreement among these several workers that the biological weight- ing, although highest in the UV-B, also contains a significant UV-A component. The accurate work of Cullen et al. (1992) provides a significant advance in our under- standing of phytoplankton action spectra and, consequently, of the influence of ozone depletion on phytoplankton. It permits more accurate comparison of biological effects produced by artificial sources with different spectral distributions and those caused by actual and/or predicted natural radiation. It also provides for the development of quanti- tative models for the prediction of future effects of ozone depletion. As will be illustrated later in this chapter, there is a need for accuracy because relatively small changes in ε(λ) can lead to large changes in biologically effective fluence.

Radiation Amplification Factor

The concept of an amplification factor, A, such that a 1% decrease in ozone may cause an A% increase in biological effect, is useful when considering the possible effect of ozone reduction on a biological system. This amplification factor has been subdivided into two components (Nachtwey and Caldwell 1975; Green et al. 1976; Rundel and Nachtwey 1978; Rundel 1983): (1) the ratio of the percentage change in biological

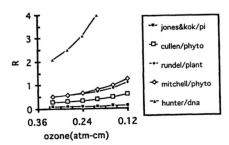

Figure 15.2. Radiation amplification factors (using percentages, equation 15.2) versus ozone level.

effective dose, $\Delta E_{Be}/E_{Be}$, to the percentage change in ozone thickness, $\Delta\omega/\omega$, i.e., the radiation amplification factor,

$$R = (\Delta E_{Be}/E_{Be}) \, / \, (\Delta\omega/\omega) \qquad (15.2)$$

and (2) the ratio of the percentage change in biological effect, $\Delta P/P$, to the percentage change in biologically effective dose, $\Delta E_{Be}/E_{Be}$, i.e., the biological amplification factor,

$$B = (\Delta P/P)/(\Delta E_{Be}/E_{Be}) \qquad (15.3)$$

so that the total amplification factor is

$$A = R \times B = (\Delta P/P)/(\Delta\omega/\omega) \qquad (15.4)$$

Since the dose-versus-ozone relationship is nonlinear, the radiation amplification factor, if calculated using equation 15.2, is not constant with ozone concentration over large changes in ozone. It has been discussed elsewhere (Madronich and Granier 1992) that a simple power law provides a more general definition of the dose-versus-ozone relationship, which results in a factor R that is relatively constant with variable ozone concentration:

$$\frac{(E_{B_\varepsilon})_2}{(E_{B_\varepsilon})_1} = \left(\frac{\omega_2}{\omega_1}\right)^{-R} \qquad (15.5)$$

so,

$$R = -\frac{\ln\left[(E_{B_\varepsilon})_2 \, / \, (E_{B_\varepsilon})_1\right]}{\ln\left(\omega_2 \, / \, \omega_1\right)} \qquad (15.6)$$

As has been shown for biological dose, the sensitivity of the radiation amplification factor to a change in ozone concentration is primarily determined by the biological weighting function and in dealing with inferences based upon a radiation factor, it is important to be aware of how the factor is derived.

A reduction in the thickness of the ozone layer leads to an increase in UV-B radiation. This will have a significant effect on biological weighting functions that are heavily weighted in the UV-B region (e.g., DNA) with $R > 1$. Conversely, biological weighting functions weighted outside the UV-B region will have smaller direct effects ($R < 1$). Figure 15.2 is a plot of the radiation amplification factor, R, for the biological weighting functions plotted in Figure 15.1 calculated using equation 15.2. This figure illustrates the nonlinearity in such a computation, whereas Figure 15.3 shows the results of using equation 15.6. A review of published radiation amplification factors is given in the United Nations Environmental Program 1991 update (UNEP 1991).

Figure 15.3. Radiation amplification factors (using power law, equation 15.6) versus ozone level. These data were computed for 64° S and represent R values appropriate for the Palmer Station area in the Antarctic during October. Low-latitude (higher-zenith-angle) R values are somewhat higher. The comparison dramatically illustrates how relatively small differences in the wavelength characteristics of biological weighting functions are translated into a measure of the relative influence of ozone-related fluence rates. The data, using most recent estimates of $\varepsilon(\lambda)$, show radiation amplification factors of approximately 0.5, which are significantly less than the R values for a DNA weighting. It should be noted that although we now have much better information on biological weighting functions for UV-B effects on phytoplankton, we continue to have very little information with respect to biological amplification factors, B. This is clearly an important area for future research.

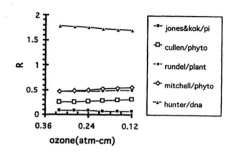

Mixing

The fluence rate received by phytoplankton, or other marine organisms within the water column, is dependent on complex physical mixing processes that determine the time spent by the phytoplankton at each depth. R. C. Smith (1989) recently reviewed this subject and pointed out that an important consideration, with respect to the dose response of phytoplankton, is the ratio of the biological dose to the photosynthetically available radiation (PAR), $E_{B\theta}/E_{PAR}$, versus biological damage. Many mechanisms, such as mixing, provide UV-B protection by reducing radiant energy but also reduce the energy available for photosynthesis. For example, if the time scale for a biological response (e.g., ultraviolet damage, photoadaptation) is shorter than that for vertical mixing, phytoplankton will exhibit a vertical gradient of this response (Lewis et al. 1984; Cullen and Lewis 1988). If mixing occurs with a time scale shorter than that of the biological response, no such gradient will be observed. The rates of various photoprocesses, as compared with that of vertical mixing, will determine the overall effect on the populations at depth.

Recent work suggests that linking the rates of the various photoprocesses to a physical mixing model may be more complex than previously thought. Work by Prézelin et al. 1986, 1992, 1993) shows strong diurnal patterns in the photosynthetic capacity and depth-dependent photosynthesis-irradiance relationships and also strong diurnal effects in ultraviolet inhibition of photosynthesis. The problem is also confounded by wavelength changes in irradiance with depth. UV-B is damaging to organisms. UV-A is known to induce both photodamage and photoreactivation processes in living cells (Hirosawa and Miyachi 1983; Caldwell 1968; Caldwell et al. 1986; Buhlmann et al. 1987; Karentz 1991; R. C. Smith et al. 1992b). PAR (400–700 nm) is essential for growth and photosynthesis. Thus, several photoprocesses are simultaneously active within the cell in these different wavelength regions, operating with different time constants and responding differently to changing irradiance levels as they are mixed within the water column. Current models permit sensitivity analysis of these various physical and biological processes, but they are inadequate for predictive modeling without further research and improvement.

Dose-Response Curves

A quantitative assessment of possible ultraviolet-related damage to an organism requires laboratory data assessing ultraviolet effects versus biologically weighted fluence rate (a relationship called a survival or dose-response curve) for comparison with current or predicted natural fluence rates. Cullen and Lesser (1991) studied the inhibition of photosynthesis by ultraviolet radiation as a function of dose (biologically weighted $J m^{-2}$) and irradiance ($W m^{-2}$). They showed that (1) photosynthesis is inhibited by UV-B radiation; (2) at a fixed time, photoinhibition is a monotonic, nonlinear function of dose (they also showed that for equal doses a relatively short exposure to high UV-B irradiance is more damaging to photosynthesis than a longer exposure to lower irradiance, hence reciprocity failure); and (3) for both nutrient-replete and nitrate-limited cultures, photoinhibition is a monotonic, nonlinear function of irradiance for time scales of 0.5 to 4 h. They echoed previous cautions with respect to the necessity to understand reciprocity relations and, for water column studies, the need to match the time scales of measurements to the time scales of vertical mixing.

Karentz et al. (1991a) studied twelve species of Antarctic diatoms for cell survival characteristics and molecular responses to UV-B radiation and determined the average fluence for cell death. Their studies, which did not simulate natural sunlight conditions and were not intended to do so, showed that (1) dose responses of population survival to ultraviolet exposure varied considerably among species and there were significant differences as a function of wavelengths available or absent for photorepair and (2) a general relationship was evident between the surface area/volume ratios of cells and the amount of damage induced by ultraviolet exposure. Smaller cells, with larger ratios, sustained greater amounts of damage per unit of DNA. They pointed out that it is difficult, in studying the cellular and molecular aspects of DNA repair in plants, to separate the synergistic effect of PAR in the metabolic processes of photoenzymatic repair and photosynthesis. They concluded from their results that the most likely consequence of ozone-related increases in UV-B will be the initiation of changes in cell size and taxonomic structure in Antarctic phytoplankton communities.

METHODOLOGICAL ISSUES

Several methodological problems are associated with estimating the implications of ozone-related UV-B increases for aquatic systems. Issues associated with the estimation of biological weighting functions, dose-response curves, and reciprocity have already been mentioned. A continuing critical factor in laboratory experiments is that radiation regimes cannot easily simulate ambient levels of solar radiation throughout the total spectrum, especially when vertical mixing through an in-water irradiance field is considered. As a consequence, although experiments may enhance the UV-B spectral region, the visible portion of the spectrum may be as much as an order of magnitude lower than that in nature, thus limiting the energy necessary for optimum photoreactivation and photorepair (Kaupp and Hunter 1981; Damkaer and Dey 1983; Worrest 1986). Furthermore, simulation of in-water irradiance is difficult to match.

Holm-Hansen and Hebling (1993) raised the issue of a possible UV-B-induced toxicity in polyethylene bags, claiming that these bags significantly lower the rate of

CO_2 assimilation in productivity experiments. This assertion runs contrary to previous tests and experience (Prézelin et al. 1992; R. C. Smith et al. 1992b; D. Karentz personal communication). In recent additional tests (Prézelin and Smith 1993; Prézelin et al. 1993; Williamson and Zagarese 1994), no toxic effects were evident from the use of polyethylene bags.

PHYSIOLOGICAL ADAPTATION AND SENSITIVITY OF PHYTOPLANKTON TO ULTRAVIOLET EXPOSURE

The work of numerous investigators, beginning with Steemann-Nielsen (1964) and continuing until today, provides evidence that exposure to UV-B reduces algal productivity. Much of this evidence is based upon comparison of rates of ^{14}C uptake in incubation bottles that transmit, or do not transmit, ultraviolet radiation, a topic that continues to be explored (El-Sayed et al. 1990; Holm-Hansen 1990; Mitchell 1990; Vernet 1990). Furthermore, there is convincing evidence that ultraviolet radiation, at levels currently incident at the surface of the ocean, influences phytoplankton productivity (Worrest et al. 1978, 1980, 1981a,b; Calkins and Thordardottir 1980; R. C. Smith and Baker 1980; Jokiel and York 1982, 1984; Worrest 1982, 1983; Dohler 1984, 1985; Häder, 1984, 1985, 1986, 1987). There is now little dispute that ultraviolet radiation damages phytoplankton in laboratory and microcosm experiments. However, extrapolation of this finding to natural populations continues to be controversial.

A recent SCOPE workshop (1992) addressed and summarized issues associated with ultraviolet impacts on global ecosystems. The SCOPE report points out that we currently lack the necessary knowledge and appropriate models to make use of laboratory results for quantitative prediction of whole-system performance at the ecosystem level. Little information is available from which to infer long-term ecological consequences from short-term observations. In short, ecosystem-level information on enhanced levels of UV-B is limited, and there is a great need for further research.

Phytoplankton have evolved a variety of protective mechanisms associated with high solar radiation in general and high ultraviolet fluence in particular. One mechanism is the synthesis of ultraviolet-absorbing compounds. Chalker and Dunlap (1990) summarize a substantial body of literature dealing with UV-B and UV-A-absorbing compounds in marine macroalgae. They point out that these UV-B-absorbing compounds, especially mycosporinelike amino acids (MAAs), have been found in many marine organisms, are frequently related to environmental levels of ultraviolet radiation (Sivalingham et al. 1974; Dunlap et al. 1986), and hence have been proposed as a physiological adaptation to ultraviolet exposure. Vernet (1990) showed that Antarctic phytoplankton exposed to ambient levels of ultraviolet radiation seem to have the ability to synthesize potentially protective ultraviolet-absorbing compounds, and that they may have the capacity to utilize some of the ultraviolet radiation in photosynthesis through pigments that absorb below 400 nm. However, the latter hypothesis remains problematical (Yentsch and Yentsch 1982). El-Sayed et al. (1990) showed changes in photosynthetic pigmentation with elevated UV-B. In laboratory studies Carreto et al. (1990) have demonstrated that MAA synthesis is stimulated by UV-A, adding further evidence for potential physiological adaptation. Karentz et al. (1991a, 1991b) surveyed 57 species (1 fish, 48 invertebrates, and 8 algae) from the vicinity of Palmer Station (Anvers Island, Antarctic

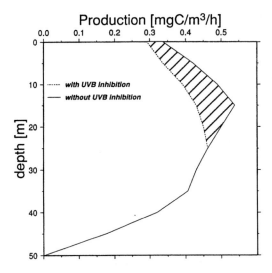

Figure 15.4. Average values for in situ phytoplankton productivity (mg cm^{-3} h^{-1}) versus depth (m) within the marginal ice zone of the Bellingshausen Sea in austral spring of 1990. Productivity inside the ozone hole (stratospheric ozone levels less than 200 DU) is compared with productivity outside the hole (stratospheric ozone levels greater than 300 DU). Higher UV-B levels (inside the ozone hole) are consistently associated with reduced levels of production (left-hand curve). Integration of these curves shows that higher UV-B levels within the ozone hole lead to a reduced water column production (6–12%). (Adapted from Smith et al. 1992b.)

Peninsula) for the presence of MAAs. They found that the majority of the species examined had absorbance peaks in the range from 315 to 335 nm, and they identified eight MAAs. They suggested that this widespread occurrence of MAAs in Antarctic marine organisms may provide some degree of natural biochemical protection from ultraviolet exposure during spring ozone depletion.

Bidigare et al. (1992) provided further direct chemical confirmation of MAAs in marine phytoplankton from the southern ocean. Their work during Icecolors '90 was undertaken to measure directly the effects of ozone diminution and ultraviolet radiation on southern ocean phytoplankton. Along a north-south transect across the marginal ice zone (MIZ), they found concentrations of diadinoxanthin (a photoprotective carotenoid found in *Phaeocystis* spp. and diatoms) highest in surface waters and decreasing with increasing depth, suggesting photoprotective adaptation to ultraviolet exposure. All of these observations lend credibility to the hypothesis that physiological adaptation to ultraviolet exposure is possible in at least some species of phytoplankton.

OZONE-RELATED UV-B EFFECTS ON NATURAL PHYTOPLANKTON POPULATIONS

Ultimately the issue of ozone-related UV-B increases must be assessed with respect to the direct impact on natural populations. R. C. Smith, Prézelin, and co-workers (Prézelin et al. 1992; R. C. Smith et al. 1992b) directly measured the increase of UV-B in, and its penetration into, Antarctic waters and provided the first conclusive evidence of a direct ozone-related effect on a natural population. Making use of the extreme change in ozone level associated with the hole—which creates a sharp gradient (or "front") in incident UV-B analogous to an atmospheric or oceanographic front—they performed comparative studies of the impact of UV-B on phytoplankton in the MIZ of the southern ocean. Their results indicate a minimum of 6–12% reduction in primary production associated with ozone depletion within the ozone hole. Figure 15.4 shows average values for in situ phytoplankton production versus depth in the MIZ of the southern ocean and offers a comparison of productivity inside the ozone hole (stratospheric ozone less than 200 Dobson units [DU]) with productivity outside the hole (stratospheric ozone levels greater than 300 DU). Higher UV-B levels (inside the hole) are consistently associated with reduced levels of production (left-hand curve).

SOUTHERN OCEAN PHYTOPLANKTON PRODUCTIVITY AND ATMOSPHERIC CO_2

For pelagic waters surrounding the MIZ of the southern ocean, current estimates of annual primary productivity are about 610×10^{12} carbon per year (El-Sayed and Turner 1977; Holm-Hansen et al. 1977; W. O. Smith and Sakshaug 1990). Productivity estimates for the MIZ were derived by W. O. Smith and Nelson (1986) using a simple model assuming ice-edge bloom genesis to be about 380×10^{12} carbon per year, or about 40% of the total MIZ plus pelagic production south of the Antarctic convergence. Thus, the MIZ potentially plays a major role in the ecological and biochemical cycles of the southern ocean.

It is possible to make an estimate of the effect of reduced ozone on primary production for the MIZ of the Bellingshausen Sea (Figure 15.3) based upon our determination of phytoplankton productivity data averaged for inside and outside the ozone hole. Based on these in situ data, a yearly estimate of production loss for the southern ocean can be made by assuming that this loss is representative of the entire MIZ and integrating production over this area and over the 3-month duration of the ozone hole during Antarctic spring. We estimate (using a 6% loss of water column productivity and conservatively assuming that a given location is outside the ozone hole half of the time) that this productivity loss to the MIZ is 7×10^{12} carbon per year, corresponding to about 2% of the estimated yearly production of the MIZ. Our assumptions are such that this is a minimum loss estimate, and values could be at least two times higher depending upon the specific space-time extent of the ozone hole. It is important to note that we have used short-term ^{14}C studies to assess changes in natural communities of phytoplankton caused by variations in the ozone hole that occurred on time scales from hours to weeks. Thus the time scale of our experimental protocol matched that of the processes observed. However, caution must be used when inferring longer-term ecological consequences from short-term observations (Smith and Baker 1980).

The interannual variability of primary production of the MIZ has been estimated to be substantial (W. O. Smith et al. 1988), such that the maximum productivity is 50% greater than the minimum. This variability associated with the annual advance and retreat of pack ice is thought to be a major physical determinant of space-time changes in the structure and function of polar biota (Ainley et al. 1986; Fraser and Ainley 1986; S. J. Smith and Vidal 1986; W. O. Smith and Nelson 1986; Walsh and McRoy 1986; Garrison et al. 1987; Ainley et al. 1988; W. O. Smith 1990). In particular, this interannual variability is likely to have a significant effect on total annual primary production, although to date these natural changes have not been accurately quantified. Thus, our estimated (2–4%) loss to MIZ productivity should be viewed in the context of a presumed natural variability of ±25%. Concern has been expressed (Voytek 1990) that ozone-induced phytoplankton loss may trigger a positive feedback with respect to atmospheric CO_2 that would exacerbate the greenhouse effect. Our estimated loss of 7×10^{12} carbon per year is about three orders of magnitude smaller than estimates of global phytoplankton production and thus is not likely to be significant in this context. However, we find that the ozone-induced loss to a natural community of phytoplankton in the MIZ is measurable, and the subsequent ecological consequences of the magnitude and timing of this early spring loss remain to be determined.

SUMMARY

Based on the evidence presented in this chapter, we may offer the following conclusions:

1. There is nearly undisputed evidence that human activities have caused a reduction in stratospheric ozone levels, especially in the south polar vortex region. The reduction has led and will continue to lead to increased UV-B at the surface of the earth.
2. There is overwhelming evidence that UV-B is an important environmental stress within aquatic ecosystems. Thus, independent of issues associated with the Antarctic ozone hole, there is a need to elucidate mechanisms and ecological consequences of natural levels of UV-B radiation on biological processes.
3. Improved atmospheric models, developed specifically for high-latitude (low-solar-elevation) situations, are now available that allow UV-B (and biological dose, provided the weighting function is known) to be computed as a function of solar elevation and atmospheric variables, including ozone concentration. Thus, UV-B and biological dose (given $\varepsilon[\lambda]$) can be accurately estimated for current and predicted reduced ozone concentrations.
4. Biooptical models of in-water optical properties allow these computations to be extended as a function of depth into the water column.
5. In spite of a number of methodological problems, the past few years have seen a significant improvement in our knowledge of $\varepsilon(\lambda)$, which, in turn, permits more realistic estimation of the radiation amplification factor.
6. We are making new advances in understanding the fundamental ultraviolet-related mechanisms underlying photodamage and have new data with respect to the dose-response characteristics of phytoplankton.
7. Recent work, using the Antarctic ozone hole as a natural laboratory, has provided the first evidence of a direct ozone-related effect on a natural phytoplankton community. This effect, although seemingly small, has a measurable impact on the marine ecosystem of the southern ocean—an impact that can be directly linked to human (primarily Northern Hemisphere) activities. Furthermore, the reduction of phytoplankton production caused by ozone-related increases in UV-B is an effect that is coincident with the warming of the earth and is likely to produce effects on the warming process in the same direction as the positive feedbacks that seem to dominate. Although the impact of humankind's large-scale UV-B experiment within the ozone hole is clear and measurable, the ultimate consequences of this unintentional global experiment remain to be seen.

REFERENCES

Ainley, D. G., W. R. Fraser, C. W. Sullivan, J. J. Torres, T. L. Hopkins, and W. O. Smith Jr. 1986. Antarctic mesopelagic micronekton: Evidence from seabirds that pack ice affects community structure. *Science* 232:847–849.

Ainley, D. G., W. R. Fraser, and K. L. Daly. 1988. Effects of pack ice on the composition of micronektonic communities in the Weddell Sea. In *Antarctic Ocean and Resources Variability*, ed. D. Sahrhage, pp. 140–146. Springer-Verlag, Berlin.

Anderson, J. G., D. W. Toohey, and W. H. Brune. 1991. Free radicals within the Antarctic vortex: The role of CFCs in Antarctic ozone loss. *Science* 251:39–46.

Bidigare, R. R., M. E. Ondrusek, and S.-H. Kang. 1992. Measurements of photosynthetic and UVB blocker pigments during the Icecolors '90 expedition. *Antarct. J. U. S.* 27:119–120.

Buhlmann, B., P. Bossard, and U. Uehlinger. 1987. The influence of longwave ultraviolet radiation (u.v.-A) on the photosynthetic activity (^{14}C-assimilation) of phytoplankton. *J. Plankton Res.* 9(5):935–943.

Caldwell, M. M. 1968. Solar ultraviolet radiation as an ecological factor for alpine plants. *Ecol. Monogr.* 38(3):243–268.

Caldwell, M. M., L. B. Camp, C. W. Warner, and S. D. Flint. 1986. Action spectra and their key role in assessing biological consequences of solar UV-B radiation change. In *Stratospheric Ozone Reduction, Solar Ultraviolet Radiation and Plant Life*, ed. R. C. Worrest and M. M. Caldwell, pp. 87–111. Springer-Verlag, Berlin.

Calkins, J., and T. Thordardottir. 1980. The ecological significance of solar UV radiation on aquatic organisms. *Nature* 283:563–566.

Carreto, J. I., M. O. Carignan, G. Daleo, and S. G. De Marco. 1990. Occurrence of mycosporine-like amino acids in the redtide dinoflagellate *Alexandrium excavatum*—UV photoprotective compounds? *J. Plankton Res.* 12:909–921.

Chalker, B. E., and W. C. Dunlap. 1990. UV-B and UV-A light absorbing compounds in marine organisms. In *Response of Marine Phytoplankton to Natural Variations in UV-B Flux: Proceedings of a Workshop*, ed. B. G. Mitchell, O. Holm-Hansen, and I. Sobolev, Appendix J. Chemical Manufacturers Association, Washington, D.C., 1990.

Cullen, J. J., and M. P. Lesser. 1991. Inhibition of photosynthesis by ultraviolet radiation as a function of dose and dosage rate-results for a marine diatom. *Mar. Biol.* 111(2):183–190.

Cullen, J. J., and M. R. Lewis. 1988. The kinetics of algal photoadaptation in the context of vertical mixing. *J. Plankton Res.* 10(5):1039–1063.

Cullen, J. J., P. J. Neale, and M. P. Lesser. 1992. Biological weighting function for the inhibition of phytoplankton photosynthesis by ultraviolet radiation. *Science* 258:646–650.

Damkaer, D. M., and D. B. Dey. 1983. UV damage and photoreactivation potentials of larval shrimp, *Pandalus platyceros*, and adult euphausiids *Thysanoessa raschii*. *Oecologia* 60:169–175.

Dohler, G. 1984. Effect of UV-B radiation on the marine diatoms *Lauderia annulata* and *Thalassiosira rotula* grown in different salinities. *Mar. Biol.* 83:247–253.

Dohler, G. 1985. Effect of UV-B radiation (290–320nm) on the nitrogen metabolism of several marine diatoms. *J. Plant Physiol.* 118:391–400.

Dunlap, W. C., B. E. Chalker, and J. K. Oliver. 1986. Bathymetric adaptations of reef-building corals at Davies Reef, Great Barrier Reef, Australia-III. UV-B absorbing compounds. *J. Exp. Mar. Biol. Ecol.* 104:239–248.

El-Sayed, S. Z., and J. T. Turner. 1977. Productivity of the Antarctic and tropical/subtropical regions: A comparative study. In *Polar Oceans*, ed. M. Dunbar, pp. 463–503. Arctic Institute of North America, Calgary. Alberta, Canada.

El-Sayed, S. Z., F. C. Stephens, R. R. Bidigare, and M. E. Ondrusek. 1990. Effect of ultraviolet radiation on Antarctic marine phytoplankton. In *Antarctic Ecosystems; Ecological Change and Conservation*, ed. K. R. Kerry and G. Hempel, pp. 379–385. Springer-Verlag, Berlin.

Fraser, W. R., and D. G. Ainley. 1986. Ice edges and seabird occurrence in Antarctica. *BioScience* 36:258–263.

Frederick, J. E., and A. D. Alberts. 1991. Prolonged enhancement in surface ultraviolet radiation during the Antarctic spring of 1990. *Geophys. Res. Lett.,* 18(10):1869–1871.

Garrison, D. L., K. R. Buck, and G. A. Fryxell. 1987. Algal assemblages in Antarctic pack ice and in ice-edge plankton. *J. Phycol.* 23:564–572.

Green, A. E. S., G. B. Findley Jr., K. F. Klenk, W. M. Wilson, and T. Mo. 1976. The ultraviolet dose dependence of non-melanoma skin cancer incidence. *J. Photochem. Photobiol. B Biol.* 24:353–362.

Häder, D.-P. 1985. Effects of UV-B on motility and photoorientation in the cyanobacterium, *Phormiduim uncinatum. Arch. Microbiol.* 140:34–39.

Häder, D.-P. 1984. Effects of UV-B on motility and photobehavior in the green flagellate, *Euglena gracilis. Arch. Microbiol.* 141:159–163.

Häder, D.-P. 1986. Effects of solar and artificial UV irradiation on motility and phototaxis in the flagellate, *Euglena gracilis. J. Photochem. Photobiol. B Biol.* 44:651–656.

Häder, D.-P. 1987. Polarotaxis, gravitaxis and vertical phototaxis in the green flagellate, *Euglena gracilis. Arch. Microbiol.* 147:179–183.

Helbling, E. W., V. Villafane, M. Ferrario, and O. Holm-Hansen. 1992. Impact of natural ultraviolet radiation on rates of photosynthesis and on specific marine phytoplankton species. *Mar. Ecol. Progr. Ser.* 80:89–100.

Hirosawa, T., and S. Miyachi. 1983. Inactivation of hill reaction by long-wavelength ultraviolet radiation (UV-a) and its photoreactivation by visible light in the cyanobacterium. *Anacystis nidulans. Arch. Microbiol.* 135:98–102.

Holm-Hansen, O. 1990. UV radiation in Antarctic waters: Effect on rates of primary production. In *Response of Marine Phytoplankton to Natural Variations in UV-B Flux: Proceedings of a Workshop,* ed. B. G. Mitchell, O. Holm-Hansen, and I. Sobolev, Appendix G. Chemical Manufacturers Association, Washington, D.C.

Holm-Hansen, O., and E. W. Helbling. 1993. Polyethylene bags and solar ultraviolet radiation. *Science* 259:534–534.

Holm-Hansen, O., S. Z. El-Sayed, G. A. Franceschini, and R. L. Cuhel. 1977. Primary production and the factors controlling phytoplankton growth in the Southern Ocean. In *Adaptations within Antarctic Ecosystems: Proceedings of the Third SCAR Symposium on Antarctic Biology,* ed. G. A. Llano, pp. 11–50. Gulf, Houston.

Jerlov, N. G. 1950. Ultraviolet radiation in the sea. *Nature* 166:111.

Jokiel, P. L., and R. H. York Jr. 1982. Solar ultraviolet photobiology of the reef coral *Pocillopora damicornis* and symbiotic zooxanthellae. *Bull. Mar. Sci.* 32:301–315.

Jokiel, P. L., and R. H. York Jr. 1984. Importance of ultraviolet radiation in photoinhibition of microalgal growth. *Limnol. Oceanogr.* 29(1):192–199.

Karentz, D. 1991. Ecological considerations of the Antarctic ozone depletion. *Antarct. Sci.* 3(1):3–11.

Karentz, D., J. E. Cleaver, and D. L. Mitchell. 1991a. Cell survival characteristics and molecular responses of Antarctic phytoplankton to ultraviolet-B radiation. *J. Phycol.* 27(3):326–341.

Karentz, D., F. S. McEuen, K. M. Land, and W. C. Dunlap. 1991b. Survey of mycosporine-like amino acid compounds in Antarctic marine organisms: potential protection from ultraviolet exposure. *Mar. Biol.* 108:157–166.

Kaupp, S. E., and J. R. Hunter. 1981. Photorepair in larval anchovy, *Engraulis mordax. J. Photochem. Photobiol. B Biol.* 33:253–256.

Lenoble, J. 1956. L'Absorption du rayonnement ultraviolet par les ions présents dans la mer. *Rev. Opt.* 35(10):526–531.

Lewis, M. R., J. J. Cullen, and T. Platt. 1984. Relationships between vertical mixing and photoadaptation of phytoplankton: Similarity criteria. *Mar. Ecol. Prog. Ser.* 15:141–149.

Lubin, D., and J. E. Frederick. 1990. Column ozone measurements from Palmer Station, Antarctica: Variations during the austral springs of 1988 and 1989. *J. Geophys. Res.* 95(D9): 13,883–13,889.

Lubin, D., and J. E. Frederick. 1991. The ultraviolet radiation environment of the Antarctic peninsula: The roles of ozone and cloud cover. *J. Appl. Meteorol.* 30:478–493.

Lubin, D., B. G. Mitchell, J. E. Frederick, A. D. Alberts, C. R. Booth, T. Lucas, and D. Neuschuler. 1992. A contribution toward understanding the biospherical significance of Antarctic ozone depletion. *J. Geophys. Res.* 97(D8):7817–7828.

Madronich, S., and C. Granier. 1992. Impact of recent total ozone changes on tropospheric ozone photodissociation, hydroxyl radicals, and methane trends. *Geophys. Res. Lett.* 19(5):465–467.

Mitchell, B. G. 1990. Action spectra of ultraviolet photoinhibition of Antarctic phytoplankton and a model of spectral diffuse attenuation coefficients. In *Response of Marine Phytoplankton to Natural Variations in UV-B Flux: Proceedings of a Workshop,* ed. B. G. Mitchell, O. Holm-Hansen, and I. Sobolev, Appendix H. Chemical Manufacturers Association, Washington, D.C.

Nachtwey D. S., and M. M. Caldwell, eds. 1975. Impacts of climatic change on the biosphere, PB 247 724. Climatic Impact Assessment Program Monograph 5, Part 1. Ultraviolet radiation effects. Department of Transportation, Washington, D.C.

National Academy of Science and National Research Council. 1979. *Stratospheric Ozone Depletion by Halocarbons: Chemistry and Transport.* Committee on Impacts of Stratospheric Change, National Academy Press, Washington, D.C.

National Academy of Science and National Research Council. 1982. *Causes and Effects of Stratospheric Ozone Reduction: An Update.* Committee on Chemistry and Physics of Ozone Depletion and Commmittee on Biological Effects of Increased Solar Ultraviolet Radiation, National Academy Press, Washington, D.C.

National Academy of Science and National Research Council. 1984. *Causes and Effects of Stratospheric Ozone Reduction: Update 1983.* Committee on Causes and Effects of Changes in Stratospheric Ozone. National Academy Press, Washington, D.C.

National Oceanic and Atmospheric Administration. 1987. *Stratospheric Ozone: The State of the Science and NOAA's Current and Future Research.* National Oceanic and Atmospheric Administration, Washington, D.C.

Prézelin, B. B., and R. C. Smith. 1993. Response: Polyethylene bags and solar ultraviolet radiation. *Science* 259:534–535.

Prézelin, B. B., M. Putt, and H. E. Glover. 1986. Diurnal patterns in photosynthetic capacity and depth-dependent photosynthesis-irradiance relationships in *Synechococus* spp. and larger phytoplankton in three water masses in the Northwest Atlantic Ocean. *Mar. Biol.* 91:205–217.

Prézelin, B. B., N. P. Boucher, and R. C. Smith. 1993. Daytime kinetics of UVA and UVB inhibition of photosynthetic activity in Antarctic surface waters. In *Current Topics in Plant Physiology: An American Society of Photosynthetic Responses to the Environment,* ed. H. Y. Yamamoto and C. M. Smith, pp. 150–155.

Prézelin, B. B., N. P. Boucher, and R. C. Smith. 1994. Marine primary production under the influence of the Antarctic ozone hole: Icecolors '90. *Antarct. Res. Ser.* 62:159–186.

Rundel, R. D. 1983. Action spectra and estimation of biologically effective uv radiation. *Physiol. Plant.* 58:360–366.

Rundel, R. D., and D. S. Nachtwey. 1978. Skin cancer and ultraviolet radiation. *J. Photochem. Photobiol. B Biol.* 28:345–356.

Schoeberl, M. R., and D. L. Hartmann. 1991. The dynamics of the stratospheric polar vortex and its relation to springtime ozone depletions. *Science* 251:46–52.

Scientific Committee on Problems of the Environment. 1992. *Effects of Increased Ultraviolet Radiation on Biological Systems. Proceedings of a Workshop.* SCOPE, Paris.

Sivalingham, P. M., T. Ikawa, Y. Yokohama, and K. Nisizawa. 1974. Distribution of a 334 UV-absorbing substance in algae, with special regard of its possible physiological roles. *Bot. Mar.* 17:23–29.

Smith, R. C. 1989. Ozone, middle ultraviolet radiation and the aquatic environment. *J. Photochem. Photobiol. B Biol.* 50(4):459–468.

Smith, R. C., and K. S. Baker. 1979. Penetration of UV-B and biologically effective dose-rates in natural waters. *J. Photochem. Photobiol. B Biol.* 29:311–323.

Smith, R. C., and K. S. Baker. 1980. Stratospheric ozone, middle ultraviolet radiation and carbon-14 measurements of marine productivity. *Science* 208(4):592–593.

Smith, R. C., and K. S. Baker. 1981. Optical properties of the clearest natural waters (200–800 nm). *Appl. Opt.* 20:177–184.

Smith, R. C., and K. S. Baker. 1989. Stratospheric ozone, middle ultraviolet radiation and phytoplankton productivity. *Oceanography* 2(2):4–10.

Smith, R. C., Z. Wan, and K. S. Baker. 1992a. Ozone depletion in Antarctica: Modeling its effect on solar UV irradiance under clear-sky conditions. *J. Geophys. Res.* 97(C5):7383–7397.

Smith, R. C., B. B. Prézelin, K. S. Baker, R. R. Bidigare, N. P. Boucher, T. Coley, D. Karentz, S. MacIntyre, H. A. Matlick, D. Menzies, M. Onderusek, Z. Wan, K. J. Waters. 1992b. Ozone depletion: Ultraviolet radiation and phytoplankton biology in Antarctic waters. *Science* 255(5047):952–959.

Smith, S. J., and J. Vidal. 1986. Variations in the distribution, abundance, and development of copepods in the southeastern Bering Sea in 1980 and 1981. *Cont. Shelf Res.* 5:215–239.

Smith, Jr., W. O., ed. 1990. *Polar Oceanography,* Academic Press, San Diego, Calif.

Smith, Jr., W. O., and D. M. Nelson. 1986. Importance of ice edge phytoplankton production in the Southern Ocean. *BioScience* 36(4):251–257.

Smith, Jr., W. O., and E. Sakshaug. 1990. Polar Phytoplankton. In *Polar Oceanography: Part B, Chemistry, Biology, and Geology,* ed. W. O. Smith, pp.477–525. Academic Press, New York.

Smith, Jr., W. O., N. K. Keene, and J. C. Comiso. 1988. Interannual variability and estimated primary productivity of the Antarctic marginal ice zone. In *Antarctic Ocean and Resources Variability,* ed. D. Sahrhage, pp. 131–139. Springer-Verlag, Berlin.

Stamnes, K., S.-C. Tsay, W. Wiscombe, and K. Jayaweera. 1988. Numerically stable algorithm for discrete-ordinate-method radiative transfer in multiple scattering and emitting layered media. *Appl. Opt.* 27(12):2502–2509.

Stamnes, K., J. Slusser, M. Bowen, C. Booth, and T. Lucas. 1990. Biologically effective ultraviolet radiation, total ozone abundance, and cloud optical depth at McMurdo Station, Antarctica, September 15 1988 through April 15 1989. *Geophys. Res. Lett.* 17(12):2181–2184.

Steemann Nielsen, E. 1964. On the complication in marine productivity work due to the influence of ultraviolet light. *J. Cons. Int. Explor. Mer.* 29:130–135.

Tsay, S.-C., and K. Stamnes. 1992. Ultraviolet radiation in the Arctic: The impact of potential ozone depletions and cloud effects. *J. Geophys. Res.* 97(D8):7829–7840.

United Nations Environment Programme. 1989. *Environmental Effects Panel Report.* United Nations Environment Programme, Nairobi, Kenya.

United Nations Environment Programme. 1991. *Environmental Effects of Ozone Depletion: 1991 Update.* United Nations Environment Programme, Nairobi, Kenya.

Vernet, M. 1990. UV radiation in Antarctic waters: Response of phytoplankton pigments. In *Response of Marine Phytoplankton to Natural Variations in UV-B Flux: Proceedings of a Workshop,* ed. B. G. Mitchell, O. Holm-Hansen, and I. Sobolev, Appendix I. Chemical Manufacturers Association, Washington, D.C.

Voytek, M. A. 1990. Addressing the biological effects of decreased ozone on the Antarctic environment. *AMBIO* 19(2):52–61.

Walsh, J. J., and C. P. McRoy. 1986. Ecosystem analysis in the southeastern Bering Sea. *Cont. Shelf Res.* 5:259–288.

Watson, R. 1988. *Ozone Trends Panel, Executive Summary.* NASA, Washington, D.C.

Williamson, C. E. and H. Zagarese, eds. 1994. *Workshop on Impacts on Pelagic Freshwater Ecosystems* (in press).

Worrest, R. C. 1982. Review of literature concerning the impact of UV-B radiation upon marine organisms. In *The Role of Solar Ultraviolet Radiation in Marine Ecosystems,* ed. J. Calkins, pp. 429–457. Plenum Press, New York.

Worrest, R. C. 1983. Impact of solar ultraviolet-B radiation (290–320 nm) upon marine microalgae. *Physiol. Plant.* 58:428–434.

Worrest, R. C. 1986. The effect of solar UV-B radiation on aquatic systems: An overview. In *Effects of Changes in Stratospheric Ozone and Global Climate, Overview,* ed. J. G. Titus, pp. 175–191. U.S. Environmental Protection Agency and United Nations Environmental Program 1.

Worrest, R. C., H. Van Dyke, and B. E. Thomson. 1978. Impact of enhanced simulated solar ultraviolet radiation upon a marine community. *J. Photochem. Photobiol. B Biol.* 17:471–478.

Worrest, R. C., D. L. Brooker, and H. Van Dyke. 1980. Results of a primary productivity study as affected by the type of glass in the culture bottle. *Limnol. Oceanogr.* 25:360–364.

Worrest, R. C., K. U. Wolniakowski, J. D. Scott, D. L. Brooks, B. E. Thomson, and H. Van Dyke. 1981a. Sensitivity of marine phytoplankton to UV-B radiation: Impact upon a model ecosystem. *J. Photochem. Photobiol. B Biol.* 33:223–227.

Worrest, R. C., B. E. Thomson, and H. Van Dyke. 1981b. Impact of UV-B radiation upon estuarine microcosms. *J. Photochem. Photobiol. B Biol.* 33:861–867.

Yentsch, C. S., and C. M. Yentsch. 1982. The attenuation of light by marine phytoplankton with specific reference to the absorption of near-UV radiation. In *The Role of Solar Ultraviolet Radiation in Marine Ecosystems,* ed. J. Calkins, pp. 691–700. Plenum Press, New York.

III

GLOBAL CARBON BUDGETS, MODELS, AND GEOPHYSICAL CONSTRAINTS

A

Terrestrial and Oceanic Interactions

There can be little question as to the emergent role of systematic modeling in contemporary science and especially in sorting through cause and effect in topics as complicated as the global carbon cycle and the warming of the earth. Each model is built on a set of assumptions as to the realm that is important and the factors that are to be considered. The models quickly devour experience and data and proceed on assumptions. Comprehensiveness is always elusive, and the models are always open to skepticism and revision. Where there is no formal model there is always an array of informal models, basic understanding, assumptions, or hypotheses as to how the world works. Competition among these informal models is acute, heavily influenced by the experience and emphasis of individuals, and the differences are difficult to resolve. The emergence of formal models and their rapid improvement in detail is one of the most promising aspects of contemporary science.

The progress in modeling is hard won. Purposes do not necessarily overlap: models of long-term changes in atmospheric CO_2 levels during the Phanerozoic Eon (the last 600 million years of the earth's history) or of changes in CO_2 levels and other ocean-atmosphere-land components on a glacial-interglacial time scale do not necessarily apply directly to the present and shed little light on the behavior of contemporary ecosystems. The types of mechanistic or physiological models that can be used to explore details of the effects of CO_2 fertilization may not take into account such important factors as the respiration of soils, the successional status of forests, or the effects of fire, disease, or drought, which become overriding considerations in determining the carbon balance of the landscape moment by moment.

Yet progress continues to be made despite these difficulties, and it is evident in both concepts and formal models. Luxmoore and Baldocchi offer a splendid review of the contemporary status of modeling around this topic, and Prentice and Sykes offer a superb example of what can now be done with a keen knowledge of vegetation and a mastery of contemporary techniques. Enting, working with a more nearly conventional approach in climatology, offers still another perspective on trace gases and the factors controlling them.

16

Modeling Interactions of Carbon Dioxide, Forests, and Climate

ROBERT J. LUXMOORE AND DENNIS D. BALDOCCHI

Atmospheric CO_2 levels are rising, forests are responding, and climate is changing. This combination of facts and premises may be evaluated on a range of temporal and spatial scales with the aid of computer simulators describing the interrelationships among forest vegetation, litter and soil characteristics, and meteorological variables. In this chapter, some insights into the effects of climate on the transfer of carbon and the converse, effects of carbon transfer on climate, are offered as a basis for assessing the significance of feedbacks between vegetation and climate under conditions of rising atmospheric CO_2 concentration.

Three main classes of forest models are reviewed: physiologically based models, forest succession simulators based on the JABOWA model of Botkin et al. (1972), and ecosystem–carbon budget models that use compartment transfer rates with empirically estimated coefficients. Some regression modeling approaches are also outlined, and energy budget models applied to forests and grasslands are reviewed. Energy exchanges at the earth's surface influence the enzyme kinetics that control the carbon metabolism of vegetation, as well as meteorological conditions within the layer of the atmosphere that interacts with the earth's surface, i.e., the planetary boundary layer. Finally, some comments are made on the scaling up of model results, and some conclusions are drawn on the positive and negative feedbacks between vegetation and global warming. This chapter presents examples of forest models; a comprehensive discussion of all available models is not undertaken.

PHYSIOLOGICALLY BASED MODELS (DIURNAL AND WET-DRY CYCLES)

Physiologically based models simulate the hourly or daily changes in plant processes and describe resource acquisition (carbon, water, and nutrient uptake), stomatal control of gas exchange, and resource utilization in tree growth and development. Physiological models simulate photosynthesis with a range of complexity, from a simple CO_2 gradient equation with empirically based pathway resistances (dependent on temperature, irradiance, and plant water status) to biochemically based formulations requiring enzymatic and leaf physiological variables for the species of interest (Farquhar and von Caemmerer

1982; Wullschleger 1993). Often physiological models are linked with leaf energy budget calculations because leaf temperature is needed to determine transpiration and to quantify enzyme kinetic rates associated with photosynthesis and respiration.

Physiological models are typically driven by external environmental variables. Philosophically, one should couple leaf physiology with canopy micrometeorology to scale leaf-level fluxes to canopy dimensions. Such coupling is desirable because environmental driving variables vary with depth in the canopy and on sunlit and shaded leaves, and the physiological response to these variables is often nonlinear. Cruder approximations of forest canopies assume that the forest is a single "big-leaf" surface. Here, evapotranspiration (ET) and surface temperature calculations are usually based on the Penman-Monteith equation, which requires meteorological variables (air temperature, dew point, wind speed, and solar radiation) and a surface resistance to vapor loss for simulation.

A significant limitation of most photosynthesis models is the lack of algorithms for feedback inhibition, even though there is abundant empirical evidence showing that lack of growth (sink) demand for photosynthate correlates with low rates of photosynthesis. Sink activity, such as cell growth, is typically much more sensitive to water and nutrient (nitrogen) stress than is photosynthesis (Luxmoore 1991). The use of photosynthesis–leaf nitrogen relationships for modeling photosynthesis may not be appropriate as a controlling mechanism in woody plants, which use leaves for nitrogen storage (Nambiar and Fife 1991). A feedforward-feedback photosynthesis framework is needed for understanding whole-plant responses to water and nutrient stresses under changing atmospheric CO_2 and climatic conditions.

The physiology of photosynthate partitioning, nutrient allocation, and the root processes of water and nutrient uptake are also represented with a variety of model formulations. Several important root-soil processes involving mycorrhizae, soil chemical and microbial reactions, and fine root dynamics are poorly represented in many physiological models. This situation is due in part to a lack of information and to preoccupation with photosynthesis and canopy processes, which has often led to an unbalanced treatment of plant processes in the context of the *whole plant–soil system*.

Forests and CO_2

Increasing atmospheric CO_2 concentrations lead to complex linkages and feedbacks among photosynthesis, respiration, stomatal conductance, and leaf temperature. For example, rising CO_2 concentration increases carboxylation rates and decreases leaf stomatal conductance. This chain of events can cause a reduction in transpiration and an increase in leaf temperature. Finally, higher leaf temperatures increase leaf respiration, and this increase moderates the influence of enhanced CO_2 on photosynthesis. Recent models of coupled physiological-micrometeorological processes (Baldocchi 1993; McMurtrie and Wang 1993) allow one to examine these linkages in detail. Calculations using a model described by Baldocchi (1993) show that a rise in CO_2 concentration from 350 to 600 µl/liter will increase the canopy photosynthesis of a deciduous temperate forest by 25% and reduce canopy stomatal conductance by 28% (Table 16.1). In turn, canopy latent heat exchange is decreased by 9% and canopy sensible heat exchange is increased by 6%.

Table 16.1. Influence of an Increase in Atmospheric CO_2 Concentration on Latent and Sensible Heat Exchange, Photosynthesis, and Stomatal Conductance of a Deciduous Forest Canopy[a]

CO_2 concentration (μl/liter)	Latent heat exchange (W m^{-2})	Sensible heat exchange (W m^{-2})	Photosynthesis (μmol m^{-2} s^{-1})	Stomatal conductance (m s^{-1})
350	331	142	36.6	0.0249
600	300	151	45.9	0.0178
Percent difference	−9.4	6.3	25.4	−28.5

[a]The canopy is assumed to be growing under typical summertime conditions (25°C air temperature and 1800 μmol m^{-2} s^{-1} of quanta). The model couples physiological and canopy micrometeorological algorithms to be able to consider feedbacks among photosynthesis, stomatal conductance, leaf temperature, humidity, and energy exchanges (Baldocchi 1993).

Even though tree models use a wide variety of differing formulations for photosynthesis, they all generally predict increased growth with elevated atmospheric CO_2 concentration. The biotic growth factor of Bacastow and Keeling (1973),

$$\beta = [(P_1 - P_0)/P_0]/(\ln C_1 - \ln C_0) \tag{16.1}$$

where P_1 and P_0 are the growth rates at atmospheric CO_2 concentrations of C_1 and C_0, respectively, is used as a means for *comparing* various simulated growth responses to CO_2 enrichment. Physiological models predict β factors in the range from 0.06 to 1.3 (Table 16.2), and these more or less correspond to the short-term growth responses obtained in CO_2 exposure experiments with seedlings (β factors in the range of 0.1–1.2: Kienast and Luxmoore 1988; Chapter 4, this volume). For comparison, recent simulations of grassland responses to CO_2 enrichment by Thornley et al. (1991) gave β factors in the range of 0.84–1.20. Gifford's (1992) recent review of vegetation responses to elevated CO_2 levels suggests that β factors in the range of 0.1–0.5 seem likely from a "bottom-up" physiological approach and also from a "top-down" global carbon pool

Table 16.2. Biotic Growth Factors (β) Calculated from Simulated Forest Growth or Carbon-Fixation Rate Responses to CO_2 Enrichment Obtained from Physiologically Based Models

Model	Reference	Atmospheric CO_2 concentration (μl/l)			Comment
		C_1	C_0	β	
BIOMASS	McMurtrie and Wang (1993)	700	350	0.36	Conifer
CANDO[a]	Reynolds et al. (1992)	680	340	0.68–0.86	Hardwood
CANOAK[a]	Baldocchi (1993)[b]	600	350	0.46	Hardwood
FOREST-BGC	Running and Nemani (1991)	680	340	1.3[c]	Conifer
GEPSI	Reynolds et al. (1993)	680	340	0.74	Short term
		680	340	0.26	Long term
MAESTRO[a]	McMurtrie and Wang (1993)	700	350	0.37	Conifer
UTM-SPL	Luxmoore et al. (1990)	600	340	0.21	Hardwood
	Rastetter et al. (1991)	680	340	0.06[c]	Hardwood

[a]Canopy models.
[b]Derived from Table 16.1.
[c]Estimated.

analysis. Reynolds et al. (1993) show that simulated β factors tend to decrease with change from short- to long-term integration (Table 16.2).

Preferential allocation of carbon to belowground processes and root growth is predicted by some models, particularly under conditions of water and nutrient limitation (e.g., Luxmoore 1981; Thornley and Cannell 1992). Several other physiologically based models, not in Table 16.1, are being developed for simulation of global change impacts. These include ECOPHYS (Rauscher et al. 1990), FORGRO (Mohren 1987), the ITE Edinburgh Forest model (Thornley and Cannell 1992), Q (G. I. Ågren personal communication), a seedling tree simulator (Webb 1991), and TREGRO (Weinstein et al. 1991). A considerable amount of effort is being given to mechanistic simulation of global change effects on forests. Physiological modeling is expected to show that forest vegetation will fix more carbon with elevated CO_2 concentration, but it is not clear if short-term physiological responses will be maintained with continuing CO_2 enrichment in a changing climate. Linkage of short-term simulation results to long-term response models is needed (see "Scaling Up and Other Aspects").

Forests, CO_2, and Precipitation

Climate change is expected to modify precipitation patterns, and a number of throughfall diversion experiments are under way in forests to evaluate the effects of more or less rainfall on forest processes. An early experiment was established at Flakaliden in central Sweden in a Norway spruce stand. Experiments are also underway in a loblolly pine (*Pinus taeda*) stand near Aiken, South Carolina, and in an eastern deciduous forest near Oak Ridge, Tennessee. Some initial simulations (R. J. Luxmoore unpublished observations) of the consequence of a ± 20% change in precipitation (without CO_2 effects) for the Oak Ridge site were conducted with the UTM-SPL model (Luxmoore et al. 1990) using local meteorological data for an average rainfall year (1372 mm/year) and for a dry year (933 mm/year). The code implements the Penman-Monteith equation for ET in a big-leaf approach and the Darcy flow equation for water flux between soil layers. The major simulated effects of more or less rainfall were corresponding changes in both drainage to groundwater and streamflow. There were slight effects on ET, forest growth, and litter decomposition for the mesic eastern Tennessee environment.

The major insight gained from this modeling was the large buffering effect of water storage offered by soil. For forests growing in mesic areas, where precipitation exceeds ET (e.g., the eastern United States) and where there are well-developed soil profiles, there may not be much impact of precipitation change on forest growth, but there could be large impacts on recharge to aquifers and drainage to streams. Such changes will greatly impact aquatic ecosystems and adjacent wetlands. Change in precipitation in forest communities where annual precipitation is less than potential ET can be expected to have a significant effect on forest growth and water budgets.

Forests, CO_2, Precipitation, and Temperature

Temperature has a direct effect on photosynthesis and respiration algorithms in physiological models, and it also modifies phenological development. The interactions of temperature and CO_2 level with physiological processes are still being actively investigated (Long 1991).

Running and Nemani (1991) reported the results of simulation with the FOREST-BGC code, which uses the Penman-Monteith big-leaf approach for ET calculations and a CO_2 gradient equation for photosynthesis. They conducted simulations for a forested region of Montana using a doubling of CO_2 concentration, a 4°C temperature rise, and a 10% increase in precipitation. In general, increases in leaf area, ET, and net photosynthesis were predicted; the duration of snow pack decreased. Their results also suggested a big decrease in streamflow. Simulations conducted for Montana (cold, dry) were compared with predictions for conditions in northern Florida (hot, wet). An increase in net primary production (NPP) of 88% and a 10% increase in ET were predicted for Montana with change in environmental conditions, whereas a decrease in NPP of 5% and a 16% decrease in ET were predicted for Florida.

Summary Comments

Physiological models with a range of differing model formulations predict increased tree growth (carbon fixation) with CO_2 enrichment. Growth could be preferentially enhanced in roots on sites with water or nutrient stress.

A β factor response in the range of 0.1–0.5 may be a consistent result for long-term CO_2 effects. Higher β values may be predicted in short-term simulations.

Soil water storage is an effective buffer to water stress on mesic sites with deep soil profiles. Wetland forests and streamflow could be significantly affected by a decline in precipitation and a warming climate.

Temperature and CO_2 interaction effects on growth processes need further quantification. Model applications show contrasting results for differing situations, suggesting that large-scale generalizations will not be defensible.

PHENOLOGICAL RELATIONSHIPS (ANNUAL CYCLES)

The influence of environmental variables on the initiation and cessation of tree development and growth in the annual cycle is generally represented by empirical functions in forest models. Cannell (1990) reviewed the concepts of cessation of shoot growth, frost hardening, dormancy release, and the onset of shoot growth in spring. Such concepts can lead to a more mechanistic basis for modeling of tree phenology; however, growth models will be dependent on empirical relationships for some time to come. Fortunately natural variation in climate has provided a range of conditions for determining empirical phenological relationships. Unfortunately we do not have adequate phenological data for the interaction effects of CO_2 enrichment and climatic variables.

Changes in temperature and precipitation have been shown to modify the empirical functions reported by Dougherty et al. (1990) for the foliar phenology of loblolly pine in the southern United States. They noted that leaf area development and senescence can vary by as much as 2 months owing to changes in climatic conditions. Large differences in leaf area duration and the absorbed radiation by a canopy can result from the influence of different meteorological conditions on foliar development. CO_2 enrichment does not seem to have much influence on leaf senescence of woody plants. Cumulative degree-days above a base temperature, often taken as 5°C, are used to determine growth rate in some models.

It is thus apparent that atmospheric CO_2 effects on the phenology of woody plants may be small, that the interaction effects of climate and CO_2 on phenology need to be investigated further, and that empirical phenological relationships will continue to be used in tree growth models.

FOREST SUCCESSION SIMULATORS (LIFE CYCLES)

Forest succession model applications, based on the JABOWA model of Botkin et al. (1972), have been developed for a wide range of forest communities. Simulation of the life cycles of individuals in a multispecies forest stand depends on species differences in temperature effects on growth rate and tolerance to shade, to drought, and in some cases to soil fertility status. Mean monthly air temperature has a direct effect on growth rate and an indirect effect on growth through change in soil water status, determined by temperature effects on ET. In a nutrient cycling version of a succession model called LINKAGES (Post and Pastor 1990), temperature controls litter decomposition rate and nitrogen availability. Succession models have been used extensively to gain insight into possible forest responses to global changes in atmospheric CO_2 concentration, temperature, and precipitation.

CO_2, Warming, and Succession

Bowes and Sedjo (1991) have conducted extensive simulations of succession in a deciduous forest in Missouri using the FORENA code of Solomon (1986). Missouri forests are on the western edge of the eastern deciduous forest range. Bowes and Sedjo modified parameter values to represent the effects of elevated CO_2 level on growth and water use efficiency. They increased the diameter growth rate factor by 5% and 10% in a sensitivity analysis. These increases are equivalent to β factor input signals of 0.15 and 0.30, respectively, based on atmospheric CO_2 concentrations of 350 and 700 μl/liter. They predicted aboveground biomass for a deciduous forest at 20 and 200 years for control and elevated CO_2 conditions using data for a 30-year-old regrowth stand as initial conditions. Mean growth rates for the 180-year period between the simulation results at 20 and 200 years were calculated, and β factors were determined for three mean annual temperature scenarios. This analysis showed very large β factors for the base case of 13.5°C, negative β factors for the 14.5°C case, and a mixed result for the 15.5°C mean annual temperature case (Table 16.3). The output β factors for the 13.5°C scenario were significantly larger than the input β factor signals. An interesting observation from Table 16.3 is the wide range of apparent growth responses to CO_2 enrichment for differing temperature scenarios.

In a second series of CO_2 simulations reported by Bowes and Sedjo (1991), β factors were essentially zero for the various temperature cases above the base case of 13.5°C. The results for the 13.5°C case in Table 16.3 are probably not robust. Thus, results from a careful simulation analysis of a Missouri forest response to CO_2 suggest that there may be no effect of elevated CO_2 on forest growth rates over the long term. A small increase

Table 16.3. β Factors Calculated from the Forest Succession
Simulation Results of Bowes and Sedjo (1991) for CO_2 Enrichment
Effects on a Missouri Deciduous Forest for Several Mean Annual
Temperature Cases

Growth enhancement (%)[a]	Biomass (Mg/ha)		Mean growth rate (Mg ha^{-1} year^{-1})	β
	Year 20	Year 200		
13.5°C case				
0	75	85	0.0556	—
5	79	100	0.1167	1.59
10	84	106	0.1222	1.73
14.5°C case				
0	60	87	0.1500	—
5	62	85	0.1278	Neg.[b]
10	70	90	0.1111	Neg.
15.5°C case				
0	37	83	0.2556	—
5	44	95	0.2833	0.16
10	50	93	0.2389	Neg.

[a]The 0%, 5%, and 10% enhancements correspond to 350, 700, and 700 µl/liter of atmospheric CO_2, respectively.
[b]<0.00.

in carbon storage was shown at year 200 (Table 16.3), but this was not found in another series of CO_2 enrichment simulations. Bowes and Sedjo (1991) showed a decline in Missouri forest biomass with CO_2-based climate change simulations, and there was little effect from inclusion of direct CO_2 effects on growth and water use efficiency in the model results. The dominant cause for forest decline in the Missouri simulations was water stress induced by temperature rise.

Succession models do not give any assurance that carbon storage will increase over the long term with CO_2 enrichment alone. This same conclusion was reached in a simulation of deciduous forests in Tennessee (Luxmoore et al. 1990; Post et al. 1992). Even though short-term (few decades) storage increases were obtained, they did not persist with an increase in stand maturity. Variability of carbon storage in vegetation owing to stochastic ingrowth and mortality algorithms masked CO_2 enrichment effects.

Climate warming can have a large effect on forests according to succession modeling results. The JABOWA family of models is very responsive to temperature effects, which operate in part through ET and soil water storage effects on growth. The Thornthwaite and Mather equation for potential ET is used in several succession models, and it is a function of air temperature. A temperature rise increases ET, inducing plant water stress and inhibition of forest growth. Inadequate simulation of ET biases model output, and this may be the case for succession models using the Thornthwaite and Mather equation. The Penman-Monteith equation would provide improved estimates of ET.

Solomon (1986) conducted an extensive series of simulations of potential climate change effects on the forested areas of eastern North America. Decline in forest growth along the western edge of the deciduous forest distribution (e.g., Missouri) was largely due to increased water stress induced by climate change.

Kienast (1991) and Kienast and Brzeziecki (1993) used a succession model to predict forest distributions in Switzerland using a scenario of linear rises in air temperature and atmospheric CO_2 over a 300-year period. The succession results were extrapolated to the whole of the Swiss forested areas with a regression-based geographical information system approach. They projected a 7–10% conversion of some forest area to steppe and a migration of deciduous forests into coniferous forest areas with the global change scenario.

Smith et al. (1992) predicted a decline of boreal forest on south-facing slopes in Alaska using a succession simulator and a linear climate projection obtained from general circulation model (GCM) output. Decline was attributed to an increase in water stress. On north-facing slopes, enhanced boreal forest growth was simulated because of warmer conditions, which favored growth without inducing water stress.

Post and Pastor (1990) and Post et al. (1992) showed that inclusion of nutrient cycling in their succession model (LINKAGES) can account for indirect climate change effects that operate through changes in carbon-nitrogen interactions and litter quality. They noted that CO_2-induced climate change has two effects, one being the direct effect of temperature and precipitation, which could lead to significant changes in forest growth and carbon storage. Some of these changes were projected to lead to changes in species composition, particularly along transition zones between biomes. Alteration in carbon-nitrogen cycling induced by changes in tree species and associated litter quality can lead to important secondary effects on forest response to climate change.

Martin (1992) reported some simulation results with a simulator called EXE, which has some similarities to the LINKAGES model but provides greater resolution of several controls on plant growth, including soil-plant water relationships and growing season length. These changes in physiological algorithms lead to second-order effects on forest succession that resulted in predictions of climate change effects on forests in Minnesota that differed from the results with LINKAGES (Martin 1992). For example, a climate change scenario for the next 100 years at Duluth, associated with a doubling of atmospheric CO_2, resulted in a large increase in aboveground biomass in LINKAGES and a decrease in biomass in EXE. These contrasting results were due in part to differences in temperature effects on growth in the parameterization of the two models.

Smith et al. (1992) summarized two regression approaches for estimating the potential changes in forest and grassland community distributions with change in mean annual temperature and precipitation. Imposition of a climate change scenario on the Holdridge lifezone relationship between vegetation types and climate variables indicated potential shifts in forest distribution. A second method for predicting distribution of vegetation was based on correlation between leaf area and ET. Simulation of changes in ET for the continent of Africa using the Penman-Monteith equation with projected changes in temperature and precipitation was used to estimate changes in leaf area. The leaf area results had some correspondence with the projections from the lifezone analysis and also with some predictions from their forest succession modeling.

Prentice et al. (1992) reported the development and application of a global vegetation distribution model that accounted for physiological attributes, dominance of species in plant communities, soil texture, and climate. The influences of climate variables and soil water status resulted in selection of plant types suited to given areas, and the dominant species within the selected plant types defined the biome for a particular region. The global vegetation patterns predicted with this approach were in agreement with a published map of natural plant ecosystems. This modeling approach can be used for assessment of vegetation distributions and carbon budgets appropriate for alternative climate change scenarios.

Pastor and Post (1993) conducted some simulations of transient climate change effects on forests with the LINKAGES succession model and obtained results that call into question the direct use of regression-based prediction of climate change effects on forests. They showed that transient responses of forests to climate change included lags in population responses and nonlinear changes in nitrogen availability that affected tree growth. These effects caused departure of projections obtained from regression relationships between climate variables and vegetation.

Succession models are very dependent on leaf area indexes for modeling the effects of competition for light, yet leaf area simulation results from succession models are rarely presented. This is unfortunate since the information could be used with biomass increment to estimate changes in solar conversion efficiency (see "Scaling Up and Other Aspects") during various stages of stand development. The simulations of coniferous forest growth and succession of Dale and Franklin (1989) provide results for aboveground production and leaf area that could be used to estimate changes in solar conversion efficiency with stand age. Additionally, succession models could be used to provide the tree heights of species at selected index ages, and these could be compared with site index values for forests from a range of climatic conditions.

Summary Comments

Forest succession models suggest that there may be little or no response to CO_2 enrichment over the long term (100 years) owing to the overriding variability introduced by the stochastic processes of establishment and mortality of individual trees in the forest community.

Conversion of forests to grassland or steppe may result from warming (e.g., south-facing slopes in Alaska, deciduous forests in Missouri, circa 10% of Swiss forests). Will this lead to a significant decrease in terrestrial carbon storage? Will change in vegetation type influence meteorological conditions in the planetary boundary layer? (See "Biophysical Effects of Forests on Climate.")

Inclusion of nutrient cycling in succession models changes the possibilities for forest growth response to CO_2 level and climate change. Litter quality and nutrient release by decomposition are directly or indirectly affected by global change variables. Two succession models that include nutrient cycling (LINKAGES, EXE) have given contrasting results in the same model application.

All insights gained from succession model applications come from essentially one model. Are the predicted responses to global change defensible? Do leaf area indexes used in simulations correspond to those of the natural communities being investigated? Is there an overdependence on competition for light? Is the model overly sensitive to temperature?

Some predicted changes in forest communities are the result of extreme parameter values that are probably not justified.

Succession modeling using transient environmental scenarios gives results that differ from predictions obtained from climate-vegetation regression relationships.

ECOSYSTEM AND CARBON BUDGET MODELS

Models of forest ecosystems sometimes incorporate physiological algorithms requiring fine-scale input variables (e.g., King et al. 1989; Rastetter et al. 1991), whereas others use aggregated transfer coefficients for fluxes between model compartments that require calibration to primary production and nutrient cycling data (e.g., McGuire et al. 1992).

McGuire et al. (1992) reported simulations for the NPP of nonwetland ecosystems in North America using over 11,000 grid cells, each 0.5° latitude by 0.5° longitude in area. Seventeen vegetation and five soil textural classes were investigated. Temperature effects on NPP were represented by direct effects on gross primary production, respiration, ET, and litter mineralization. The results suggested that soil water stress affected NPP more strongly through nitrogen availability than directly through CO_2 uptake. The importance of coupling carbon and nitrogen was shown in simulations for temperate mixed forests. In one case temperate forest simulations were conducted without nitrogen cycling, and these showed a decrease in NPP with a 2°C increase in mean temperature caused largely by increased respiration. When nitrogen cycling was included the same temperature increase resulted in higher NPP owing to enhanced growth induced by increased nitrogen availability—an effect that offset increased respiration. It has been noted that soil has a large store of nitrogen (a low C:N ratio) and that a small shift of nitrogen to vegetation could support a significant increase in biomass that has a high C:N ratio (Luxmoore 1981; Rastetter et al. 1991).

Simulation of the potential CO_2-enhanced carbon storage was conducted by Polglase and Wang (1992) with a 10-biome model that combined input estimates of NPP with biotic growth factors to represent CO_2 and temperature effects on biome productivity. Partition and residence time coefficients were used to estimate changes in carbon storage in vegetation components, and the results were linked with the Rothamsted soil carbon model to estimate changes in soil carbon storage. A biotic growth factor value of 0.3 was used to relate NPP to atmospheric CO_2 for the reference case. Adjustments were made for temperature effects such that β increased at higher temperature. Temperature adjustments to β were based on photosynthetic rate responses to temperature.

The most important biomes for sequestering carbon shown by Polglase and Wang (1992) for 1990 meteorological conditions were the tropical humid forest, savanna, and temperate forest biomes. Most of the increase in carbon storage was due to accumulation of undecomposed plant litter. The global estimate of CO_2-enhanced carbon storage for 1990 was 1.26 petagrams (Pg) ($=10^{15}$ g). Polglase and Wang also estimated that the annual CO_2-enhanced carbon storage in tropical forests (0.72 Pg) was considerably less than the annual carbon release from tropical deforestation. Long and Hutchin (1991) suggest that models based on the extrapolation of primary production data obtained from measurements of the standing crop may be of limited value owing to high error rates in the data. Nevertheless, models based on primary production measurements provide some guidance.

Kauppi et al. (1992) conducted a carbon budget analysis of European forests for the period 1971–1990 and estimated that the fertilization effect of atmospheric pollutants along with increased forest planting gave an increasing carbon storage pool that could account for a large proportion of the "missing carbon" in the global carbon budget (Chapter 21, this volume). The use of forest plantings as a means of sequestering carbon has been modeled by Dewar and Cannell (1992) for conditions in the United Kingdom.

They showed that long-term (100-year) carbon storage can be increased with new plantations of conifer or hardwood species. This finding reinforces the analysis of Kauppi et al. (1992), who have documented increased forest plantings in Europe.

Thus, we conclude that ecosystem modeling shows increased carbon storage with CO_2 enrichment. Physiological and carbon budget models both suggest that forests have the capacity to moderate the rate of atmospheric CO_2 rise through an increase in carbon storage. How transient will this effect be? Succession models suggest that storage may increase for about 100 years in some cases.

Linkage of water, carbon, and nitrogen (phosphorus) cycles is critical for effective prediction of ecosystem responses to global change impacts.

Increases in forest planting may be having a greater effect on current global carbon budgets than CO_2 enrichment. Will increases in forest planting induce changes in climate variables?

BIOPHYSICAL EFFECTS OF FORESTS ON CLIMATE

Models of surface energy budgets account for the inflows, outflows, transformations, and energy storage changes of vegetation-soil systems. Application of these models reveals important differences in the energy budgets of forests and grasslands, and these are needed for interpretation of the biophysical consequences of predicted changes in forest distribution from succession models. Between 70% and 90% of incoming solar radiation (R_g) over forests is converted into net radiation (R_n), whereas less than 70% of R_g is converted into R_n over croplands (Figure 16.1). A different proportion of R_n is converted into sensible (H) and latent (LE) energy for the two vegetation types; grassland has a lower component of sensible heat loss. The relationships in Figure 16.1 apply to daytime conditions. Forest has a component of sensible heat storage (G) that builds up during the day and declines at night. Soil heat fluxes (S) are greater in the grassland than in the forest, resulting in greater diurnal and seasonal maximum and minimum soil temperatures in the grassland. Nevertheless, the mean annual soil temperature in the upper meter of soil is the same in both vegetation types in a given environment, even though the ranges of temperatures about the mean differ for different vegetation types. Any increase in mean global air temperature will lead to a similar increase in the mean annual temperature in the litter and root zones of terrestrial ecosystems. This relationship has a very important bearing on soil respiration and decomposition processes, which can lead to positive or negative feedbacks to global warming (cf. McGuire et al. 1992 and Raich and Schlesinger 1992).

Albedo

Broadleaf and coniferous forests are visibly darker and have lower albedos than croplands, deserts, and grasslands. Thus, under identical solar radiation loads more R_n is available to evaporate water and heat the air over a forest than over a cropland, grassland, or desert landscape. Charney (1975) hypothesized that deforestation of the Sahel region by overgrazing was a prime factor in desertifying the edges of that region. Denuding the surface of the vegetation increased its surface albedo, causing a net

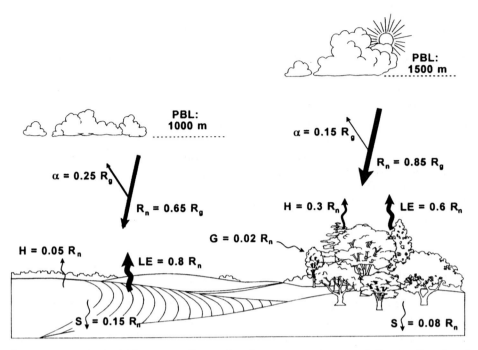

Figure 16.1. Typical midday energy fluxes of surface energy components for a well-watered deciduous forest (Rauner 1976; Verma et al. 1986; D. D. Baldocchi unpublished observations) and a grassland canopy (Ripley and Redman 1976; Verma et al. 1989).

radiative loss relative to nearby forested regions. In turn, an atmospheric circulation was suggested to develop that imported heat from aloft and forced the air column to subside. These processes act to suppress convective precipitation, giving a positive feedback to desertification. Schlesinger et al. (1990) offered a biological feedback mechanism for desertification associated with an increase in the spatial heterogeneity of nutrient distribution as well as a change in energy budgets as vegetation becomes increasingly patchy.

Heat Storage

Energy exchange models often ignore the role of heat storage by vegetation. Yet the appreciable depth and bulk of forest canopies allow incoming solar and terrestrial energy to be stored as heat in the canopy's air space and in the biomass (McCaughey 1985; Moore and Fisch 1986).

The current neglect of canopy heat storage may partly explain why GCM modules do not mimic the time course of convective precipitation. GCM hydrologic calculations of tropical precipitation distribute precipitation over the course of the day instead of during discrete afternoon events that occur because of convective thunderstorms (Dickinson 1989). Correct timing of surface heating and planetary boundary layer (PBL) growth is needed to calculate the development of clouds at the top of the PBL and the onset of convective precipitation. In turn, the temporal distribution of precipitation affects the duration of canopy wetness and the subsequent partitioning of energy into sensible and latent heat exchange.

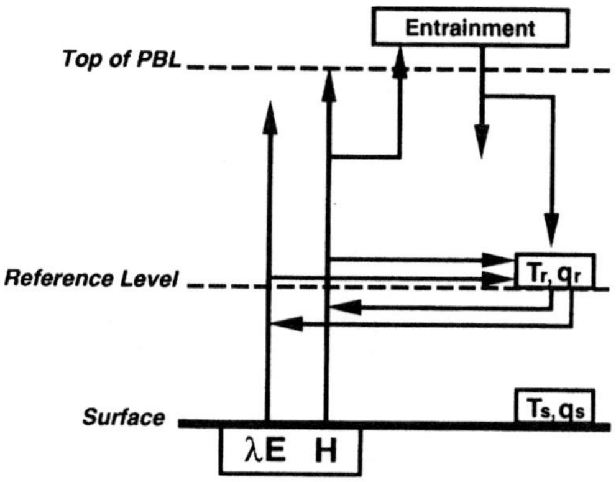

Figure 16.2. Relationships between sensible (H) and latent (E) heat fluxes at the surface and the temperature (T) and humidity (q) conditions in the planetary boundary layer are influenced by entrainment, which may be modified by climate change. (After Jacob and deBruin 1989.)

Planetary Boundary Layer

The PBL is the layer of air in contact with and influenced by the earth's surface. Its height is defined by the inversion level of the atmosphere's temperature profile. Visually, its height is associated with the base height of fair weather cumulus clouds, and this height is influenced by the surface vegetation. Computations by de Bruin and Jacob (1989) show that the PBL height over Thetford forest (dry sparse canopy, Bowen ratio 2–3) can reach 2200 m. In contrast, that over Les Landes forest (Bowen ratio 1) can reach 1400 m and that over an actively transpiring grassland can be 900 m. What effect do feedbacks between surface vegetation and the PBL have on surface energy, mass, and momentum fluxes and meteorological variables (Figure 16.2)? Jacob and de Bruin (1992) examined this problem with a detailed model that included feedbacks among surface evaporation, PBL growth, and climatic forcing. They concluded that the sensitivity of transpiration to the surface and aerodynamic resistances is decreased by PBL feedbacks. However, they showed that PBL feedbacks increase the sensitivity of LE to net radiation and albedo changes. Considering feedbacks between surface layer and PBL, Jacob and de Bruin predicted that a change from forest to grassland would increase LE by 29 W m^{-2}. In contrast, an 87-W m^{-2} increase was estimated if PBL feedback effects were neglected.

Surface Roughness

Forests are much taller and rougher than croplands and grasslands. Consequently, any transition from a forest to a cropland or grassland will impact the efficacy by which heat, momentum, water vapor, and CO_2 are exchanged between the biosphere and the atmosphere. This impact is greatest for entities that have a negligible surface resistance, such as momentum and sensible heat transfer. In contrast, altering surface roughness will have a relatively small impact on canopy CO_2 exchange because the canopy's aerodynamic resistance is a minor component of the bulk canopy exchange coefficient. Water vapor exchange is also dependent upon surface resistance. However, evaporation rates are coupled to available energy and sensible heat exchange, so one cannot conclude that evaporation rates will be as minimally affected by a change in surface roughness as will canopy CO_2 exchange.

Uptake of CO_2

It is unlikely that PBL feedback effects between canopy photosynthesis and the CO_2 concentration in the PBL will significantly influence photosynthesis. Experimental measurements by Wofsy et al. (1988) and theoretical calculations by McNaughton (1988) show that the CO_2 drawdown in a well-mixed PBL is only 10–15 μl/liter during the photosynthetically active portion of the day. CO_2 buildup at night and in the early morning under stable inversions can be over 500 μl/liter. Forests are not expected to take up much of this CO_2 because of low light conditions (stomata closed) and the rapid mixing that occurs within 2 h after sunrise.

Reforestation Effects on Climate

Miller (1992) anticipates very little effect of regional reforestation on regional temperature and humidity owing to negative feedback mechanisms that operate in the PBL. Expansion of the PBL during the day tends to buffer any increase in vapor pressure from ET. Little credence is now given to the idea that ET from temperate forests contributes to local precipitation. For example, midlatitude precipitation in the United States comes primarily from frontal storms originating in subtropical ocean regions (Miller 1992). Local convective storms are a significant component of rainfall received in tropical rain forests.

Summary Comments

De Bruin and Jacob (1989) cite some important features concerning the micrometeorological behavior of forests relative to grasslands: (1) Because of their greater aerodynamic roughness the evaporation rate of intercepted precipitation is higher over forests than over crops and grassland. (2) Transpiration rates over forests can be lower than those over crops and grasslands because the surface conductance of forest canopies is less than that of shorter vegetation. (3) Forests have a lower albedo than shorter vegetation, causing R_n to be 20% greater during summer and 10% greater during winter than for grasslands.

Is heat storage in forests a significant factor in diurnal weather dynamics? How much do changes in energy budgets with conversion from forest to grassland impact the conditions in the PBL?

Negative feedback of PBL processes on conditions controlling water vapor fluxes from vegetation will reduce the direct effects of climate change on water budgets of terrestrial ecosystems.

SCALING UP AND OTHER ASPECTS

The need for scaling up (integration, extrapolation) from short-time and small-area processes to long-time and large-area approaches has been addressed by some modeling groups. Reynolds et al. (1992) have developed a hierarchical set of simulators that provide a means for examining climate change effects on vegetation on a number of different scales. Martin (1992) has bridged a range of scales from physiology to forest succession with his EXE simulator.

On short time scales computations of canopy photosynthesis require detailed evaluation of the canopy light environment to evaluate the nonlinear response curve on sunlit and shaded leaves (Baldocchi 1993). On longer time scales such effects are ignored. Gutschick and Wiegel (1988) show that daily photosynthesis is insensitive to leaf angle, contrary to earlier predictions.

An alternative to scaling up from the small to the large scale is to model the large scale directly by the judicious choice of aggregated parameters that capture the attributes of lower-scale responses to environmental variables that impact the large scale. The use of these codes presents the challenge of estimating aggregated parameter values, and this is often accomplished by tuning model parameters such that predictions match empirical observation. This can be done for some variables, such as temperature effects, but not for others, such as atmospheric CO_2 level, because of the lack of experiments on the ecosystem scale.

Scaling Up by Linking Models

Temporal integration provided by simulation is a means for scaling up processes defined within the model framework. In other cases scaling up leads to new phenomena that are not present on the lower scale. These new phenomena can be accounted for by linkage of lower-scale model results to a larger-scale model that includes the new phenomena. One application of this technique was the propagation of a CO_2 enrichment response from the physiological scale to the scale of forest succession (Luxmoore et al. 1990).

Linkage of physiology, succession, and forest management models (Luxmoore 1992) offers one means for incorporating extensive forest inventory data into the analysis of global change impacts on forests and carbon budgets. In this approach a succession model is used to calculate the height of dominant and codominant trees at an index age (this gives the site index) that can be utilized as input in forest management models. The approach can also be adapted to link simulation of changes in tree height with age (e.g., from a JABOWA-type succession model) into forest management models such as PTAEDA (Burkhart et al. 1987) and NE-TWIGS (Hilt and Teck 1989). An equivalent approach could be adapted to calculate the yield classes (growth rates) in the carbon budget modeling reported by Dewar and Cannell (1992). The hypothesis that site index or yield class will change with rising CO_2 concentration and changing climate is the basis for this linked modeling approach.

Handling Heterogeneity

A wide range of plant and soil variable values exist within a forested region. A procedure called Latin hypercube sampling is an efficient means for propagating variability through a simulator. Examples of its use are given by King et al. (1989) and Luxmoore et al. (1991). The latter authors showed that annual photosynthate production for three atmospheric CO_2 levels had overlapping frequency distributions when a range of normally and lognormally distributed input variables were included in the simulation. Nevertheless, the photosynthesis results were statistically different for the three CO_2 cases. The Latin hypercube sampling method provides output variables in the form of frequency distributions that can be statistically compared for alternative modeling scenarios.

Solar Conversion Efficiency

The partitioning of net primary production into two terms (Monteith 1977) has been actively investigated in recent years.

$$\text{NPP [g m}^{-2}\text{ year}^{-1}] = \text{Solar conversion efficiency [g/MJ]} \times$$
$$\text{Absorbed radiation [MJ m}^{-2}\text{ year}^{-1}] \qquad (16.1)$$

Global changes could influence water and nitrogen availability to vegetation and induce changes in leaf area; leaf area adjustment is a major plant response to these variables. A secondary effect operates through water and nutrient stress effects on assimilation, which may reduce solar conversion efficiency. Linder (1985) found one conversion efficiency value for aboveground growth of *Pinus radiata* in a fertilization experiment showing the dominant effect of leaf area adjustment and a lack of a response in solar conversion efficiency to fertilization for the field conditions used.

   Remote sensing is one means for estimating leaf area of vegetation, and calculation of absorbed radiation by the canopy is feasible. NPP could then be predicted if suitable values of solar conversion efficiency were available. A number of existing models could be modified to determine solar conversion efficiency values, as has been done by McMurtrie and Wang (1993). They showed in simulation of *P. radiata* that the relationship between annual carbon gain and annual absorbed radiation had linear slopes that increased with increase in atmospheric CO_2 concentration. It would be useful to adapt forest succession models so that solar conversion efficiency could be calculated with change in stand maturity. An increase in forest biomass with stand age leads to a decline in solar conversion efficiency owing to an increase in respiratory load and perhaps to age effects (Saldarriaga and Luxmoore 1991).

Model Testing

Forest modeling has provided a number of ideas on the impacts of rising CO_2 level and changing climate on forests and the carbon budgets of forested land, but there has been very limited testing and validation. Although this work has been valuable, continued modeling without testing may become counterproductive.

   Micrometeorological measurements of canopy CO_2 fluxes provide a means of testing short-term physiological models of CO_2 exchange. Results from a recent test of a model that computes photosynthesis of a broadleaf deciduous forest suggest that it is possible to calculate short-term CO_2 fluxes of a broadleaf forest by scaling leaf-level information up to the canopy scale (Baldocchi 1993). However, similar exercises need to be conducted with conifer, savanna, boreal, and tropical forests in order for us to have reasonable confidence in our ability to scale leaf-level physiological relationships up to the canopy scale in forests with different architecture.

LINKAGE OF VEGETATION MODELS WITH GCMs

The attributes of a suitable vegetation model formulation that can be coupled to GCMs need to be determined. Can we abstract the canopy as a big leaf in order to get reasonable and defensible estimates of carbon exchange? Some big-leaf models are uncoupled from soil processes and apply to stress-free conditions (e.g., Sellers et al. 1992). Other big-leaf models incorporate plant responses to soil variables (e.g., Dixon et al. 1978; Martin 1992). It has been shown that soil water storage and litter nutrient dynamics can be critical factors in determining plant response to variations in temperature, precipitation, and soil heterogeneity. A big-leaf simulator with a source-sink framework will be robust for applications to heterogeneous landscapes having a range of stressors that impact above- and below-ground processes of vegetation.

Martin (1992) reviewed models of vegetation and climate and noted that the coupling of climate and vegetation interactions goes beyond the "piecing together" of existing models. The space and time basis of vegetation and climate processes must be constrained by the requirement that atmospheric information not be transferred faster through space than through time in climate models. In addition, ecological information on species migration in space must be transmitted fast enough to have an impact on landscape responses.

CONCLUSIONS

Results from a range of computer simulators suggest that some additional carbon storage is feasible over the next 50 years in response to rising atmospheric CO_2 concentration. Succession model results suggest that any forest growth response to CO_2 could be limited to the stand development phase and not to final stand production. A growth response to rising atmospheric CO_2 concentration acts as a negative feedback on global warming.

Global change may result in changes in vegetation distribution patterns, but there is no clear consensus on the projected outcome of global change for the distribution of vegetation. Any conversion of forest into grassland will change the energy budgets at the land surface, and these effects will affect the surface boundary conditions needed in GCMs.

It is not clear if rising global temperature will act through forests as a positive or a negative feedback to warming. Less carbon storage will result if respiration exceeds carbon fixation and a positive feedback effect will result. Ecosystem simulators that include nitrogen cycling suggest that warming will enhance carbon storage through increased forest growth in response to increased nitrogen mineralization in soil. Some critical experiments are required to evaluate the impacts of soil warming on soil respiration, mineralization of organic matter, and plant growth. The outcomes of these processes are identified from ecosystem modeling to be critical determinants of whether global warming will beget further warming.

ACKNOWLEDGMENTS

We thank Nicole Lepoutre-Baldocchi for drafting Figure 16.1. Work performed by D.D.B. was supported by USDOE, NOAA, and the RAISA project of the Consiglio Nazionale delle Ricerche, Italy. Research was sponsored in part by the Carbon Dioxide Research Program of the Office of Health and Environmental Research, U.S. Department of Energy, under contract DE-AC05–84-OR21400 with Martin Marietta Energy Systems, Inc., and in part by the Southern Global Change Program of the USDA Forest Service.

REFERENCES

Bacastow, R., and C. D. Keeling. 1973. Atmospheric carbon dioxide and radiocarbon in the natural carbon cycle. II. Changes from AD 1700 to 2070 as deduced from a geochemical model. In *Carbon and the Biosphere*, ed. G. M. Woodwell and E. V. Pecan, pp. 86–135. U.S. Atomic Energy Commission, Washington, D.C.

Baldocchi, D. D. 1993. Scaling water vapor and carbon dioxide exchange from leaves to a canopy: Rules and tools. In *Scaling Physiological Processes: Leaf to Globe*, ed. J. Ehleringer and C. Field, pp. 77–114. Academic Press, London.

Botkin, D. B., J. F. Janik, and J. R. Wallis. 1972. Some ecological consequences of a computer model of forest growth. *J. Ecol.* 60:849–872.

Bowes, M. D., and R. A. Sedjo. 1991. Processes for identifying regional influences of and responses to increasing atmospheric CO_2 and climate change—The MINK project. Report III—Forest resources. DOE/RL/01830T-H9. Carbon Dioxide Research Program, U.S. Department of Energy, Washington, D.C.

Burkhart, H. E., K. D. Farrar, R. L. Amateis, and R. F. Daniels. 1987. Simulation of individual tree growth and stand development in loblolly pine plantations on cutover, site-prepared areas. Report FWS-1-87, Virginia Polytechnic Institute and State University, Blacksburg, Va.

Cannell, M. G. J. 1990. Modelling the phenology of trees. *Silva Carel.* 15:11–27.

Charney, J. G. 1975. Dynamics of deserts and drought in the Sahel. *Q. J. R. Meteorol. Soc.* 101:193–202.

Dale, V. H., and J. F. Franklin. 1989. Potential effects of climate change on stand development in the Pacific Northwest. *Can. J. For. Res.* 19:1581–1590.

De Bruin, H. A. R., and C. M. J. Jacob. 1989. Forests and regional scale processes. *Phil. Trans. R. Soc. London B* 324:393–406.

Dewar, R. C., and M. G. R. Cannell. 1992. Carbon sequestration in the trees, products, and soils of forest plantations: An analysis using UK examples. *Tree Physiol.* 11:49–71.

Dickinson, R. E. 1989. Modeling the effect of Amazonian deforestation on regional surface climate: A review. *Agric. For. Meteorol.* 47:339–347.

Dixon, K. R., R. J. Luxmoore, and C. L. Begovich. 1978. CERES—A model of forest stand biomass dynamics for predicting trace contaminant, nutrient, and water effects. I. Model description. *Ecol. Mod.* 5:17–38.

Dougherty, P. M., P. Oker-Blom, T. C. Hennessey, R. E. Witter, and R. O. Teskey. 1990. An approach to modelling the effects of climate and phenology on the leaf biomass dynamics of a loblolly pine stand. *Silva Carel.* 15:133–143.

Farquhar, G. D., and S. von Caemmerer. 1982. Modelling of photosynthetic response to environmental conditions. In *Physiological Plant Ecology II: Water Relations and Carbon Assimilation, Encyclopedia of Plant Physiology, New Series*, Vol. 12B, ed. O. L. Lange, P. S. Nobel, C. B. Osmond, and H. Ziegler, pp. 549–587. Springer-Verlag, Berlin.

Gifford, R. M. 1992. Interactions of carbon dioxide with growth-limiting environmental factors in vegetation productivity: Implications for the global carbon cycle. *Advances in Bioclimatology*, with contributions by R. L. Desjardins, R. M. Gifford, T. Nelson, and E. A. N. Greenwood, pp. 24–58. Springer-Verlag, New York.

Gutschick, V. P., and F. W. Wiegel. 1988. Optimizing the canopy photosynthetic rate by patterns of investment in specific leaf mass. *Am. Nat.* 132:67–86.

Hilt, D. E., and R. M. Teck. 1989. NE-TWIGS: An individual-tree growth and yield projection system for the northeastern United States. *Compiler* 7(2):10–16.

Jacob, C. M. J., and H. A. R. de Bruin. 1992. The sensitivity of regional transpiration to land-surface characteristics: Significance of feedback. *J. Climate* 5:683–698.

Kauppi, P. E., K. Mielikainen, and K. Kuusela. 1992. Biomass and carbon budget of European forests, 1971 to 1990. *Science* 256:70–74.

Kienast, F. 1991. Simulated effects of increasing atmospheric CO_2 and changing climate on the successional characteristics of alpine forest ecosystems. *Land. Ecol.* 5:225–238.

Kienast, F., and B. Brezeziecki. 1993. Potential temporal and spatial responses of forest communities to climate change: Application of two simulation models for ecological risk assessment. *IUFRO World Series* 4:20–21. IUFRO, Vienna.

Kienast, F., and R. J. Luxmoore. 1988. Tree-ring analysis and conifer growth responses to increased atmospheric CO_2 levels. *Oecologia* 76:487–495.

King, A. W., R. V. O'Neill, and D. L. DeAngelis. 1989. Using ecosystem models to predict regional CO_2 exchange between the atmosphere and the terrestrial biosphere. *Glob. Biogeochem. Cycles* 3:337–361.

Linder, S. 1985. Potential and actual production in Australian forest stands. In *Research for Forest Management,* ed. J. J. Landsberg and W. Parsons, pp. 11–35. CSIRO, Canberra, Australia.

Long, S. P. 1991. Modification of the response of photosynthetic productivity to rising temperature by atmospheric CO_2 concentrations: Has its importance been underestimated? *Plant Cell Environ.* 14:729–739.

Long, S. P., and P. R. Hutchin. 1991. Primary production in grassland and coniferous forests with climate change: An overview. *Ecol. Appl.* 1:139–156.

Luxmoore, R. J. 1981. CO_2 and phytomass. *BioScience* 31:626.

Luxmoore, R. J. 1991. A source-sink framework for coupling water, carbon, and nutrient dynamics of vegetation. *Tree Physiol.* 9:267–280.

Luxmoore, R. J. 1992. An approach to scaling up physiological responses of forests to air pollutants. In *Responses of Southern Commercial Forests to Air Pollution,* ed. R. Flagler, pp. 321–330. Air and Waste Management Association, Pittsburgh.

Luxmoore, R. J., M. L. Tharp, and D. C. West. 1990. Simulating the physiological basis of tree-ring responses to environmental changes. In *Process Modeling of Forest Growth Responses to Environmental Stress,* ed. R. K. Dixon, R. S. Meldahl, G. A. Ruark, and W. G. Warren, pp. 393–401. Timber Press, Portland, Oreg.

Luxmoore, R. J., A. W. King, and M. L. Tharp. 1991. Approaches to scaling up physiologically based soil-plant models in space and time. *Tree Physiol.* 9:281–292.

McCaughey, J. H. 1985. Energy balance storage terms in a mature mixed forest at Petawawa, Ontario—A case study. *Bound. Layer Meteorol.* 31:89–101.

McGuire, A. D., J. M. Melillo, L. A. Joyce, D. W. Kicklighter, A. L. Grace, B. Moore III, and C. J. Vorosmarty. 1992. Interactions between carbon and nitrogen dynamics in estimating net primary production for potential vegetation in North America. *Glob. Biogeochem. Cycles* 6:101–124.

McMurtrie, R. E., and Y.-P. Wang. 1993. Mathematical models of the photosynthetic response of tree stands to rising CO_2 concentrations and temperatures. *Plant Cell Environ.* 16:1–13.

McNaughton, K. G. 1988. Regional interactions between canopies and the atmosphere. In *Plant Canopies: Their Growth, Form and Function,* ed. G. Russell, B. Marshall, and P. G. Jarvis, pp. 63–81. Cambridge University Press, Cambridge.

Martin, P. 1992. EXE: A climatically sensitive model to study climate change and CO_2 enhancement effects on forests. *Aust. J. Bot.* 40:717–735.

Martin, P. 1993. Vegetation responses and feedbacks to climate: A review of models and processes. *Climate Dyn.* 8:201–210.

Miller, D. R. 1992. Regional climate changes from increasing forest area. In *Forests and Global Change.* Vol. 1: *Opportunities for Increasing Forest Cover,* ed. R. N. Sampson and D. Hair, pp. 11–21. American Forests, Washington D.C.

Mohren, G. M. J. 1987. Simulation of forest growth, applied to Douglas fir stands in the Netherlands. Doctoral thesis. Agricultural University, Wageningen, The Netherlands.

Monteith, J. L. 1977. Climate and the efficiency of crop production in Britain. *Phil. Trans. R. Soc. London B* 281:277–294.

Moore, C. J., and G. Fisch. 1986. Estimating heat storage in Amazonian tropical forest. *Agric. For. Meteorol.* 38:147–169.

Nambiar, E. K. S., and D. N. Fife. 1991. Nutrient retranslocation in temperate conifers. *Tree Physiol.* 9:185–207.

Pastor, J., and W. M. Post. 1993. Linear regressions do not predict the transient responses to eastern North American forests to CO_2-induced climate change. *Climat. Change* 23: 111–119.

Polglase, P. J., and Y.-P. Wang. 1992. Potential CO_2-enhanced carbon storage by the terrestrial biosphere. *Aust. J. Bot.* 40:641–656.

Post, W. M., and J. Pastor. 1990. An individual-based forest ecosystem model for projecting forest response to nutrient cycling and climate changes. In *Forest Simulation Systems, Bull. 1927,* ed. L. C. Wensel and G. S. Biging, pp. 61–74. Berkeley, University of California.

Post, W. M., J. Pastor, A. W. King, and W. R. Emanuel. 1992. Aspects of the interaction between vegetation and soil under global change. *Water Air Soil Pollut.* 64:345–363.

Prentice, I. C., W. Cramer, S. P. Harrison, R. Leemans, R. A. Monserud, and A. M. Solomon. 1992. A global biome model based on plant physiology and dominance, soil properties and climate. *J. Biogeog.* 19:117–134.

Raich, J. W., and W. H. Schlesinger. 1992. The global carbon dioxide flux in soil respiration and its relationship to vegetation and climate. *Tellus* 44B:81–90.

Rastetter, E. B., M. G. Ryan, G. R. Shaver, J. M. Melillo, K. J. Nadelhoffer, J. E. Hobbie, and J. D. Aber. 1991. A general biogeochemical model describing the responses of the C and N cycles in terrestrial ecosystems to changes in CO_2, climate, and N deposition. *Tree Physiol.* 9:101–126.

Rauner, J. L. 1976. Deciduous forests. In *Vegetation and the Atmosphere,* Vol. 2, ed. J. L. Monteith, pp. 241–264. Academic Press, London.

Rauscher, H. M., J. G. Isebrands, G. E. Host, R. E. Dickson, D. I. Dickmann, T. R. Crow, and D. A. Michael. 1990. ECOPHYS: An ecophysiological growth process model for juvenile poplar. *Tree Physiol.* 7:255–281.

Reynolds, J. F., D. W. Hilbert, J.-L. Chen, P. C. Harley, P. R. Kemp, and P. W. Leadley. 1992. Modeling the response of plants and ecosystems to elevated CO_2 and climate change. DOE/ER-60490T-H1. Carbon Dioxide Research Program, U.S. Department of Energy, Washington, D.C.

Reynolds, J. F., D. W. Hilbert, and P. R. Kemp. 1993. Scaling ecophysiology from the plant to the ecosystem: A conceptual framework. In *Scaling Physiological Processes: Leaf to Globe,* ed. J. Ehleringer and C. Field, pp. 127–140. Academic Press, New York.

Ripley, E. A., and R. E. Redmann. 1976. Grassland. In *Vegetation and the Atmosphere,* Vol. 2, ed. J. L. Monteith, pp. 351–398. Academic Press, London.

Running, S. W., and R. R. Nemani. 1991. Regional hydrologic and carbon balance responses of forests resulting from potential climate change. *Climat. Change* 19:349–368.

Saldarriaga, J. G., and R. J. Luxmoore. 1991. Solar energy conversion efficiencies during succession of a tropical rain forest in Amazonia. *J. Trop. Ecol.* 7:233–242.

Schlesinger, W. H., J. F. Reynolds, G. L. Cunningham, L. F. Huenneke, W. M. Jarrell, R. A. Virginia, and W. G. Whitford. 1990. Biological feedbacks in global desertification. *Science* 247:1043–1048.

Sellers, P. J., J. A. Berry, G. J. Collatz, C. A. Field, and F. G. Hall. 1992. Canopy reflectance, photosynthesis, and transpiration. III. A reanalysis using improved leaf models and a new canopy integration scheme. *Remote Sensing Environ.* 42:187–216.

Smith, T. M., H. H. Shugart, G. B. Bonan, and J. B. Smith. 1992. Modeling the potential response of vegetation to global climate change. *Adv. Ecol. Res.* 22:93–116.

Solomon, A. M. 1986. Transient response modeling of forests to CO_2-induced climate change: Simulation modeling experiments in eastern North America. *Oecologia* 68:567–579.

Thornley, J. H. M., and M. G. R. Cannell. 1994. Nitrogen relations in a forest plantation—Soil organic matter ecosystem model. *Ann. Bot.* 70:137–151.

Thornley, J. H. M., D. Fowler, and M. G. R. Cannell. 1991. Terrestrial carbon storage resulting from CO_2 and nitrogen fertilization in temperate grasslands. *Plant Cell Environ.* 14:1007–1011.

Verma, S. B., D. D. Baldocchi, D. E. Anderson, D. R. Matt, and R. J. Clement. 1986. Eddy fluxes of CO_2, water vapor, and sensible heat over a deciduous forest. *Boundary-Layer Meteorol.* 36:71–91.

Verma, S. B., J. Kim, and R. J. Clement. 1989. Carbon dioxide, water vapor and sensible heat fluxes over a tallgrass prairie. *Boundary-Layer Meteorol.* 46:53–67.

Webb, W. L. 1991. Atmospheric CO_2, climate change, and tree growth: A process model. I. Model structure. *Ecol. Mod.* 56:81–107.

Weinstein, D. A., R. M. Beloin, and R. D. Yanai. 1991. Modeling changes in red spruce balance and allocation in response to interacting ozone and nutrient stresses. *Tree Physiol.* 9:127–146.

Wofsy, S. C., R. C. Harris, and W. A. Kaplan. 1988. Carbon dioxide in the atmosphere over the Amazon basin. *J. Geophys. Res.* 93:1377–1387.

Wullschleger, S. D. 1993. Biochemical limitations to carbon assimilation in C_3 plants—A retrospective analysis of the A/C_i curves from 109 species. *J. Exp. Bot.* 44:907–920.

17

Vegetation Geography and Global Carbon Storage Changes

I. COLIN PRENTICE AND MARTIN T. SYKES

The terrestrial biosphere currently stores $\approx$ 560 petagrams (Pg) ($=10^{15}$ g) carbon in plant biomass and a further $\approx$ 1500 Pg carbon in soil organic matter (these and other figures in this paragraph are from W. H. Schlesinger 1991 and are representative of values in the current literature). Residence times of this carbon range widely, from <1 year for deciduous leaves and fine roots to >1000 years for refractory soil compounds. About 120 Pg carbon are taken up from the atmosphere annually through (gross) primary production, and a similar amount is returned to the atmosphere through autotrophic and heterotrophic respiration. These annual fluxes amount to a considerable fraction of the 720 Pg carbon currently in the atmosphere. Atmospheric CO_2 takes centuries to equilibrate with the large reservoir of carbon in the ocean; thus, any consistent trend in the annual balance of primary production and respiration (net ecosystem production [NEP]) can significantly influence the evolution of atmospheric CO_2 over the time scales of global change.

Over the long term (thousands of years), global NEP might be expected to be balanced by the flux of carbon from the land to the sea in rivers, estimated at circa 0.4 Pg carbon year^{-1} (W. H. Schlesinger 1991). This flux provides an approximate estimate of the long-term average NEP. But there is no direct way to measure annual or decadal changes in NEP. One hypothesis, supported by some biogeochemical model results, attributes the "missing sink" for anthropogenic CO_2 emissions to CO_2 fertilization producing a transient increase in NEP. According to this hypothesis, increasing atmospheric CO_2 concentration will produce an increase in global net primary production (NPP), which in turn would produce a transient increase in NEP (Taylor and Lloyd 1992). The magnitude of this increase in NEP might be 1–2 Pg carbon year^{-1}, sufficient to account for the missing sink (Keeling et al. 1989; Kohlmaier et al. 1989; Gifford 1992; Taylor and Lloyd 1992). This hypothesis implies that the terrestrial biosphere is already providing a strong negative feedback to CO_2-induced climatic change.

However, there are several other terrestrial biospheric response mechanisms to consider, including mechanisms that could cause either positive or negative feedbacks. These mechanisms can be roughly classified by time scale:

MODERN

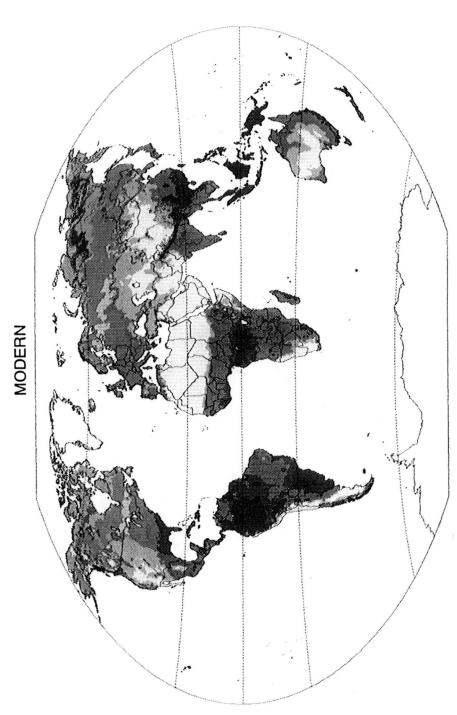

Figure 17.1. Biome distributions from modern climate data and AGCM scenarios for a doubling of CO_2 concentration.

OSU BIOMES

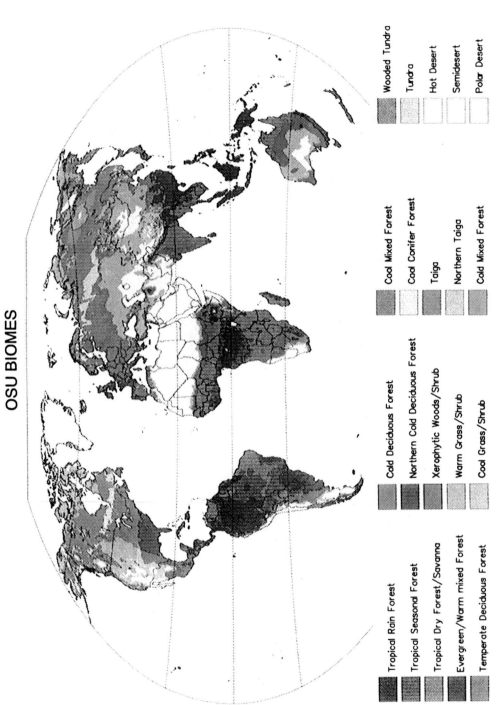

Tropical Rain Forest
Tropical Seasonal Forest
Tropical Dry Forest/Savanna
Evergreen/Warm mixed Forest
Temperate Deciduous Forest

Cold Deciduous Forest
Northern Cold Deciduous Forest
Xerophytic Woods/Shrub
Warm Grass/Shrub
Cool Grass/Shrub

Cool Mixed Forest
Cool Conifer Forest
Taiga
Northern Taiga
Cold Mixed Forest

Wooded Tundra
Tundra
Hot Desert
Semidesert
Polar Desert

Figure 17.1. *(Continued).*

GFDL BIOMES

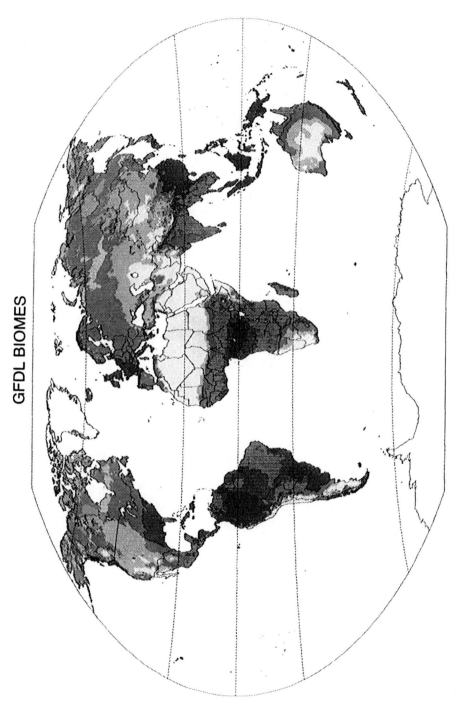

Figure 17.1. (*Continued*).

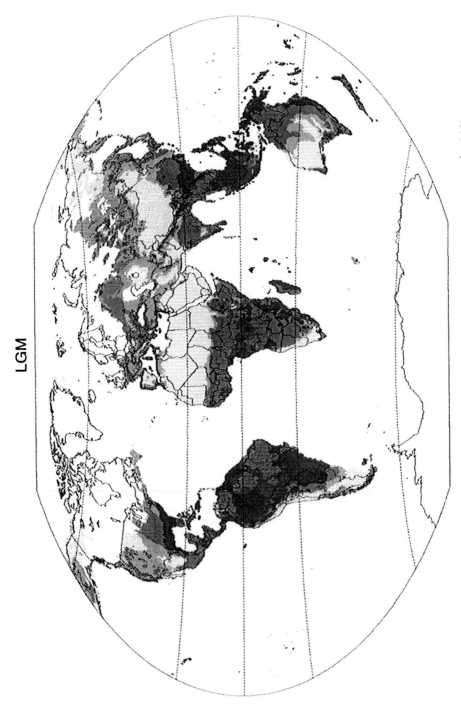

LGM

Figure 17.2. Biome distributions from an AGCM scenario for the last glacial maximum (I. C. Prentice et al. 1993).

1. On a time scale of years, CO_2 level and climatic changes can affect NPP and NEP through differential effects on primary production and respiration within existing ecosystems (Esser 1992). Such effects are not confined to CO_2 fertilization. For example, rising temperatures may increase the respiration flux from high-latitude, carbon-rich ecosystems such as tundra and boreal peatlands, implying a positive feedback (Oechel et al. 1993; Woodwell 1983).

2. On a time scale of centuries, climatic changes can cause shifts in the distribution of broad ecosystem types (biomes). Total terrestrial carbon storage will then change according to the balance of carbon uptakes and releases implied by these shifts (Esser 1987; I. C. Prentice 1994). Again there is no prior reason to expect these changes to imply generally positive or negative feedback. What happens will depend on the spatial structure of the climate change and the relationship between changes in temperature and precipitation on a regional and seasonal basis (I. C. Prentice and Sarnthein 1994). More rapid changes in biome distributions also cannot be ruled out, for example, large areas of forests may be transformed into grasslands through a combination of drought and drought-induced increases in the incidence of natural fires. Such events would cause a transient net release or "pulse" of carbon into the atmosphere (King and Neilson 1992; see also Chapters 1 and 6, this volume).

Here we are concerned with terrestrial carbon storage responses of type (2), brought about through climate-induced changes in the distribution of biomes. We use the global biome model of I. C. Prentice et al. (1992) to analyze the consequences of climatic changes simulated by atmospheric general circulation models (AGCMs) under a $2 \times CO_2$ climate. Our approach focuses on changes in equilibrium carbon storage implied by the biome shifts; we do not address the dynamics of the changes explicitly. The usefulness of our approach for examining future (high-CO_2) climate scenarios is qualitative or semiquantitative, indicating the direction of vegetation change on a regional basis and the general trend of changes in carbon storage associated with these changes.

The Holdridge (Emanuel et al. 1985; Henderson-Sellers 1990; Leemans 1990; K. C. Prentice 1990; K. C. Prentice and Fung 1990; Smith et al. 1992a,b), Köppen (Guetter and Kutzbach 1990), and Budyko (Monserud et al. 1994) schemes have previously been applied in a similar way, to estimate changes in potential biome distributions. The biome model may provide more accurate results than these earlier schemes because of its more physiologically mechanistic basis. Environmental limits to biomes in the model are inferred from the physiological and phenological requirements of different plant types; appropriate bioclimatic variables are used to define these limits, including a moisture index derived from a physically based algorithm similar to those commonly used in AGCMs. Changes in carbon storage are, however, still treated empirically, by assigning typical modern carbon storage values to biomes, as in previous studies (K. C. Prentice and Fung 1990; Smith et al. 1992a; I. C. Prentice et al. 1993).

The same modeling procedures can be applied to past climatic conditions, for which the forcing is understood, for testing against palaeodata. We illustrate this approach with an example for the last glacial maximum (LGM). The LGM has been the subject of many climate model simulations and is also a key time for the reconstruction of past environments.

METHODS

Half-degree-resolution biome simulations were derived from AGCM scenarios using the method described in I. C. Prentice et al. (1993). Baseline climate data were specified from Leemans and Cramer (1991) and soil texture data (for soil water capacity estimates) were taken from Zobler (1986). Sea-level and ice sheet distributions were kept as at present for the 2×CO$_2$ simulations and were specified from geological evidence (I. C. Prentice et al. 1993) for the LGM simulations.

AGCM anomalies (differences between "experiment" and control run for monthly mean temperatures, precipitation, and cloud cover) were first interpolated to the 0.5° grid, then added to the baseline data. This method goes some way toward eliminating AGCM biases and creates a more informative map than would be produced by confining the analysis to the coarser (AGCM) spatial resolution. However, it must be recognized that some of the resulting spatial detail may be spurious because the map resolution exceeds the resolution of the AGCM.

Global carbon storage values were calculated as in I. C. Prentice et al. (1994). Each biome was assigned "low," "medium," and "high" biomass carbon storage values from Olson et al. (1983) and soil carbon values (mean ± 2 standard errors to represent low, medium, and high) from data summarized in Zinke et al. (1984).

POTENTIAL BIOME DISTRIBUTIONS IN A 2×CO$_2$ CLIMATE

A set of four 2×CO$_2$ equilibrium climate change scenarios is in wide circulation in the climate impact modeling community. These scenarios were derived from simulations with the OSU (Schlesinger and Zhao 1989), GFDL (Manabe and Wetherald 1987), GISS (Hansen et al. 1988), and UKMO (Mitchell 1983) AGCMs. They were made with roughly comparable AGCMs incorporating a full seasonal cycle, a simple (mixed-layer) ocean model, and a simple ("bucket" type) land surface model. They all simulated the effects of increasing CO$_2$ to twice its preindustrial level. None of these model runs allowed for any feedback effects of possible vegetation changes on the atmosphere (e.g., through changing surface roughness, vegetation phenology, or canopy conductance), which could further alter climate.

There are general, qualitatitative similarities among the biome shifts simulated using these four sets of model results, but differences in their magnitude. Two scenarios were selected for carbon storage analysis (Figure 17.1). The UKMO results show biome shifts of similar magnitude to the GFDL results, whereas the GISS results show shifts of similar magnitude to the OSU results, but the former two model results show markedly greater shifts. Common to all four scenarios are a poleward shift of the taiga, cool conifer, cool mixed, temperate deciduous, and warm-temperate evergreen–warm mixed forest belts in the northern high and middle latitudes; a marked inland (eastward) shift of these forest belts in western Eurasia; a poleward shift of "warm" (probably C4-dominated) grass and shrublands; and a slight but consistent expansion of the areas occupied by tropical rain and seasonal forests. Close examination on a regional basis, however, shows inconsistencies, especially in the direction of shift in moisture-controlled boundaries in semiarid regions.

Table 17.1. Estimated Present (Potential Natural) Carbon
Storages, and Changes Caused by Greenhouse (2×CO$_2$) and
Ice Age (Last Glacial Maximum) Climates[a]

	Low	Medium	High
Present (from modern climate)			
Biomass	451	749	1042
Soil	1057	1333	1610
Total	1508	2082	2652
Changes caused by CO$_2$ level doubling (OSU model)			
Biomass	68	110	135
Soil	23	−7	−38
Total	91	103	97
Changes caused by CO$_2$ level doubling (GFDL model)			
Biomass	51	72	93
Soil	−2	−53	−104
Total	49	19	−11
Changes caused by glacial conditions (ECHAM model)			
Biomass	−78	−129	−168
Soil	−136	−185	−233
Total	−214	−314	−401

[a]Low, medium, and high refer to the set of carbon storage estimates applied to each biome.
All figures are in Pg carbon.

CHANGES IN POTENTIAL CARBON STORAGE CAUSED BY BIOME SHIFTS

The implications of these changes for equilibrium carbon storage are not immediately obvious; they depend on a balance of gains and losses over different areas. To give some indication of the uncertainty caused by carbon storage variability within biomes, Table 17.1 shows values obtained when all "low," all "medium," or all "high" carbon storage values were applied. Note that the reference point for these simulated changes in global carbon storage is the control run (in which the modern climate data were used to drive the biome model), i.e., an estimate of the carbon that would be stored by potential natural vegetation, rather than the actual present carbon storage, which is somewhat less. We do not consider the impact of present or future land use on these figures.

A key finding from Table 17.1 is that the potential for additional long-term carbon storage caused by biome shifts, if climate remained at its 2×CO$_2$ state, could be as much as circa 100 Pg carbon (equivalent to 50 μl liter^{-1} atmospheric CO$_2$), as shown in the OSU scenario. However, this potential would be less (perhaps nil) if climate change proceeded as far as is indicated by the GFDL scenario. This result suggests that the equilibrium carbon storage potential of the terrestrial biosphere may respond nonlinearly to the magnitude of global warming. Beyond a certain point, climate-induced biome shifts might turn the terrestrial biosphere from a sink into a source of carbon, if high-latitude warming proceeds far enough to reduce the taiga and tundra soil carbon inventories, as shown for the GFDL scenario.

The results summarized in Table 17.1 differ from those obtained by K. C. Prentice and Fung (1990), who calculated a larger carbon storage potential of 250–350 Pg. The reasons for this difference are unknown. Comparable calculations by Smith et al. (1992a) yielded carbon storage potentials somewhat greater than ours, but not as great as those indicated by K. C. Prentice and Fung. The difference between our figures and those of Smith et al. is probably due to the greater expansion of tropical moist forests predicted by the Holdridge scheme.

TRANSIENT EFFECTS

So far we have considered only differences in equilibrium carbon storage among climatic states. However, further nonlinearities in the biospheric reponse to climate change imply that the potential for ecosystems to act as a carbon sink on time scales of decades or centuries cannot be determined simply from these equilibrium values.

To estimate the carbon sink potential on these time scales one could perhaps consider only changes in biomass, treating soil carbon as constant (W. H. Schlesinger 1990). Table 17.1 then indicates a carbon storage potential of 60–140 Pg carbon for both climate models. According to this reasoning the soil carbon losses implied by the GFDL scenario would be deferred, owing to the long residence time of much soil carbon.

But this "optimistic" reasoning (in the sense that it implies a net terrestrial carbon sink, and therefore a negative feedback to the CO_2 level rise, on a time scale of centuries) is flawed because carbon sequestration and oxidation have very different time scales. The rate of carbon gain in ecosystems is limited by primary production (and in practice can be only be a small fraction of NPP), whereas carbon may be lost much more rapidly from both live vegetation and soils, especially when, for example, the transition from one biome to another is accelerated by wildfire. A "pessimistic" calculation would estimate the total *loss* of carbon from areas predicted to have lower carbon storage than at present (King and Neilson 1992). This value would represent a maximum loss of carbon, part of which would subsequently be taken up in areas predicted to have higher carbon storage than at present. In other words, there might be a net release of carbon on a time scale of decades, even if equilibrium carbon storage were set to increase. Smith and Shugart (1993) provided support for this prediction, using a simple compartment model with prescribed time constants for the kinetics of biomass and soil carbon during different types of biome transition. A more rigorous analysis awaits the development of a mechanistic global model incorporating vegetation dynamics.

PEATLANDS AS A POSSIBLE CARBON SOURCE

Peatlands occupy about a third of the total area of the taiga biome ($\approx 3.5 \times 10^6$ km^2) and can store several times more carbon per unit area than is implied by conventional soil carbon storage data (Gorham 1991; Malmer 1992). Gorham (1991) estimated a total storage of about 450 Pg carbon in peatlands, mostly within the taiga and cold deciduous forest biomes, as shown in Figure 17.1.

If we accept Gorham's estimate and assume that the total *equilibrium* carbon storage in peat is proportional to the combined area of these two biomes, then the simulated reduction in boreal forest areas with global warming would imply a small further loss, on the order of 10 Pg carbon, according to the OSU scenario, but a much larger loss, on the order of 100 Pg carbon, according to the GFDL scenario. These very rough calculations suggest that terrestrial carbon storage may be particularly sensitive to the magnitude of high-latitude warming. Furthermore, peatland accumulation in climatically favorable areas is expected to be much slower than peatland destruction in climatically unfavorable areas (Gorham 1991), implying that peatlands may additionally contribute to a transient net release of carbon if climate change proceeds too far or too fast.

VEGETATION AND CARBON STORAGE AT THE LAST GLACIAL MAXIMUM

It is somewhat reassuring to turn from the future (where uncertainty rules) to the past, where there is a large body of data on the distribution of vegetation at all time intervals from the LGM (18,000 years B.P. on the radiocarbon time scale; circa 21,000 years B.P. by astronomical time) to the present (COHMAP members 1988). The LGM was characterized by extensive ice sheets in northern midlatitudes, a sea level 120 m lower than that at present, sea-surface temperatures generally colder than those at present, more extensive sea ice, and low atmospheric CO_2 levels ($\approx$185 μl liter^{-1}) (Bartlein 1988). The earth's orbital configuration was similar to that at present, the state of glaciation being a consequence of antecedent insolation conditions. These "glacial boundary conditions" can be provided to AGCMs as a means of simulating the glacial climate.

Although the LGM represents a cold, low-CO_2 world, it does not represent a mirror image of the future warm, high-CO_2 world. The purpose of comparing data and simulations for the LGM is to test the performance of the whole modeling procedure used to project climate, biome distributions, and carbon storage into the future. The problem is simplified because on this time scale the equilibrium approximation is reasonable for biome distributions (I. C. Prentice 1992; Webb 1992), biomass, and soil carbon storage. Simulated biome distributions for the LGM can be compared with vegetation reconstructions based on pollen data from lake sediment cores, and the implied net change in terrestrial carbon storage can be compared with ocean-carbon isotopic ratio reconstructions based on measurements on benthic foraminifera from deep-sea cores (Crowley 1991; I. C. Prentice and Sarnthein 1994).

Figure 17.2 shows results obtained from an LGM simulation made in conjunction with the Max Planck Institute for Meteorology's ECHAM 1 (T21) AGCM (Lautenschlager and Herterich 1990), a model of comparable structure and resolution to those used for the 2×CO_2 simulations in this chapter. A systematic comparison of the biome shifts shown by Figure 17.2 will have to await the development of a global pollen-based biome data set for 18,000 years B.P., a project actively being planned in the palynological community. Existing map reconstructions such as that of Adams et al. (1990) rely too much on subjective inference and extrapolation to be used in a quantitative way. However, Figure 17.2 shows features that are strikingly consistent with the palaeorecord in a qualitative sense, including poleward shifts and compression of the northern midlatitude vegetation belts, major reduction of the taiga biome, and fragmentation of the African rain forest (Maley 1991). The changes shown for Amazonia are consistent with data suggesting encroachment by dry forest and savanna to the north and east and temperate tree taxa to the north (van der Hammen 1991; I. C. Prentice et al. 1994). The simulation also shows expanded tropical rain forests on the exposed continental shelves;

it is not known whether such forests existed or not because the direct palynological evidence has been obliterated.

Carbon storage calculations (Table 17.1) suggest that the continental biosphere stored about 400 Pg carbon less at the LGM than in preagricultural time (Friedling-stein et al. 1992). The "extra" carbon stored on land now inundated was not enough to compensate for the shortfall of carbon in the areas that either were then glaciated (and are now occupied by relatively carbon-rich biomes such as taiga and tundra) or were occupied by biomes with lower carbon storage than today, e.g., drier types of tropical forest instead of present-day rain forest. (Again there is a puzzling discrepancy from the findings of K. C. Prentice and Fung [1990], who found no change from LGM to the present.)

Peatlands were less extensive at the LGM than today. If we make the same rough calculation as for the 2×CO$_2$ simulation, again using Gorham's (1991) estimates of present peatland carbon storage, we find that the ice age world was missing a further 300 Pg carbon. Even with more conservative assumptions about total peatland carbon storage, I. C. Prentice et al. (1994) found a shortfall of 100 Pg carbon, bringing the total carbon deficit to 300–700 Pg carbon depending on the figures used. This is consistent with the well-documented increase in the δ^{13}C of deep ocean waters from the LGM to present (Shackleton 1977; Duplessy et al. 1988), which implies that terrestrial carbon storage increased by 400–500 Pg carbon after the LGM (Crowley 1991; I. C. Prentice et al. 1993). So there is, at least, a rough correspondence between our model-based calculations and calculations based on independent geochemical data. In terms of the long-term dynamics of the carbon cycle, it seems probable that the terrestrial biosphere was a net sink for several hundred petagrams of carbon released from the ocean during the several thousand years following the LGM and that this sink arose (in spite of opposite effects caused by sea-level rise) as a result of climate-induced biome shifts (I. C. Prentice 1994).

REFERENCES

Adams, J. M., H. Faure, L. Faure-Denard, J. M. McGlade, and F. I. Woodward. 1990. Increases in terrestrial carbon storage from the Last Glacial Maximum to the present. *Nature* 348:711–714.

Bartlein, P. J. 1988. Late-Tertiary and Quaternary palaeoenvironments. In *Vegetation History,* ed. B. Huntley and T. Webb III, pp. 113–152. Kluwer, Dordrecht, The Netherlands.

COHMAP members. 1988. Climatic changes of the last 18,000 years: Observations and model simulations. *Science* 241:1043–1052.

Crowley, T. J. 1991. Ice age carbon. *Nature* 352:575–576.

Duplessy, J. C., N. J. Shackleton, R. G. Fairbanks, L. Labeyrie, D. Oppo, and N. Kallel. 1988. Deepwater source variations during the last climatic cycle and their impact on the global deepwater circulation. *Paleoceanography* 3:343–360.

Emanuel, W. R., H. H. Shugart, and M. P. Stevenson. 1985. Climate change and the broad-scale distribution of terrestrial ecosystem complexes. *Climat. Change* 7:29–43.

Esser, G. 1987. Sensitivity of global carbon pools and fluxes to human and potential climatic impacts. *Tellus* 39B:245–260.

Esser, G. 1992. Implications of climate change for production and decomposition in grasslands and coniferous forests. *Ecol. Appl.* 2:47–54.

Friedlingstein, P., C. Delire, J. F. Müller, and J. C. Gérard. 1992. The climate induced variation of the continental biosphere: A model simulation of the last glacial maximum. *Geophys. Res. Lett.* 19:897–900.

Gifford, R. M. 1992. Implications of CO_2 effects on vegetation for the global carbon budget. In *The Global Carbon Cycle,* ed. M. Heimann, pp. 159–190. Wiley, New York.

Gorham, E. 1991. Northern peatlands: Role in the carbon cycle and probable responses to global warming. *Ecol. Appl.* 1:182–195.

Guetter, P. J., and J. E. Kutzbach. 1990. A modified Köppen classification applied to model simulations of glacial and interglacial climates. *Climat. Change* 16:193–215.

Hansen, J., I. Fung, A. Lacis, D. Rind, G. Russell, S. Lebedeff, and R. Ruedy. 1988. Global climate changes as forecast by the GISS-3-D model. *J. Geophys. Res.* 93:9341–9364.

Henderson-Sellers, A. 1990. Predicting generalized ecosystem groups with the NCAR GCM: First steps towards an interactive biosphere. *J. Climate* 3:917–940.

Keeling, C. D., R. B. Bacastow, A. F. Carter, S. C. Piper, T. P. Whorf, M. Heimann, W. G. Mook, and H. Roeloffzen. 1989. A three-dimensional model of atmospheric CO_2 transport based on observed winds: 1. Analysis of observational data. In *Aspects of Climate Variability in the Pacific and the Western Americas,* ed. D. H. Peterson, pp. 165–236. American Geophysical Union, Washington, D.C.

King, G. A., and R. P. Neilson. 1992. The transient response of vegetation to climate change: A potential source of CO_2 to the atmosphere. *Water Air Soil Pollut.* 65:365–383.

Kohlmaier, G. H., E. Siré, A. Janecek, C. D. Keeling, S. C. Piper, and R. Revelle. 1989. Modelling the seasonal contribution of a CO_2 fertilization effect on the terrestrial vegetation to the amplitude increase in atmospheric CO_2 at Mauna Loa Observatory. *Tellus* 41B:487–510.

Lautenschlager, M., and K. Herterich. 1990. Atmospheric response to ice age conditions—Climatology near the earth's surface. *J. Geophys. Res.* 95:22,547–22,557.

Leemans, R. 1990. *Possible Changes in Natural Vegetation Patterns due to a Global Warming.* IIASA, Laxenburg, Austria.

Leemans, R., and W. Cramer. 1991. *The IIASA Climate Database for Mean Monthly Values of Temperature, Precipitation and Cloudiness on a Terrestrial Grid.* IIASA, Laxenburg, Austria.

Maley, J. 1991. The African rain forest vegetation and palaeoenvironments during the Late Quaternary. *Climatic Change* 19:79–98.

Malmer, N. 1992. Peat accumulation and the global carbon cycle. *Catena* (Suppl.) 22:97–110.

Manabe, S., and R. T. Wetherald. 1987. Large scale changes in soil wetness induced by an increase in carbon dioxide. *J. Atmos. Sci.* 44:1211–1235.

Mitchell, J. F. B. 1983. The seasonal response of a general circulation model to changes in CO_2 and sea temperature. *Q. J. R. Meteorol. Soc.* B109:113–152.

Monserud, R. A., N. M. Tchebakova, and R. Leemans. 1994. Global vegetation change predicted by the modified Budyko model. *Climatic Change.* In press.

Oechel, W. C., S. J. Hastings, G. Vourlitis, M. Jenkins, G. Riechers, and N. Gruike. 1993. Recent change of Arctic tundra ecosystems from a net carbon dioxide sink to a source. *Nature* 361:520–523.

Olson, J. S., J. A. Watts, and L. J. Allison. 1983. *Carbon in live vegetation of major world ecosystems.* Oak Ridge National Laboratory, Oak Ridge, Tenn.

Prentice, I. C. 1992. Climate change and long-term vegetation dynamics. In *Plant Succession: Theory and Prediction,* ed. D. C. Glenn-Lewin, R. A. Peet, and T. Veblen, pp. 293–339. Chapman and Hall, London.

Prentice, I. C. 1994. Biome modelling and the carbon cycle. In *The Global Carbon Cycle,* ed. M. Heimann. Wiley, New York.

Prentice, I. C. and M. Sarnthein. 1994. Self-regulatory processes in the biosphere in the face of climate change. In *Global Changes in the Perspective of the Past,* ed. J. Eddy and H. Oeschger, pp. 29–38. Wiley, New York.

Prentice, I. C., W. Cramer, S. P. Harrison, R. Leemans, R. A. Monserud, and A. M. Solomon. 1992. A global biome model based on plant physiology and dominance, soil properties and climate. *J. Biogeog.* 19:117–134.

Prentice, I. C., M. T. Sykes, M. Lautenschlager, S. P. Harrison, O. Denissenko, and P. J. Bartlein. 1993. Modelling global vegetation patterns and terrestrial carbon storage at the last glacial maximum. *Glob. Ecol. Biogeog. Lett.* 3:67–76.

Prentice, K. C. 1990. Bioclimatic distribution of vegetation for general circulation model studies. *J. Geophys. Res.* 95(D8):11,811–11,830.

Prentice, K. C., and I. Y. Fung. 1990. The sensitivity of terrestrial carbon storage to climate change. *Nature* 346:48–50.

Schlesinger, M. E., and Z. C. Zhao. 1989. Seasonal climatic changes induced by doubled CO_2 as simulated by the OSU atmospheric GCM/mixed-layer ocean model. *J. Climate* 2:459–495.

Schlesinger, W. H. 1990. Evidence from chronosequence studies for a low carbon-storage potential of soils. *Nature* 348:232–234.

Schlesinger, W. H. 1991. *Biogeochemistry: An Analysis of Global Change.* Academic Press, New York.

Shackleton, N. J. 1977. Carbon-13 in Uvigerina: Tropical rain forest history and the equatorial Pacific carbonate dissolution cycles. In *The Fate of Fossil Fuel Carbon in the Ocean,* ed. N. R. Anderson and A. Malahoff, pp. 401–428. Plenum Press, New York.

Smith, T. M., and H. H. Shugart. 1993. The transient response of terrestrial carbon storage to a perturbed climate. *Nature* 361:523–526.

Smith, T. M., R. Leemans, and H. H. Shugart. 1992a. Sensitivity of terrestrial carbon storage to CO_2 induced climate change: Comparison of four scenarios based on general circulation models. *Climatic Change* 21:367–384.

Smith, T. M., H. H. Shugart, G. B. Bonan, and J. B. Smith. 1992b. Modelling the potential response of vegetation to global climate change. *Adv. Ecol. Res.* 22:93–116.

Taylor, J., and J. Lloyd. 1992. Sources and sinks of atmospheric CO_2. *Aust. J. Bot.* 40:407–418.

van der Hammen, T. 1991. Palaeoecological background: Neotropics. *Climatic Change* 19:37–47.

Webb, T., III. 1992. Past changes in vegetation and climate: Lessons for the future. In *Global Warming and Biological Diversity,* ed. R. L. Peters and T. E. Lovejoy, pp. 59–75. Yale University Press, New Haven, Conn.

Woodwell, G. M. 1983. Biotic effects on the concentration of carbon dioxide: A review and projection. In *Changing Climate,* pp. 216–241. National Academy of Science Press, Washington, D.C.

Zinke, P. J., A. G. Stangenberger, W. M. Post, W. R. Emanuel, and J. S. Olson. 1984. *Worldwide Organic Soil Carbon and Nitrogen Data.* Oak Ridge National Laboratory, Oak Ridge, Tenn.

Zobler, L. 1986. *A World Soil File for Global Climate Modeling.* NASA, Washington, D. C.

18

CO$_2$-Climate Feedbacks: Aspects of Detection

I. G. Enting

The IPCC (1990) report includes a number of projections of CO$_2$ concentrations resulting from various possible CO$_2$ release scenarios. A common feature of these projections is the assumption that the processes determining atmospheric CO$_2$ levels remain unchanged. In particular, various possible feedbacks between CO$_2$ and climate are neglected. This approximation is probably adequate for short-term projections. However, some study of feedback processes is necessary for longer-term studies and even for defining what is actually short-term versus long-term (see also Broecker 1987).

Although the calculations in the IPCC report did not take feedbacks into account, a number of possible feedbacks were listed for consideration. These were as follows (using the numbering corresponding to the subsections of Section 1.2.7 [page 16] of the IPCC report):

Oceanic effects

1.1. Sea-surface temperature—PCO$_2$'s being elevated by the increase in sea-surface temperature (SST) of surface waters.

1.2. Ocean circulation—Changes in ocean circulation as part of an overall change in atmosphere-ocean dynamics could affect atmospheric CO$_2$ levels through changes in uptake of anthropogenic CO$_2$ and also by influencing the biogeochemical cycling of carbon in the oceans (see 1.4). Ocean circulation changes will reflect climate change through wind stress, temperature, and the effect of hydrologic changes on salinity.

1.3. Effect of winds on gas exchange rates—This is regarded as a minor effect since air-sea gas exchange plays little part in determining the rate of oceanic uptake of CO$_2$.

1.4. Ocean biogeochemical cycling—Changes could occur from either circulation changes or the direct effects of temperature changes.

1.5. UV-B—The effect of UV-B radiation on marine productivity could affect the biological pumping of ocean carbon and thus affect the atmospheric CO$_2$ concentration.

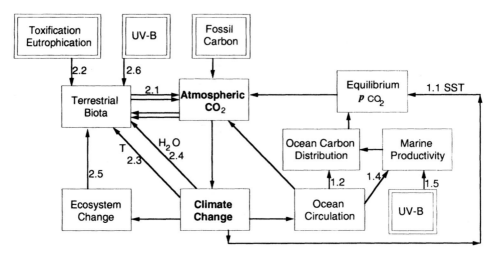

Figure 18.1. Feedbacks affecting atmospheric CO_2, as listed in IPCC (1990). Numbers refer to subsections of Section 1.2.7 of that report. The double arrows denote the carbon cycle feedback, which does not require climate change in order to operate. Double boxes signify direct and indirect anthropogenic forcings.

Effects on the terrestrial biota

2.1. CO_2-enhanced growth—The effect of increased CO_2 levels in enhancing plant growth has been extensively demonstrated in laboratory studies and commercial operations. The extrapolation of this effect to the ecosystem scale is subject to very great uncertainties.

2.2. Eutrophication and toxification—Anthropogenic releases of a multitude of compounds, either as nutrients promoting increased growth or as toxins inhibiting growth, may have direct effects on terrestrial ecosystems.

2.3. Temperature—Temperature increases will have a range of effects on biological processes. The IPCC (1990) report suggests that the dominant effect will be an increased rate of respiration with increased temperature.

2.4. Water—Changes in the hydrologic cycle will have a direct effect on many terrestrial ecosystems.

2.5. Shifts in ecosystems—Climatic changes will lead to changes in the spatial distribution of ecosystem types. During periods of rapid change the adaptation may not be fast enough to sustain the biomass at current levels.

2.6. UV-B—Increased levels of UV-B may be harmful to many terrestrial plants.

Figure 18.1 gives a schematic representation of these feedbacks. One important point is that, within the context of the CO_2-climate problem, three of the processes are not really feedbacks but are instead external forcings. The eutrophication effect and the effects of UV-B radiation on marine and terrestrial ecosystems are the results of anthropogenic disturbances not directly related to CO_2. However, like the actual feedbacks, these processes are still indirect consequences of anthropogenic change, and the issues involved in modeling and/or detecting these processes are similar to the issues involved in studying the feedbacks. A consideration of biotic feedbacks on the global carbon cycle must also take account of a range of other anthropogenic influences—deforestation, land degradation, hydrologic changes, and toxification—through their effects on the terrestrial biota.

A second important distinction is that the possible effect of CO_2 in inducing enhanced growth of the terrestrial biota does not involve climatic change as an intermediate step. This means that the feedback can occur without being subject to the delays inherent in the climatic response. (The directness of the interaction also simplifies the process of modeling this feedback, and indeed many carbon cycle modeling studies have taken it into account.) Because this is a special case, it seems worthwhile to introduce a distinguishing terminology. We suggest the term "carbon cycle feedback" for the feedback processes in which excess CO_2 leads to an increase in net biotic CO_2 uptake and the term "CO_2-climate feedback" to refer to the remaining feedback processes in which climatic change is an intermediate stage of the feedback loop. Another distinction between the carbon cycle feedback and CO_2-climate feedback is that the carbon cycle feedback is apparently a significant negative feedback, whereas many of the processes identified as having potential for CO_2-climate feedback produce positive feedback.

For most of the processes involved in the feedback loops, there is little scope for direct measurement of the interactions on a global scale. Furthermore, although some preliminary modeling has been undertaken (e.g., Lashof 1989), the information required for such modeling is often imprecise. Therefore an important part of the study of such feedbacks will be the detection of those that produce significant changes (or inhibit significant changes). Part of such detection will be quantitative—detection of discrepancies between observed carbon cycle behavior and that predicted in the absence of feedbacks. In such detection of anomalous behavior, the quantification of uncertainties will be extremely important in order to determine whether apparently anomalous behavior is in fact anomalous or whether it is consistent with the absence of feedbacks. In such studies, techniques such as those of Enting (1992a) (a consistency analysis of disparate carbon budget information using linear programming techniques) and Enting (1993) (quantification of the range of uncertainty in regional budgets derived from transport modeling) will be essential.

In some cases it will be possible to obtain qualitative indications of the presence of feedback processes. For example the analysis by Hansen et al. (1985) is quite general and not restricted to climatic systems. They show that the characterisic response times of a system will increase as the amount of positive feedback increases. Other generic relations may involve isotopic information. The section on "CO_2-Climate Feedback" describes an analysis based on a violation of generic properties of linear steady-state systems.

In the remainder of this chapter I consider carbon cycle feedback, review an analysis of possible detection of a past CO_2-climate feedback, and discuss the analysis of spatial distributions of CO_2.

CARBON CYCLE FEEDBACK

The most common characterization of the global carbon cycle is in terms of the atmospheric carbon budget. This is usually expressed in terms of five main components:

Table 18.1. Scope of Actual and Potential Independent
Determination of Components of the Atmospheric Carbon Budget

Component	Measurement possibility	Modeling possibility
Atmospheric increase	Good	No[a]
Fossil	Not practical; Using economic records	
Deforestation	Remote sensing	Using agricultural records
Oceanic uptake	P_{CO_2}	Box models, ocean general circulation models
CO_2 fertilization	Not practical	Difficult to validate

[a]The atmospheric CO_2 concentration cannot be modeled independently of a knowledge of the other components.

fossil fuel use, oceanic uptake, deforestation, CO_2 fertilization, and atmospheric increase. The precise specification of each of these requires more careful description. In particular:

Fossil—This includes the CO_2 release from both fossil fuel use and cement production.

Oceanic uptake—The work of Smith and Mackenzie (1987), Wollast and Mackenzie (1989), and Sarmiento and Sundquist (1992) has brought out the fact that the net air-sea gas exchange differs from net carbon storage because of the role of rivers in transporting carbon into the oceans (see also Chapter 2).

Deforestation—The estimates produced by Houghton et al. (1983 and subsequent work) refer to the net carbon release due to land-use change and not simply the gross release (see also Chapter 19).

CO_2 fertilization—To produce a closed budget, this component must include all biotic processes not associated with land use change. Although CO_2-induced growth may be the largest of these, other effects will be those of nitrogen fertilization and degradation due to toxic pollutants (see Chapters 3, 4, and 11).

Atmospheric increase—This change can be measured directly.

The CO_2 fertilization component provides the carbon cycle feedback. Wigley and Raper (1992) and many others have analyzed (see Chapters 3, 4, and 16, this volume) the importance of this component in projections of future CO_2 concentrations.

The same five components provide a basis for discussing regional budgets except that the measured atmospheric increase must be replaced by a component representing the net regional source as deduced from transport modeling. Potentially, there are three main ways in which each component can be estimated: direct measurement, process modeling, and deduction from the other four components. Table 18.1 lists some of the possibilities for estimating the fluxes by direct measurement and modeling. It is also important that, for looking at the components of past budgets, direct measurement is generally precluded. The most notable exception is the recovery of past atmospheric CO_2 concentrations from bubbles in polar ice. However, on the longer time scales, progress has been made in reconstructing past biotic distributions and ocean carbon contents from paleo-records.

Enting (1992a) has reviewed the various estimates of the components of the atmospheric carbon budget, both on a global scale and in terms of a three-region division. The

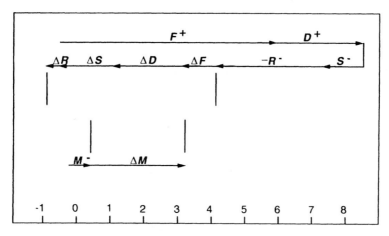

Figure 18.2. Vector representation of the atmospheric carbon budget for 1980–1990, based on the IPCC (1990) report. The upper range is the "worst-case" uncertainty for the imbalance, M, obtained by summing uncertainties in components. The lower range ΔM is the range presented in the IPCC report: a root-sum-of-squares combination of the uncertainties.

estimates and associated ranges of uncertainty were combined using linear programming to combine the various constraints. The constraints can be represented in terms of vector sums, as shown in Figure 18.2. For the fossil component, F, we define the maximum and minimum as F^+ and F^- with the range $\Delta F = F^+ - F^-$. A similar notation is used for deforestation, CO_2-enhanced growth, and air-sea flux: D, G, and S, respectively. When adding source-sink components with specified ranges, the sums are reordered so that the uncertainties (e.g., ΔF) are brought together to give a combined range of uncertainty. Figure 18.2 shows the budget from the IPCC (1990) report presented as

$$\text{Fossil} + \text{Deforestion} - \text{Air-sea flux} - \text{Net release} = \text{Missing sink} \qquad (18.1)$$

where the net release, R, is equal to the observed atmospheric increase.

The same form of vector representation is used to present the regional budget in the section on "Regional Carbon Budgets."

CO₂-CLIMATE FEEDBACK

Detection of any CO_2-climate feedback will need to take the form of a departure from the behavior expected in the absence of such a feedback. As noted previously, the requirement for balance of the atmospheric carbon budget provides one constraint. This can serve to help determine the most poorly defined component—currently the carbon cycle feedback. However, if the atmospheric carbon budget is being used to determine the strength of the carbon cycle feedback then it cannot be used to detect any additional CO_2-climate feedbacks until their effects become so large as to lie outside the range of possible CO_2-fertilization, i.e., outside a range of roughly 0–3 petagrams (Pg) (=10^{15} g) carbon per year. Detection of less extreme CO_2-climate feedbacks will require identifying special aspects of behavior that provide a signature of the feedback.

The studies by Enting and Mansbridge (1987a) and Enting (1992b) give an example of how the special properties of the carbon system without feedback can be used as a basis for detecting feedbacks on the basis of violations of these properties. The special property is the linear steady-state property. This implies that the oceanic uptake can be expressed in terms of a response function, $R(t)$, that gives the amount of carbon remaining in the atmosphere at a time t after a unit input. This allows the CO_2 concentrations resulting from inputs $S(t)$ to be written as

$$C(t) = C\left(t_0\right) + \int_{t_0}^{t} R(t - t')\, S(t')\, dt' \qquad (18.2)$$

Forward modeling involves using a model, or equivalently a specification of $R(t)$, and a release history, $S(t)$, to deduce concentrations, $C(t)$. Inverse modeling involves using the model and the concentration history to deduce the sources, $S(t)$ (e.g., Siegenthaler and Oeschger 1987). The approach used by Enting and Mansbridge (1987a) was to use the concentration history, $C(t)$, from ice core data and source estimates, $S(t)$ (from Houghton et al. 1983; Rotty 1987) to attempt to deduce the model response, $R(t)$. Their result was that they could not find any response function that had the following properties:

1. The fit to the CO_2 data was acceptable.
2. $R(t) \geq 0$.
3. $dR(t)/dt \leq 0$.
4. $d^2R(t)/dt^2 \geq 0$.

The actual calculation was set up as a linear programming problem that showed that the constraints could not be simultaneously satisfied. It was the set of constraints as a whole that could not be fitted. No particular constraint could be identified as being uniquely anomalous. However, one aspect that contributed to the inconsistency was that the ratio of increase to emission rate was greater in the 19th century than the 20th.

The calculation by Enting and Mansbridge (1987a) did not include the possibility of the "carbon cycle feedback," i.e., CO_2-enhanced growth, except to the extent that the carbon cycle feedback can be modeled in a linear way. For a linear description of the carbon cycle feedback, it is possible to define a combined response function that specifies the net effect of oceanic CO_2 uptake and CO_2-enhanced growth. (The formalism for combining response functions is given by Enting [1991]; see also Appendix B.) The later study (Enting 1992b) treated the carbon cycle feedback separately. (It also refined the mathematical treatment in minor ways.) The separation made it possible to enforce the requirement that the so-called β factor be positive. The result was that β factors greater than zero gave greater discrepancies than were seen for the case with no carbon cycle feedback (i.e., $\beta = 0$).

In the original study, Enting and Mansbridge (1987a) reviewed the possible reasons for the discrepancy and favored errors in the biotic source estimates as the most likely explanation. The more recent study by Enting (1992b) suggested that the discrepancy could be due to a climatic influence on the carbon cycle, possibly associated with the Little Ice Age. There is a small amount of additional evidence supporting that view, as shown in Figure 18.3. The CO_2 concentrations measured in air extracted from bubbles in polar ice suggest that prior to the anthropogenic increase, there was a slight decline in concentration from around 1600 to 1750 (although the timing is poorly defined). On the

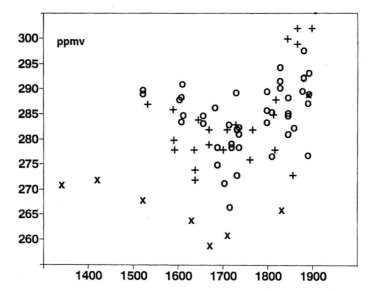

Figure 18.3. Composite CO_2 record from ice cores. +, Data from Wahlen et al. (1991); ∘, data from Etheridge et al. (1988); ×, data from Raynaud and Barnola (1985), which appear to have a systematic offset in concentration (see Raynaud 1993). Although none of the records taken alone shows any statistically significant decline from 1600 to 1750, the agreement among all the cases suggests the reality of such a feature.

basis of this interpretation, the anomalously high rates of increase of CO_2 during the last century would reflect a recovery from this decline, superimposed on the effects of anthropogenic CO_2 emissions.

DETECTION FROM ATMOSPHERIC DATA

The majority of recent studies of the atmospheric carbon budget have drawn heavily on the constraints implied by atmospheric transport modeling (e.g., Keeling et al. 1989; Tans et al. 1990; Sarmiento and Sundquist 1992). The principle is that the spatial distribution of atmospheric constituents can be used to deduce the spatial distribution of sources and sinks, which in turn constrains the possible source-sink processes. Enting and Pearman (1993) and Tans (1992) have described the global networks that have been established to obtain the requisite data.

Given estimates of the net source-sink strengths on a regional basis, the approaches described previously for global budgeting can be applied to regional budgeting. The form of the deduction will vary from region to region depending on which components are best determined. For example, Tans et al. (1990) used PCO_2 data to estimate oceanic CO_2 uptake in northern and tropical regions but used the results of transport modeling to deduce oceanic uptake in southern regions. The approach can be extended by the use of ^{13}C data to give an extra class of constraint.

The detection of any CO_2-climate feedback from atmospheric CO_2 data will have to be in terms of anomalous departures from the regional budgets expected on the basis of no feedback. Therefore, the detection of such anomalies requires a quantification of the uncertainties in the budgeting. In particular, it is necessary to estimate the range of uncertainty in the net sources determined by transport modeling. This is a matter of some considerable difficulty, and it is discussed, together with some preliminary results from Enting (1993), in the following section.

A more qualitative illustration of the difficulties in identifying the onset of feedbacks comes from the work of Enting and Mansbridge (1991). They noted that their inversion calculation gave a relatively small southern sink, in common with most other recent modeling studies. However, the earlier study by Pearman and Hyson (1986) indicated a larger sink. Although different models were involved, Enting and Mansbridge noted that the main reason for the different interpretation seemed to be differences in the CO_2 data that were used. They were unable to determine whether this difference represented a real change in the carbon cycle or whether it reflected the poor degree of intercalibration in the earlier data.

MATHEMATICAL DETAILS

The problem of deducing source-sink distributions from the spatial distributions of atmospheric CO_2 concentration is an ill-conditioned inverse problem. These problems have the characteristic that arbitrarily small errors in the concentration data can imply arbitrarily large errors in the estimated sources. The difficulties have long been recognized—Bolin and Keeling (1963) noted that "no details in the distribution of sources and sinks are reliable." The most important consequences of this ill-conditioning are:

The small-scale aspects of the source distribution are highly attenuated by atmospheric mixing—a k^{-1} dependence on meridional wave number applies.

Thus the effect of small-scale aspects of the source distribution become progressively more difficult to distinguish from the various "noise" contributions present in the observational data.

Consequently there is a trade-off between resolution and variance—the greater the spatial detail required, the less certain the estimates. (More recently ^{13}C data have been used in several different ways to obtain conflicting estimates of the oceanic uptake of CO_2. See Broecker and Peng 1993.)

Enting (1993) has presented some preliminary estimates of this trade-off. Figure 18.4 shows some of the estimates that can be derived from that approach. The calculation uses two approximations that require improvement—the transport model uses only diffusion and the data error is assumed to be independent for each site. The estimates are for the integrated Southern Hemisphere source (south of 15°S) as derived (using the diffusive model) from the annual mean data quoted by Tans et al. (1990), assuming that independent 0.4-ppmv errors apply to each data point.

In Figure 18.4, the resolution is specified in terms of the number of singular vectors used to construct the estimate, or equivalently the dimension of the subspace onto which the source distribution is projected.

The uncertainty increases significantly at about the point where the resolution that is sought drops below about twice the mean spatial separation of the data.

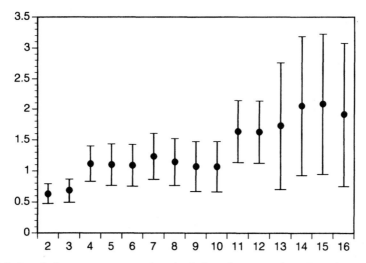

Figure 18.4. Resolution versus uncertainty in deduced sources, based on the approximate modeling of Enting (1993). The ranges show estimates of the Southern Hemisphere CO$_2$ source (south of 15°S) together with the 1-standard-deviation range as estimated from a purely diffusive model, plotted against the resolution defined by the number of singular vectors used to construct the estimate.

REGIONAL CARBON BUDGETS

Figure 18.5 demonstrates the extent to which transport modeling can constrain the atmospheric carbon budget. It gives a graphical representation of the analysis from Enting (1992a) with refined estimates for the sources derived from transport modeling. It uses a division of the earth into three zones, a tropical zone from 12°S to 12°N, with northern and southern zones outside this range. The source-sink processes used the same $F, D, G,$ and S description as in the section on "Carbon Cycle Feedback." The fossil data are from Rotty (1987) with the regional division as given by Enting and Mansbridge (1991). The range +10% to −5% is used. The deforestation data are from Houghton et al. (1983 and personal communication) with the regional distribution based on Houghton et al. (1987). The ocean data are based on the PCO$_2$ data presented by Tans et al. (1990). These are assigned a range of uncertainty reflecting the uncertainty in the air-sea gas exchange coefficient. The ranges are then corrected for the skin-temperature effect (Robertson and Watson 1992) and the role of carbon transport through rivers (Sarmiento and Sundquist 1992). The CO$_2$-enhanced growth is based on the work of Kohlmaier et al. (1987). The regional distribution reflects the distributions of ecosystems used in that study. On the basis of work by Polglase and Wang (1992), we have reduced the tropical contribution to reflect a smaller capacity for soil carbon storage in tropical regions.

The ranges of uncertainty for the regional budgets calculated from these "direct" estimates are compared to the net sources deduced from transport modeling. The midpoints of the range are the net surface *carbon* fluxes deduced by Enting and Mansbridge (1991), i.e., the role of CO has been included. The ranges are based on the simplified, purely diffusive, model of the atmosphere used by Enting (1993). Both the estimates and the ranges are subject to a number of serious qualifications, as described by Enting (1993).

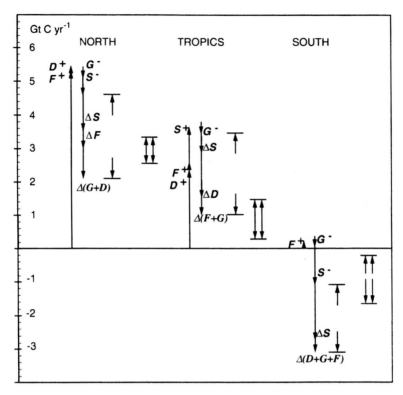

Figure 18.5. Atmospheric carbon budgets for northern, tropical, and southern regions. The symbols F, D, G, and S denote fossil carbon, deforestation, CO_2-induced growth, and seas, respectively. The + and − superscripts denote the upper and lower ends (in magnitude) of the range. The symbols of the form ΔF give the ranges based on direct estimates. The combined uncertainty in the net flux derived from direct estimates is the range shown by the single arrows. The double arrows show a range whose spread is taken from diffusive model calculations but whose midpoint is taken from the inversion calculation by Enting and Mansbridge (1991).

The points to note from these results are as follows:

Transport modeling can provide useful constraints on regional carbon budgets.

Because the global carbon budget involves cancellations between the regions, much greater precision would be required to constrain the global budget significantly.

Therefore the use of transport modeling in detecting the signals of feedback processes will require that the predicted anomalies be expressed in regional terms.

CONCLUDING REMARKS

The information contained in the spatial distribution of atmospheric CO_2 has the potential for revealing anomalies in the carbon cycle resulting from feedback processes. Spatial information about such anomalies may help identify the processes involved. This potential is subject to the very considerable limitations on resolution that arise from the nature of the inversion problem.

An additional aspect that has not previously been discussed is the use of ^{13}C data. The atmospheric carbon budget of Keeling et al. (1989) has made considerable use of ^{13}C. The resulting budget differs from that of Sarmiento and Sundquist (1992). Differences of this type could be indicators of the occurrence of feedback processes, if such differences were significant compared to the uncertainties in the individual estimates. Thus we cannot assess the significance of apparent anomalies until we can assign realistic estimates of uncertainty to current budget estimates. (More recently ^{13}C data have been used in several different ways to obtain conflicting estimates of the oceanic uptake of CO_2—see Broecker and Peng, 1993.)

The long-term trend in atmospheric $\delta^{13}C$ may provide a basis for detecting some classes of feedback. Enting and Pearman (1986, 1987) noted that their model produced a rate of change in good agreement with observations, under a wide range of assumptions about the current atmospheric carbon balance. Thus any observation of a trend differing from that implied by models without feedbacks could imply the onset of some feedback process.

One important point is that this discussion has been oriented toward CO_2. The uncertainties for the atmospheric CH_4 budget are even greater (Fung et al. 1991).

ACKNOWLEDGMENTS

The CSIRO atmospheric transport modeling project was funded in part by the State Electricity Commission of Victoria. The author wishes to thank Graeme Pearman and Tom Wigley for comments on the manuscript.

APPENDIX A: REFINING ATMOSPHERIC TRANSPORT MODELING

Modeling of the spatial distribution of CO_2 has been a major source of input to most recent estimates of the atmospheric carbon budget. However, as described in the section on "Mathematical Details," there are very considerable uncertainties in such modeling. This appendix describes some of the areas in which improvements seem both desirable and possible.

The inversions rely on the use of models of atmospheric transport. However, the approach to the inversion involves broader modeling questions concerning the objectives. One way to frame these broader aspects is in statistical terms. A Bayesian statistical approach seems to offer the best basis for a comprehensive description. In these terms, one has a source distribution about which one has some prior knowledge that needs to be expressed in statistical terms. One has a set of observations, also with statistical uncertainties, and a model with its own uncertainties. The inversion problem becomes one of determining the extent to which the observations, as interpreted by the model, serve to refine the knowledge of the sources beyond the information embodied in the prior distributions. (A comprehensive account of inversion formalisms from such a Bayesian point of view is given by Tarantola [1987].)

Inversion Techniques

Questions of inversion techniques involve both the formal approach used and the computational implementation. In principle, the optimal inversion formalism will be determined by the underlying statistical model. In practice there is much scope for further study of inversion schemes that are nearly optimal. Desirable additional qualities are robustness (schemes should remain nearly optimal for a range of statistical models) and feasibility of implementation without requiring excessive computing resources.

Some of the computational techniques for use with numerical models of atmospheric transport are the following:

Running a model with the concentrations on the lower boundary specified by observations (Enting and Mansbridge 1989, 1991; Tans et al. 1989). This is practical in two dimensions; in three dimensions the sparseness of the observational data limits the possibility of applying this technique (see, however, some preliminary studies by Law et al. 1992). The direct inversion using specified surface concentrations is restricted to the case of deducing surface sources using only surface data. Sources in the free atmosphere need to be modeled explicitly (e.g., Enting and Mansbridge 1991), and the formalism is unable to utilize data from above the earth's surface. Another limitation of this approach is that it does not produce estimates of the uncertainties in the sources.

The synthesis approach (Keeling et al. 1989; Tans et al. 1990). The various source components are modeled explicitly and the transport model is used to calculate the atmospheric response for each source individually. The source strengths are then estimated by seeking linear combinations of sources such that the combined response matches observations.

The Green's function approach. This is similar in concept to the synthesis approach except that the source components represent a geometric discretization of the source distribution rather than a process-based representation. Obvious choices of basis are either a separate source component for each grid point or a basis defined by a set of spherical harmonics. The problem of fitting observations will be severely underdetermined, so that prior constraints will play a key role (either explicitly or implicitly) in determining the form of the solution.

Adjoint modeling. The use of the adjoint of a transport model running with specified surface concentrations would allow the use of adjoint sensitivity analysis to give estimates of the sensitivity of the results to uncertainties in the specified surface concentrations.

Kalman filtering. Hartley and Prinn (1992) have investigated this approach. The difficulty lies in constructing an appropriate "state-space model" to specify the system.

Model Refinement

The most fruitful path for improving models of atmospheric transport would seem to be through an integrated approach to chemical and dynamic modeling using general circulation models in conjunction with a climate modeling program.

Error Analysis

The problem of error analysis in inversion problems is rather complex, and only a few preliminary results have been obtained. We suggest that an appropriate starting point is a statistical characterization of the prior estimates of the source distribution. The reasons for this suggestion are as follows:

Some constraints on the source distribution are needed given the possibility of unbounded error amplification in the inversion processes. Simple constraints such as boundedness (or even continuity) are usually too weak to provide useful information. Projection of the solution onto predetermined distributions such as smooth mathematical functions or assumed source distributions introduces assumptions that can only be justified in the context of a (statistical) model of characteristics of the sources.

The statistical characteristics of the observed concentration data are needed. Since much of the so-called noise in the observational data reflects unresolved sources, this is best treated as part of an overall statistical modeling of source characteristics.

Similarly, the model error will depend on the characterisitics of the sources. As a simple example, the effect of an error in the seasonal transport in a model will be most serious for those tracers that have strong seasonal variations in the sources.

Such a "statistical" approach has recently been implemented as a "Bayesian synthesis" inversion by Enting et al. (1993). This produced estimates of regional carbon fluxes with associated uncertainty derived from the uncertainties in observational data and prior estimates of sources.

APPENDIX B: CHARACTERIZING CARBON CYCLE FEEDBACK

The strength of carbon cycle feedback is commonly expressed in terms of the so-called β factor. This expresses the degree of enhancement of the biotic uptake of CO_2. However, a more useful characterization for carbon budgeting studies would give a measure of the rate of carbon *storage*. In this appendix, we define such a measure.

The objective is to define a concept that characterizes the amount of carbon cycle feedback in a similar way to the airborne fraction description of CO_2 concentrations, i.e., a convenient numerical index that gives a good approximation to the behavior of the system under circumstances of continually growing anthropogenic sources.

The atmospheric carbon budget is expressed in terms of an anthropogenic source, $S(t)$, representing the combination of fossil plus deforestation releases. The biotic uptake from carbon cycle feedback is denoted by $B(t)$ and the oceanic uptake by $\Phi(t)$. The amount of excess atmospheric carbon relative to preindustrial equilibrium is denoted by $Q(t)$, using units consistent with the sources (e.g. Pg carbon and Pg carbon per year).

The atmospheric carbon budget is

$$\dot{Q} = S - \Phi - B \qquad (18.B.1a)$$

In a linear approximation, the oceanic uptake can be expressed using a response function,

$$Q(t) = \int_{t_0}^{t} [S(t') - B(t')] R_s(t - t') dt' \qquad (18.B.1b)$$

Defining a similar response function formalism for the CO_2-enhanced growth gives

$$Q(t) = \int_{t_0}^{t} [S(t') - \Phi(t')] R_b(t - t') dt' \qquad (18.B.1c)$$

Following Enting and Mansbridge (1987b) and Enting (1990), we use the corresponding lowercase symbols to denote Laplace transforms of the fluxes and the excess carbon. The three relations above become

$$pq = s(p) - \varphi(p) - b(p) \tag{18.B.2a}$$

$$q = [s(p) - b(p)]r_s(p) \tag{18.B.2b}$$

$$q = [s(p) - \varphi(p)]r_b(p) \tag{18.B.2c}$$

The Laplace transform for argument p actually defines the amplitude for the case of exponentially growing fluxes and carbon content, i.e., $Q(t) = q(p)\exp(pt)$ and so on (strictly only in the limit $t_0 \to -\infty$).

For the case of exponential growth, one can define the ratio of the increases in atmospheric and oceanic carbon as the ratio

$$\kappa_s = \dot{Q}(t) / \Phi(t) = \left[\frac{1}{pr_s(p)} - 1\right]^{-1} \tag{18.B.3a}$$

a quantity that we call the ocean-atmosphere growth ratio. Similarly we define a biota-atmosphere growth ratio

$$\kappa_b = \dot{Q}(t) / B(t) = \left[\frac{1}{pr_b(p)} - 1\right]^{-1} \tag{18.B.3b}$$

In these terms, the airborne fraction α, defined as the ratio of atmospheric carbon increase to anthropogenic source is

$$d = (1 + \kappa_b^{-1} + \kappa_s^{-1})^{-1} \tag{18.B.4}$$

Note that each of the variables α, κ_b, κ_s is a function of p and so can only be regarded as a constant for the situation in which the combined anthropogenic source has a single exponential growth.

Combining relations (B.2a–c) gives

$$pq(p) + s(p) = 2s(p) - \varphi(p) - b(p) = \frac{q(p)}{r_s(p)} + \frac{q(p)}{r_b(p)} \tag{B.5}$$

whence

$$q(p) = r(p)s(p) \tag{B.6a}$$

with the combined response function $R(t)$ having a Laplace transform

$$r(p) = \left(\frac{1}{r_s} + \frac{1}{r_b} - p\right)^{-1} \tag{18.B.6b}$$

It will be seen that this result is consistent with the relation

$$\alpha = pr(p) \tag{18.B.7}$$

presented by Enting (1990) for the case of an $\exp(pt)$ growth.

Enting (1991) considered the simple model defined for the excess biomass, D, as

$$\dot{D} = \gamma Q(t) - D(t) / \tau \qquad (18.B.8a)$$

where $\gamma = \beta P_0/N_0$, with P_0 the initial biotic production, N_0 the initial atmospheric carbon content, and τ as a biotic lifetime. The Laplace transform is

$$pd(p) = \gamma q(p) - d(p) / \tau = b(p) \qquad (18.B.8b)$$

whence

$$d(p) = \frac{\gamma q(p)}{p + (1/\tau)}$$

and

$$s - \varphi = \frac{q(p)}{r_b} = pq(p) + b(p) = pq(p) + pd(p)$$

whence

$$r_b = p^{-1}\left[\frac{\gamma}{p + (1/\tau)}\right]^{-1} \qquad (18.B.9)$$

This result implies that $R_b(t)$ approaches a constant (representing an equilibrium partition of carbon between the atmosphere and biota) as $t \rightarrow \infty$. For finite t, $R_b(t)$ has an exponential approach to its limit. The biota-atmosphere growth ratio is

$$\kappa_b = \frac{\gamma}{p + (1/\tau)} \qquad (18.B.10)$$

REFERENCES

Bolin, B., and C. D. Keeling. 1963. Large-scale atmospheric mixing as deduced from the seasonal and meridional variations of carbon dioxide. *J. Geophys. Res.* 68:3899–3920.

Broecker, W. S. 1987. Unpleasant surprises in the greenhouse? *Nature* 328:123–126.

Broecker, W. S. and T.-H. Peng. 1993. Evaluation of the ^{13}C constraint on the uptake of fossil fuel CO₂ by the ocean. *Global Biochem. Cycles* 7:619–626.

Enting, I. G. 1990. Ambiguities in the calibration of carbon cycle models. *Inv. Prob.* 6:L39–L46.

Enting, I. G. 1991. Calculating future atmospheric CO₂ concentrations. Tech. Pap. No. 22. Division of Atmospheric Research, CSIRO, Australia.

Enting, I. G. 1992a. Constraining the atmospheric carbon budget: A preliminary assessment. Tech. Pap. No. 25. Division of Atmospheric Research, CSIRO, Australia.

Enting, I. G. 1992b. The incompatibility of ice-core CO₂ data with reconstructions of biotic CO₂ sources (II). The influence of CO₂-fertilised growth. *Tellus* 44B:23–32.

Enting, I. G. 1993. Inverse problems in atmospheric constituent studies: III. Estimating errors in surface sources. *Inv. Probl.* 9:649–665.

Enting, I. G., and J. V. Mansbridge. 1987a. The incompatibility of ice-core CO₂ data with reconstructions of biotic CO₂ sources. *Tellus* 39B:318–325.

Enting, I. G., and J. V. Mansbridge. 1987b. Inversion relations for the deconvolution of CO₂ data from ice cores. *Inv. Probl.* 3:L63–L69.

Enting, I. G., and J. V. Mansbridge. 1989. Seasonal sources and sinks of atmospheric CO_2: Direct inversion of filtered data. *Tellus* 41B:111–126.

Enting, I. G., and J. V. Mansbridge. 1991. Latitudinal distribution of sources and sinks of CO_2: Results of an inversion study. *Tellus* 43B:156–170.

Enting, I. G., and G. I. Pearman. 1986. The use of observations in calibrating and validating carbon cycle models. In *The Changing Carbon Cycle: A Global Analysis,* ed. J. Trabalka and D. Reichle, pp. 423–458. Springer-Verlag, New York.

Enting, I. G., and G. I. Pearman. 1987. Description of a one-dimensional carbon cycle model calibrated using techniques of constrained inversion. *Tellus* 39B:459–476.

Enting, I. G., and G. I. Pearman. 1992. Average global distributions of CO_2. In *Proc. NATO ASI,* ed. M. Heimann, pp. 31–64. Springer-Verlag, Il Ciocco, Italy.

Enting, I. G., C. M. Trudinger, R.-J. Francey, and H. Cranek. 1993. Synthesis inversion of atmospheric CO_2 using the GISS tracer transport model. Tech. Paper No. 29. Division of Atmospheric Research, CSIRO, Australia.

Etheridge, D. M., G. I. Pearman, and F. de Silva. 1988. Atmospheric trace-gas variations as revealed by air trapped in an ice core from Law Dome, Antarctica. *Ann. Glaciol.* 10:28–33.

Fung, I., J. John, J. Lerner, E. Matthews, M. Prather, L. P. Steele, and P. J. Fraser. 1991. Three-dimensional model synthesis of the global methane cycle. *J. Geophys. Res.* 96D: 13033–13065.

Hansen, J., G. Russell, A. Lacis, I. Fung, D. Rind, and P. Stone. 1985. Climate response times: Dependence on climate sensitivity and ocean mixing. *Science* 229:857–859.

Hartley, D., and R. Prinn. 1992. On the feasibility of determining surface emissions of trace gases using an inverse method in a three-dimensional chemical transport model. *J. Geophys. Res.* 98D:5183–5197.

Houghton, R. A., J. E. Hobbie, J. M. Melillo, B. Moore, B. J. Peterson, G. R. Shaver, and G. M. Woodwell. 1983. Changes in the carbon content of terrestrial biota and soils between 1860 and 1980: A net release of CO_2 to the atmosphere. *Ecol. Monogr.* 53:235–262.

Houghton, R. A., R. D. Boone, J. R. Fruci, J. E. Hobbie, J. M. Melillo, C. A. Palm, B. J. Peterson, G. R. Shaver, G. M. Woodwell, B. Moore, D. L. Skole, and N. Myers. 1987. The flux of carbon from terrestrial ecosystems to the atmosphere in 1980 due to changes in land use: Geographic distribution of the global flux. *Tellus* 39B:122–139.

Intergovernmental Panel on Climate Change. 1990. *Climate Change: The IPCC Scientific Assessment,* ed. J. T. Houghton, G. J. Jenkins, and J. J. Ephraums. Cambridge University Press, Cambridge.

Keeling, C. D., S. C. Piper, and M. Heimann. 1989. A three-dimensional model of atmospheric CO_2 transport based on observed winds: 4. Mean annual gradients and interannual variations, ed. D. H. Peterson, pp. 305–363. In *Aspects of Climate Variability in the Pacific and Western Americas.* Geophysical Monograph 55. American Geophysical Union, Washington, D.C.

Kohlmaier, G. H., H. Bröhl, E. O. Siré, M. Plöchl, and R. Revelle. 1987. Modelling stimulation of plants and ecosystem response to present levels of excess atmospheric CO_2. *Tellus* 39B: 155–170.

Lashof, D. A. 1989. The dynamic greenhouse: Feedback processes that may influence future concentrations of atmospheric trace gases and climate change. *Climatic Change* 14: 213–242.

Law, R., I. Simmonds, and W. F. Budd. 1992. Application of an atmospheric tracer model to high southern latitudes. *Tellus* 44B:358–370.

Pearman, G. I., and P. Hyson. 1986. Global transport and inter-reservoir exchange of carbon dioxide with particular reference to stable isotope distribution. *J. Atmos. Chem.* 4:81–124.

Polglase, P. J., and Y.-P. Wang. 1992. Potential CO_2-enhanced carbon storage by the terrestrial biosphere. *Austr. J. Bot.* 40:641–656.

Raynaud, D. 1993. Ice core records as a key to understanding the history of atmospheric trace gases. In *The Biogeochemistry of Global Change: Radiative Trace Gases,* ed. R. S. Oremland. Chapman and Hall, New York.

Raynaud, D., and J. M. Barnola. 1985. An Antarctic ice core reveals atmospheric CO_2 variations over the past few centuries. *Nature* 315:309–311.

Robertson, J. E., and A. J., Watson. 1992. Thermal skin effect of the surface ocean and its implications for CO_2 uptake. *Nature* 358:738–740.

Rotty, R. M. 1987. A look at 1983 CO_2 emissions from fossil fuels (with preliminary data for 1984). *Tellus* 39B:203–208.

Sarmiento, J. L., and E. T. Sundquist. 1992. Revised budget for the oceanic uptake of anthropogenic carbon dioxide. *Nature* 356:589–593.

Siegenthaler, U., and H. Oeschger. 1987. Biospheric CO_2 emissions during the past 200 years reconstructed by deconvolution of ice core data. *Tellus* 39B:140–154.

Smith, S. V., and F. T. Mackenzie. 1987. The ocean as a net heterotrophic system: Implications from the carbon biogeochemical cycle. *Global Biogeochem. Cycles* 1:187–198.

Tans, P. P., ed. 1992. *Environmental Pollution Monitoring and Research Program.* Publ. No. 77. World Meteorological Organization, Geneva.

Tans, P. P., T. J. Conway, and T. Nakazawa. 1989. Latitudinal distribution of the sources and sinks of atmospheric carbon dioxide derived from surface observations and an atmospheric transport model. *J. Geophys. Res.* 94D:5151–5172.

Tans, P. P., I. Y. Fung, and T. Takahashi. 1990. Observational constraints on the global atmospheric CO_2 budget. *Science* 247:1431–1438.

Tarantola, A. 1987. *Inverse Problem Theory: Methods for Data Fitting and Model Parameter Estimation.* Elsevier, Amsterdam.

Wahlen, M., A. B. Deck, and A. Herchenroder. 1991. Initial measurements of CO_2 concentrations (1530 to 1940 AD) in air occluded in the GISP 2 ice core from central Greenland. *Geophys. Res. Lett.* 18:1457–1460.

Wigley, T. M. L., and S. C. B. Raper. 1992. Implications for climate and sea level of revised IPCC emissions scenarios. *Nature* 357:293–300.

Wollast, R., and F. T. Mackenzie. 1984. Global biogeochemical cycles and climate. In *Climate and Geo-Sciences,* ed. A. Berger, S. Schneider, and J.-Cl. Duplessy, pp. 453–510. Kluwer, Dordrecht, The Netherlands.

B

Modeling the Carbon Sink

The scientific community has had difficulty evaluating the global carbon cycle for more than 15 years, since the possibility was widely recognized in 1977 that deforestation globally was large enough to constitute a significant further release of carbon into the atmosphere beyond the release from combustion of fossil fuels. During most of the 1970s, terrestrial ecosystems were assumed by most scientists either to be at equilibrium with the atmosphere or to be absorbing atmospheric carbon globally at some low rate. The latter assumption was an outgrowth of the definition and application of the biotic growth factor in models used by C. D. Keeling and his colleagues, beginning with their contribution to the Brookhaven Conference of 1972. The term was a correction designed to express the effect of CO_2 fertilization on carbon storage in long-lived terrestrial plants. The evidence from the middle and late 1970s suggested that the terrestrial vegetation, far from acting as a net absorber of CO_2 globally, was probably a net source of additional carbon for the atmosphere. The topic has been discussed in Chapter 1 by Woodwell and is covered in this section by R. A. Houghton. The global carbon cycle could no longer be "balanced." Where was the error?

The answer is elusive, although Tans et al. believe it must lie in an enhanced storage of carbon in the midlatitude forests of the Northern Hemisphere. Houghton calls attention to the fact that different approaches to the appraisal are measuring different attributes of forests and that it is possible that both are correct, although he also expresses puzzlement that we cannot measure an increase in carbon storage if the annual increment in forests is as high as 1–2 petagrams. The problem is made more difficult if chronic disturbance, especially a rapid warming, may be expected to release additional carbon from forests and soils. The explanation of the global carbon cycle remains puzzling, especially to those most intensively involved in appraising the contemporary changes in forests. Those changes, if they involve the rate of absorption of carbon suggested by the modeling summarized here by Tans et al., must hang on a series of tenuous circumstances, certain to change as the earth warms and as human influences spread. The change can be in one direction only, unless the warming can be controlled.

19

Effects of Land-Use Change, Surface Temperature, and CO₂ Concentration on Terrestrial Stores of Carbon

R. A. Houghton

Two questions are addressed in this chapter. First, are there feedbacks between the earth's environment and the amount of carbon held in terrestrial ecosystems? More specifically, will a warmer earth with higher concentrations of CO_2 in its atmosphere store more or less carbon in vegetation and soil than the current earth? The time frame of interest is years to decades, and consideration is given to metabolic processes (productivity and respiration) rather than to migration of species or ecotones. The question is addressed from the perspective of the global carbon balance over the last 140 years. The answer is tentative: it appears as though there may have been both positive and negative biotic feedbacks.

The second question is whether or not there is direct evidence for an accumulation of carbon in northern midlatitude terrestrial ecosystems. Recent analyses based on forest growth seem to have demonstrated a significant net accumulation, but the analyses are incomplete. Rather than demonstrating a net terrestrial sink, these analyses confirm the unlikelihood that the global carbon imbalance will be found in the vegetation of terrestrial ecosystems.

Concerning the first question, feedbacks, the assumption made here is that the imbalance in the global carbon cycle (the so-called "missing" carbon) is accounted for by an as yet undetected accumulation of carbon on land. Analyses based on models of oceanic uptake and atmospheric transport provide indirect evidence for this terrestrial sink. The assumption could be incorrect, however. The imbalance might be the result of errors in calculation of one or more of the terms in the global carbon balance. But if estimates of the terrestrial release of carbon from land-use change and estimates of oceanic uptake are correct, and the imbalance does represent a terrestrial sink, the magnitude of the sink is probably determined by environmental variables affecting the storage of terrestrial carbon. In other words, the imbalance in the global carbon budget may be related to biotic feedbacks between the carbon cycle and climate. The processes that control the accumulation of carbon on land—photosynthesis, growth, respiration, decomposition, and mineralization—are all affected by temperature, moisture, and nutrients, and some of them are affected by the concentration of CO_2 in air. Thus changes in these environmental variables may be expected to cause rapid changes in the density of carbon per unit area of land.

The amount of carbon held in terrestrial ecosystems is also changed through direct human effects on land use, such as conversion of forests to agricultural lands. In this case the changes in carbon result from changes in the area of ecosystems and changes in the stocks of carbon per unit area that result from management. Changes in terrestrial carbon that result from changes in land use can be determined with more confidence than changes in the density of carbon where no change in land use has occurred. This is so because of the large difference in carbon stocks between forests and the agricultural ecosystems that replace them (forests hold 20–100 times more carbon in vegetation per unit area), and because changes in agricultural lands and rates of deforestation and reforestation are, to some extent, measured and recorded. Most estimates of the flux of carbon between terrestrial ecosystems and the atmosphere have been based on these changes in land use. In contrast, changes in the density of carbon within ecosystems are likely to be small relative to the amounts of carbon held in vegetation and soil, and, hence, difficult to measure.

TOTAL NET FLUX OF CARBON FROM TERRESTRIAL ECOSYSTEMS

One method used to estimate the net flux of carbon between terrestrial ecosystems and the atmosphere over the last 100–200 years is that based on an inverse technique, sometimes referred to as deconvolution (Siegenthaler and Oeschger 1987; Keeling et al. 1989; Sarmiento et al. 1992). Normally, ocean models of carbon uptake are used to calculate atmospheric concentrations of CO_2 from specified annual emissions of carbon from fossil fuels and deforestation. The models can be inverted, however. If the historic pattern of atmospheric concentrations of CO_2 is known, the same models can be used to calculate the annual emissions and accumulations required to generate that pattern. Subtraction of the annual emissions of fossil fuel carbon from the total emissions calculated from these deconvolutions yields a residual flux, a nonfossil flux of carbon. The flux need not be a terrestrial flux, but it is often assumed to be because the other terms in the budget are specified. Furthermore, short-term evidence from $^{13}C/^{12}C$ ratios suggests that the changes are of terrestrial rather than oceanic origin (Siegenthaler and Oeschger 1987; Keeling et al. 1989).

A recent deconvolution by Sarmiento et al. (1992) shows the residual (presumably terrestrial) flux to have been a net release of about 0.4 petagrams (Pg) ($=10^{15}$ g) carbon per year in 1800, rising gradually to 0.6 Pg carbon per year by 1900, remaining at 0.6 Pg carbon per year until about 1920, and then falling to zero by the late 1930s (Figure 19.1). The flux continued to fall, becoming a net accumulation of carbon in terrestrial ecosystems after 1930, reaching a maximum sink of 0.8 Pg carbon per year in the 1970s, and then abruptly returning to zero by the early 1980s. It has remained near zero since 1982 or so. The cumulative net flux for the period 1850–1990 was a release of 25 Pg carbon according to this deconvolution (Sarmiento et al. 1992). This estimate is about half that calculated by Siegenthaler and Oeschger (1987) with their box-diffusion model, but the temporal patterns of the two estimates are almost identical. Since 1940

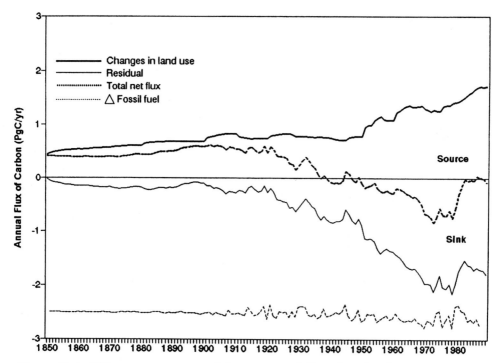

Figure 19.1. Different components of the net annual flux of carbon between terrestrial ecosystems and the atmosphere. The heavy solid curve is the flux of carbon from changes in land use. The heavy dashed curve is the net biotic flux, based on deconvolution (Sarmiento et al. 1992). The thin solid curve is the residual flux of carbon, the difference between the net flux and the flux from land-use change. This curve represents the global carbon imbalance or the annual "missing" carbon. The fact that it is negative is assumed here to indicate that carbon is accumulating in ecosystems undisturbed by land use. The thin dashed curve at the bottom is year-to-year variation in worldwide combustion of fossil fuels. Most of the short-term variation in the net flux (from deconvolution) and in the residual flux is an artifact due to the fact that atmospheric CO_2 concentrations used in the deconvolution were smoothed. Short-term variations are related to fossil fuel use and should not be interpreted as representing metabolic processes on land.

the cumulative net flux calculated by Sarmiento et al. (1992) has been an accumulation of 13.5 Pg carbon in terrestrial ecosystems.

FLUX OF CARBON FROM CHANGES IN LAND USE

Ecologists have used a different approach to calculate the flux of carbon between terrestrial ecosystems and the atmosphere. Theirs is based on changes in land use and associated changes in vegetation and soil. The conversion of forests to agricultural lands, for example, releases carbon to the atmosphere through burning and decay and accumulates (seasonally) a small amount of carbon in the agricultural crop. Conversely, the regrowth of forests on abandoned agricultural lands withdraws carbon from the atmosphere and stores it again on land, in trees and soil. Current estimates of the flux of carbon from changes in land use indicate a net release into the atmosphere of between 0.6 and 2.5 Pg carbon for 1980 (Houghton et al. 1987; Hall and Uhlig 1991; Houghton 1991) and between 1.1 and 3.6 Pg carbon for 1990 (Houghton 1991), almost entirely from the tropics. It is important to note that this net release of carbon includes the accumulation of carbon in forests recovering after logging and after abandonment of agriculture. Land use includes reforestation as well as deforestation.

The high end of the range for 1990 now seems almost certainly too high. The increase between 1980 and 1990 was in large part due to an apparent increase in the rate of deforestation in the Brazilian Amazon (Myers 1991). Myers (1991) based his estimate of Brazilian deforestation on a study by Setzer and Pereira (1991), who used Advanced Very High Resolution Radiometer (AVHRR) data from the NOAA-7 satellite to determine the number of fires burning in Legal Amazonia (an area of 500×10^6 ha including most of the Brazilian Amazon) during the dry season (mid-July through September, 1987). They accounted for the fact that some fires burned for more than one day (and should not be counted twice), and that a small, hot fire would saturate the entire pixel (1 km square) and overestimate the area actually burned. With these adjustments, Setzer and Pereira (1991) estimated that about 20×10^6 ha of fires burned in the Brazilian Amazon in 1987, about 60% of which, or 12×10^6 ha, were on lands that had already been deforested. Their estimate of deforestation was 8×10^6 ha.

Myers (1991) reduced their estimate for 1987 to 5×10^6 ha, to account for other factors. Nevertheless, even this reduced rate seems high according to more recent studies. Using data from Landsat (80-m resolution rather than 1-km resolution), the National Institute for Space Research (INPE) in Brazil found the rate of deforestation of closed forests in Brazil's Legal Amazonia to have averaged about 2.1×10^6 ha/year between 1978 and 1989 (Fearnside et al. 1990), about one-fourth the rate initially determined by Setzer and Pereira (1991). The actual rate probably increased between 1978 and 1987, but fell substantially after 1987 to 1.8×10^6 ha in 1988–1989, to 1.38 in 1989–1990, and to 1.11 in 1991. Recent work by Skole and Tucker (1993) shows an even lower average rate than that reported by INPE (1.52×10^6 ha/year between 1978 and 1988).

When the recent data from INPE are substituted for those reported by Myers (1991), the estimated rate of deforestation for all Latin America is revised from 7.7 to 4.5×10^6 ha/year. Over the 10-year period 1980–1990, the rate of deforestation throughout the entire humid tropics now appears to have increased by about 40% rather than 90%. The revised estimate for the late 1980s is 10.66×10^6 ha/year for closed forests. An independent estimate of the total rate for closed and open tropical forests is 15.4×10^6 ha/year (FAO 1993).

These revised rates of deforestation were used here with the data initially used by Houghton et al. (1991) to recalculate emissions of carbon from Latin America. The reanalysis omitted the conversion of forests to degraded land because that conversion is now thought to be redundant with conversion of forest to shifting cultivation already included. The effect of both of these revisions was to reduce the calculated emissions from Latin America from about 0.7 Pg carbon per year (Houghton et al. 1991) to about 0.5 Pg carbon per year in 1980, and from about 0.9 to 0.7 Pg carbon per year in 1990.

Revisions were also made here to the data used to calculate the flux of carbon from tropical Asia and Africa. Rates of deforestation in Southeast Asia were revised upward recently by FAO (1993), and historical rates of degradation in the region were reassessed by Flint and Richards (1994). These revisions gave an estimate of flux for South and Southeast Asia that was about 0.7 Pg carbon in 1990 (Houghton and Hackler 1994). A reanalysis of land-use change in Africa, using improved estimates of biomass (Brown et al. 1989), gave an estimate of flux there of 0.35 Pg carbon per year in 1990.

As a result of these revisions, the global net flux of carbon from changes in land use is estimated to have been about 1.4 Pg carbon in 1980 (1.3 Pg from the tropics and 0.1 Pg from outside the tropics) and 1.7 Pg carbon in 1990 (essentially all of it from the tropics). The cumulative flux over the period 1850–1990 was 120 Pg carbon. The estimate is about 15% lower than an earlier one (Houghton 1993), with most of the difference occurring between 1985 and 1990. For South and Southeast Asia, the estimate is about 30% lower than an independent study (Flint and Richards 1994), and, for all the tropics in 1980, it is about 45% higher than the recent estimate by Hall and Uhlig (1991). The uncertainty is estimated here to be better than ±30%.

The revised estimate of flux from changes in land use (120 Pg carbon over the period 1850–1990) is considerably higher than the estimate obtained from deconvolution (25 Pg carbon), and the temporal patterns of annual flux do not bear much resemblance (Figure 19.1).

DIFFERENCE BETWEEN THE TWO ESTIMATES

The difference between these two estimates of flux (one from deconvolution, one from land-use change) may represent an actual flux of carbon between terrestrial ecosystems and the atmosphere because the two approaches address different processes (Table 19.1). Assuming that the entire flux obtained through deconvolution is terrestrial, the deconvolution approach provides an estimate of the net terrestrial flux, including both the land-use flux and a flux due to other changes in terrestrial ecosystems. The flux of carbon from land-use change, on the other hand, includes only changes on lands directly and deliberately managed. Thus, the difference between the net flux obtained through deconvolution and the flux obtained through analysis of land-use change may define the flux of carbon from terrestrial ecosystems not directly modified by humans. There has, as yet, been no direct measurement of even the direction of such a change. Possible causes include changes in climate, in atmospheric concentrations of CO_2, and in the deposition of nutrients or toxins. The effect of these factors, either singularly or together, on the storage of carbon in the world's terrestrial ecosystems is unknown.

The errors involved in calculating this flux of carbon from the difference between the deconvolution technique and the land-use approach are potentially large, but the gross trends may indicate which environmental factors were important in the past and which, perhaps, may be important over the next century. What does the pattern of this residual flux suggest about feedbacks?

The residual flux, or the difference between the net biotic flux (from deconvolution) and the land-use flux, is shown in Figure 19.1 (thin solid line). This residual flux has always been negative and is interpreted here to mean that some terrestrial ecosystems

Table 19.1. Factors Included and Not Included in Analyses of the Flux of Carbon from Changes in Land Use

Included	Not included
Change in the area of forests	Unmanaged, undisturbed ecosystems assumed to be in steady state
Deforestation	
Croplands	
Pastures	
Degradation of land	Forest decline or enhancement
Reforestation	
Agricultural abandonment	CO_2 fertilization
Afforestation	
	Eutrophication
Change in biomass within forests	
	Frequency of fires
Logging and regrowth	
	Climatic change
Shifting cultivation	
	Desertification
Degradation of forests (reduction of biomass)	

have been accumulating carbon independent of land-use change. The magnitude of the accumulation, worldwide, is given by the residual flux. Positive differences, if they occurred, would indicate that some terrestrial ecosystems were releasing carbon to the atmosphere in addition to that released from changes in land use.

The curve of this residual flux exhibits three features: first, a period before 1920 showing a small terrestrial sink not different from zero, with little variation; second, a period between about 1920 and 1975, when some of the world's terrestrial ecosystems were apparently accumulating carbon at an increasing rate; and, finally, a period since the 1970s in which the rate of accumulation has decreased. The reader is reminded not to interpret this residual flux as showing a net terrestrial sink for carbon. The net terrestrial flux has generally been a small net source of carbon to the atmosphere, except between 1940 and 1980 (Sarmiento et al. 1992) (Figure 19.1). The residual flux is assumed here to result from changing environmental conditions. The accumulation of carbon in ecosystems not directly disturbed by changes in land use was about 95 Pg carbon between 1850 and 1990 (net flux: 25 Pg carbon; flux from land-use change: 120 Pg carbon).

POSSIBLE FEEDBACK MECHANISMS

The residual flux is the imbalance in the global carbon budget, or the "missing" carbon, a term initially coined by Broecker et al. (1979) to refer to the difference between the amount of carbon released to the atmosphere from combustion of fossil fuels and the amount that could be accounted for in the atmosphere (measured) and oceans (modeled). Estimates of a release of carbon from deforestation and other changes in land use increased the size of the "missing" carbon. If terrestrial ecosystems were required as a sink for carbon, an even larger sink was needed to compensate for the additional release from deforestation and reforestation. This missing carbon has, of course, never been measured; it has only been defined as the difference among three "knowns" (emissions from fossil fuels, accumulation in the atmosphere, and uptake by the oceans). In the analysis described here, it is being defined as the difference among four "knowns," the fourth being emissions of carbon from changes in land use.

The amount of carbon held in terrestrial ecosystems is determined by the balance between photosynthesis and respiration, the latter including decay of soil organic matter. Both of these metabolic processes are affected by many environmental factors, but surface air temperature and concentrations of CO_2 have received the most attention. Arguments are made, on the one hand, that terrestrial ecosystems will amplify a global warming (positive feedback) by releasing additional carbon to the atmosphere as a result of a warming-enhanced increase in respiration rates, and, on the other hand, that they will reduce the warming (negative feedback) by accumulating carbon in response to increasing concentrations of CO_2 in the atmosphere. The historical variation in the residual flux, calculated here, allows us to ask whether the variation is related to temperature, atmospheric CO_2, or some other environmental factor. A correlation would not demonstrate a cause-effect relationship, of course, but it might reveal which environmental factor had been most important in the past, and it might suggest the direction of response of terrestrial ecosystems to global change in the future. Will they act as a net source or sink for carbon?

Linear regressions were used here to explore correlations over the period 1880–1990 between the annual residual flux and two environmental factors: concentrations of CO_2 in the atmosphere and mean global surface air temperatures.

Concentrations of CO_2 in the Atmosphere

The progressive increase in the residual flux of carbon after 1920 (Figure 19.1) parallels the trend in industrial activity, and one cause of the increasing sink might be increasing concentrations of CO_2 in the atmosphere. Elevated concentrations of CO_2 have been shown to enhance photosynthesis and growth in many plants (Strain and Cure 1985; Allen et al. 1987) and may increase the storage of carbon in terrestrial ecosystems if the CO_2-enhanced growth is not respired (Grulke et al. 1990; Mooney et al. 1991).

The linear regression equation (with 11-year running averages for both variables)

$$\text{Residual flux [Pg carbon per year]} = 10.104 - 0.0355 \, (CO_2) \, \text{[ppmv]} \quad (19.1)$$

showed that annual concentrations of CO_2 explained 75% of the variation in the residual flux of carbon over the period 1850–1990 (regression 1, Table 19.2). CO_2 concentrations over this period were obtained from Neftel et al. (1985), Friedli et al. (1986), and Keeling et al. (1989). According to the linear assumptions of the regression, 0.0355 Pg carbon per year are stored per ppmv increase in CO_2.

A term more commonly used to express the effect of CO_2 fertilization on carbon storage is the biotic growth factor (or β factor). It relates the response of net primary productivity (NPP) (the annual net uptake of carbon by green plants) to elevated concentrations of CO_2 (C) and was defined initially by Bacastow and Keeling (1973) as

Table 19.2. Results of Regression Analyses

Independent variable	Period	R^2	Effect	Flux in late 1980s (Pg carbon per year)	Biotic[a] growth factor	Q_{10}[b]	Net effect of doubled CO_2 levels and 2.5°C increase (Pg carbon per year)
1. CO_2 (no lag)	1880–1988[c]	0.75	Sink: 0.0355 Pg carbon/ppmv	Sink: 2.3	0.19		Sink: 15
2. CO_2 (no lag)	1940–1988	0.84	Sink: 0.043 Pg carbon/ppmv	Sink: 2.8	0.14	1.6	Sink: 10
temperature deviation (no lag)			Source: 3.4 Pg carbon/°C	Source: 0.7			
3. CO_2 (no lag)	1940–1988	0.94	Sink: 0.004 Pg carbon/ppmv	Sink: 1.4	0.026		
temperature deviation (7-year lag)			Source: 6.4 Pg carbon/°C	Source: 0.9		2.1	Source: 13

[a]These biotic growth factors assume that all of the increase in NPP is stored. If only a fraction of an increase in NPP is stored, the calculated biotic growth factor would be larger.
[b]These values for Q_{10} are calculated for a global respiration of 100 Pg carbon per year (autotrophic and heterotrophic respiration). If temperature affects only heterotrophic respiration, the resulting Q_{10}s are 2.5 and 3.2 for regressions 2 and 3, respectively. In both cases, temperature is assumed to have no effect on photosynthesis. If higher temperatures increase the rate of photosynthesis as well as the rate of respiration, Q_{10}s for respiration alone will be higher.
[c]Temperature deviation not statistically significant.

$$\beta = (\text{NPP/NPP}_0) - [1 / \ln (C/C_0)] \tag{19.2}$$

where subscripts 0 refer to the initial (preindustrial) rates or concentrations. Pre-industrial NPP was assumed to have been 50 Pg carbon per year (range: 45–60); NPP in 1990 was 50 + 2.3 Pg carbon per year (the increase of 2.3 was calculated: 0.0355 Pg carbon per year per ppmv $\times$ 65 ppmv difference between 1850 and 1985); and C_0 and C were 285 ppmv and 350 ppmv, respectively. The biotic growth factor was calculated to be 0.19 assuming all of the increase in NPP accumulated on land and was not subsequently oxidized.

The biotic growth factor is useful for comparative purposes, but it is of limited use biologically because production of organic matter by green plants (NPP) need not be related to the accumulation of carbon (net ecosystem production [NEP]). In mature ecosystems, NPP may be large while NEP is close to zero. In such systems, the organic matter produced annually is balanced by the decomposition of organic matter accumulated over many years. In young systems, some fraction of NPP accumulates. The amount of carbon held currently in vegetation (about 560 Pg) and soil (about 1500 Pg) is considerably greater than annual global NPP (about 50 Pg/year) and represents an accumulation of carbon over a long time. The problem with calculating a biotic growth factor from annual accumulations is that there is no way to know what fraction of an increment in NPP has been stored. If half of an increase in NPP is accumulated in either wood or soil, the increase in NPP will be twice the increase in storage, or 5.0 Pg carbon/year according to the results of this regression. The corresponding biotic growth factor would then be 0.45 rather than 0.19. If a smaller fraction of the increment in NPP is stored, the increase in NPP will have been even larger. There are no data to suggest what fraction of an increment in growth will accumulate in the ecosystem, or even whether an increase in NPP is required for an accumulation of carbon. Carbon would accumulate from a reduction in the rate of respiration or decomposition, as well as from an enhancement of NPP.

The use of linear regression coefficients (from equation 19.1) in a nonlinear equation (19.2) is not strictly correct. It may be justified here, however, because the error it introduces is small relative to other errors and unknowns in the analysis. The increase in atmospheric CO_2 concentration has been relatively small, about 25% since 1850.

The apparent increase in carbon storage between 1850 and 1990 may not have been driven by increased concentrations of CO_2 but by some other factor associated with increased industrial activity. One such factor is the increased availability of nitrogen, often the mineral element thought to limit productivity. Analyses of the increased loading of nitrogen and phosphorus on land and in coastal waters suggest that 0.2 (Peterson and Melillo 1985) to 0.8 Pg carbon per year (Chapter 2, this volume) may be sequestered as a result of these additional nutrients. The estimates are considerably less than the rate of storage suggested here (2.3 Pg carbon per year), but nitrogen may also have become more available through a warming-caused increase in the rate of mineralization (decomposition).

Surface Temperature

The role of temperature in the accumulation and release of terrestrial carbon is uncertain. On the one hand, rates of respiration (including decomposition) are temperature sensi-

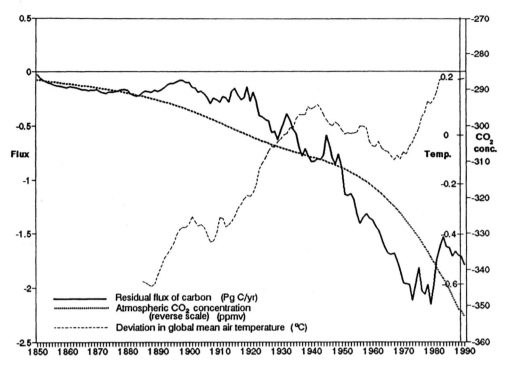

Figure 19.2. The residual flux of terrestrial carbon (heavy solid line); the concentration of CO_2 in the atmosphere (from Neftel et al. 1985; Friedli et al. 1986; Keeling et al. 1989) (dotted line) (plotted with the ordinate reversed so that the overall trend in concentration parallels the trend in the "missing" sink); and mean global land surface temperature deviations from the 1951–1980 baseline (Hansen and Lebedeff 1987) (thin dashed line).

tive, and higher temperatures should increase the rates (Schleser 1982; Woodwell 1989; Raich and Schlesinger 1992; Chapter 8, this volume). On the other hand, respiration-mineralization not only releases CO_2, it also releases nitrogen and other nutrients. Because the ratio of carbon to nitrogen is so much higher in woody tissue than in soil organic matter, the same amount of nitrogen can sequester much more carbon in wood than in soil (Rastetter et al. 1992; Shaver et al. 1992). For such an accumulation to occur, the nitrogen released during decomposition must be taken up by the plant and used in the production of additional wood. If the processes are decoupled in time, as they might be in winter and spring (Chapter 11, this volume), or if higher temperatures increase the rate of denitrification, nitrogen may be lost from the system and not contribute to enhanced growth.

If the net effect of warmer temperatures is to release carbon, the pattern of the residual flux should parallel the trend in temperature (Figure 19.2): a decreasing sink (increasing source) of carbon with warming between 1880 and 1940, an increasing sink with the cooling between 1940 and 1970, and then an abruptly decreasing sink again after 1970. If the net effect of warming is to store carbon (changing its form from soil to wood), the expected flux would be reversed. Comparison of the temperature trend with the residual flux is ambiguous. The flux shows no response to the early warming between 1880 and 1940. After 1940, however, the increasing sink of carbon matches the decrease in global temperature until the 1970s, and the decreased sink after the 1970s matches the recent warming (Figure 19.2). The results before 1940 show no effect; the results after 1940 suggest a warming-caused release of carbon, a positive feedback. The difference in the response to warming might be related to the different rates of warming in the two periods, 1880–1940 and 1970–1990. There is no evidence for a negative feedback between temperature and carbon.

The magnitude of a warming-caused release is difficult to obtain from these analyses. Linear regressions of the following form were used with different lag times between temperature deviation and flux:

$$\text{Residual flux [Pg carbon/year]} = \text{Constant} + A(\text{temp. dev.}) \, [°C] \qquad (19.3)$$

Annual deviations in global mean temperature explained essentially none of the variance in the residual flux over the period 1880–1983. The same was true over the period 1940–1983 if the lag between temperature and response was ignored. However, if the residual flux was assumed to lag behind temperature by 7 or 8 years, temperature deviations explained about 60% of the variation in flux. With these lags, the shapes of the dependent and independent variables were coincident, but the increase in the sink with cooling before the 1970s was greater than the decrease in the sink with warming after the 1970s (Figure 19.2).

More of the variance was explained over the period 1940–1983 when both CO_2 concentration and temperature deviation were used as independent variables in multiple regressions of the following form:

$$\text{Residual flux [Pg carbon/year]} = \text{Constant} + A(CO_2 \text{ concentration}) \, [\text{ppmv}] + B(\text{temp. dev.}) \, [°C] \qquad (19.4)$$

Of the variation in flux, 84% was explained with no lag, and 94% was explained with a 7-year lag (Table 19.2). Unfortunately, the coefficients of the two analyses are quite different. With no lag (regression 2), the effect of warming is to release 3.4 Pg carbon per °C, and the effect of CO_2 is to store 0.043 Pg carbon per ppmv. With the 7-year lag (regression 3), the effects are 6.4 Pg carbon per °C and 0.004 Pg carbon per ppmv. The latter analysis shows temperature to be about twice as effective, and CO_2 concentrations about one-tenth as effective, as the results obtained from regression 2. The relative effects of the two independent variables in explaining the residual flux are shown in Figure 19.3.

The regression coefficients for the temperature-flux relationship suggest that the 0.5°C warming between 1880 and 1990 may have been responsible for an additional release that is currently 2.4 (regression 2) to 3.7 (regression 3) Pg carbon per year. These values allow computation of a temperature coefficient (Q_{10}), defined as the factor by which the respiration rate increases when temperature is increased by 10°C. By assuming an exponential relationship between respiration and temperature, one can derive the equation

$$Q_{10} = \left(\frac{k_2}{k_1}\right)^{\frac{10}{T_2 - T_1}} \qquad (19.5)$$

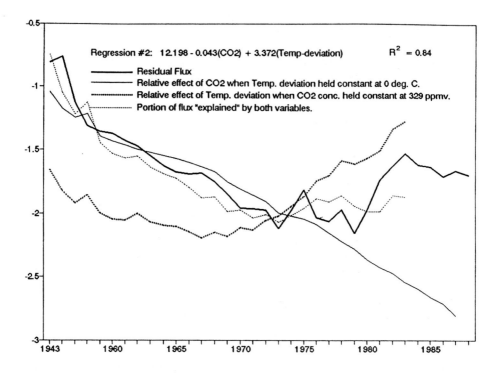

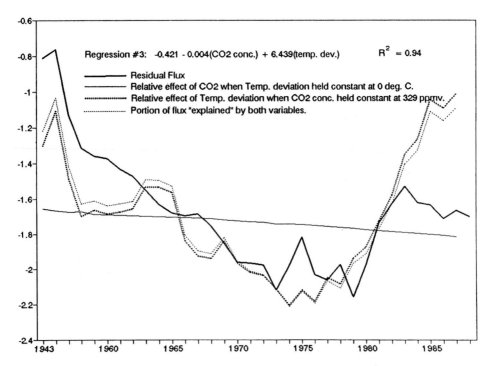

Figure 19.3. Relative ability of CO_2 concentrations and temperature deviations to account for variations in the residual flux of carbon over the period 1940–1983.

where k_1 and k_2 are rates of carbon flux at temperatures T_1 and T_2, respectively. Total respiration of terrestrial ecosystems, globally, is about 100 Pg carbon per year. Heterotrophic respiration (respiration of animals and microbes) is about 50 Pg carbon per year. If temperature is assumed to have affected only the rate of heterotrophic (largely soil) respiration ($k_1 = 50.0$, $k_2 = 52.4$–53.7), then the Q_{10} is between 2.5 and 3.2, somewhat higher than the value of 2 commonly measured in the laboratory and field but within the range observed (Chapter 8, this volume). If the change in temperature is assumed to affect also the rate of plant respiration (autotrophic + heterotrophic = total respiration of the ecosystem), the values of Q_{10} range between 1.6 and 2.1 (Table 19.2). These values of Q_{10} assume that only respiration is affected by temperature. If rates of photosynthesis have also been increased by warmer temperatures, the Q_{10} for respiration alone is higher.

If increasing concentrations of CO_2 in the atmosphere and variation in surface temperature have been the "causes" of variation in the annual flux of carbon to undisturbed ecosystems, their effects have largely canceled each other thus far. They may not continue to do so in the future, however. Assuming that the linearity of the relationships persists into the future, and it almost certainly will not, one can calculate the magnitude of these feedbacks on a warmer earth with doubled CO_2 concentration. If CO_2 exerts the only feedback (negative), a doubled concentration (700 ppmv) would sequester 15 Pg carbon per year in terrestrial ecosystems. Considering the effects of both temperature (assuming a global mean temperature increase of 2.5°C) and a doubled CO_2 concentration, feedbacks could range from a net annual storage of 10 Pg carbon (regression 2) to a net release of 13 Pg carbon per year (regression 3) (Table 19.2). These crude extrapolations demonstrate the ambiguity of the results to date. Even the net direction of the feedbacks is uncertain, to say nothing of the magnitudes possible under environmental conditions very different from those of the present day. However, the effects of future changes in land use seem less ambiguous. Deforestation in the tropics is about 15×10^6 ha (almost 1%) year^{-1} at present and increasing. The net effect of this conversion will be a reduction in the capacity of ecosystems to accumulate carbon in the future even if they have in the past.

In summary, if the net biotic flux was about zero throughout the late 1980s (Sarmiento et al. 1992), it may have been the result of three offsetting fluxes: a flux from changes in land use, releasing 1.4–1.7 Pg carbon per year to the atmosphere; a CO_2- or nutrient-enhanced storage, annually storing 1.4–2.8 Pg carbon per year on land; and a warming-enhanced respiration, releasing on the order of 0.7–0.9 Pg carbon per year (Table 19.2). These calculated sources and sinks of carbon and the factors responsible are, of course, only suggestive. The analysis has not considered changes in precipitation, which could well be the dominating influence on the storage of terrestrial carbon.

The abrupt decline (about 0.5 Pg carbon per year) in the global (terrestrial?) sink for carbon since the end of the 1970s is, perhaps, important. It has been observed by others (Keeling et al. 1989; Sarmiento et al. 1992), but the cause remains uncertain. Clearly, there has been an increased release of carbon from tropical deforestation over the last decade. It appears, from the analysis reported here, that there has also been an increased release (or reduced accumulation) elsewhere, perhaps caused by a warming-enhanced increase in respiration and decomposition.

IS THERE A TERRESTRIAL CARBON SINK IN NORTHERN MIDLATITUDES?

If the ocean model of Sarmiento et al. (1992) is correct, and the net flux of carbon from terrestrial ecosystems has been close to zero since the early 1980s, then the releases of carbon from deforestation in the tropics (between 1.4 and 1.7 Pg in the 1980s) must be balanced by the accumulation of carbon elsewhere. The analysis by Tans et al. (1990) suggests that the missing sink is in the northern midlatitudes or the northern temperate zone. Although the analysis presented here cannot distinguish geographical variations in the sources and sinks of carbon, two recent studies claim to have found direct evidence for a major sink in northern temperate and boreal forests (Kauppi et al. 1992; Sedjo 1992). Kauppi et al. (1992) calculated that European forests were accumulating 0.085–0.120 Pg carbon per year. If the same rate of accumulation applies to the temperate and boreal forests of North America and Asia, the total accumulation rate for northern regions could be 0.85–1.2 Pg carbon per year. Extrapolation of the European rate to all northern forests is probably not valid. Unlike large areas of Canadian and Russian forests, European forests have had a long history of intensive use and can be expected to be growing throughout the region.

Using similar data from forest inventories to compute the carbon balance for Europe, the United States, Canada, and the former USSR, Sedjo (1992) calculated that as much as 0.7 Pg carbon per year may be sequestered in northern regions. Sedjo, furthermore, pointed out that this sink represents a major departure from earlier studies that found such regions to be approximately in balance with respect to carbon (Houghton et al. 1987; Melillo et al. 1988).

The two recent analyses (Kauppi et al. 1992; Sedjo 1992) do not, however, provide sufficient evidence to revise previous estimates of the role of northern forests in the global carbon balance. On the contrary, the most remarkable aspect of these recent studies is that they confirm the results of the earlier studies. All four analyses show northern temperate and boreal zones to be accumulating between 0.6 and 0.8 Pg carbon per year in growing forests (Table 19.3). Different estimates for individual regions often vary by more than a factor of 2, but the differences are somewhat compensating and absolutely small. The new estimates of a carbon sink do not change those calculated earlier by more than a few tenths of a petagram of carbon per year.

The major difference between the earlier (Houghton et al. 1987; Melillo et al. 1988) and later (Kauppi et al. 1992; Sedjo 1992) studies arises not from their consideration of forest growth, but from the methods used to account for the rest of the carbon originally held in the vegetation and soil of forests logged, cleared, or otherwise disturbed. The rest of the carbon includes (1) that left to decay in the forest at harvest: stumps, roots, branches, twigs, leaves, and other debris; (2) that burned either as fuelwood or during site preparation before replanting; (3) that removed in wood products but gradually oxidized at rates depending on end use, such as lumber, paper, or veneer; and (4) that released to the atmosphere as a result of oxidation of soil carbon following clearing of forests for agriculture. Because of the low decay constants for soil organic matter, the release of carbon continues long after the initial clearing and cultivation of forest soil. Thus, although agricultural land may not have expanded significantly in northern regions over the last few decades, or may even have reverted to forests, some amount of carbon may still be being lost from soils placed under cultivation decades ago. Carbon may also be lost from soils immediately after logging, but this loss is not always

Table 19.3. Annual Rates of Accumulation (–) and Loss (+) of Carbon (Tg Carbon per Year) from Changes in the Major Forested Regions of the Northern Temperate Zone

Process	Reference	Former USSR	North Ameria	Europe	Total
Accumulation: growth of forests	Kauppi et al. (1992)			–85 to –120	–850 to –1200[a]
	Sedjo (1992)	–416	–179	–91	–686
	Melillo et al. (1987)	–154	–407	–225	–786
Loss	Melillo et al. (1987)				
Decay of logging debris		82	301	143	526
Burning[b]		24	6	16	46
Oxidation of wood products		46	96	51	193
Expansion and abandonment of agricultural lands		39	–4	–26	9
Total (accumulation and loss)		37	–8	–41	–12[c]

[a]The rest of the world's northern temperate and boreal forests are assumed to be accumulating carbon at a rate similar to that of European forests.

[b]Burning associated with logging and replanting, and with burning of fuelwood. Emissions do not include those resulting from wildfires.

[c]Houghton et al. (1987) calculated a net release of 32 Tg carbon per year for these regions. The major difference between the analyses by Houghton et al. and Melillo et al. was that the former assumed a 20% reduction in soil carbon with logging, followed by recovery during regrowth. Thus, the uncertainty associated with the response of soils to logging contributes an uncertainty of about 50 Tg carbon per year to the net flux.

observed (Johnson 1992) and is generally balanced by the redevelopment of soil carbon with subsequent regrowth of the forest. For example, the assumption of a 20% reduction in soil carbon with logging, followed by a recovery during regrowth, added less than 0.05 Pg carbon per year to the calculated flux of carbon for the temperate and boreal forests of North America, Europe, and the former USSR (Houghton et al. 1987). Thus, the net losses of carbon from soil are generally important for agriculture but not for logging.

When the four components listed in the preceding paragraph are included in carbon budgets, the net flux of carbon from changes in land use is quite different from that calculated on the basis of forest regrowth alone (Table 19.3). Had Kauppi et al. (1992) or Sedjo (1992) accounted for the initial carbon held in the disturbed ecosystems, they too might have found a net flux close to zero.

If the recent studies confirm that the accumulation of carbon in northern forests is taking place at a rate expected from past and current logging, what does this imply for the missing carbon? If additional carbon were accumulating in the northern temperate zone, could we measure it? The global carbon imbalance is about 1.5 Pg carbon per year (Houghton et al. 1990; Sarmiento and Sundquist 1992). If all of that carbon were accumulating in the trees of the northern temperate zone (area of temperate zone forests = 600×10^6 ha [Ajtay et al. 1979]), the accumulation would be about 2.5 Mg carbon ha^{-1} year^{-1} (1 Mg = 10^6 g = 1 metric ton). In comparison, the accumulation of carbon in European forests was 0.4–0.6 Mg carbon ha^{-1} year^{-1} (Kauppi et al. 1992). Thus, the accumulation of missing carbon would be about five times higher than that observed. It would be difficult to miss in the data on growth rates. Including boreal forests in the calculation lowers the average accumulation rate required to balance the global carbon equation to 1.0 Mg carbon ha^{-1} year^{-1}, but it is still high relative to observed growth rates. If the missing carbon were accumulating throughout the temperate zone, not just in forests but in woodlands, grasslands, and cultivated lands, the average accumulation rate would be about 0.5 Mg carbon ha^{-1} year^{-1}, about the rate observed in European forests. This rate of accumulation in the vegetation of grasslands and cultivated lands is high relative to the standing stocks of carbon there, and it would be obvious. Finally, if the missing carbon were accumulating in trees throughout the world, the rate would, again, average about 0.5 Mg carbon ha^{-1} year^{-1}. But this is the rate observed for regrowing European forests, forests with a long history of intensive use. That an additional accumulation of this magnitude, representing the missing sink, could be consistently overlooked is unlikely. That such an accumulation could be occurring in the rest of the forests of the world, many of which have not been logged and, hence, are not regrowing (for example, large regions of Canada, central and eastern Russia, and Brazil) seems equally unlikely. Regrowing forests are young; mortality is low and little carbon is lost with mortality. Even if mature forests around the world are not in steady state but are accumulating carbon, they would be unlikely to be growing at rates that approximate rates of regrowth immediately after logging. Overall, it seems unlikely that carbon is accumulating in vegetation anywhere at the rate required to balance the global carbon budget.

Perhaps the missing carbon is accumulating in the soils of the temperate zone. If so, the chances of measuring it are small. Annual carbon accumulation in soils is not readily available, and the standing stock of organic carbon in soils is approximately 1500 ± 300 Pg. The global carbon imbalance is two to three orders of magnitude smaller than the error associated with estimates of global soil carbon. Also, observed rates of carbon accumulation in soils not affected by land-use change are too low to account for the missing carbon even if the accumulation were spread over the entire land surface (Schlesinger 1990 and Chapter 8, this volume). Rates of accumulation in abandoned agricultural soils may be large enough, but these accumulations have already been included in the net flux of carbon from change in land use. Finally, if carbon were accumulating in undisturbed soils, what is its pathway? Have increasing concentrations of CO_2 enhanced belowground production (Norby et al. 1992), not measured in forestry surveys of growth? It is difficult to believe that increased belowground production would not lead to increased aboveground growth as well.

The calculations in this chapter are based on the best estimates of carbon emissions and accumulations. The errors associated with these estimates are about ±0.5 Pg carbon per year for emissions of carbon from land-use change and for uptake by the ocean. Thus, the missing sink might be 0.5 Pg carbon per year, considerably smaller than 1.5 Pg carbon per year, and it could be accumulating in vegetation and soil without detection.

ACKNOWLEDGMENTS

I thank J. Hackler for his help with analysis and graphics and J. L. Sarmiento for providing the results of his deconvolution. The research was supported by the U.S. Department of Energy, Carbon Dioxide Research Program.

REFERENCES

Ajtay, G. L., P. Ketner, and P. Duvigneaud. 1979. Terrestrial primary production and phytomass. In *The Global Carbon Cycle,* ed. B. Bolin, E. T. Degens, S. Kempe, and P. Ketner, pp. 129–182. SCOPE 13, Wiley, New York.

Allen, L. H., K. J. Boote, J. W. Jones, P. H. Jones, R. R. Valle, B. Acock, H. H. Rogers, and R. D. Dahlman. 1987. The response of vegetation to rising carbon dioxide: Photosynthesis, biomass, and seed yield of soybean. *Global Biogeochem. Cycles* 1:1–14.

Bacastow, R., and C. D. Keeling. 1973. Atmospheric carbon dioxide and radio-carbon in the natural carbon cycle. II. Changes from A.D. 1700 to 2070 as deduced from a geochemical model. In *Carbon and the Biosphere,* ed. G. M. Woodwell and E. V. Pecan, pp. 86–135. U.S. Atomic Energy Commission, Symp. Series 30, National Technical Information Service, Springfield, Va.

Broecker, W. S., T. Takahashi, H. H. Simpson, and T.-H. Peng. 1979. Fate of fossil fuel carbon dioxide and the global carbon budget. *Science* 206:409–418.

Brown, S., A. J. R. Gillespie, and A. E. Lugo. 1989. Biomass estimation methods for tropical forests with applications to forest inventory data. *For. Sci.* 35:881–902.

Fearnside, P. M., A. T. Tardin, and L. G. M. Filho. 1990. Deforestation rate in Brazilian Amazonia. National Secretariat of Science and Technology, Brasilia, Brazil.

Flint, E. P., and J. F. Richards. 1994. Trends in carbon content of vegetation in south and southeast Asia associated with changes in land use. In *Effects of Land Use Change on Atmospheric CO_2 Concentrations: South and Southeast Asia as a Case Study,* ed. V. H. Dale, pp. 201–209. Springer-Verlag, New York.

Food and Agriculture Organization. 1993. Forest Resources Management 1990 Program: tropical countries. FAO Forestry Paper 112. FAO, Rome, Italy.

Friedli, H., H. Lotscher, H. Oeschger, U. Siegenthaler, and B. Stauffer. 1986. Ice core record of the $^{13}C/^{12}C$ ratio of atmospheric CO_2 in the past two centuries. *Nature* 324:237–238.

Grulke, N. E., G. H. Riechers, W. C. Oechel, U. Hjelm, and C. Jaeger. 1990. Carbon balance in tussock tundra under ambient and elevated atmospheric CO_2. *Oecologia* 83:485–494.

Hall, C. A. S., and J. Uhlig. 1991. Refining estimates of carbon released from tropical land-use change. *Can. J. For. Res.* 21:118–131.

Hansen, J., and S. Lebedeff. 1987. Global trends of measured surface air temperature. *J. Geophys. Res.* 92:13345–13372.

Houghton, J. T., G. J. Jenkins, and J. J. Ephraums, eds. 1990. *Climatic Change. The IPCC Scientific Assessment.* Cambridge University Press, New York.

Houghton, R. A. 1991. Tropical deforestation and atmospheric carbon dioxide. *Climatic Change* 19:99–118.

Houghton, R. A. 1993. Changes in terrestrial carbon over the last 135 years. In *The Global Carbon Cycle,* ed. M. Heimann, pp. 139–157. Springer-Verlag, Berlin.

Houghton, R. A., and J. L. Hackler. 1994. The net flux of carbon from deforestation and degradation in South and Southeast Asia. In *Effects of Land Use Change on Atmospheric CO_2 Concentrations: South and Southeast Asia as a Case Study,* ed. V. H. Dale, pp. 301–327. Springer-Verlag, New York.

Houghton, R. A., R. D. Boone, J. R. Fruci, J. E. Hobbie, J. M. Melillo, C. A. Palm, B. J. Peterson, G. R. Shaver, G. M. Woodwell, B. Moore, D. L. Skole, and N. Myers. 1987. The flux of carbon from terrestrial ecosystems to the atmosphere in 1980 due to changes in land use: Geographic distribution of the global flux. *Tellus* 39B:122–139.

Houghton, R. A., D. L. Skole, and D. S. Lefkowitz. 1991. Changes in the landscape of Latin America between 1850 and 1980. II. A net release of CO_2 to the atmosphere. *For. Ecol. Man.* 38:173–199.

Johnson, D. W. 1992. Effects of forest management on soil carbon storage. *Water Air Soil Pollut.* 64:83–120.

Kauppi, P. E., K. Mielikainen, and K. Kuusela. 1992. Biomass and carbon budget of European forests, 1971–1990. *Science* 256:70–74.

Keeling, C. D., R. B. Bacastow, A. F. Carter, S. C. Piper, T. P. Whorf, M. Heimann, W. G. Mook, and H. Roeloffzen. 1989. A three-dimensional model of atmospheric CO_2 transport based on observed winds: 1. Analysis of observational data. In *Aspects of Climate Variability in the Pacific and the Western Americas,* ed. D. H. Peterson, pp. 165–236. Geophysical Monograph 55. American Geophysical Union, Washington, D.C.

Melillo, J. M., J. R. Fruci, R. A. Houghton, B. Moore, and D. L. Skole. 1988. Land use change in the Soviet Union between 1850 and 1980: Causes of a net release of CO_2 to the atmosphere. *Tellus* 40B:116–128.

Mooney, H. A., B. G. Drake, R. J. Luxmoore, W. C. Oechel, and L. F. Pitelka. 1991. Predicting ecosystem responses to elevated CO_2 concentrations. *BioScience* 41:96–104.

Myers, N. 1991. Tropical forests: Present status and future outlook. *Climatic Change* 19:3–32.

Neftel, A., E. Moor, H. Oeschger, and B. Stauffer. 1985. Evidence from polar ice cores for the increase in atmospheric CO_2 in the past two centuries. *Nature* 315:45–47.

Norby, R. J., C. A. Gunderson, S. D. Wullschleger, E. G. O'Neill, and M. K. McCracken. 1992. Productivity and compensatory responses of yellow-poplar trees in elevated CO_2. *Nature* 357:322–324.

Peterson, B.J., and J.M. Melillo. 1985. The potential storage of carbon caused by eutrophication of the biosphere. *Tellus* 37B:117–127.

Raich, J. W., and W. H. Schlesinger. 1992. The global carbon dioxide flux in soil respiration and its relationship to vegetation and climate. *Tellus* 44B:81–99.

Rastetter, E. B., R. B. McKane, G. R. Shaver, and J. M. Melillo. 1992. Changes in C storage by terrestrial ecosystems: How C-N interactions restrict responses to CO_2 and temperature. *Water Air Soil Pollut.* 64:327–344.

Sarmiento, J. L., and E. T. Sundquist. 1992. Revised budget for the oceanic uptake of anthropogenic carbon dioxide. *Nature* 356:589–593.

Sarmiento, J. L., J. C. Orr, and U. Siegenthaler. 1992. A perturbation simulation of CO_2 uptake in an ocean general circulation model. *J. Geophys. Res.* 97:3621–3645.

Schleser, G. H. 1982. The response of CO_2 evolution from soils to global temperature changes. *Z. Naturforsch. A* 37:2037/1–2037/5.

Schlesinger, W. H. 1990. Evidence from chronosequence studies for a low carbon-storage potential of soils. *Nature* 348:232–234.

Sedjo, R. A. 1992. Temperate forest ecosystems in the global carbon cycle. *Ambio* 21:274–277.

Setzer, A. W., and M. C. Pereira. 1991. Amazonia biomass burnings in 1987 and an estimate of their tropospheric emissions. *Ambio* 20:19–22.

Shaver, G. S., W. D. Billings, F. S. Chapin, A. E. Giblin, K. J. Nadelhoffer, W. C. Oechel, and E. B. Rastetter. 1992. Global change and the carbon balance of arctic ecosystems. *BioScience* 42:433–441.

Siegenthaler, U., and H. Oeschger. 1987. Biospheric CO_2 emissions during the past 200 years reconstructed by deconvolution of ice core data. *Tellus* 39B:140–154.

Skole, D. L., and C. J. Tucker. 1993. Tropical deforestation and habitat fragmentation in the Amazon: Satellite data from 1978 to 1988. *Science* 260:1905–1910.

Strain, B. R., and J. D. Cure, eds. 1985. Direct effects on increasing carbon dioxide on vegetation. DOE/ER-0238. U.S. Department of Energy, Washington, D.C.

Tans, P. P., I. Y. Fung, and T. Takahashi. 1990. Observational constraints on the global atmospheric CO_2 budget. *Science* 247:1431–1438.

Woodwell, G. M. 1989. The warming of the industrialized middle latitudes 1985–2050: Causes and consequences. *Climatic Change* 15:31–50.

20

Storage versus Flux Budgets: The Terrestrial Uptake of CO₂ during the 1980s

Pieter P. Tans, Inez Y. Fung, and Ian G. Enting

Measurements of the atmospheric CO_2 mixing ratio have played a major role in the discussion during recent years of the magnitudes and whereabouts of the major sources and sinks of atmospheric CO_2. Any hypothesis about sources of global significance must satisfy the observed spatial distribution of CO_2 and its rate of growth, today and in recent history. The link between spatial and temporal source-sink patterns and the global distribution of CO_2 mixing ratios in the atmosphere is provided by numerical models of atmospheric diffusion and transport. The models calculate concentration patterns resulting from specified source distributions. The major transport characteristics of the models have been independently calibrated with relatively well-known input functions, such as those describing the anthropogenic chlorofluorocarbons and radioactive [85]Kr.

First we will briefly review two approaches to the problem, those taken by Keeling et al. (1989a,b) and Tans et al. (1990). Various revisions proposed since then—by Enting and Mansbridge (1991), Robertson and Watson (1992), and Sarmiento and Sundquist (1992)—will also be discussed. In all of these attempts, agreement with the atmospheric observations is obtained only when it is assumed that terrestrial ecosystems at temperate latitudes are, on average, currently functioning as a large sink of CO_2. However, there is not yet direct and conclusive evidence of increased carbon storage of the required magnitude in these ecosystems. (See recent papers by Kauppi et al. (1992), Wofsy et al. (1993) and Dixon et al. (1994). See also Chapter 19.)

Our convention in this chapter will be to call fluxes out of the atmosphere positive, and those when CO_2 is entering the atmosphere, negative. The reason for this choice is that we shall end by describing storage in reservoirs other than the atmosphere, i.e., the oceans and terrestrial ecosystems. An increase of carbon storage in the oceans will then have a positive sign, and fossil fuel combustion carries a negative sign because the fossil fuel reservoir is being depleted. The budget is closed by requiring that the sum of all components (oceans, terrestrial, fossil fuel, deforestation, and atmospheric increase) equal zero.

Table 20.1 summarizes the global CO_2 budgets for the 1980s of Keeling et al. (1989b), designated KPH 1984, and Tans et al. (1990), called TFT 1981–1987. The

Table 20.1. Proposed Contemporary Atmospheric Carbon Budgets
(Pg Carbon per Year)

	Oceans	Terrestrial	Fossil	Deforestation
KPH 1984[a]				
16–90°N	+2.3	+1.0	–4.8	–0.4
Equator	–1.1	+0.9	–0.2	–1.2
16–90°S	+1.1	+0.3	–0.2	–0.2
	+2.3	+2.3[b]	–5.2	–1.8
TFT 1981–1987[c]				
15–90°N	+0.6	+2.3	–4.9	–0.0
Equator	–1.3	+0.5	–0.2	–1.0
15–90°S	+1.1	+0.1	–0.2	–0.0
	+0.4	+2.9	–5.3	–1.0

[a] Standard case scenario of Keeling et al. (1989b) for the year 1984. Total deforestation is 1.8 Pg carbon per year.
[b] This sum appears not to add up because of rounding errors.
[c] Scenario 7 of Tans et al. (1990), with "empirical" air-sea transfer and 1 Pg carbon per year of tropical deforestation, and "CO_2 fertilization" proportional to net primary productivity.

primary constraint used by Keeling et al. (1989a) is the time history of CO_2 and its $^{13}C/^{12}C$ isotopic ratio during the last century as established from measurements in ice cores and in the atmosphere. A box model of the global carbon cycle incorporating the history of fossil fuel combustion, estimates of past and present deforestation, ocean CO_2 uptake, and postulated increased terrestial plant uptake ("CO_2 fertilization") provided satisfactory agreement with the observed CO_2 and $^{13}C/^{12}C$ time series. As a secondary constraint use was made of the latitudinal gradient (north pole to south pole) of the CO_2 mixing ratio during the last decades and the gradient of its $^{13}C/^{12}C$ ratio in more recent years. However, it should be noted that in their standard case budget for 1984, representative of the most recent period, the atmospheric increase was only 2.4 petagrams (Pg) ($=10^{15}$ g) per year. The observed increase was close to 3.1 Pg carbon per year. This discrepancy could be remedied by increasing the postulated rate of deforestation in the tropics by 0.7 Pg carbon per year. Such a remedy would provide the minimum upset to the modeled latitudinal gradient of CO_2, as it would increase the modeled annual average CO_2 mixing ratios at the tropical air sampling sites by approximately only 0.2 ppmv. However, such an adjustment might upset the long-term trade-off between deforestation and CO_2 fertilization in their simulation of historic CO_2 levels.

The observed $^{13}C/^{12}C$ isotopic ratio of CO_2 decreases from the equator to the Arctic just as one would expect from the burning of fossil fuels, which have a lower $^{13}C/^{12}C$ ratio. A net flux of CO_2 into the oceans does not have a significant impact on the atmospheric isotopic ratio, whereas terrestrial uptake discriminates against ^{13}C, enriching the $^{13}C/^{12}C$ ratio of the CO_2 left behind in the atmosphere. Terrestrial uptake would thereby partially compensate for the $^{13}C/^{12}C$ depletion caused by fossil fuel burning. The analysis of the atmospheric $^{13}CO_2/^{12}CO_2$ observations by Keeling et al. (1989a, b) suggests that there is no need for such compensation and implicates the oceans as the major uptake reservoir in the Northern Hemisphere.

A complicating factor in the $^{13}C/^{12}C$ argument of the previous paragraph is isotopic exchange with seawater. Note that here we are not talking about *net* uptake, because in the pure exchange process a molecule of $^{12}CO_2$ leaves the water for every one of $^{13}CO_2$ entering, and vice versa (see Tans et al. [1993] for a detailed treatment). The exchange will try to establish thermodynamic equilibrium between the isotopic ratio in the air and that in the water locally. The equilibrium fractionation factor between gaseous and dissolved CO_2 depends on temperature. Thus CO_2 in the atmosphere will tend to have a low $^{13}C/^{12}C$ ratio at high latitudes (above cold waters) in both hemispheres compared to the equator. The magnitude of this tendency in the atmosphere is significant compared to the isotopic signal produced by fossil fuel burning. It depends mostly on the speed of air-sea exchange relative to atmospheric mixing, once the isotopic ratios in ocean surface water are constrained by observations.

It has been shown by Broecker and Peng (1992) from chemical oceanographic data that North Atlantic deep water (NADW) transports about 0.6 Pg more inorganic carbon each year from the Northern to the Southern Hemisphere than the return flow carries northward. This is a part of the "conveyor belt" circulation of the oceans. They postulate that the CO_2 thus transported is released to the atmosphere in southern circumpolar waters. If this process had been occurring in preindustrial times, it could have led to a reverse (i.e., southern concentrations higher than northern) atmospheric gradient of circa 1.2 ppmv, as was postulated by Keeling et al. (1989a, 1989b) to explain the low value of the gradient observed today. However, there is currently no direct evidence from ΔP_{CO_2} data (corrected for the fossil fuel increase of atmospheric CO_2) that most of this excess CO_2 indeed was leaving the oceans around Antarctica. (ΔP_{CO_2} is the difference in CO_2 partial pressure between the ocean surface and the atmosphere.) It is possible that a large fraction of the excess carbon could have been carried along farther with the conveyor belt circulation to end up in the North(!) Pacific.

The approach used by Tans et al. was somewhat different. The primary constraints were the observed latitudinal gradient of atmospheric CO_2 and its global annual rate of increase from 1981 to 1987. Their calculations led to the conclusion that there had to be quite a large sink of CO_2 in the Northern Hemisphere because the observed CO_2 gradient was much smaller than that expected owing to fossil fuel combustion alone. To constrain the problem further they used a compilation of ΔP_{CO_2}, the thermodynamic driving force of net air-sea fluxes. Combined with different wind speed parameterizations of the piston velocity describing the kinetics of air-sea exchange, this led to estimates of the net CO_2 flux into the oceans in the Northern Hemisphere. The net air-sea fluxes derived in this way were rather small, suggesting that the larger part of the CO_2 sink in the Northern Hemisphere had to be on the land. The ΔP_{CO_2} values appeared to be too small for the large uptake of CO_2 in the North Atlantic required by Keeling et al. and by Broecker and Peng.

The scenario presented in Table 20.1 has a "reasonable" amount of carbon loss from tropical ecosystems, which is estimated to be 1.6 ± 1.0 Pg carbon per year (IPCC 1991). A higher rate of tropical deforestation than 1 Pg carbon per year could be accommodated by TFT 1981–1987 in the comparison of the model transport with the observations, but only if additional stimulation of CO_2 uptake (more than 0.5 Pg carbon per year^{-1}) was assumed in tropical latitudes at the same time. The observed atmospheric gradients constrain the *net* amount of carbon loss in an area. If the direct losses due to land use changes were 3.0 Pg carbon per year, they would have to be compensated by CO_2 absorption in the same area of approximately 2.5 Pg carbon per year.

A CO_2 SURFACE FLUX BUDGET

Two types of corrections must be applied to carbon budgets inferred from observed CO_2 gradients. First of all, the gradients themselves are not caused exclusively by surface sources and atmospheric transport. The oxidation of CO to CO_2 in the atmosphere makes a small contribution to the concentration gradient, which should be taken into account. Second, surface fluxes of CO_2 do not translate directly into storage at the same location; part of the carbon fixed from CO_2 by photosynthesis may escape again in the form of reduced gaseous compounds (e.g., terpenes and isoprene) and as dissolved organic and inorganic carbon in rivers. A small part of the carbon storage involves sedimentary reservoirs instead of the more labile oceans and terrestrial ecosystems.

In this section we will treat the corrections needed to obtain the direct CO_2 fluxes between the earth's surface and the atmosphere. Enting and Mansbridge (1991) discussed adjustments to the carbon budget due to the fluxes of CO and CH_4. Here we will only include the correction due to CO oxidation by OH radicals in the atmosphere. This actually includes most of the CH_4 budget because about 90% of the atmospheric oxidation of CH_4 proceeds via CO as an intermediate (Warneck 1988). Using a two-dimensional atmospheric model (Tans et al. 1989), we first calculate an estimate for the CO_2 production from this source and the resulting CO_2 distribution at the surface. The CO mixing ratios are defined as follows: observed annual mean and seasonal surface values (Novelli et al. 1992), observed (sparse) values for the upper troposphere (Logan et al. 1981) with no seasonal cycle, linear interpolation for the rest of the troposphere, and an assumed CO mixing ratio of 40 ppb in the stratosphere. For the OH fields we use those calculated by Spivakovsky et al. (1990) in the troposphere, augmented by values in the lower stratosphere calculated from a stratospheric photochemical model (S. Solomon personal communication). The total resulting CO_2 production is 0.92 Pg carbon per year (Figure 20.1). If we then invert the calculated resulting CO_2 surface distribution in the usual way, i.e., assuming that all the sources and sinks causing the distribution are located at the surface, we find for the apparent surface sources a somewhat different distribution (dashed line in Figure 20.1). The difference arises because the CO_2 produced in the atmosphere is transported toward the surface through eddy diffusion and through the large-scale advective structure incorporated into the two-dimensional model. These calculated apparent surface sources—0.4, 0.3, 0.2 Pg carbon per year for the northern, equatorial, and southern latitudes, respectively—originating from atmospheric CO oxidation are subtracted from the terrestrial CO_2 flux in the equatorial and northern regions (Table 20.2A), or, equivalently, added to the terrestrial storage at these latitudes. The correction is not applied to the ocean fluxes in these regions because they are based on observed ΔP_{CO_2}. However, the CO correction is applied to the estimate of the southern ocean uptake because it was derived from the observed atmospheric CO_2 distribution, not from ΔP_{CO_2}.

Another minor flux adjustment is applied at this stage. A small part of the fossil fuel usage results in the emission of CH_4 and CO. Ratios between excess CO_2, CH_4, and CO concentrations at Arctic sites remote from source regions in air masses contaminated by anthropogenic sources suggest that the total amount of excess CH_4 and CO relative to excess CO_2 is circa 1.5% for each (Conway et al. 1994). We take account of this observation by subtracting 0.1 Pg carbon per year from the fossil fuel CO_2 emissions in the Northern Hemisphere because CO was already included in the

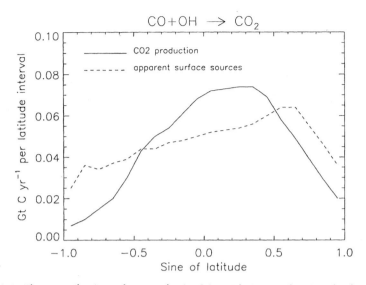

Figure 20.1. The contribution of atmospheric CO oxidation to the CO_2 budget. The CO_2 from this source (solid line; for assumptions see main text) gives rise to a concentration pattern in the atmosphere. The dashed line gives the equivalent surface source resulting in the same concentration pattern in the model surface layer (900–1000 mb).

emission estimates that are based on consumption. In order not to change the latitudinal atmospheric CO_2 gradient, the terrestrial sink of CO_2 is decreased by 0.1 Pg carbon per year at the same latitudes. The CH_4 emissions are related to the mining and transportation of the fossil fuels and have not been counted as consumption. We will include these later. At this point the total sinks from the atmosphere have been increased by 0.9 Pg carbon per year to compensate exactly for the photochemical production of CO_2 from CO in the atmosphere. We will account later for where the CO comes from.

Robertson and Watson (1992) pointed out that a correction should be applied to the measured ΔP_{CO_2} values in the derivation of air-sea fluxes. The CO_2 partial pressure is measured in water a few meters below the surface. However, the atmosphere "feels" only the skin of the ocean, which is, on average, a few tenths of a degree cooler than the bulk of the water. Since the solubility of CO_2 is higher at lower temperatures, the amount of dissolved CO_2 in equilibrium with the atmosphere is a little higher at the cooler skin temperature. The same amount of dissolved CO_2 at a few meters depth would, at the higher bulk water temperature, result in a P_{CO_2} several microatmospheres higher than that in equilibrium with the atmosphere.

We have used the skin temperature corrections estimated by Robertson and Watson to correct the CO_2 surface flux budget (Table 20.2A). The adjustment is applied to the northern and equatorial ocean fluxes because those estimates were based on observed ΔP_{CO_2}, and a corresponding opposite correction is given to the terrestrial fluxes to keep the atmospheric latitudinal gradient unchanged. Robertson and Watson arrived at their estimates by using a parameterization of the skin temperature difference based on the heat flux, radiation flux, and wind speed proposed by Hasse (1971). However,

Table 20.2. Corrections to the TFT 1981–1987 Budget[a]

A. CO_2 surface flux budget

	TFT(7)	CO	Fossil fuel	Skin temperature	Total
Oceans					
15–90°N	+0.6			+0.2	+0.8
Equator	−1.3			+0.2	−1.1
15–90°S	+1.1	+0.2			+1.3
Land					
15–90°N	+2.3	+0.4	−0.1	−0.2	+2.4
Equator	+0.5	+0.3		−0.2	+0.6
15–90°S	+0.1				+0.1
Fossil					
15–90°N	−4.9		+0.1		−4.8
Equator	−0.2				−0.2
15–90°S	−0.2				−0.2
Land use					
15–90°N	0.0				
Equator	−1.0				−1.0
15–90°S	0.0				

B. Carbon storage budget

	CO_2 surface fluxes	$C_{organic}$	$CaCO_3$	Reduced gas escape	Ocean currents	Total
Oceans						
15–90°N	+0.8	+0.1 (+0.3)	+0.2		−0.7 (−0.8)	+0.4 (+0.5)
Equator	−1.1	+0.2 (+0.4)	+0.1		+1.1 (+1.0)	+0.3 (+0.4)
15–90°S	+1.3		−0.1		−0.4 (−0.2)	+0.8 (+1.0)
Land						
15–90°N	+2.4	−0.1 (−0.3)	−0.1	−0.3		+1.9 (+1.7)
Equator	+0.6	−0.2 (−0.4)	−0.1	−0.4		−0.1 (−0.3)
15–90°S	+0.1			−0.1		0.0
Fossil						
15–90°N	−4.8			−0.1		4.9
Equator	−0.2					−0.2
15–90°S	−0.2					−0.2
Land use						
15–90°N						
Equator	−1.0					−1.0
15–90°S						

[a]The sign convention is positive for fluxes out of the atmosphere and for increased storage in carbon reservoirs other than the atmosphere. Units are Pg carbon per year. TFT scenario 7 is adjusted for the oxidation of CO to CO_2 in the atmosphere, the fraction of fossil fuel consumption not initially released as CO_2, and the ΔP_{CO_2} correction due to the cool skin of the oceans, respectively. The CO_2 surface flux budget in the last column of (A) together with CO oxidation determines the spatial pattern of the atmospheric CO_2 mixing ratio. This budget is balanced as follows: −5.2 (fossil fuels) + 3.0 (atmospheric increase) + 1.0 (ocean uptake) + 3.1 (land uptake) − 1.0 (deforestation) − 0.9 (CO oxidation) = 0.0. The carbon storage budget, the last column of (B), determines the long-term changes of carbon storage in reservoirs. It is derived from the CO_2 flux budget by accounting for fluxes other than CO_2 gas, namely the transport of organic carbon and bicarbonate in rivers, the escape of reduced gases from terrestrial ecosystems, and the redistribution of carbon in the oceans. The numbers in parentheses correspond to the river fluxes of Wollast and Mackenzie (1989).

Schluessel et al. (1990) found that the Hasse parameterization was only moderately successful in reproducing their skin temperature measurements.

At this point we have derived a budget of surface CO_2 fluxes that is, together with the oxidation of CO, fully responsible for and consistent with the observed latitudinal distribution of the atmospheric CO_2 mixing ratio.

A CARBON STORAGE BUDGET

If one is looking for changes in the amounts of carbon stored in various reservoirs, several corrections need to be applied to the surface flux budget of CO_2. This section will consider fluxes of carbon in forms other than gaseous CO_2. Part of the carbon fixed via photosynthesis on the land is not stored there, but is carried by rivers to the oceans in the form of dissolved and particulate organic matter. This flux was estimated by Sarmiento and Sundquist (1992), based on various sources, as 0.4 Pg carbon per year. As a first approximation, we divide the total of 0.4 Pg carbon per year equally between the northern and the equatorial latitudes because of the location of continents and rivers. This is done in the atmospheric budget by moving a sink from the terrestrial biosphere to the oceans. Note that we define ocean storage of carbon here as organic and inorganic carbon combined.

Part of the organic carbon in the oceans, about 0.1 Pg carbon per year, is ultimately buried in the sediments each year, mostly on the continental shelves (Berner 1982). We have subtracted it from the ocean storage in the section with the largest coastal area, the northern oceans. The buried organic material is eventually returned to the atmosphere as CO_2 during the erosion of sedimentary rocks, which would represent a source of CO_2 to the atmosphere that has been neglected thus far. To account for it, the storage in terrestrial ecosystems derived by us from the atmospheric concentration should be increased by the same amount, 0.1 Pg carbon per year. We have added it to the terrestrial uptake north of 15°N. The actual fate of the organic fraction in sedimentary rocks is very uncertain. Kerogen (the collective name of this organic material) is quite refractory after its passage through the sedimentary cycle.

The "steady-state" deposition of $CaCO_3$ was estimated by Sarmiento and Sundquist (1992) to be 0.2 Pg carbon per year. On the land the same amount of carbon (neglecting silicate weathering) is needed to dissolve the limestone, via the reaction $CO_2 + CaCO_3 + H_2O \Leftrightarrow Ca^{2+} + 2HCO_3^-$ (and also dolomite, $MgCa[CO_3]_2$). We subtract this amount from the storage in terrestrial ecosystems at northern and equatorial latitudes because it represents a CO_2 sink on the land that does not contribute to terrestrial storage. The products of the erosion of sedimentary rock on the continents—calcium, magnesium, silicate, and 0.4 Pg carbon per year in the form of bicarbonate—are delivered to the oceans via rivers. We again divide this input of carbon to the oceans equally between the northern and equatorial oceans. A large part of the $CaCO_3$ removal takes place in the building of coral reefs (Opdyke and Walker 1992). To a first order of approximation, we take account of the $CaCO_3$ deposition by removing 0.1 Pg carbon per year each from the equatorial and southern portions of the oceans.

Finally, part of the CO_2 taken up in photosynthesis is lost in the form of emissions of reduced gases by plants, mostly isoprene and terpenes. The direct production of CO_2

from isoprene and terpene oxidation in the atmosphere has already been taken into account implicitly in the CO_2 gradient method. One group of oxidation products, the organic acids, tend to dissolve in water, where they form a part of the organic carbon loading of natural waters. Another important group is particles, which have generally short atmospheric lifetimes. The latter group could possibly constitute another significant route of transfer of organic carbon from the land to the oceans, but we will neglect it here. In that case we only have to consider the contribution of the long-lived gaseous intermediate oxidation product CO. Direct emissions of CO by vegetation and biomass burning are further sources of CO that must be subtracted from terrestrial storage. We adopt the estimates of WMO 1991 for the latter two CO sources and J. Logan's estimate for CO production from the oxidation of nonmethane hydrocarbons (NMHC), also presented in WMO 1991. The vegetation and NMHC source are assumed to be distributed in proportion to net primary productivity.

Furthermore, the emissions of CH_4 also must be subtracted from terrestrial storage. We take the estimates, including their spatial distribution, of the preferred scenario 7 of Fung et al. (1991) for the CH_4 emissions from animals, peat bogs and tundra, swamps, rice agriculture, landfills, biomass burning, and termites. The sum of these CH_4 emissions and the CO sources mentioned in the previous paragraph leads to a decrease of terrestrial storage by 0.3 Pg carbon per year north of 15°N, 0.4 Pg carbon per year in equatorial latitudes, and 0.1 Pg carbon per year south of 15°S (Table 20.2B). The escape of 0.1 Pg carbon per year of CH_4 associated with fossil fuel mining operations and transport must be subtracted from the fossil fuel reservoir. Total carbon uptake by the oceans is now 1.5 Pg carbon per year, whereas the global net uptake by terrestrial ecosystems is 0.8 Pg carbon per year.

Until this point the ocean estimates are still based on where the inputs take place. The inventory of carbon in the ocean basins does not change strictly in proportion to those inputs because in the steady state the ocean circulation transports carbon from one basin to another. Adopting Broecker and Peng's estimate of 0.6 Pg carbon per year for the steady-state flux carried by NADW, the inventory increase in the northern basins would decrease to 0.5 Pg carbon per year. The carbon loss from equatorial waters is balanced by influx from both north and south. Sarmiento et al. (1992) have simulated the uptake of the anthropogenic atmospheric CO_2 perturbation by the oceans with an ocean general circulation model. In the last column of Table 20.2B, we have adopted a distribution of ocean carbon storage proportional to the total excess carbon inventory simulated by Sarmiento et al. (their Figure 5). This would imply the movement of 0.7 and 0.4 Pg carbon per year from the northern and southern oceans, respectively, to the equator. The Broecker and Peng NADW flux of 0.6 Pg carbon per year is a part of this. If indeed all of the 0.6 Pg left the ocean in Antarctic waters, that would imply a large asymmetry in any further redistribution of the anthropogenic CO_2. Respectively 0.1 and 1.0 Pg carbon per year would have to be moved from the northern and the southern waters to the equatorial zone because all of the 0.6 Pg carbon per year would initially contribute to an increase of storage in the southern waters.

In our treatment of carbon sources and sinks, transport by rivers is a mechanism to move carbon into the oceans independent of air-sea exchange and ΔPco_2. Today's river flux of carbon could be significantly larger because of mankind's influence, in which case our shifting of the carbon sink from terrestrial ecosystems to the oceans would be larger as well. Wollast and Mackenzie (1989) estimate the present-day flux of organic carbon in rivers as 0.8 Pg carbon per year. This alternative scenario is presented within parentheses in Table 20.2B. In it the rivers carry 0.4 Pg carbon per year each from the land to the oceans in the equatorial and northern zones. Total ocean uptake would increase to 1.9 Pg carbon per year, and net carbon storage on land would decrease to 0.4 Pg carbon per year. If we assume that the NADW flux of 0.6 Pg carbon per year completely outgasses in Antarctic waters, this scenario would imply that there should be fluxes of 0.2 and 0.8 Pg carbon per year from the northern and southern oceans, respectively, to the equator, a somewhat less asymmetrical situation than that described in the previous paragraph.

Opdyke and Walker (1992) argue that the calcite deposition-erosion cycle is not in a long-term steady-state mode, but that it is likely that calcite deposition during the Holocene has been larger than the input to the oceans. Anthropogenic perturbations would come on top of the average Holocene flux. The river flux of inorganic carbon may not have increased significantly in recent times (Wollast and Mackenzie 1989). In the budget of Table 20.2B we do not consider such possible modifications of the calcium carbonate cycle.

DISCUSSION

Our starting point for revisions to the TFT 1981–1987 budget was scenario 7 (Tans et al. 1990). We could have used another scenario, for instance number 6, which matches equally well the constraints imposed by the atmospheric CO_2 mixing ratios and the ocean ΔPCO_2 data. Carrying the corrections through to a storage budget for scenario 6 similar to that presented in Table 20.2B, the total uptake by the oceans would increase from 1.5 to 1.8, and the total storage on land would decrease from 0.8 to 0.5 Pg carbon per year. Scenario 6 would imply a net loss of carbon from the tropical forests of 0.6 Pg year in addition to the initially postulated 1.0 Pg carbon per year from deforestation activities.

It has been assumed thus far that the entire anthropogenically perturbed river flux of organic carbon contributes to ocean storage. That assumption would be correct only if we neglected sedimentation on the shelves and if the global coverage of ΔPCO_2 measurements (actually we would need direct air-sea flux measurements) would incorporate coastal and shelf waters, because any escape in the form of CO_2 gas would then be included. This is not the case. Our CO_2 surface flux budget is based on open ocean ΔPCO_2 measurements, and any CO_2 fluxes in coastal waters have been missed. The fate of organic carbon delivered by rivers to coastal waters is uncertain, and much of it may be transformed into CO_2 and lost to the atmosphere relatively rapidly, never making it to the open oceans. Suppose, as a thought experiment, that all organic carbon delivered by rivers escapes from the coastal waters to the atmosphere as CO_2 gas. For this to happen coastal ΔPCO_2 values would have to be quite high. Their inclusion in the ocean surveys would then lead to a lower estimate of global net ocean uptake of CO_2 via air-sea gas exchange, with the change precisely balancing the river import of organic carbon. Therefore, the ocean storage in the case of the estimate of the river flux of organic carbon by Wollast and Mackenzie (1989) is an upper limit, assuming that the figure of 0.8 Pg carbon per year is correct. Sarmiento and Sundquist (1992) do not include an anthropogenically enhanced river flux of organic carbon in their budget because they only include the fraction that turns into inorganic carbon in the oceans, which is small in their estimation. Note that our storage of carbon in the oceans includes both inorganic carbon and organic matter, which eventually has to either be turned into inorganic carbon or be buried in sediments. The amount and age of organic material in the oceans produced by marine organisms is subject to much debate (e.g., Sugimura and Suzuki 1988; Martin and Fitzwater 1992). Knowledge of the fate of terrestrial organic carbon and nutrients delivered to coastal and shelf waters is mostly lacking.

The CO_2 surface flux corrections and the carbon storage corrections we have applied to the TFT 1981–1987 budget do not apply in the same way to the KPH 1984 budget because the primary constraint in the latter case was the long-term increase of atmospheric CO_2. The question of how the CO_2 entered the oceans or the terrestrial biosphere was not addressed. The KPH 1984 budget should be compared with Table 20.2B before the ocean currents are taken into account. For the northern region, the remaining difference consists of a sink of about 1 Pg carbon per year that is attributed to either the land or the oceans. However, the differences between the two budgets for the equatorial region have widened compared to the values given in Table 20.1.

The $^{13}C/^{12}C$ latitudinal gradient of atmospheric CO_2 provides a constraint on the balance between terrestrial and oceanic sources and sinks. The isotopic budget calculation for KPH will have to be modified because originally it was assumed that the CO_2 ending up in the oceans had arrived there via air-sea exchange, without significant isotopic fractionation. The carbon entering the oceans in rivers has a terrestrial isotopic signature, except for the fraction originating from $CaCO_3$. The isotopic effects of CO oxidation and gas escape also have to be taken into account properly. In order to estimate the consequences of both budgets for the latitudinal gradient of $^{13}C/^{12}C$, we first need to transform KPH 1984 from a storage budget into a CO_2 surface flux budget. This is done in Table 20.3. A total of 0.6 Pg carbon per year (1.0 in the case of the river flux of organic carbon as estimated by Wollast and Mackenzie 1989) is shifted from the oceans to the land as this flux is initially assimilated on the land. It consists of 0.4 Pg as organic carbon and 0.2 Pg as CO_2 respired in soils that is used for the chemical weathering of $CaCO_3$.

The annual mean latitudinal CO_2 gradients generated by the flux budgets of Tables 20.2A (TFT) and 20.3 (KPH) are shown in Figure 20.2. The two-dimensional (latitude, height) atmospheric transport model used (Tans et al. 1989) was based on the transport fields generated by Plumb and Mahlman (1987) from the GFDL general circulation model. The two-dimensional model was initialized on January 1, 1980, with observed CO_2 concentrations and run with estimated fossil fuel input (Marland and Boden 1991) through 1987. The calculated mixing ratios in the lowest model layer (900–1000 mb) for 1987 are shown in Figure 20.2. The CO_2 gradient is almost indistinguishable between the TFT and KPH budgets, but the calculated values for KPH are uniformly too low because the global budget was not balanced in the latter case (see the introduction). The CO_2 mixing ratios calculated by the two-dimensional model are higher at northern midlatitudes than the observations. This result has been ascribed to the fact that the observations are not representative of true averages over longitude since they are primarily from ocean sites far away from major source regions (Tans et al. 1989). Three-dimensional transport models are fundamentally better in this respect because they simulate longitudinal differences in mixing ratios.

Table 20.3. KPH 1984 Surface Flux Budget[a]

	KPH	River fluxes	CO	Fossil fuel	Total
Oceans					
15–90°N	+2.3	−0.3 (−0.5)			+2.0 (+1.8)
Equator	−1.1	−0.3 (−0.5)			−1.4 (−1.6)
15–90°S	+1.1		+0.2		+1.3
Land					
15–90°N	+1.0	+0.3 (+0.5)	+0.4	−0.2	+1.5 (+1.7)
Equator	+0.9	+0.3 (+0.5)	+0.3		+1.5 (+1.7)
15–90°S	+0.3				+0.3
Fossil					
15–90°N	−4.8			+0.2	−4.6
Equator	−0.2				−0.2
15–90°S	−0.2				−0.2
Land use					
15–90°N	−0.4				−0.4
Equator	−1.2				−1.2
15–90°S	−0.2				−0.2

[a]The first column is the KPH budget from Table 20.1. The last column together with the atmospheric oxidation of CO determines the latitudinal gradients of CO_2 and $^{13}C/^{12}C$. The numbers in parentheses are the corrections for the case of the larger river flux estimates of organic carbon of Wollast and Mackenzie (1989).

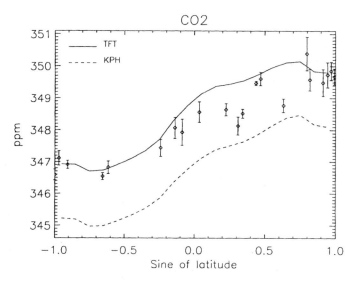

Figure 20.2. Model calculated annual mean CO_2 mixing ratios in 1987 for the TFT 1981–1987 budget (solid line) and KPH 1984 budget (dashed line), as defined in Tables 20.2A and 20.3. The data (diamonds) are from Tans et al. (1990). Error bars are based on the interannual variability (one sigma) of the annual mean mixing ratios at each site relative to the global mean.

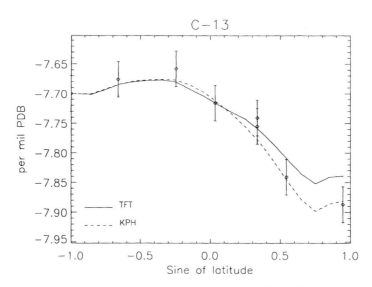

Figure 20.3. Observed and model calculated annual mean $^{13}CO_2/^{12}CO_2$ isotopic ratios for 1987, expressed in the standard notation $\delta^{13}C$ relative to PDB. The observed value at the South Pole is from a linear fit to the data for 1978–1986 (Keeling et al. 1989). The other observed values are multiyear differences relative to the South Pole plotted (diamonds) as differences from the South Pole. The error bars are based on the interannual variations of the differences from the South Pole of the entire data set. The calculated curves are for TFT 1981–1987 (solid line) and KPH 1984 (dashed line) as defined in Tables 20.2A and 20.3. Only the depicted latitudinal gradient is significant, as the calculated values were forced to go through the observed South Pole value.

The calculated 1987 annual mean latitudinal $^{13}CO_2/^{12}CO_2$ gradients are shown in Figure 20.3. Several assumptions about isotopic ratios went into these calculations. The isotopic ratio of fossil fuels was set at –27.3‰ (Tans 1981), that of terrestrial plants at –25‰ (Craig 1953), and that of CO at –27‰ (Stevens et al. 1972), and the isotopic fractionation for CO + OH → CO_2 was set at –4‰ (Stevens et al. 1980). As boundary values the sea surface isotopic ratios were estimated for 1980 from Kroopnick (1977, 1987), and they were assumed to decrease uniformly by 0.02‰ per year. There was no seasonal cycle in the surface water $\delta^{13}C$ values, but seasonal cycles in the sea-surface temperature (Shea et al. 1992) and wind-speed-dependent transfer coefficient (Erickson 1989) were included. Initial isotopic ratios for the atmosphere were taken from Keeling et al. (1989a). The observations from Keeling et al. (1989a) are also plotted for comparison. Only the spatial gradient can be usefully compared because the calculated values are shifted up and down nearly uniformly by assuming different initial values or a different trend for the ocean isotopic ratios, which are at present not constrained sufficiently by observations. Therefore, the calculated values are plotted as deviations from the annual mean calculated for the South Pole, which is made to coincide with the observations.

Terrestrial "memory" effects were neglected in the calculations, i.e., it was assumed that the isotopic ratio of the photosynthetically fixed carbon was identical to that of the respired carbon. In principle these two values should be different by a few tenths per mil because some of the respired CO_2 is old. The atmospheric isotopic ratio is decreasing because of the addition of fossil fuel CO_2 and old biospheric material fixed when the atmosphere was less depleted in ^{13}C. Memory effects will tend to decrease the isotopic gradient because isotopically heavier CO_2 is being respired, especially in the Northern Hemisphere. Another way to look at this effect is to consider that the isotopic anomaly caused by fossil fuel burning in the atmosphere is slowly being absorbed by the terrestrial biosphere. As a result the atmosphere changes its isotopic composition more slowly. Since we do not know very well the magnitude of this isotopic exchange, we have made the calculated South Pole values coincide with the observations.

The calculated isotopic difference between the KPH and TFT budgets is significant, but it is only about 0.05‰. If the river fluxes of organic carbon of Wollast and Mackenzie (1989) are accepted, the difference in the isotopic gradient between the two scenarios would be cut in half, making it insignificant. However, if most of the organic carbon carried by the rivers escapes from the coastal waters as CO_2 gas, the difference in isotopic gradient between the KPH and TFT budgets would be about 0.07‰.

The partitioning of CO_2 sinks between oceans and terrestrial systems has also been estimated by Quay et al. (1992) using $^{13}C/^{12}C$ isotopic ratios in the oceans and atmosphere. The average ocean uptake over the last two decades was 2.1 Pg carbon per year according to their analysis. Their budget is clearly a storage budget because it is immaterial how the isotopic anomaly caused by the combustion of fossil fuels finds its way into the oceans. It was pointed out by Tans et al. (1993) that the isotopic data can be analyzed in a different way. Instead of estimating the decadal changes in the ocean total column inventory of the isotopic anomaly, one can estimate from the measurements the isotopic disequilibrium across the air-sea boundary, which also implicates an isotopic flux. The latter method gives a very different result from the inventory method, an estimated ocean uptake of only a few tenths of 1 Pg carbon. The disequilibrium method is a flux budget. The adjustments required to convert from a flux budget to a storage budget, as discussed in this chapter, were not considered by Tans et al. (1993). Assuming that an amount of 0.6 Pg year of organic and inorganic carbon carried by rivers indeed ends up in the open oceans, we estimate a resulting steady-state isotopic disequilibrium between the ocean surface and the atmosphere of about 0.11‰, the oceans being "lighter," or more depleted in ^{13}C, and not in equilibrium with the atmosphere. The skin temperature correction on the isotopic equilibrium had already been included. If we take account of an overall steady-state disequilibrium of 0.1‰, the estimate of the ocean uptake of fossil fuel carbon by that method would be raised from 0.2 to 0.7 Pg carbon per year.

CONCLUSIONS

The conclusion that there was substantial uptake of CO_2 by northern terrestrial ecosystems during the early 1980s seems fairly solid. It is based on the observation that the CO_2 gradient from north to south was small. The uncertainty of the atmospheric transport of numerical models does not leave enough room to explain away the observation of the low CO_2 gradient. The major features of the transport have been calibrated by observations of the levels of CFC-11 and ^{85}Kr, for which the sources are known. This conclusion was reached independently by studying two different CO_2 data sets and several different transport models.

By making a distinction between CO_2 flux budgets and storage budgets, we have clarified some of the differences between the approaches taken by Keeling et al. (1989a, 1989b) and Tans et al. (1990). In the latter case the estimate of ocean uptake is increased and land uptake is decreased. The remaining difference at northern latitudes is that the KPH estimate of ocean uptake is about 1 Pg carbon per year larger than that of TFT, and the land uptake is 1 Pg carbon per year smaller at these latitudes.

The relatively few $^{13}C/^{12}C$ data published thus far are not yet decisive in partitioning the CO_2 sinks between the oceans and the terrestrial biosphere. Modeling suggests that the differences to be distinguished by the measurements are small.

ACKNOWLEDGMENTS

This research has been supported in part by the Atmospheric Chemistry and Ocean-Atmosphere Carbon Flux projects of the Climate and Global Change Program of the National Oceanic and Atmospheric Administration, and by the Atmospheric Research and Exposure Assessment Laboratory of the Environmental Protection Agency.

REFERENCES

Berner, R. A. 1982. Burial of organic carbon and pyrite sulfur in the modern ocean: Its geochemical and environmental significance. *Am. J. Sci.* 282:451–473.

Broecker, W. S., and T.-H. Peng. 1992. Interhemispheric transport of carbon dioxide by ocean circulation. *Nature* 356:587–589.

Conway, T. J., L. P. Steele, and P. C. Novelli. 1993. Correlations among atmospheric CO_2, CH_4, and CO in the Arctic, March 1989. *Atmos. Environ.* 27A:2881–2894.

Craig, H. 1953. The geochemistry of the stable carbon isotopes. *Geochim. Cosmochim. Acta* 3:53–92.

Dixon, S. Brown, R. A. Houghton, A. M. Solomon, M. C. Trexler, and J. Wisniewski. 1994. Carbon pools and flux of global forest ecosystems. *Science* 263:185–190.

Enting, I. G., and J. V. Mansbridge. 1991. Latitudinal distribution of sources and sinks of CO_2: Results of an inversion study. *Tellus B* 43:156–170.

Erickson, D. 1989. Variations in the global air-sea transfer velocity field of CO_2. *Global Biogeochem. Cycles* 3:37–41.

Fung, I., J. John, J. Lerner, E. Matthews, M. Prather, L. P. Steele, and P. J. Fraser. 1991. Three-dimensional model synthesis of the global methane cycle. *J. Geophys. Res.* 96: 13033–13065.

Hasse, L. 1971. The sea surface temperature deviation and the heat flow at the sea-air interface. *Boundary-Layer Meteorol.* 1:368–379.

IPCC. 1991. *Climate Change, The IPCC Scientific Assessment,* ed. J. T. Houghton, G. J. Jenkins, and J. J. Ephraums. Cambridge University Press, Cambridge.

Kauppi, P. E., K. Mielikäinen, and K. Kuusela. 1992. Biomass and carbon budget of European forests, 1971 to 1980. *Science* 256:70–74.

Keeling, C. D., R. B. Bacastow, A. F. Carter, S. C. Piper, T. P. Whorf, M. Heimann, W. G. Mook, and H. Roeloffzen. 1989a. A three-dimensional model of atmospheric CO_2 transport based on observed winds: 1. Analysis of observational data. In *Aspects of Climate Variability in the Pacific and Western Americas,* ed. D. H. Peterson, pp. 165–236. Geophysical Monograph 55. American Geophysical Union, Washington, D.C.

Keeling, C. D., S. C. Piper, and M. Heimann. 1989b. A three-dimensional model of atmospheric CO_2 transport based on observed winds: 4. Mean annual gradients and interannual variations. In *Aspects of Climate Variability in the Pacific and Western Americas,* ed. D. H. Peterson, pp. 305–363. Geophysical Monograph 55. American Geophysical Union, Washington, D.C.

Kroopnick, P. M. 1987. Carbon-13 in dissolved carbon dioxide (ΣCO_2). In *Geosecs Atlantic, Pacific, and Indian Ocean Expeditions,* Vol. 7: *Shorebased Data and Graphics.* National Science Foundation, Washington, D.C.

Kroopnick, P. M., S. V. Margolis, and C. S. Wong. 1977. $\delta^{13}C$ variations in marine carbonate sediments as indicators of the CO_2 balance between the atmosphere and the oceans. In *The Fate of Fossil Fuel in the Oceans,* ed. N. Anderson and A. Malahoff, pp. 296–312. Plenum Press, New York.

Logan, J., M. J. Prather, S. C. Wofsy, and M. B. McElroy. 1981. Tropospheric chemistry: A global perspective. *J. Geophys. Res.* 86:7210–7254.

Marland, G., and T. Boden. 1991. CO_2 emissions—modern record. In *Trends '91: A Compendium of Data on Global Change,* eds. T. A. Boden, R. J. Sepanski, and F. W. Stoss. ORNL/CDIAC-46. Oak Ridge Laboratory, Oak Ridge, Tenn.

Martin, J. H., and S. E. Fitzwater. 1992. Dissolved organic carbon in the Atlantic, Southern and Pacific oceans. *Nature* 356:699–700.

Novelli, P. C., L. P. Steele, and P. P. Tans. 1992. Mixing ratios of carbon monoxide in the troposphere. *J. Geophys. Res.* 97:20,731–20,750.

Opdyke, B. N., and J. C. G. Walker. 1992. Return of the coral reef hypothesis: basin to shelf partitioning of $CaCO_3$, and its effect on atmospheric CO_2. *Geology* 20:733–736.

Plumb, R. A., and J. D. Mahlman. 1987. The zonally averaged transport characteristics of the GFDL general circulation/transport model. *J. Atmos. Sci.* 44:298–327.

Quay, P. D., B. Tilbrook, and C. S. Wong. 1992. Oceanic uptake of fossil fuel CO_2: Carbon-13 evidence. *Science* 256:74–79.

Robertson, J. E., and A. J. Watson. 1992. Thermal skin effect of the surface ocean and its implications for CO_2 uptake. *Nature* 358:738–740.

Sarmiento, J. L., J. C. Orr, and U. Siegenthaler. 1992. A perturbation simulation of CO_2 uptake in an ocean general circulation model. *J. Geophys. Res.* 97:3621–3645.

Sarmiento, J. L., and E. T. Sundquist. 1992. Revised budget for the oceanic uptake of anthropogenic carbon dioxide. *Nature* 356:589–593.

Schluessel, P., W. J. Emery, H. Grassl, and T. Mammen. 1990. On the bulk-skin temperature difference and its impact on satellite remote sensing of sea surface temperature. *J. Geophys. Res.* 95:13341–13356.

Shea, D. J., K. E. Trenberth, and R. W. Reynolds. 1992. A global monthly sea surface temperature climatology. *J. Climate* 5:987–1001.

Spivakovsky, C. M., R. Yevitch, J. A. Logan, S .C. Wofsy, and M. B. McElroy. 1990. Tropospheric OH in a three-dimensional chemical tracer model: An assessment based on observations of CH_3CCl_3. *J. Geophys. Res.* 95:18441–18471.

Stevens, C. M., L. Krout, D. Walling, A. Venters, A. Engelkemeir, and L. E. Ross. 1972. The isotopic composition of atmospheric carbon monoxide. *Earth Planet. Sci. Lett.* 16:147–165.

Stevens, C. M., L. Kaplan, R. Gorse, S. Durkee, M. Compton, S. Cohen, and K. Bielling. 1980. The kinetic isotope effect for carbon and oxygen in the reaction CO+OH. *Int. J. Chem. Kinet.* 12:935–948.

Sugimura, Y., and Y. Suzuki. 1988. A high temperature catalytic oxidation method of non-volatile dissolved organic carbon in seawater by direct injection of liquid samples. *Mar. Chem.* 24:105–131.

Tans, P. 1981. $^{13}C/^{12}C$ of industrial CO_2. In *Carbon Cycle Modelling,* Vol. 16, ed. B. Bolin, pp. 127–129. SCOPE, Wiley, Chichester, U.K.

Tans, P. P., T. J. Conway, and T. Nakazawa. 1989. Latitudinal distribution of sources and sinks of atmospheric carbon dioxide derived from surface observations and an atmospheric transport model. *J. Geophys. Res.* 94:5151–5172.

Tans, P. P., I. Y. Fung, and T. Takahashi. 1990. Observational constraints on the global atmospheric CO_2 budget. *Science* 247:1431–1438.

Tans, P. P., J. A. Berry, and R. F. Keeling. 1993. Oceanic 13C/12C observations: A new window on ocean CO_2 uptake. *Global Biogeochem. Cycles* 7:353–368.

Warneck, P. 1988. *Chemistry of the Natural Atmosphere.* Academic Press, San Diego, Calif.

WMO 1991. Scientific Assessment of Ozone Depletion: 1991. Global Ozone Research and Monitoring Project. Report No. 25. World Meteorological Organization, Geneva.

Wofsy, S. C., M. L. Goulden, J. W. Munger, S.-M. Fan, P. S. Bakwin, B. C. Daube, S. L. Bassow, and F. A. Bazzaz. 1993. Net exchange of CO_2 in a midlatitude forest. *Science* 260:1314–1317.

Wollast, R., and F. T. Mackenzie. 1989. Global biogeochemical cycles and climate. In *Climate and Geosciences: A Challenge for Science and Society in the 21st Century,* ed. A. Berger, S. Schneider, and J.-Cl. Duplessy, pp. 453–473. Kluwer, Dordrecht, The Netherlands.

21

Perturbations to the Biospheric Carbon Cycle: Uncertainties in the Estimates

Inez Fung

Renewed focus on the contemporary CO_2 budget, especially on the partitioning of the anthropogenic CO_2 sink between land and sea, has greatly expanded our understanding of both terrestrial and oceanic carbon dynamics. Nevertheless, many uncertainties remain about the processes that maintain the present-day carbon balance and about how those processes may change in the future. In this chapter, I briefly review estimates of biospheric carbon fluxes derived from atmospheric observations and models, as well as the processes that may be responsible.

INFERRED PERTURBATIONS TO THE BIOSPHERIC CARBON CYCLES

Mass balance determines the total strength of the CO_2 sink. Information available to deduce the locations and magnitudes of the terrestrial and oceanic components of the sink includes (1) the observed latitudinal gradient of CO_2 in the atmosphere, (2) the observed latitudinal gradient of $\delta^{13}C$ in the atmosphere, and (3) the compiled distribution of ΔP_{CO_2}, the difference of CO_2 partial pressures across the air-sea interface.

Table 21.1 shows the terrestrial sink implied by two atmospheric three-dimensional model calculations. The Tans et al. (1990) budget violates the constraint of atmospheric $\delta^{13}C$, whereas the Keeling et al. (1989) budgets ignore the ΔP_{CO_2} compilation. Two cases are cited from Keeling et al. (1989). Their standard case yields an atmospheric increase of 2.8 petagrams (Pg) ($=10^{15}$ g) carbon per year for 1980, smaller than the observed 3.6 Pg carbon per year in the atmosphere. Their double deconvolution case satisfies all the mass balance constraints for ^{12}C as well as ^{13}C. It is noteworthy that a midlatitude sink is required in all three studies.

Extending the Tans et al. (1990) study to include $\delta^{13}C$ does not eliminate the need for a midlatitude land sink. Figure 21.1 shows the modeled and observed north-south gradient of $\delta^{13}C$ at the observing stations (Fung 1993). Fossil fuel and land-use sources release "light" carbon (circa $-25\permil$) into the atmosphere, the $\delta^{13}C$ of which was circa $-7.5\permil$ in the 1980s. Because 96% of the fossil fuel source is in the Northern Hemi-

Table 21.1. Biospheric Sources and Sinks (Pg Carbon per Year) from
Three-Dimensional Atmospheric Transport Models

	Tropics		Midlatitudes		Global: net land sink
	Source	Sink	Source	Sink	
Tans et al. (1990) scenario 5	0	0 0	2.0	2.0	
Keeling et al. (1989) Standard case	1.8	2.0	0	2.1	2.3
Keeling et al. (1989) double deconvolution	2.8	2.0	0	2.1	1.3

sphere, the atmospheric $\delta^{13}C$ signature due to the anthropogenic sources alone would be lower in the Northern than in the Southern Hemisphere (Figure 21.1). The signature due to air-sea exchange of ^{13}C is sensitive only to sea-surface temperature and the magnitude of the gross fluxes and not to the magnitude of ΔP_{CO_2}. Thus the difference between the observed and combined anthropogenic and oceanic signatures implies a significant land sink. The estimated magnitude of the land sink depends on $\Delta\delta^{13}C_{ab}$, the magnitude of the ^{13}C disequilibrium between the atmosphere and biosphere. $\Delta\delta^{13}C_{ab}$ is dependent on the age of the CO_2 returned to the atmosphere from the biosphere, which in turn is dependent on the fraction of CO_2 respired from litter and soil carbon pools, their respective turnover times, and $\delta^{13}C$ histories. Thus an understanding of the dynamics of these pools and the latitudinal variation of $\Delta\delta^{13}C_{ab}$ is a prerequisite to a reliable estimate of the magnitude of the land sink.

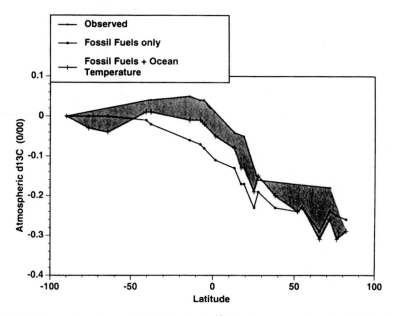

Figure 21.1. Observed and modeled latitudinal $\delta^{13}C$ in the atmosphere for 1980, relative to the $\delta^{13}C$ at the South Pole. The shaded region denotes the signal from the terrestrial carbon sink required to match the observations.

The mechanism of net CO_2 uptake by the terrestrial biosphere has not been established unequivocably. Tans et al. (1990) state that the sink mechanism is unknown. The terrestrial sink of Keeling et al. (1989) is made proportional to the annual net primary productivity (NPP) of natural ecosystems and hence attributed to CO_2 fertilization. Approximately half of the global NPP is in the tropics and half is in the northern midlatitudes. Nutrient limitation and other competing mechanisms may alter ecosystem response to CO_2 fertilization (e.g., Korner and Arnone 1992; Norby et al. 1992), so that, if climate is unchanging, wet tropical forests, coniferous boreal forests, and arctic tundra may have low potential to respond to CO_2 fertilization (Mooney et al. 1991).

DIRECT ESTIMATES OF PERTURBATIONS TO THE BIOSPHERIC CARBON CYCLE

In this section, I discuss some of the uncertainties in the estimation of CO_2 fluxes from land-use modification and the biotic feedbacks that may perturb terrestrial carbon dynamics and alter the growth rate of CO_2 in the atmosphere.

Anthropogenic Sources and Sinks in the Tropics

The major biotic source of CO_2 is land-use modification, principally as a result of deforestation in the tropics. Estimation of this source requires information on (1) the area modified, (2) the amount of carbon (above and below ground) prior to modification, and (3) the fate of the land after modification.

Despite advances in remote sensing of the areal extent of deforestation, the availability of archived satellite data limits the use of this approach to small regions in the Amazon (e.g., Woodwell et al. 1986; Skole and Tucker 1993). Economics and agricultural statistics of unverified quality are relied upon for information for the period prior to the satellite era and for the rest of the tropics, where satellite data with <1-km resolution have not been either collected or analyzed for land conversion rates (e.g., Flint and Richards 1991; Houghton et al. 1991a). Uncertainties in this term cannot be quantified without access to the satellite data.

The amount of standing biomass prior to modification is not known, except in very limited regional studies. Brown et al. (1991) demonstrate that, for South and Southeast Asia, the total aboveground biomass (TAGB) actually in the forest may be as much as a factor of 2 smaller than the potential TAGB of a climax vegetation. Such a TAGB survey has not been carried out elsewhere in the tropics. There is also little information on the amount of biomass below ground and its fate at and after disturbance. The use of a lower biomass figure would obviously lower the estimates of CO_2 source strength from this term.

Modifications of the landscape by shifting cultivation, sustained conversion to pasture or croplands, or logging and abandonment have very different impacts on the timing and magnitudes of carbon exchange. As with TAGB prior to modification, there is little record of carbon stocks after modification and of the oxidation rate of residual necromass, soil carbon, and harvested products after disturbance. Models such as that of Houghton et al. (1987, 1991b) assume that, across the whole tropics, 20% of the top 1 m of forest soil carbon is lost in the first 5 years after clearing and 5% is lost over the next

20 years. Although the values used are within the range of 2–69% reported by Schlesinger (1986) and 12–34% estimated for the top meter by Detweiler (1986), it is not clear if they are representative of soil oxidation in the whole tropics. Resolving the geographic distribution of land-use practices, above- and belowground carbon stocks, and carbon loss rates in the models would provide at least the possibility of spot-checking the assumptions at a few sites.

Degradation (i.e., undocumented or illicit reduction of biomass in a forest) would result in a lower TAGB; it may also be an unaccounted-for CO_2 source as the removed biomass decays or is burned. Although the TAGB may be included in surveys, there is no way to evaluate its contribution to the net flux.

Recent estimates that take into account many of the factors just discussed range from 0.6 Pg carbon per year (Hall and Uhlig 1991) to 1.1–3.6 Pg carbon per year (Houghton 1991b). This term is the largest source of uncertainty in the carbon budget, and hence the magnitude of the total sink itself.

Natural Sources and Sinks in the Tropics

In their three-dimensional atmospheric model calculation, Keeling et al. (1989) inferred a sink of CO_2 from the natural tropical biota in the 1980s (see Table 21.1). The same conclusion obtains if the small net flux from the tropical biosphere in Tans et al. (1990) and Enting and Mansbridge (1991) is viewed as the difference between a land-use source and a natural sink. The mechanism of this natural sink is not clear. Phosphorus limitation leads to low potential for CO_2 fertilization of the wet tropical forests (Mooney et al. 1991). In addition, experiments with artifical tropical ecosystems found vigorous growth enhancement but little net carbon accumulation in response to elevated CO_2 levels (Korner and Arnone 1992). Mechanisms for increasing carbon accumulation in the tropics remain unclear. However, temperature and precipitation fluctuations in the past 50 years may have resulted in enhanced NPP over soil respiration during that period (Dai and Fung 1993). There are few data to substantiate or refute any hypothesis about changing carbon dynamics in the tropical biosphere.

Anthropogenic Sources and Sinks in Northern Hemisphere Midlatitudes

What is the contribution of forest management and regrowth in midlatitudes to the global carbon budget? Contemporary carbon accumulation is found in European trees (Kauppi et al. 1992). Sedjo (1992) also estimated that changes in forest stocks in northern midlatitudes may account for a carbon sink of 0.7 Pg carbon per year. However, these studies include changes only in TAGB and represent incomplete views of carbon exchange. Changes in the litter and belowground carbon pools may enhance or diminish the sink strengths obtained. Recovery of forests in midlatitudes, such as after abandonment of agriculture, may represent a sink for atmospheric CO_2, as confirmed by the recent multiyear measurements of ecosystem CO_2 exchange in the northeastern United States (Wofsy et al. 1993). Houghton et al. (1987), Melillo et al. (1988), and Harmon et al. (1990) cautioned, however, that oxidation of wood products, detritus, and soil carbon may balance the CO_2 absorbed by regrowing forests, and that land-use modification in northern midlatitudes may result in near-zero flux of CO_2 to the atmosphere.

Natural Sources and Sinks in Northern Hemisphere Middle to High Latitudes

Productivity of midlatitude ecosystems, where temperature, nutrients, and other factors may not be totally limiting, may increase as a result of elevated CO_2 levels (Mooney et al. 1991) and of nitrogen deposition and may be at a maximum in midlatitudes in the Northern Hemisphere (e.g., Jenkinson et al. 1991). Whether net carbon accumulation may result will depend on the degree of cancellation by other processes, such as soil respiration, that may increase with litter and root biomass. In their experiments with yellow poplar saplings, Norby et al. (1992) found that changing carbon allocation patterns under elevated CO_2 conditions resulted in little increase in whole-plant carbon storage. Changes in climate may also play a role in affecting imbalances between carbon uptake and release in midlatitude ecosystems (Dai and Fung 1993).

At high latitudes, nutrient limitation prevents additional carbon accumulation in response to elevated CO_2 levels (Billings et al. 1984; Tissue and Oechel 1987). However, climate warming may liberate nutrients now frozen in soils and permit ecosystem production to respond to CO_2 fertilization (Billings et al. 1984; Shaver et al. 1992). However, drying may increase the aerobicity of the soil, and CO_2 effluxes may replace CH_4 effluxes (Billings et al. 1984), a hypothesis supported by the measurements of Oechel et al. (1993). Although high-latitude ecosystems may have experienced large perturbations to carbon processing as a result of large climate excursions (e.g., Dai and Fung 1993), assessment of their contribution to the global carbon budget must include evaluation of compensatory processes and the areal extent of the response.

What Is "Natural"?

The debate over detectability of the greenhouse warming signal in the 200-year instrumental record of surface air temperature underlies the importance of quantifying transient interdecadal fluctuations in a time series whose long-term trend is being sought. Models of the contemporary carbon cycle focus on perturbations to a background state with zero net carbon exchange with the atmosphere. Field measurements, on the other hand, yield total (steady state plus perturbation) pool sizes or fluxes. Reconciliation between modeled results and measurements thus requires information about the background state itself. Does steady state for a carbon reservoir imply net zero exchange with the atmosphere? Over what period is steady state defined? What is the natural variability of the fluxes?

Over long time scales, the steady-state terrestrial and oceanic carbon reservoirs may each be in carbon flux imbalance, by circa 0.6 Pg carbon per year, with respect to the atmosphere, the magnitude of the imbalance being equal to the riverine carbon transport to the oceans (Smith and Mackenzie 1987; Wollast and Mackenzie 1989; Sarmiento and Sundquist 1992). The continual accumulation of soil organic carbon since the last deglaciation is a steady biospheric carbon sink of 0.4 Pg carbon per year (Schlesinger 1986, 1990). Whether there is a similar recovery of the oceanic carbon reservoir is not known. This would mean that, without anthropogenic perturbations, there is a net flux of 0.6–1.0 Pg carbon per year from the atmosphere to the biosphere, and a net flux of at least 0.6 Pg carbon per year from the oceans to the atmosphere. These net fluxes are two orders of magnitude smaller than the bidirectional gross fluxes between the atmosphere and the biosphere and between the atmosphere and the ocean. However, they are of a magnitude comparable to the unidirectional perturbation fluxes. This nonzero background flux does not alter the magnitude of the sink for anthropogenic CO_2, as the latter is estimated from annual sources and year-to-year differences in atmospheric CO_2 abundance. However, the perturbation to the oceanic sink estimated from compilations of wind speeds and air-sea differences in CO_2 partial pressures may be underestimated by the net background flux out of the ocean, and, correspondingly, the terrestrial sink strength may be overestimated by the net background flux into the biosphere. The geographic distributions of the background land sink and presumed ocean source, and their implications for the preindustrial atmospheric CO_2 gradient, are yet to be investigated.

Even without these long-term fluxes, the definition of a steady-state biosphere may be not straightforward. In middle to high latitudes, there are decades-long turnover times of litter and upper soils as well as cycles of natural disturbance (e.g., fires) and recovery. In these systems, equilibrium (in terms of carbon fluxes or carbon stocks) has to be defined over several cycles, or centuries. Similarly, climate has not remained constant over the past century. There is no clear-cut choice of a period for which climate-induced net fluxes average to zero. For annually averaged atmosphere-biosphere exchange of CO_2, a climate-induced natural variability of ± 0.3 Pg carbon per year is estimated for the natural biosphere (Dai and Fung 1993). It is thus crucial to determine if the terrestrial sink for the 1980s as implied by the atmospheric studies or by ecosystem models (e.g., Bonan 1991) is a manifestation of natural variability or of a secular trend.

CONCLUDING REMARKS

The term "biotic feedbacks" is a shorthand expression for ecosystem responses to a myriad of interacting and competing external perturbations. The richness of the literature attests to the progress that has been made in identifying the external perturbations as well as in setting up the building blocks for quantifying and verifying the system response. This chapter has tried to highlight the gaps in our understanding of the external perturbations as well as the system response to them. Because changes in atmospheric CO_2 abundance reflect changes in whole ecosystems rather than individual components of them, it is crucial to adopt a holistic approach to the problem, to investigate the response to each external change as well as to the entire suite of external changes, including temperature and precipitation variability, CO_2 levels, and nitrogen deposition. With the response to human perturbations, end-to-end analysis of the fate of carbon at the disturbed site as well as that removed from the site must be carried out. Similarly, changes in all components of the system, both above- and belowground stocks, and fluxes in as well as fluxes out, must be included before whole-system responses can be evaluated. Nonlinearity in the system guarantees that the total response will not be equal to the sum of the individual responses, as seen in the case of CO_2 fertilization in arctic ecosystems.

ACKNOWLEDGMENTS

This work is supported by the DOE Carbon Dioxide Research Program, NASA Geochemistry and Geophysics Branch, and by the NASA EOS-IDS Program.

REFERENCES

Billings, W. D. 1983. Increasing atmospheric carbon dioxide: Possible effects on arctic tundra. *Oecologia* 58:286–289.

Billings, W. D., J. O. Lukeen, D. A. Mortensen, and K. M. Peterson. 1984. Arctic tundra: A source or sink for atmospheric carbon dioxide in a changing environment? *Oecologia*, 53:7–11.

Bonan, G. B., 1991. Atmosphere-biosphere exchange of carbon dioxide in boreal forests. *J. Geophys. Res.* 96:7301–7312.

Brown, S., A. J. R. Gillespie, and A. E. Lugo. 1991. Biomass of tropical forests of south and southeast Asia. *Can. J. For. Res.* 21:111–117.

Dai, A. G., and I. Fung. 1993. Can climate variability contribute to the "missing" CO_2 sink? *Glob. Biogeochem. Cycles* 7:599–609.

Detweiler, R. P. 1986. Land use change and the global carbon cycle: The role of tropical soils. *Biogeochemistry* 2:67–93.

Enting, I. G., and J. V. Mansbridge. 1991. Latitudinal distribution of sources and sinks of CO_2: Results of an inversion study. *Tellus* 43B:156–170.

Flint, E. P., and J. F. Richards. 1991. Historical analysis of changes in land use and carbon stock of vegetation in south and southeast Asia. *Can. J. For. Res.* 21:91–110.

Fung, I. 1993. Models of oceanic and terrestrial sinks of anthropogenic CO_2: A review of the contemporary carbon cycle. In *The Biogeochemistry of Global Change: Radiative Trace Gases,* ed. R. Oremland, pp. 166–189. Chapman and Hall, New York.

Hall, C. A. S., and J. Uhlig. 1991. Refining estimates of carbon released from tropical land-use change. *Can. J. For. Res.* 21:118–131.

Harmon, M. E., W. K. Ferrell, and J. F. Franklin. 1990. Effects on carbon storage of conversion of old-growth forests to young forests. *Science* 247:699–702.

Houghton, R. A. 1991a. Releases of carbon to the atmosphere from degradation of forests in tropical Asia. *Can. J. For. Res.* 21:132–142.

Houghton, R. A. 1991b. Tropical deforestation and atmospheric carbon dioxide. *Climatic Change* 19:99–118.

Houghton R. A., R. D. Boonem, J. R. Fruci, J. E. Hobbie, J. M. Melillo, C. A. Palm, B. J. Peterson, G. R. Shaver, G. M. Woodwell, B. Moore, D. L. Skole, and N. Myers. 1987. The flux of carbon from terrestrial ecosystems to the atmosphere in 1980 due to changes in land use: Geographic distribution of the global flux. *Tellus* 39B:122–139.

Houghton, R. A., D. S. Lefkowitz, and D. L. Skole. 1991a. Changes in the landscape of Latin America between 1850 and 1985. I. Progressive loss of forests. *For. Ecol. Manage.* 38: 143–172.

Houghton, R. A., D. L. Skole, and D. S. Lefkowitz. 1991b. Changes in the landscape of Latin America between 1850 and 1985. II. Net release of CO_2 to the atmosphere. *For. Ecol. Manage.* 38:173–199.

Jenkinson, D. S., D. E. Adams, and A. Wild. 1991. Model estimates of CO_2 emissions from soil in response to global warming. *Nature* 351:304–306.

Kauppi, P. E., J. Mielikinen, and K. Kusela. 1992. Biomass and carbon budget of European forests, 1971–1990. *Science* 256:70–74.

Keeling, C. D., S. Piper, and M. Heimann. 1989. A three-dimensional model of atmopsheric CO_2 transport based on observed winds. 4. Mean annual gradients and interannual variations. In *Aspects of Climate Variability in the Pacific and Western America,* ed. D. H. Peterson, pp. 305–363. Geophysical Monograph 55, American Geophysical Union, Washington, D.C.

Korner, C., and J. A. Arnone III. 1992. Responses to elevated carbon dioxide in artificial tropical ecosystems. *Science* 257:1672–1675.

Melillo, J. M., J. R. Fruci, R. A. Houghton, B. Moore, and D. L. Skole. 1988. Land-use change in the Soviet Union between 1850 and 1980: Causes of a net release of CO_2 to the atmosphere. *Tellus* 40B:116–128.

Mooney, H. A., B. G. Drake, R. J. Luxmoore, W. C. Oechel, and L. F. Pitelka. 1991. Predicting ecosystem responses to elevated CO_2 concentrations. *BioScience,* 41:96–104.

Norby, R. J., C. A. Gunderson, S. D. Wullschleger, E. G. O'Neill, and M. K. McCracken. 1992. Productivity and compensatory responses of yellow-poplar trees in elevated CO_2. *Science* 357:322–324.

Oechel, W. C., S. J. Hastings, G. Vouritis, M. Jenkins, G. Reichers, and N. Grulke. 1993. Recent change of arctic tundra ecosystems from a net carbon dioxide sink to a source. *Nature* 361:520–523.

Sarmiento, J. L., and E. T. Sundquist. 1992. Revised budget for the oceanic uptake of anthropogenic carbon dioxide. *Nature* 356:589–593.

Schlesinger, W. H. 1986. Changes in soil carbon storage and associated properties with disturbance and recovery. In *The Changing Carbon Cycle: A Global Analysis,* ed. J. R. Trabalka and D. E. Reichle, pp. 194–220. Springer-Verlag, New York.

Schlesinger, W. H. 1990. Evidence from chronosequence studies for a low carbon storage potential of soils. *Nature* 348:232–234.

Sedjo, R. A. 1992. Temperate forest ecosystems in the global carbon cycle. *Ambio* 21:274–277.

Shaver, G. R., W. D. Billings, F. S. Chapin III, A. E. Giblin, K. J. Nadelhoffer, W. C. Oechel, and E. B. Rastetter. 1992. Global change and the carbon balance of Arctic ecosystems. *Bioscience* 42:433–441.

Skole, P., and C. J. Tucker. 1993. Topical deforestation and habitat fragmentation in the Amazon: satellite data from 1978 to 1988. *Science* 260:1905–1910.

Smith, S. V., and F. T. Mackenzie. 1987. The ocean as a net heterotrophic system: Implications from the carbon biogeochemical cycle. *Global Biogeochem. Cycles* 1:187–198.

Tans, P. P., I. Y. Fung, and T. Takahashi. 1990. Observational constraints on the global atmospheric CO_2 budget. *Science* 247:1431–1438.

Tissue, D. T., and W. C. Oechel. 1987. Response of *Eriophorum vaginatum* to elevated CO_2 and temperature in an Alaskan artic tundra. *Ecology* 68:401–410.

Wofsy, S. C., J. E. Munger, P. S. Bakwin, B. C. Daube, and T. R. Moore. 1993. Net CO_2 uptake by northern woodlands. *Science* 260:1314–1317.

Wollast, R., and F. T. Mackenzie. 1989. Global biogeochemical cycles and climate. In *Climate and Geosciences: A Challenge for Science and Society in the 21st Century,* ed. A. Berger, S. Schneider, and J.-Cl. Duplessy, pp. 453–473. Kluwer, Dordrecht, The Netherlands.

Woodwell, G. M., R. A. Houghton, T. A. Stone, and A. B. Park. 1986. Changes in the area of forests in Rondonia, Amazon Basin, measured by satellite imagery. In *The Changing Carbon Cycle: A Global Analysis,* ed. J. R. Trabalka and D. E. Reichle, pp. 242–257. Springer-Verlag, New York.

IV

BIOTIC FEEDBACKS IN THE GLOBAL CLIMATIC SYSTEM

The final two chapters summarize where we are now in this continuous search for insights into the workings of the world, and where we might go.

The broad dimensions of the threat of a global warming posed by the accumulation of heat-trapping gases in the atmosphere have been recognized for a century. These dimensions have not changed and are unlikely to change as a result of further research: adding heat-trapping gases to the atmosphere will warm the earth. The speed with which that problem might develop has not been known and remains in question. And the geographic and temporal details of the climatic changes associated with a warming are also much in doubt.

It has been a central caveat of ecology that whatever the climatic regime, the vegetation of the earth at any moment integrates the effects of climate and soil and the interactions of populations of plants and animals to produce communities that constitute at once the most general and specific expression of what is possible at that moment on any site. The relationships between these communities provide insight into what has happened in the past and what will probably happen in the future as climates change. This now classical form of "reading the landscape" provides a check on inferences from contemporary physiological studies and process models that have been so thoroughly useful and provocative in defining how nature works to build these same communities.

W. Dwight Billings has decades of experience in reading landscapes globally. He has spanned in that experience the limits of forests from deserts to tundra, from equator to the poles. He offers here in a penultimate chapter a perspective on research that would be appropriate in examining and predicting the transitions in vegetation that can be expected as the earth warms, continental centers become more arid, and climatic patterns become appreciably less stable than they have been in the past.

In the final chapter the authors have picked up threads of data and experience summarized in the earlier chapters to weave a contemporary tapestry of causes and effects associated with the climatic and other environmental changes anticipated from the warming. The emphasis is on the short term of years to decades, a period in which the responses of terrestrial ecosystems is large in proportion to other factors, and that of marine systems remains much in doubt. There is no escape from the need to limit releases of heat-trapping gases, if the warming of the earth is to be controlled.

22

What We Need to Know: Some Priorities for Research on Biotic Feedbacks in a Changing Biosphere

W. D. Billings

There are two main sources of facts on biotic feedbacks in a changing world: (1) constant monitoring of whole ecosystems at all levels, both physical and biological, and (2) quantitative measurement and experimentation, both in nature and in controlled environments, using the main biological components of ecosystems.

Of paramount importance to understanding biotic feedbacks globally is the use of improved and continuous instrumental monitoring. Continuous remote sensing and *in situ* measurements of the structure and fluxes within the biosphere and among its component biomes are now possible. Such monitoring from space must include mapping the extent of green vegetation and the amounts of chlorophyll through the year. Without such data, it is difficult, if not impossible, to measure, model, and predict feedbacks between biological communities and the rapidly changing physical environments of the earth, its continents, and its oceans. The sites for such monitoring are, appropriately, ecosystems of the oceans, the atmosphere, and continental ice, as well as those on land. On land and in the seas, Biosphere Reserves, the Long-Term Ecological Research Programs of the National Science Foundation, and preserved natural areas in marine environments are invaluable for such studies. Additionally, croplands within the earth's major climates should also be principal research sites.

Among the useful techniques, including remote and *in situ* sensing by appropriate instruments, should be mapping, repeat photography, and biological censusing at the community and population levels, both floristically and faunistically. Such techniques would be informative regarding changes in community structure, biological invasions, and biodiversity. At the physiological level, badly needed are more automated measurements of photosynthesis and respiration rates in C3 and C4 plants and studies of their water-use efficiencies, with special attention to the possibility of increasing water yields on critical watersheds in semiarid and other marginal climates and ecosystems.

Monitoring should also provide advance warning, in terms of years and decades, of the effects of disturbances, both biological and physical. Such disturbances are often the triggers that set off future ecosystemic and biospheric change. We have been surprised far too often by the delayed effects of changes, such as increases in radioactivity, atmospheric CO_2 concentration, UV-B radiation, destruction of stratospheric ozone, and biological invasions that have occurred when nobody was looking or expecting the consequences.

DEFORESTATION

Among the most urgent tasks of global vegetational and chlorophyll monitoring is the establishment of the extent of deforestation, especially in the rainy tropics. This process now accounts for the most serious loss of ecosystemic biodiversity and biotic carbon capture since full-glacial times, and it is largely caused by, and is within the control of, human populations. In the more temperate climates of Eurasia and North America during the last two or three centuries, the deciduous forests have suffered a similar fate. The change has been followed in this century by the loss of large areas of the subarctic coniferous forests and the very old temperate-zone coniferous rain forests of the west coasts of North America, New Zealand, and South America. In addition to species loss, the flux of CO_2 from the biota and soils after deforestation exceeded that from the combustion of fossil fuels until about 1960 (Woodwell et al. 1983). In spite of the global increase in use of fossil fuels in recent decades, the biotic feedback of CO_2 from deforestation is still significant, even though the area of forest has continued to decrease rapidly on a global scale because of lumbering and fire.

Biotic feedback to the atmosphere is not confined to gases derived from soil and decomposition of biotic debris in those barren lands once forested. Also important are the reflected visible and infrared wavelengths entering the atmosphere and space that can alter the energy budget of the earth. Shukla et al. (1990), in their models, show that when Amazonian tropical rain forests are replaced by overgrazed and deteriorating pasture grasses, there is a significant increase in soil temperature (thus radiating infrared) and a decrease in evapotranspiration and precipitation over the region—and a lengthening of the dry season. Henderson-Sellers (1993) also predicts that, in the Amazon region, the climatic effects of removing all the tropical forest would be so marked that in only 1 of the 18 deforested grid elements could the new mesoclimates of the deforested region sustain new tropical forest vegetation.

The extinction rates of dependent species of plants and animals following destruction of tropical rain forests are probably very high, but precise figures are rare. Lovejoy (1989) cites two extreme examples. The first, based on several centuries of observation and data, is the coastal Atlantic rain forest of the state of São Paulo, Brazil. It is estimated that only 2% of this species-rich forest remains after the destruction of the last 150 years. One can only guess at the number of species, including endemics, lost to extinction in that time. Lovejoy estimates that the original species complement has probably been reduced to 75% of what it was two centuries ago. The second example is the highly endemic fauna and flora of Madagascar (with 90% of its forests gone), which must have lost about half of its species, according to Lovejoy.

INCREASING GREENHOUSE GASES IN THE ATMOSPHERE

Concentrations of the various natural greenhouse gases in the atmosphere (especially CO_2, CH_4, water vapor, and N_2O) have increased, decreased, and increased again during the Pleistocene and Holocene as climates and vegetation have changed (Dansgaard and Oeschger 1989; Khalil and Rasmussen 1986). CO_2 dioxide concentrations varied during the glacials and interglacials of the Pleistocene from circa 190 to 300 ppmv. The preindustrial benchmark of atmospheric CO_2 during the 18th and early 19th centuries, as measured directly and in ice cores, was 275–280 ppmv (Friedli et al. 1986; Siegenthaler and Oeschger 1987). In the last 200 years atmospheric CO_2 concentrations have risen curvilinearly at a steadily increasing rate.

The evidence for this CO_2 increase is, of course, straightforward monitoring. This continuous monitoring began on the summit of alpine Mauna Loa on Hawaii in the mid-Pacific in 1958 as part of studies by C. D. Keeling and his co-workers (1982). The Mauna Loa curve showing the annual cycle in CO_2 concentration and the year-by-year accumulation has been a standard of excellence for 35 years; in 1958 the concentration was about 316 ppmv; at present it is about 360 ppmv. The Mauna Loa measurements have now been augmented by those from many other CO_2 monitoring stations. Such careful monitoring, although frequently scorned as "routine," is obviously essential and must be continued without interruption. Continued monitoring of synthetic chlorofluorocarbons must be added to the monitoring of the natural atmospheric gases. The chlorofluorocarbons are two-edged swords: not only are they greenhouse gases but they also break down the stratospheric ozone layer that shields the earth from much of the biologically damaging solar UV-B radiation (Cicerone 1987; Rowland 1989). The radiation poses a serious problem for both human populations (Kopf 1984) and plants and their vegetation (Caldwell et al. 1982; Billings 1984).

Some of the chlorofluorocarbons have long residence times in the atmosphere: more than a century according to Cicerone (1987). Other fluorinated synthetic gases, such as CF_4, have a strong global warming potential (Ravishankara et al. 1993). Cicerone (1979) concluded that the inert CF_4 has a residence time in the atmosphere of more than 10,000 years. Global warming may thus continue for a long time indeed and not just because of CO_2, CH_4 and water vapor.

The increases in concentrations of CO_2, CH_4, water vapor, and chlorofluorocarbons are at the root of the estimates of global warming and studies of the role of temperature in biospheric change. One does not have to emphasize the importance not only of globally monitoring the atmospheric concentrations of these gases through time, but also of monitoring the inputs of these gases into and through the atmosphere. More research must be initiated on strategies that can be employed to lower the input rates and how to augment the sinks.

MIGRATIONAL RATES, DISTANCES, AND RESISTANCES

At present, the most readily available information for northward migrational rates of trees, shrubs, and herbaceous plants in a warming climate is that based on studies of Holocene paleoecological macrofossils and microfossils, such as pollen grains and spores (Jacobson et al. 1987; Davis 1989; Overpeck et al. 1991; Davis and Zabinski 1992). This kind of research is extremely valuable as an analogue for what may happen in the next century or two if the climates of the Northern Hemisphere warm even more dramatically than they did during parts of the Holocene. Such paleoecological research is needed and must be continued. Additionally, however, calculation of migrational rates, distances, and directions based on quantitative measurements using living plants actually *in* the vegetation is absolutely necessary. Such data would conceivably use aerial and/or space photography. Quantitative ground truth data based on plainly marked permanent transects, plots, and mapping will also be required from vegetation ecologists who are competent in plant identification under field conditions. Such ecologists will have to be educated, trained, and supported in appropriate graduate ecology programs.

Of comparable importance to the migrational rates of species of plants and animals are the death and possible extinction of local populations of species in biological communities that are changing floristically or disintegrating ecologically due to climatic change or loss of soil. Simultaneously with climatic change, there often is increased competition from native and/or exotic species of plants and animals as ecosystems become warmer, drier, and more prone to wildfires. A good example is the invasion into Great Basin ecosystems by the European annual grass *Bromus tectorum* in the early part of this century and the susceptibility of this grass to extensive wildfires during the dry summers. The result has been the destruction of great parts of the sagebrush (*Artemisia tridentata*) and pinyon-juniper (*Pinus monophylla–Juniperus osteosperma*) ecosystems (Billings 1990, 1993). The lowered biomass of the postfire annual grassland and its high reflectance in the visible spectrum during the dry, hot summers are excellent examples of positive feedback to the atmosphere and to space occasioned by a drastic and unexpected change in vegetation.

Similarly, the piedmont forests of the southeastern United States are already changing rapidly as tree species native farther south (for example, *Magnolia grandiflora* and *Ilex vomitoria,* both of which are planted as ornamentals in the piedmont) are invading the surrounding native deciduous and pine forests. Such invasions are due to a combination of vectors, including people, birds, and squirrels, carrying viable seeds of the introduced trees into seemingly sustainable and stable native forests. Is climatic warming also involved? Such forest invasions are not limited to northward migration of native North American tree species. Many species of trees, shrubs, and woody vines from eastern Asia, introduced as ornamentals to the American South in the 19th and 20th centuries, are also invading native American piedmont forests in fair numbers. These include *Camellia sasanqua, Acer palmatum, Paulownia tomentosa, Ilex cornuta,* and *Lonicera japonica* (Billings manuscript in preparation).

Present models of how vegetation zones could move northward or longitudinally in the CO_2-rich world of the next century are relatively simplistic. Most models assume that there are almost no resistances in the north to invasions or migrations from the south. Actually, there are many such resistances, including

Glacially scoured rocks or thin soils
Acid podzols with low supplies of nutrients
Boggy and wet peatlands
Existing forests that do not tolerate, or shade out, competition
The lack of appropriate bird or mammal seed vectors

Day-length problems with leaf dormancy mechanisms
Increased forest fire frequency
Increased chance of severe frost during the growing season
Lack of sufficient local seed sources

Such information must be incorporated into the models.

The present vegetation of Canada has been produced only during the last several thousand years by migrations from southern refugia since deglaciation of the Wisconsinan and other ice sheets. Even after these thousands of years, the floristic richness of eastern Canada is relatively low as compared with the relatively rich floras of the unglaciated southeastern quarter of the North American continent. These past migrations northward from the southern refugia, however, have been natural ones that took place under relatively low atmospheric CO_2 concentrations during relatively steady and slow climatic changes. The next two centuries will probably see much faster warming and certainly much higher levels of atmospheric CO_2 and CH_4. The migrations of plant populations and species will probably not keep up with such warming, in which the only constant will be the unchanged length of day, as it cycles through the year from winter to summer and back to winter again, characteristic of each latitude. This lack of climatic and soil synchrony may lead to sparsely vegetated areas where the old original vegetation has died but more southerly species have not yet arrived. One other prospect is that adventive species from other continents, including herbaceous weeds, may occupy the floristically poor gaps. It is quite likely that wholly new vegetation types will result in such places. In some cases, these might be poor habitats for the native birds and mammals of the northern ecosystems. The problems posed by migrations and invasions thus call for a great deal of thought, planning, and broadly based ecological research.

PERMAFROST WARMING AND THERMOKARST

All of the arctic tundra, parts of the boreal taiga, and some alpine areas above timberline in the Northern Hemisphere are underlain with permanently frozen ground (permafrost). In some cases, the permafrost is as deep as several hundred to a thousand meters (Brown et al. 1980). Such permafrost has existed for thousands of years, long before Holocene warming commenced. It can be "dry" permafrost (frozen rock or dry soil) or "wet" permafrost, characterized by white ice lenses or ice wedges below shallow peaty wet soils and vegetation that insulate the permafrost during the cold but sunny tundra summers. The "active" layer of soil above the permafrost thaws briefly during the short arctic summers to depths of 20–50 cm. It is in the active layer where plant root growth and nutrient uptake occur. The roots of some plants, notably various species of *Eriophorum,* follow the thaw down to the top of the permafrost, where they grow horizontally along the impervious surface at temperatures of 0°C or even lower (Billings et al. 1978).

Permafrost is a fundamental and characteristic part of tundra ecosystems. In the wet coastal arctic tundras, the large white ice wedges embedded in permafrost have formed extensive polygonal patterns during the thousands of years since the land rose above sea level. As tundra plants invaded these patterned and very cold soils, they formed vegetations that are themselves patterned on the ice wedge-polygoned substratum. Such patterned vegetation changes through several thousand years in a cyclic sequence (the thaw lake cycle) that never reaches any stage that is completely stable (Billings and Peterson 1980, 1992). This cycle is completely dependent on permafrost and its patterned ice wedges. If the ice wedges warm to temperatures above 0°C, either as a result of increasing soil temperatures or by exposure to warm air or flowing water due to climatic change or disturbance by vehicles, the result is accelerated erosion caused by permafrost melting. This erosion is called thermokarst. It is significant that Lachenbruch and Marshall (1986) have reported increasing permafrost temperatures in the Alaskan Arctic tundra. Oechel et al. (1993) have concluded, based on weather data from Barrow and Prudhoe Bay, that summer air temperatures over the North Slope tundra have increased through the last 25 years.

Monitoring of arctic tundra soil and permafrost temperatures is urgently necessary and of very high priority in understanding biotic feedback relationships in possible global warming. The Arctic is predicted to be the geographic region in which temperature changes will be most pronounced and that will provide the earliest warning. Preservation of the permafrost by protection of the overlying vegetation is absolutely necessary if thermokarst and loss of wet tundra are to be prevented. This necessity has been demonstrated in wet tundra near Point Barrow, Alaska, by Billings and Peterson (1992). If the upper 2 m of permafrost are lost by melting for whatever reason, the coastal tundra is lost. Tundra and permafrost have been inseparable for more than 10,000 years. So, conversely, if the tundra is damaged first by vehicular traffic or other disturbance, the underlying permafrost is doomed.

The thermokarst along Footprint Creek, near Barrow, Alaska, is an example of what warmer waters can do to a pristine tundra ecosystem. Much of this previously undisturbed wet tundra of circa 14 ha, the Voth Tundra, has been destroyed by thermokarst during the last 30–40 years because of the artificial drainage of slightly warmer water still flowing by ditch into Footprint Creek from four thaw lakes artificially drained in 1950 to protect the natural gas wells east of the village. Billings and Peterson (1992) measured the loss of 1050 metric tons of soil, ground ice, and vegetation washed out into the Chukchi Sea from the Voth Tundra through Footprint Creek alone during the 11 summers between 1977 and 1988. This erosion is just a small sample of what can happen when such tundra ecosystems and their permafrost are disturbed. The Voth Tundra is gone.

ECOSYSTEM CARBON BALANCE

Since it is carbon compounds that make up the majority of the greenhouse gases in the atmosphere, considerably more research is needed on the cycling of these gases through the biosphere and their relative fluxes and carbon balances in the biomes. The principal gas in this regard is, of course, CO_2, with lesser metabolic roles played by CH_4 and CO. The synthetic chlorofluorocarbons and other fluorinated gases, although not metabolically active, are also greenhouse gases, but they are not directly involved within terrestrial or aquatic ecosystems.

Simply put, the carbon balance of an ecosystem is the difference between the carbon capture as CO_2 in photosynthesis by the green vegetation at the primary producer trophic level and the respiratory loss of CO_2 at *all* trophic levels, including green plants, herbivores, carnivores, and decomposers. Inorganic import of carbon and mass export of carbon compounds in water or wind are also parts of the equation.

The measurement of ecosystemic carbon balance with any precision is far from simple, and in most cases it is very difficult. Oechel and Lawrence (1985) put it succinctly: "Calculating the carbon balance for a single plant is difficult; attempting to estimate the carbon balance for an ecosystem such as the taiga is rudimentary at best." When one considers doing this for a single hectare of tropical rain forest, compiling the raw data alone is a staggering task. Obtaining a rough carbon budget for any ecosystem requires, at a minimum, quantitative data on the following parameters:

1. Gross and net photosynthesis of the whole plant community.
2. Loss of carbon by dark respiration and photorespiration of the whole biotic community, including the green vegetation; its leaves, stems, flowers, fruits, and seeds with their export rates and decomposition; and all belowground roots, rhizomes, bulbs, tubers, and other vascular plant carbon storage sinks.
3. Carbon fluxes and sinks of all other trophic levels, including decomposers and pathogens, such as bacteria and fungi, herbivores, and carnivores. We performed such an analysis for all trophic levels in the wet coastal tundra ecosystem at Barrow, Alaska (Chapin et al. 1980). At that time (the early 1970s), the measurements showed a carbon budget in which photosynthetic carbon capture in g carbon m^{-2} $year^{-1}$ was +214, respiratory and export carbon loss was −105 g carbon m^{-2} $year^{-1}$, and carbon balance was positive into the system at +109 g carbon m^{-2} $year^{-1}$ (see Billings 1987). Most of this positive balance of carbon went into the cold soil as live and dead roots, rhizomes, and litter. Live roots, rhizomes, and stem bases high in stored carbohydrates are harvested and stored by lemmings, which eat them under the snow during the winter. In turn, these rodents are eaten by some of the carnivores (e.g., foxes, snowy owls, and jaegers). The migratory shorebirds (e.g., sandpipers, plovers, and phalaropes) were not incorporated into the balance except as "export," nor were the migratory grazing caribou, which pass through the Barrow tundra only briefly during the peak summer season.

Most of the carbon in the wet, cold soil eventually becomes peat in the "active layer." In the past, the insulating effect of the dead litter and live vegetation has allowed the permafrost table to rise into the peat and preserve the peat carbon for centuries until sporadic warming initiates thermokarst. Then decomposition is renewed aerobically with an efflux of CO_2 (Peterson and Billings 1975; Billings and Peterson 1992).

In the geologic and historical past, the Alaskan arctic tundra has been a net sink for CO_2 and biologically produced carbon compounds (Chapin et al. 1980; Billings 1987). A large taiga bog of black spruce (*Picea mariana*) and *Sphagnum* moss in central Alaska also was a net sink for carbon into the early 1980s (Billings 1987). The result was a carbon-rich peat maintained for millennia in cold frozen soils. In estimates of global soil carbon pools, the combined arctic tundra–boreal forest estimates range from 21% (Raich and Schlesinger 1992) to about 27% (Post et al. 1982) of the total global soil carbon pool, or about one-quarter of the earth's soil carbon. This sum is large under any circumstances. It represents a ticking time bomb of potential atmospheric CO_2 that has been stored in tundra and muskeg for several centuries (Billings 1987). If the arctic and

Sub-Arctic should become warmer, this frozen carbon could become a large source of CO_2 and CH_4.

Billings et al. (1982, 1983, 1984), in experimental prediction experiments with intact frozen tundra cores from Barrow, have shown that the postulated increase in summer temperatures at Barrow will turn that wet tundra from a sink into a source of CO_2. In addition, if the depth of thaw increases, the water table in the tundra will drop from the surface to depths of 10–20 cm or more. As we have shown, if this happens, efflux of decompositional CO_2 will increase dramatically (Billings et al. 1983). This transition may be under way now in the tundra of the North Slope and perhaps elsewhere. Oechel et al. (1993) have reported recent warming of this tundra region, with concomitant loss of CO_2 to the atmosphere from the tundra ecosystem. Their results suggest that the process may be a positive feedback on the concentration of atmospheric CO_2 and increased global warming, as we predicted that it might be (Billings et al. 1982). But we desperately need many more measurements in other biomes on a global basis. Policy-makers at all levels should take note and be forewarned.

DIRECT EFFECTS OF ATMOSPHERIC CO_2 ON VEGETATION

Beyond the indirect interactions of CO_2 with vegetation through climatic effects, it must be remembered that CO_2 is one of the raw materials of photosynthesis. One must know how increasing atmospheric concentrations of this gas will affect the metabolism not only of individual plants and populations of the large number of plant species on earth but also of plant communities, and how the health and productivity of ecosystems will be affected. Are there feedbacks to the biosphere and its climatic systems from vegetations that have been directly affected by the documented increases of atmospheric CO_2 concentration? The answers are not yet clear.

Strain (1985, 1987, and subsequent work) has reported from his experimental studies under controlled phytotron conditions, and also from very large field experiments with individual trees and whole forests, that the water use efficiency of individual plants increases as CO_2 level increases. The increase is largely the result of stomatal closure that reduces transpiration while net photosynthesis is increasing. This metabolic inter-action allows greater plant growth during droughts. Strain hypothesizes that water yield will rise on some watersheds, particularly in the mountain ranges of western North America. In addition, the direct effects of increasing atmospheric CO_2 concentrations on photosynthesis tend to favor C3 plants over those with C4 metabolic pathways (Pearcy and Björkman 1983).

At the ecosystem level, direct effects of atmospheric CO_2 are even less clear in most cases. Bazzaz (1990) concludes that the intricate interactions between CO_2 levels and other environmental factors make it difficult to extrapolate from the individual plant to the ecosystem in regard to direct effects. It is to be expected, with the great diversity of plants, vertebrates, and insects in tropical rain forests, that some species populations will benefit more than others from increases in atmospheric CO_2 concentration.

Körner and Arnone (1992), in a controlled experiment using small ecosystems put together from a mixture of tropical plants of different life forms under different CO_2 concentrations, emphasize the inadequacy of scaling up from the physiological baselines of individual plants to whole ecosystems. If complex biospheric CO_2 problems are to be solved, one really must *start* with ecosystems, as indeed Billings et al. (1982, 1983, 1984), Tissue and Oechel (1987), and Shaver et al. (1992) have done with the tundra ecosystems of northern Alaska. This approach assumes, of course, that some understanding has been acquired of the physiological ecology of direct CO_2 effects on individual plants. Strain's research group has now begun to tackle this ecosystematic problem by using large forest communities in which the experimental CO_2 is supplied to a whole forest patch in the freely moving atmosphere (the FACE Project). It has been demonstrated in crop ecosystems (Hendrey 1993) that this kind of unconstrained experiment should go a long way toward answering the CO_2-ecosystem problems as they occur in tropical forests, temperate forests, the floristically rich warm deserts such as the Sonoran Desert, and the circumboreal taiga.

Direct CO_2 effects in the floristically poor arctic tundra are likely to be quite different from the effects in midlatitude or tropical ecosystems. In addition to the tundra's floristic simplicity, the tundra angiosperms are all C3 plants (Teeri and Stowe 1976), which are favored by higher atmospheric CO_2 concentrations as compared with the mixtures of C4, C3, and CAM plants in the vegetations of the warmer parts of the world. In a preliminary attempt to establish the relative importance of direct CO_2 effects on tundra ecosystems, we used natural vegetation-soil microcosms (8 cm diameter × 55 cm long) that we extracted whole and frozen from the tundra during the Arctic winter by coring through the snow with a motorized ice auger. After keeping the microcosms frozen until the start of our experiments, we measured whole ecosystem CO_2 exchange by infrared gas analyzer as thaw proceeded under controlled conditions that simulated both the present time and the 21st century in the Duke Phytotron (Billings et al. 1982, 1983, 1984). We arrived at the following results and conclusions using live cores from Barrow, Alaska, at 71°N:

1. Increasing summer temperature from 4°C to 8°C reduced net ecosystem carbon uptake by half.
2. Lowering the water table by 5 cm and increasing temperature by 4°C greatly lowered ecosystem carbon storage.
3. Doubling ambient atmospheric CO_2 concentration from 400 ppmv to 800 ppmv had little effect on total ecosystem carbon balance.
4. The addition of nitrogen (the present limiting chemical factor in most arctic tundras) had a far greater effect on ecosystemic net carbon gain than simultaneously increasing CO_2 cocentration.
5. In conclusion, the most probable influence of higher CO_2 levels on the Alaskan tundra will be through the gas's indirect effects on temperature, lowering water table levels and thus allowing oxygenation of the peat and increasing soil organic matter decomposition.

CO_2 is not now limiting to tundra ecosystem productivity, but soil nutrients and length of growing season do and will continue to limit the productivity of this cold arctic ecosystem.

Our results as stated previously have been confirmed under field conditions in tussock tundra at Toolik Lake, north of the Brooks Range in Alaska, by Tissue and Oechel (1987). They conclude that growth of the principal dominant graminoid there

(*Eriophorum vaginatum*) is limited more by soil nutrient supply than by carbon availability for photosynthesis. The tundra plants growing there in the field under small
"greenhouses" at elevated CO_2 concentrations and those at ambient CO_2 concentrations
had similar photosynthetic rates to plants of the same species. The reduction of photo-
syntheic capacity for plants growing at elevated CO_2 levels did not appear to be due to
stomatal closure or end-product inhibition. Shaver et al. (1992) also conclude that
changes in atmospheric CO_2 concentrations have relatively little long-term effect on
whole-system carbon budgets and the productivity of wet and moist arctic tundras. They
present a conceptual model of carbon-nutrient interactions in terrestrial ecosystems,
using carbon and nitrogen, that should be very useful in studying the effects of climate
change on other terrestrial nutrient-limited ecosystems.

For these tundras, perhaps the most important factor is the potential for redistribution
of the large pool of nitrogen in soils with low C:N ratios and its uptake by vegetation
with higher C:N ratios. If climatic warming should be the cause of soil CO_2 efflux in
northern ecosystems, those carbon losses may be offset by the release of soil nutrients by
increasing depth of thaw, resulting in plant carbon gains in photosynthesis (Schimel et
al. 1990).

DROUGHT, DESERTIFICATION, AND WATER SUPPLIES

One very likely result of climatic warming in the lower latitudes (equator to circa 40°
north and south latitudes) is an increase in drought frequency and possible desertification. In continental climates, gradients from forests to grasslands to the present deserts
extend poleward from tropical forests to extreme deserts, as in Africa. In the midlatitudes, gradients toward aridity at any given latitude are, in general, across longitudes
toward the middle of the continent, as in North America and in Eurasia. This pattern is
complicated by mountain rain-shadow effects, as between the Sierra Nevada of California and the Rocky Mountains. (This arid intermontane region is actually in a double
rain-shadow, with the Sierra Nevada blocking Pacific air on the west and the Rocky
Mountains blocking moist air from the Gulf of Mexico on the east.)

Any trend toward desertification within the last century is not due solely to climatic
changes, or to sequences of dry and wet years. Human population increases are also
involved, as is the dependence of these peoples on plant and animal foods. Shifting
cultivation in the tropical savannas and overgrazing of the savannas on their desert
margins have allowed the Sahara to expand southward into the Sahel (Cloudsley-
Thompson 1974; Hare 1977; Le Houérou 1989; Schlesinger et al. 1990). The combination of people, their livestock, overgrazing, and possible regional climatic change has
resulted in ecosystem degradation, with a strong possibility of the repetition of the
human famine episodes that are now occurring in Africa.

The loss of savanna vegetation from a combination of these various causes leads to
increased albedo that results in positive feedback and loss of ecosystem stability. Thus
desertification and drought proceed apace in both intensity and total area.

One result on a global basis is the loss of water supplies by evaporation to the
atmosphere as water vapor. Since this is another greenhouse gas, such input only adds to
the global warming problem. A considerable part of this atmospheric water does return

to the biosphere as precipitation in one form or another. But this liquid water, in turn, may evaporate or sublime more rapidly than before. More research is needed on global and ecosystem water balance in those deserts of colder climates that may become warmer with an accompanying increase in vegetational cover and total leaf area.

People in all biomes are dependent on abundant water supplies for agriculture and domestic use. Ecologists must have knowledge of total water inputs, circulation, and losses from city ecosystems, particularly the great metropolitan centers in semiarid and arid regions. The water sources being used at present are local precipitation, runoff from regional mountains, aquifers and other groundwater, and reservoirs and their canals. These sources are likely to be subject to considerable stress and change in a warming and drier climate.

For the last century, the lowlands, including cities and irrigation agriculture, in western North America have been dependent upon mountain snowfall and resultant snowpack for much of their urban and agricultural water supplies. The science of snow hydrology is extremely important to human populations in the western half of North America, where drought years are frequent. Snow meltwater runoff predictions will be even more important to the survival of people, cities, and agriculture in that region during the next few decades and through the 21st century.

The roles of vegetational patterns in the catchment and short-term maintenance of alpine and subalpine snowdrifts, snowpacks, and their meltwater contents cannot be ignored in the consideration of any scenarios of regional warming effects in mountainous areas of the earth (Billings 1969, 1973, 1993; Walker et al. 1993).

They are of particular importance in those mountain ranges bordering the semiarid and arid regions that are so dependent on mountain meltwater runoff as their principal water source. Further research on these snow ecology problems is very much needed.

CONCLUSION: ATTEMPTING TO PREDICT BIOSPHERIC FUTURES

The key word in this discussion of research needs on biotic feedbacks in an unstable biosphere is "change". To paraphrase the distinguished ecologist William S. Cooper (1926): change in vegetation and ecosystems is like a braided glacial river flowing through time. The complexities of the separations, rejoinings, and new combinations of plant and animal species are almost limitless and filled with surprises.

One cannot view the biosphere provincially in either space or time. There are no crosswalls or barriers to biotic migrations or environmental changes. This is also true of any regional ecosystem. Factors limiting to the successful growth and reproduction of biotic populations exist and are unique to specific ecotypic populations. Within ecosystems, a carrying capacity for people and animals, both regionally and globally, is eventually reached. Carrying capacity is determined by primary productivity of the vegetation and the nutrient availability of the soil.

Any change in a limiting factor may set off long-lasting chain reactions in an ecosystem or in the biosphere. The initiator of such a chain reaction was designated a "trigger factor" by Billings (1952). A trigger factor can be either a changed limiting factor, such as the addition of water to a desert, or an entirely new environmental factor, a "surprise." Examples of the latter are fallout from a nuclear accident or the destruction of stratospheric ozone by chlorofluorocarbons; both are anthropogenic. The ultimate future is unknown in both cases.

The biosphere has changed continuously throughout the past in both space and time, and it will continue to do so. Some of these changes have been very slow and have occurred over long periods. Other changes have been quite rapid, such as the onset and termination of the Younger Dryas in the Arctic and North Atlantic region, as shown by both marine and ice core data. Both the onset and the termination of this late-glacial event occurred within an elapsed time of 10–20 years from near-glacial conditions to relatively warm temperatures and vice versa (Mayewski et al. 1993). Such sudden changes in temperature and atmosphere in the mid-Holocene were natural and, of course, preindustrial. Many natural events are episodic and present major problems. Earthquakes, floods, droughts, and volcanic eruptions are examples. Mt. Pinatubo's eruption in the Philippines in 1991 changed temperatures in the northern midlatitudes and also detrimentally affected the stratospheric ozone layer in those same latitudes. Global average total ozone in 1992 was 2–3% lower than in any previous year (Gleason et al. 1993), and it was lower over a wide range of latitudes in both the Northern and Southern hemispheres. The atmospheric ash, dust, and aerosols from Mt. Pinatubo are thought to be the cause.

In the last two to three centuries, within the span of the Industrial Revolution, there has been an acceleration of environmental and biospheric change, both physical and biological, that has been, in many cases, almost logarithmic in progression. Much of this acceleration has been caused by well-documented changes in the composition and temperature of the atmosphere. Levels of extinction and migration of the biota are also increasing. These latter centuries have been filled with biospheric surprises that were not predicted and that have changed our world.

In the last decade, much attention has been devoted to the question "Are ecosystems and the biosphere sustainable?" (Clark and Munn 1986; Lubchenco et al. 1991). This is not a trivial question. The future of the earth depends upon the answer. Lubchenco and her committee propose three research priorities: (1) the ecological causes and consequences of global change in all its aspects, (2) the ecological determinants and consequences of biological diversity and the effects of global and regional changes on such diversity, and (3) the management of sustainable ecological systems and the interface between ecological processes and human social systems. The implementation of this very important and complex research program in earth ecology is spelled out in detail in Lubchenco's committee report.

The most succinct approach to the question of the principles of ecosystem sustainability is that of Chapin (1993). He defines "a sustainable ecosystem as one . . . that maintains its characteristic diversity of major functional groups, productivity, and rates of biogeochemical cycling in the face of some disturbance." Such a system occurs when negative feedbacks tend to maintain the current characteristics of an ecosystem; positive feedbacks hasten change away from stability.

Can we predict future biospheric characteristics and interactions? At this point in time, estimates of approximate values and fluxes can be attempted by monitoring changes of all kinds on a global basis, by experiments, and by mathematical modeling. Facts are necessary to all of these approaches. We must conserve all sources of these facts ("information") before they are lost forever: genetic, environmental, populational, ecosystemic, human—all in the broadest sense.

Beyond all this, cooperation is needed within and among all the sciences (including ecology and economics), between representatives of the sciences and of politics (governments), and among all nations. All of these problems concern the future of a livable earth. The "braided stream" flows on, and ever faster.

REFERENCES

Bazzaz, F. A. 1990. The response of natural ecosystems to the rising global CO_2 levels. *Annu. Rev. Ecol. Syst.* 21:167–196.

Billings, W. D. 1952. The environmental complex in relation to plant growth and distribution. *Q. Rev. Biol.* 27:251–265.

Billings, W. D. 1969. Vegetational pattern near alpine timberline as affected by fire-snowdrift interactions. *Vegetatio* 19:192–207.

Billings, W. D. 1973. Arctic and alpine vegetations: Similarities, differences, and susceptibility to disturbance. *BioScience* 23:697–704.

Billings, W. D. 1984. Effects of UV-B radiation on plants and vegetation as ecosystem components. In *National Research Council, 1984. Causes and Effects of Changes in Stratospheric Ozone: Update 1983*, ed. L. C. Harber, pp. 206–217. National Academy Press, Washington, D.C.

Billings, W. D. 1987. Carbon balance of Alaskan tundra and taiga ecosystems: Past, present, and future. *Quat. Sci. Rev.* 6:165–177.

Billings, W. D. 1990. *Bromus tectorum*, a biotic cause of ecosystem impoverishment in the Great Basin. In *The Earth in Transition: Patterns and Processes of Biotic Impoverishment*, ed. G. M. Woodwell, pp. 300–322. Cambridge University Press, New York.

Billings, W. D. 1994. Ecological impacts of *Bromus tectorum* and resultant fire on ecosystems in the western Great Basin. In *Ecology, Management, and Restoration of Intermountain Annual Rangelands*, ed. S. B. Monsen. U.S. Forest Service, Ogden, Utah.

Billings, W. D., K. M. Peterson, and G. R. Shaver. 1978. Growth, turnover, and respiration rates of roots and tillers in tundra graminoids. In *Vegetation and Production Ecology of an Alaskan Arctic Tundra*, Ecological Studies 29, ed. L. L. Tieszen, Chap. 18, pp. 415–434. Springer-Verlag, New York.

Billings, W. D., and K. M. Peterson. 1980. Vegetational change and ice-wedge polygons through the thaw-lake cycle in arctic Alaska. *Arct. Alp. Res.* 12:413–432.

Billings, W. D., and K. M. Peterson. 1992. Some possible effects of climatic warming on arctic tundra ecosystems of the Alaskan North Slope. In *Climatic Warming and Biological Diversity*, R. Peters and T. E. Lovejoy, pp. 233–243. Yale University Press, New Haven, Conn.

Billings, W. D., J. O. Luken, D. A. Mortensen, and K. M. Peterson. 1982. Arctic tundra: A source or sink for atmospheric carbon dioxide in a changing environment? *Oecologia* 53:7–11.

Billings, W. D., J. O. Luken, D. A. Mortensen, and K. M. Peterson. 1983. Increasing atmospheric carbon dioxide: Possible effects on arctic tundra. *Oecologia* 58:286–289.

Billings, W. D., K. M. Peterson, J. O. Luken, and D. A. Mortensen. 1984. Interaction of increasing atmospheric carbon dioxide and soil nitrogen on the carbon balance of tundra microcosms. *Oecologia* 65:26–29.

Brown J., P. C. Miller, L. L. Tieszen, and F. L. Bunnell, eds. 1980. *An Arctic Ecosystem: The Coastal Tundra at Barrow, Alaska*. Dowden, Hutchinson, and Ross, Stroudsburg, Pa.

Caldwell, M. M., R. Robberecht, R. S. Nowak, and W. D. Billings. 1982. Differential photosynthetic inhibition by ultraviolet radiation in species from the arctic-alpine life zone. *Arct. Alp. Res.* 14:195–202.

Chapin, F. Stuart, III. 1993. Principles of ecosystem sustainability. *Bull. Ecol. Soc. Am.* 74(2):189–190.

Chapin, F. Stuart, III, P. C. Miller, W. D. Billings, and P. I. Coyne. 1980. Carbon and nutrient budgets and their control in coastal tundra. In *An Arctic Ecosystem: The Coastal Tundra at Barrow, Alaska,* ed. J. Brown, P. C. Miller, L. L. Tieszen, and F. L. Bunnell, pp. 458–482. Dowden, Hutchinson, and Ross, Stroudsburg, Pa.

Cicerone, R. J. 1979. Atmospheric carbon tetrafluoride: A nearly inert gas. *Science* 206:59–61.

Cicerone, R. J. 1987. Changes in stratospheric ozone. *Science* 237:35–42.

Clark, W. C., and R. E. Munn, eds. 1986. *Sustainable Development of the Biosphere.* IIASA and Cambridge University Press, New York.

Cloudsley-Thompson, J. L. 1974. The expanding Sahara. *Environ. Cons.* 1:5–13.

Cooper, W. S. 1926. The fundamentals of vegetational change. *Ecology* 7:391–413.

Dansgaard, W., and H. Oeschger. 1989. Past environmental long-term records from the Arctic. In *The Environmental Record in Glaciers and Ice Sheets,* ed. H. Oeschger and C. C. Langway, Jr. pp. 287–318. Wiley, New York.

Davis, M. B. 1989. Lags in vegetation response to greenhouse warming. *Climatic Change* 15:75–82.

Davis, M. B., and C. Zabinski. 1992. Changes in geographical range resulting from green-house warming: Effects on biodiversity in forests. In *Global Warming and Biological Diversity,* ed. R. L. Peters and T. E. Lovejoy, pp. 297–308. Yale University Press, New Haven, Conn.

Friedli, H., H. Lotscher, H. Oeschger, U. Siegenthaler, and B. Stauffer. 1986. Ice core record of 13C/12C ratio of atmospheric CO_2 in the past two centuries. *Nature* 324:237–238.

Gleason, J. F., P. K. Bhartia, J. R. Herman, R. McPeters, P. Newman, R. S. Stolarski, L. Flynn, G. Labow, D. Larko, C. Seftor, C. Wellemeyer, W. D. Komhyr, A. J. Miller, W. Planet. 1993. Record low global ozone in 1992. *Science* 260:523–526.

Hare, F. K. 1977. Connections between climate and desertification. *Environ. Cons.* 4:81–90.

Henderson-Sellers, A. 1993. Continental vegetation as a dynamic component of a global climate model: A preliminary assessment. *Climatic Change* 23:337–377.

Hendrey, G. R., ed. 1993. *FACE: Free-Air CO_2 Enrichment for Plant Research in the Field.* CRC Press, Boca Raton, Fla.

Jacobson, G. L., Jr., T. Webb III, and E. C. Grimm. 1987. Patterns and rates of vegetation change during the deglaciation of eastern North America. In *North America and Adjacent Oceans during the Last Deglaciation. The Geology of North America,* Vol. K-3, ed. W. F. Ruddiman and H. E. Wright, Jr., pp. 277–288. Geological Society of America, Boulder, Colorado.

Keeling, C. D., R. B. Bacastow, and T. P. Whorf. 1982. Measurements of the concentration of carbon dioxide at Mauna Loa Observatory, Hawaii. In *Carbon Dioxide Review: 1982,* ed. W. C. Clark, pp. 377–385. Oxford University Press, New York.

Khalil, M. A. K., and R. A. Rasmussen. 1986. Interannual variability of atmospheric methane: Possible effects of the El Nino–Southern Oscillation. *Science* 232:56–58.

Kopf, A. W. 1984. Malignant melanoma in humans. In *National Research Council, 1984. Causes and Effects of Changes in Stratospheric Ozone: Update 1983,* ed. L. C. Harber, pp. 168–190. National Academy Press, Washington, D.C.

Körner, C., and J. A. Arnone III. 1992. Responses to elevated carbon dioxide in artificial tropical ecosystems. *Science* 257:1672–1675.

Lachenbruch, A. H., and B. V. Marshall. 1986. Changing climate: Geothermal evidence from permafrost in the Alaskan Arctic. *Science* 234:689–696.

Le Houérou, H. N. 1989. *The Grazing Land Ecosystems of the African Sahel.* Ecological Studies 75. Springer-Verlag, Heidelberg and New York.

Lovejoy, T. E. 1989. Deforestation and the extinction of species. In *Changing the Global Environment,* ed. D. B. Botkin, pp. 91–98. Academic Press, Boston.

Lubchenco, J., A. M. Olson, L. B. Brubaker, S. R. Carpenter, M. M. Holland, S. B. Hubbell, S. A. Levin, J. A. MacMahon, P. A. Matson, J. M. Melillo, H. A. Mooney, C. H. Peterson, H. R. Pulliam, L. A. Real, P. J. Regal, and P. G. Risser. 1991. The sustainable biosphere initiative: An ecological research agenda. A report from the Ecological Society of America. *Ecology* 72:371–412.

Mayewski, P. A., L. D. Meeker, S. Whitlow, M. S. Twickler, M. C. Morrison, R. B. Alley, P. Bloomfield, and K. Taylor. 1993. The atmosphere during the Younger Dryas. *Science* 261:195–197.

Oechel, W. C., and W. T. Lawrence. 1985. Taiga. In *Physiological Ecology of North American Plant Communities,* ed. B. F. Chabot and H. A. Mooney, pp. 66–94. Chapman and Hall, New York.

Oechel, W. C., S. J. Hastings, G. Vourlitis, M. Jenkins, G. Riechers, and N. Grulke. 1993. Recent change of arctic ecosystems from a net carbon dioxide sink to a source. *Nature* 361:520–523.

Overpeck, J. T., P. J. Bartlein, and T. Webb III. 1991. Potential magnitude of future vegetation change in eastern North America: Comparisons with the past. *Science* 254:692–695.

Pearcy, R. W., and O. Björkman. 1983. Physiological effects. In *CO$_2$ and Plants: The Response of Plants to Rising Levels of Atmospheric Carbon Dioxide.* ed. E. R. Lemon. pp. 65–105. AAAS Selected Symposium 84. Westview Press, Boulder, Colorado.

Peterson, K. M., and W. D. Billings. 1975. Carbon dioxide flux from tundra soils and vegetation as related to temperature at Barrow, Alaska. *Am. Midl. Nat.* 94:88–98.

Post, W. M., W. R. Emanuel, P. J. Zinke, and A. G. Stangenberger. 1982. Soil carbon pools and world life zones. *Nature* 298:156–159.

Raich, J. W., and W. H. Schlesinger. 1992. The global carbon dioxide flux in soil respiration and its relationship to vegetation and climate. *Tellus* 44B:81–99.

Ravishankara, A. R., S. Solomon, A. A. Turnipseed, and R. F. Warren. 1993. Atmospheric lifetimes of long-lived halogenated species. *Science* 259:194–198.

Rowland, F. S. 1989. Chlorofluorocarbons and the depletion of stratospheric ozone. *Am. Sci.* 77:36–45.

Schimel, D. S., W. J. Parton, T. G. F. Kittel, D. S. Ojima, and C. V. Cole. 1990. Grassland biogeochemistry: Links to atmospheric processes. *Climatic Change* 17:13–25.

Schlesinger, W. H., J. F. Reynolds, G. L. Cunningham, L. F. Huenneke, W. M. Jarrell, R. A. Virginia, and W. G. Whitford. 1990. Biological feedbacks in global desertification. *Science* 247:1043–1048.

Shaver, G. R., W. D. Billings, F. S. Chapin III, A. E. Giblin, K. J. Nadelhoffer, W. C. Oechel, and E. B. Rastetter. 1992. Global change and the carbon balance of arctic ecosystems. *BioScience* 42:433–441.

Shukla, J., C. Nobre, and P. Sellers. 1990. Amazon deforestation and climate change. *Science* 247:1322–1325.

Siegenthaler, U., and H. Oeschger. 1987. Biospheric CO$_2$ emissions during the past 200 years reconstructed by deconvolution of ice core data. *Tellus* 39B:140–154.

Strain, B. R. 1985. Physiological and ecological controls on carbon sequestering in terrestrial ecosystems. *Biogeochemistry* 1:219–232.

Strain, B. R. 1987. Direct effects of increasing atmospheric CO$_2$ on plants and ecosystems. *Trends Ecol. Evol.* 2:18–21.

Teeri, J. A., and L. G. Stowe. 1976. Climatic patterns and the distribution of C4 grasses in North America. *Oecologia* 23:1–12.

Tissue, D. T., and W. C. Oechel. 1987. Response of *Eriophorum vaginatum* to elevated CO_2 and temperature in the Alaskan tussock tundra. *Ecology* 68:401–410.

Walker, D. A., J. C. Halfpenny, M. D. Walker, and C. A. Wessman. 1993. Long-term studies of snow-vegetation interactions. *BioScience* 43:287–301.

Woodwell, G. M., J. E. Hobbie, R. A. Houghton, J. M. Melillo, B. Moore, B. J. Peterson, and G. R. Shaver. 1983. Global deforestation: Contribution to atmospheric carbon dioxide. *Science* 222:1081–1086.

23

Will the Warming Speed the Warming?

George M. Woodwell, Fred T. Mackenzie, R. A. Houghton, Michael J. Apps, Eville Gorham, and Eric A. Davidson

A significant body of experience, much of it summarized in the preceding chapters, suggests that there are mechanisms entrained by a change in global climate that tend to increase the trend of temperature change, whatever the trend and its original cause may be. In the case of a warming, there is a possibility that the warming itself may cause a series of further changes in the earth that will speed the warming. If so, what are the mechanisms involved and how seriously might our current appraisals of the speed and severity of the warming be in error? Will the warming speed the warming? If so, for how long and by how much?

The global climatic system involves a complicated array of primary and secondary causes that influence global temperature and many other climatic factors. The secondary effects are considered feedbacks insofar as they influence the primary effect. In the case of the warming of the earth, feedbacks are positive if they tend to enhance a warming trend and negative if they diminish the warming. Kellogg (1983) and Lashof (1989) explored this topic in detail for the climatic system as a whole, and M. Schlesinger and Mitchell (1985) discussed the physical atmospheric feedbacks. The general circulation models (GCMs) used in estimating the responses of climate to various disturbances usually contain various physical feedback systems (see J. T. Houghton et al. 1990, 1992).

The biotic feedbacks are complicated and have not commonly been incorporated into such models or into the general calculus of the warming of the earth, despite their potential for significant influence. The most widely recognized biotic feedback is the enhancement of photosynthesis by elevated levels of CO_2 in air, the so-called CO_2 fertilization effect. It has been incorporated for many years into carbon cycle models by Keeling and colleagues (Bacastow and Keeling 1973) as the β factor (see the discussion in Wullschleger et al., Chapter 4) on the assumption that the increased photosynthesis results in increased carbon storage on land. The issue is complicated by the fact that carbon storage (net ecosystem production [NEP]) is determined by the balance between gross production (total photosynthesis) and total respiration of the ecosystem as defined by Woodwell and Whittaker (1968) and discussed in Chapter 1. Many factors apart from CO_2 concentration limit photosynthesis. Similarly, many factors influence the respiration of plants and also affect the rates of decay of organic matter in soils. NEP is determined not only by net primary production (NPP) but also by the rates of respiration of the organisms that feed on plants and the organisms of decay in soils. These processes are influenced by the availability of water, nitrogen and other nutrient elements, and sunlight; temperature; the successional status of ecosystems; and other ecological factors, such as disturbance by fire, disease, toxins, and storms. NEP is always less than NPP, commonly by 50% or more (Woodwell and Whittaker 1968). It fluctuates around zero in a quasi-stable climax vegetation and may become negative under chronic disturbance.

Any of the factors just mentioned may have an influence on NEP that may vary from time to time and may be considerably greater than the direct effects observed so far from changes in the concentration of CO_2 in the atmosphere. The topic has been examined in detail by R. A. Houghton in Chapter 19 and by Woodwell (1983, 1989). Wullschleger et al. in Chapter 4 and Allen and Amthor in Chapter 3 show that although there may be a stimulation of photosynthesis due to the increase in CO_2 concentration in the atmosphere, there will not necessarily be an equivalent increase in the storage of carbon in terrestrial ecosystems. The major reason is the sensitivity of respiration to temperature: a warming increases rates of respiration, including the rate of decay of organic matter in soils. In addition the loss of carbon due to disturbance is commonly much more rapid than its accumulation through regrowth of the disturbed ecosystems, as outlined by Kurz et al. in Chapter 6 in a discussion of the implications of fire and other past disturbances for NEP in forests.

Direct experience with global biotic feedbacks is limited and will remain so until the warming progresses over the next several years. Even then we shall have no basis for experimentation with the earth as a whole, and analyses will have to be based, then as now, on observation, on knowledge of present and past climate and vegetation, and on local experience with natural ecosystems. Modeling offers clear advantages, and progress in the development of models in support of such analyses is evident (see Section III, especially Chapters 16 by Luxmoore and Baldocchi and 17 by Prentice and Sykes). The models offer apparent detail and the possibility of exploring mechanisms. They are, however, only a tool; they offer a system of record keeping, a never-ending series of new questions about how nature works, and insights into the logic of current evaluations. They do not offer new data, nor do they offer answers to all questions. They depend heavily on assumptions of a new stability of climate and interpretations of the next years as "transient" periods, but it is instability that is the dominant characteristic of the next decades and the subject of the analysis. Although progress in modeling remains impressive, it does not yet reflect the experience of the Vostok Core, the Little Ice Age, and the contemporary correlations of CO_2 level with temperature outlined in Chapter 1 by Woodwell and later in this chapter. These data appear to establish a set of limits within which we must assume that the world operates. We must further assume that these limits are of primary importance in determining the course of the response to the warming now underway.

CRITICAL CONSTRAINTS ON THIS ANALYSIS

Time and Stability

Our interest is in the next several years to a century, not millennia. During this period of the span of our lives and those of our children, we expect profound changes in the human circumstance quite apart from the warming of the earth. We seek the best appraisal possible of the implications of the warming for the earthly habitat while the human population is involved in yet another doubling beyond the nearly six billion humans that now crowd the planet and scramble for access to as wide a variety of earthly resources as possible.

We observe that the mean global warming over two decades, 1970–1990, has been about 0.2°C per decade. In the mid-high latitudes the future warming is expected to be greater, as much as twice the global average. A 1°C change in temperature is equivalent to a latitudinal migration in the midlatitudes of 60–100 miles or 100–160 km. It is reasonable to consider the possibility of a warming in the middle and higher latitudes of 0.5°C or more per decade or a migration of climatic zones on the order of 30–50 miles per decade. Such a change in the prairie-forest border in Minnesota or Wisconsin would be rapid by any standard. The difficulty is, however, that the climatic change is universal and affects the entire landscape and its vegetation, not simply the prairie-forest border. Each individual of each perennial species finds itself progressively maladapted to its environment under conditions of rapidly changing climate.

The warming would be especially rapid by comparison with the life cycle of most trees and the time required for terrestrial ecosystems, especially forests, to attain equilib-rium with climate. Changes of tenths of a degree centigrade per decade can be expected to outrun the capacity of forest trees to respond in very few decades. This point has been made in various forms by Woodwell (1983, 1989), Davis and Zabinski (1992), and Solomon and Cramer (1993) among others. The effect is a transition from forest to shrublands or grasslands, or to more severe impoverishment (Woodwell 1990). Such transitions in the structure and pattern of vegetation will be accelerated, as warming progresses, by increased frequency and intensity of fires, the spread of diseases and pests into new ranges, and the increased frequency of human disturbance. The transitions in climate are not transitions to a new stability, but to continuous instability marked by a progressive, open-ended global warming in the patterns defined by the GCMs and summarized by the IPCC in J. T. Houghton et al. (1990, 1992).

Recent Glacial History: The Record from Ice

The glacial record over the past 160,000 years has provided important insights that are at once revealing and confusing (Barnola et al. 1987; Lorius et al. 1988; Raynaud et al. 1993). Throughout that period atmospheric CO_2 and CH_4 concentrations have been correlated with temperature. As temperature has risen, so have CO_2 and CH_4 concentrations; as temperature has dropped, CO_2 and CH_4 concentrations have also dropped. N_2O follows the same pattern (Khalil and Rasmussen 1989). Although the trends were reversed several times in that period and the causes of the reversals are not clear, the pattern is consistent with a positive feedback once the warming or cooling has begun. Temperature appears to have led the changes, especially during the cooling periods. The record suggests that a change of 1°C in this period was equivalent to a change of about 10 ppmv of CO_2 or 20 petagrams (Pg) ($=10^{15}$ g) carbon in the global atmosphere (Table 23.1).

The topic has been addressed recently by Raynaud et al. (1993), who confirmed the correlation and explored the question of causes of the variations in temperature and atmospheric trace gases during the cycles of glaciation over the past 160,000 years.

Table 23.1. Changes in CO_2 and CH_4 Concentrations in the Atmosphere per Degree Centigrade Change in Temperature as Reported in Recent Data[a]

Period	$CO_2/°C$ (ppm)	Carbon as $CO_2/°C$ (Pg)	$CH_4/°C$ (ppb)	Reference
Glacial records (~160,000 years)	~10	~20	~60	Raynaud et al. 1993 (from Figure 5)
1940–1990		3.4–6.4		R. A. Houghton, Chapter 19
1958–1990	~2.9	5.8		Woodwell, Chapter 1 (from Figure 1.2)

[a]Data from the glacial record report what must be considered "equilibrium" conditions arrived at over a millennium or more. The data from this century report the short-term changes observed. Both sets, however, confirm an overall positive feedback that applies to CO_2 and CH_4. Although not shown here, N_2O concentrations follow the same pattern.

Their conclusion was that about 50% of the variation in temperature was controlled by the concentration of trace gases and that little is known of the factors determining the trace gas concentrations. Changes in trace gas concentrations involved not only changes in stocks of carbon on land, but also large changes in oceanic circulation and in the oceanic carbon cycle. It is possible that oceanic changes may be quite rapid, difficult to predict, and in themselves a feedback into climatic changes (Broecker 1987).

The Vostok and Byrd ice cores record what may have been equilibrium conditions, achieved over centuries to millennia. The resolution of time in these cores is, unfortunately, not fine enough to feed our need for addressing the global warming anticipated over the next several decades. More recent data from ice cores taken in Greenland bring the resolution down to a few years (Alley et al. 1993). These records in ice were established in periods when the CO_2, CH_4, and N_2O concentrations in the atmosphere were substantially lower than those in the postindustrial world. The Greenland records establish that, even during inter-glacial periods, the climatic system has been open to rapid changes. There is no reason to assume either that the correlations that occurred at that time are less valid now or that the rapid changes that marked earlier periods could not occur again in this interglacial period.

The Little Ice Age

The temperature–trace gas correlation holds, at least crudely, during the shorter-term cooling of the Little Ice Age, although details of the data and the correlation are less precise. The temperature decline began during the 14th century in Europe, replaced a medieval warm period of about 300 years, and lasted, with fluctuations, well into the 19th century. The data on CO_2 concentrations have been obtained from various ice cores and tabulated by Enting (Chapter 18, page 319). The concentrations follow the same pattern observed in the Vostok Core: the decline in temperature corresponds to a decline in the concentration of CO_2 in the atmosphere, and the late-19th-century warming is reflected in the familiar increase in CO_2 concentration that has now continued and become the focus of our concern.

Contemporary Observations of Temperature and Trace Gases

Limited contemporary local or regional data, summarized here by Woodwell (Chapter 1, page 7) from Kuo et al. (1990), Keeling et al. (1989), and Marston et al. (1991), show a similar pattern with temperature leading changes in the CO_2 concentrations by weeks to a few months. The data of Keeling et al. (1989), which are the basis of all the analyses, suggest that a change of approximately 3 ppm in CO_2 concentration occurs per degree centigrade change in temperature, equal to about 6 Pg carbon in the atmosphere if the data can be interpreted globally (Table 23.1). The pattern is consistent both with a release of CO_2 from surface water of the ocean in response to warming and with a release from land through increases in rates of respiration of plants and in the rate of decay of organic matter in soils. Increases in temperature speed respiration, including the respiration of decay, and speed the release of both CO_2 and CH_4.

CH_4 concentrations in the glacial record follow CO_2 concentrations closely (Raynaud et al. 1993), a relationship that seems to tie the fluctuations to the land as opposed to the sea because of the low solubility of CH_4 in water. CH_4 is produced principally through anaerobic decay in wetlands. Production and emission are heavily influenced by the moisture content of soils and peats (see Gorham, Chapter 9, and Nisbet and Ingham, Chapter 10). Large quantities of CH_4 occur as clathrates in coastal oceans, are vulnerable to release through warming, may have been large (perhaps major) sources of carbon for the atmosphere in the past, and may be such sources for the future (MacDonald 1990). Because CH_4 has very low solubility in seawater, the major sink is oxidation to CO_2 in the atmosphere. The uncertainties concerning the magnitude of the effects of the decomposition of clathrates through warming simply add to the potential for an accentuation of the positive feedback systems already identified.

The combination of the glacial record and the contemporary data appears to set overall limits on the patterns of responses possible. There is no basis for an assumption that the combination of changes inherent in a greenhouse gas–driven warming will stop the warming through accelerated storage of carbon on land in much-enriched forests and other terrestrial ecosystems, as suggested by some (see Idso et al. 1991 and discussions by Allen and Amthor in Chapter 3 and Wullschleger et al. in Chapter 4). On the contrary, the data provide an envelope of experience that seems to limit the overall response of temperature and trace gases to a positive feedback in which a warming, however caused, results in the further accumulation of CO_2 and CH_4 (and probably N_2O) in the atmosphere within weeks to a few months. The net effect must be a further contribution to the warming. Within that context, there are undoubtedly a host of interactions, some of which add to the positive feedback while others, as negative feedbacks, reduce it.

The factors that initiated the periods of glacial advance and retreat are not known. The reversals of trends capped periods of warming and cooling that lasted many thousands of years. During those millennia the climate appeared to be under the influence of a positive feedback. We cannot say whether that feedback system has limits that caused the reversals. The probability seems greater that other outside factors, such as the amount of energy reaching the surface of the earth, changed at those times. Our interest is in the next years to decades, when we expect the earth to continue in its most recent postglacial warming as heat-trapping gases accumulate further in the atmosphere.

THE "MISSING CARBON" PROBLEM

The factors that determine the CO_2 and CH_4 content of the atmosphere would appear to be few, easily measured, and open to control by deliberate interference. There are but two large reservoirs of carbon in addition to the atmosphere: the oceans and the terrestrial biota, especially forests and soils, including peat. The imbalance that has resulted in an unprecedented accumulation of CO_2 in the atmosphere has been caused by the combustion of fossil deposits of hydrocarbons as fuels for industrialization and by long-continued and accelerating deforestation associated with human activities.

Measurement of the stocks and flows of carbon among these major pools has proven unsatisfactory, at least from the standpoint of providing a global accounting for the carbon cycle. The difficulty lies in reconciling the major flows. We know, or think we know, with considerable accuracy the amount of carbon released through burning fossil fuels. In 1992 it is thought to have been about 6 Pg. To that we have added the release from deforestation globally, estimated most recently by R. A. Houghton as 1.6 ± 1 Pg. This release is the net release from changes in land uses. It includes the direct release from deforestation, burning, and the decay of organic matter in soils stimulated by changes in land use, as well as the storage of carbon resulting from the successional recovery of vegetation after human disturbance. It does not include a correction for any stimulation of carbon storage on land caused by a stimulation of photosynthesis from additional CO_2 in the atmosphere or of NEP from fertilization by atmospheric deposition of nitrogen mobilized by human activities, as discussed later in this chapter. These feedbacks are cited by several authors as factors that resolve the "missing carbon" problem. Nor does it include any assumption concerning forests or other vegetation globally beyond the basic assumption that undisturbed forests globally are late successional and approximately stable, and therefore have a net ecosystem production of zero.

Some of the carbon released into the atmosphere accumulates by an amount that varies from year to year. Between 1988 and 1991 the accumulation ranged between about 5 and 2 Pg carbon annually and averaged about 3. In the 1990s the rate of accumulation has been dropping abruptly. The difference between what is released (a sum commonly thought to include the 6.0 Pg carbon from fossil fuels plus 1.6 Pg carbon from changes in land use, plus any further increment from the warming itself) and the amount accumulating in the atmosphere would be expected to be the amount absorbed into the oceans, more than 4.5 Pg carbon. The difficulty is that the amount accumulating in the oceans, as estimated by several techniques, seems to be limited to about 2 ± 1 Pg (see Mackenzie, Chapter 2; Sarmiento and Sundquist 1992 and Sarmiento 1993). The difference has been called "missing carbon," as discussed by Woodwell (Chapter 1) and R. A. Houghton (Chapter 19).

The possibilities for explaining the discrepancy are limited. One possibility, of course, is that the appraisals of oceanic absorption are too low. The difficulty is that several techniques for appraising the absorption have been applied and all come to approximately the same magnitude for the current annual absorption (Woodwell, Chapter 1, and Mackenzie, Chapter 2). Broecker and Peng (1992) call attention to evidence of the magnitude of the transfer of carbon in cold oceanic currents from the Northern Hemisphere to the Southern Hemisphere as confirming the analyses of Tans et al. (1990) discussed later in this chapter. The overall appraisal of oceanic absorption and transport of carbon seems well established, although not at all beyond the need for continued scrutiny. It is also possible that the absorption varies significantly from year to year in response to temperature and other factors.

Could the appraisals of terrestrial releases be too high? Those involved in studies of forests and land use do not believe so. They call attention to the host of changes currently underway in forests, including the toxification of extensive areas by industrial wastes, ozone, and acid rain; the continued rapid deforestation and impoverishment of land in the tropics; the increasing rates of deforestation or harvesting of primary forests and their replacement by young, secondary forests with low carbon stocks in the Northern Hemisphere; and the increasing frequency of forest fires over the past decade in northern forests (see discussions by Woodwell in Chapter 1 and by Kurz et al. in Chapter 6). Warming over the past century may also have increased the decay of soil organic matter, releasing CO_2 into the atmosphere in a positive feedback, as suggested by Woodwell (1983 and Chapter 1) and Jenkinson et al. (1991) among others (see, for example, W. H. Schlesinger 1986). The direct experience with land use leads to the conclusion that releases from forests are increasing, not diminishing. If midlatitude forests are accumulating 1–2 Pg carbon annually above the amounts allowed for successional recovery in the models used for terrestrial systems, as suggested by Tans et al. (1990) (see Tans et al., Chapter 20, and Fung, Chapter 21), that accumulation would be expected to be conspicuous and measurable over a few years, as shown by R. A. Houghton (Chapter 19).

The data of Wofsy et al. (1993) showing that a forest in central Massachusetts is successional and accumulating carbon (has a positive NEP) do not change the earlier interpretations, which accommodated this successional storage, according to R. A. Houghton's discussion in Chapter 19. Houghton has also shown that the analysis of carbon storage in European forests by Kauppi et al. (1992) is incomplete and, if corrected for the decay of organic matter, confirms earlier analyses and offers no evidence that the forests are accumulating carbon at a rate sufficient to resolve the issue of imbalance in current accounts.

No part of this discussion of the global carbon budget suggests that errors or oversights are not possible or have not occurred. These claims of a solution, however, have been based on partial analyses and have brought no insight or data not considered in the earlier global estimates that defined the problem. A forested region with an NEP of 500 g carbon per year requires about 2×10^6 km^2 of midsuccessional forest to store 1 Pg carbon annually. A change of 1–2 Pg carbon per year in the storage of carbon in forests is a large change and difficult to effect in a way that will not be conspicuous. Recent analyses provide reason to examine the topic further.

Schindler and Bayley (1993), for instance, have presented an analysis that suggests that enough nitrogen is being mobilized by human activities in the Northern Hemisphere over land to account on a stoichiometric basis for an additional carbon storage of 1–2 Pg annually in the midlatitudes. This possibility, taken with the conclusions of physiological ecologists (Chapters 3 and 4) and other modeling efforts such as that of Post et al. (1992), provides a basis for a significant increase in carbon storage on land but does not reverse or negate the overall correlation between CO_2 concentration and temperature. Moreover, other fates of mineral nitrogen, such as immobilization by microbial biomass and loss via leaching and denitrification, as outlined by Davidson in Chapter 11, may prevent realization of the full potential of the increased carbon storage in plant biomass from the excess of nitrogen now circulating. The suggestions of Lugo and Brown (1986) that forests do not in fact reach an equilibrium but continue to store carbon indefinitely are not consistent with the pattern of carbon flows that must have occurred prior to human intervention in global cycles (see Chapter 2 and Meybeck 1981). It is possible that the mechanism advanced by Schindler and Bayley (1993)—alone or in combination with other adjustments, such as the high NEP observed for the stand examined by Wofsy et al. (1993) and an adjustment for large areas of suburban forests hitherto overlooked—will prove significant. If so, improved field measurements should soon resolve the issue.

The fact is that the global carbon budget is "balanced" in nature, which knows nothing of our problems in maintaining global accounts. The contemporary cycle is novel, the product of human tinkering, in ignorance, with global cycles on a grand scale. In postglacial time there was probably a net transfer of carbon from the seas to the land as forests were restored on land exposed from under glacial ice and as peat deposits accumulated (Gorham 1991; Gorham and Janssens 1992). More recently, prior to the period of human tinkering, the last two centuries, there was probably a net movement of carbon from the atmosphere to the land and into the seas (as particulate and dissolved organic matter) and a net release of that carbon back into the atmosphere from the oceans, as outlined by Mackenzie in Chapter 2 (see also Meybeck 1981). Otherwise the atmosphere would have been depleted rapidly of carbon to levels far below those recorded in glacial ice.

Human intervention released a total of about 220 Pg carbon into the atmosphere from fossil fuels between 1860 and 1990, an amount equal to about one-third of the total in the atmosphere in 1860. Additional carbon has been released from destruction of forests (about 120 Pg according to R. A. Houghton in Chapter 19), from drainage of wetlands, and from the accelerated decay of organic matter in soils. Enough CO_2 has accumulated in the atmosphere to raise its partial pressure of CO_2 significantly above that in the surface layer of the ocean. The oceans have shifted to become a net accumulator of carbon. Their capacity for accumulating atmospheric carbon at any time, however, appears limited because of the slow exchange of CO_2 between surface water and abyssal waters. As warming progresses, the capacity of the surface water for absorbing CO_2 will decline, a process that amplifies the positive feedbacks outlined here.

These processes do not necessarily proceed in a simple linear fashion (see Chapter 6 by Kurz et al.). Temperature changes, so important in determining climatic features and the responses of biotic systems, have not followed closely the accumulation of heat-trapping gases over the past century. Presumably, in due course, the accumulation of these gases will reach a point at which their effect will become conspicuous. In the interim, we may expect the diversity of responses outlined in the various chapters of this book. Nevertheless, if the earth warms at the rates projected (tenths of a degree centigrade per decade), we can expect the capacity of existing terrestrial ecosystems for storing carbon to diminish; they can be expected ultimately to release carbon into the atmosphere as the warming proceeds. The pool of carbon is large, about 2200 Pg (Table 23.2D), with a little more than two-thirds of it in soils and peats. This stock is maintained by a continuous flow from the gross production of plant communities that are vulnerable to any disturbance.

It is at this point of rapid change that the longer-lived species, and the communities of which they are a part, become vulnerable. Species have an intrinsic genetic variability that is usually higher near the margin of the distribution of the species (Cain 1944). Environment works on that variability to select combinations of genes that are appropriate for the particular circumstances of any place. The selection occurs over generations, but few generations are required to refine ecotypes to specificity for any place. There are many spectacular examples of such selection. Its importance and the generality of the phenomenon are not in question (Woodwell, Chapter 1). Each individual, including each tree or herb of the forest and each animal, carries a set of genes selected over generations for the particular environment of that place. A rapid, continuous, and persistent change in the environment moves it out from under the species, which becomes progressively maladapted to the new environment and vulnerable to disease and other disturbances. The process can, if the disturbance is sudden and severe, lead to the progressive impoverishment of the forest, with an immediate and probably continuing release of carbon into the atmosphere. At the prairie-forest margin, for example, forest is transformed first into savanna and then into grassland by the combination of drought, fire, and disease.

If the transition were to a newly stable climate, there would be a new selection of ecotypes adapted to the new environment, and carbon storage might again increase. But if the changes are continuous, the selection is for species that survive under chronic disturbance. These are small-bodied organisms with short life cycles and rapid reproduction. The carbon stocks drop as forests make the transition to grasslands. Such transitions can be rapid, far more rapid than the development of forests in tundra or other regions where climate is ameliorated. Woodwell (Chapter 1) calls attention to the sharp, genetically fixed differences in ecotypes within larch of the forests of northern Siberia. Although forest trees may invade bog and tundra under a warming climate, the rate of invasion is likely to be far slower than that required by current estimates of global warming (Gear and Huntley 1991; MacDonald et al. 1993). Such invasions will usually require a selection of ecotypes appropriate for the new environment; the assumption that such ecotypes are uniformly available is clearly false. Continuous change is the enemy of such developments.

The "missing carbon" problem is indicative of the frailty of our understanding of the interactions between climate and the terrestrial vegetation and its soils. Although a solution to the conundrum is clearly desirable, the solution is not necessary to an appraisal of biotic feedbacks and their potential for affecting the course of a warming. What is clear is that the potential for a positive feedback is large enough to be of significant concern. The seriousness of the problem grows rapidly with the speed of the warming.

TEMPERATURE AND ATMOSPHERIC TRACE GASES DURING THE LAST FEW HUNDRED YEARS: A REINTERPRETATION

Since the late 1700s atmospheric CO_2 levels have increased by nearly 30%, from 280 to about 360 ppmv. If we estimate the total effect of all heat-trapping gases (except water vapor) as though it were due to CO_2, the CO_2 levels would have risen 40%, from 310 ppmv 100 years ago to 430 ppmv in 1992. With this increase in heat-trapping capacity the GCMs predict several changes in climate. The changes include an average warming globally of at least 1°C with a reduction in the range of diurnal temperature, an increase in precipitation and cloud cover; a slight decrease in stratospheric temperatures; a small decrease in the area of snow, glacial ice, and sea ice; and a slight rise in sea level. To some degree, all these changes are present in the observational record of climate and sea level.

Climate is not regulated by the concentration of heat-trapping gases alone. The amount of warming of the planet globally is less than that predicted for an increase in

CO_2 concentration equivalent to 120 ppmv. In addition, much of the warming took place prior to the 1940s, before the most substantial increase in atmospheric trace gas concentrations. Between 1940 and the mid-1970s the observational record shows a very small decrease in temperature. This trend has been followed by increasing global temperatures with eight of the warmest years on record in the 1980s and 1990s. Furthermore, in contrast to most model calculations, the Southern Hemisphere has warmed more than the Northern Hemisphere, particularly since 1950. These observations suggest that the temperature of the earth over the past century and more has been influenced heavily, possibly controlled, by factors other than the accumulation of trace gases in the atmosphere.

One interpretation of the temperature history of the earth since the late 1800s is that the early part of the record still represents in part a recovery from the Little Ice Age of the 15th through mid-19th centuries. Temperatures during this cold period at times were 0.5°C cooler than a global average of 15°C. It should be recalled that recovery from the last glacial stage of the Wisconsin has not been continuous and unidirectional. The recovery has involved a series of temperature fluctuations, including the cool periods of the Younger Dryas and Little Ice Age and the warm intervals of the Holocene Climatic Optimum and the Medieval Warm Period. Of special interest here is the observation that during the Little Ice Age, when temperatures fell, atmospheric CO_2 levels also appear to have fallen, as summarized by Enting in Chapter 18. Furthermore, it appears that the decline in temperature may have preceded the decline in atmospheric CO_2 concentration. The observation is consistent with the Vostok Core, which shows that during cooling phases changes in atmospheric CO_2 and CH_4 concentrations lagged behind changes in temperature (see the discussion in Chapter 1). The point is that the recovery from the Little Ice Age, whatever its cause, clearly extended into the latter part of the 19th century and perhaps into part of the 20th. As CO_2 and other heat-trapping gases accumulated with the warming, their contribution to the warming increased as well. Nevertheless, although they contributed to the increase of about 0.34°C in global mean temperature between the late 1800s and 1940, they probably did not dominate that change.

Beginning in 1940 and through the mid-1970s there was an erratic cooling of about 0.1°C. The cause of this cooling is uncertain. Several mechanisms have been suggested, including decreased solar activity, increased levels of stratospheric volcanic dust and sulfate aerosol in the high atmosphere, and, most recently, increased tropospheric sulfate aerosol, particularly in the Northern Hemisphere, derived from anthropogenic SO_2 emissions (see the discussion by Charlson in Chapter 14). The last mechanism is most appealing because of the increased global rate of emission of anthropogenic SO_2 that has occurred in the years after World War II, primarily in the Northern Hemisphere. This SO_2 forms sulfate aerosol in the troposphere that can trigger a cooling by stimulating the formation of brighter clouds that reflect solar radiation back to space, and by particulate backscattering of incoming solar radiation. Whatever the actual mechanism, the effect of greenhouse gases on temperature was masked by other factors for the period between 1940 and the mid-1970s.

Between 1974 and 1990 global mean surface temperature rose at a rate of 0.02°C per year, or 0.32°C for the period. For comparison, the rate between 1910 and 1939 was 0.014°C per year, or 0.41°C for this period of time. The rate of temperature increase between 1974 and 1990 was about 1. 3 times the rate prior to the midcentury cooling that began in 1940.

The recent rate of increase in temperature leads to concern that we are entering a new phase in climate, one in which the enhanced greenhouse effect is emerging as the dominant influence on the temperature of the earth. A host of observational and theoretical arguments support the conclusion that the accelerated warming will have important effects on the NEP of terrestrial ecosystems. R. A. Houghton's analysis (Chapter 19) suggests that the recent warming coincided with a release (or a reduced uptake) of carbon from land, whereas the earlier, more gradual warming before 1940 did not. Before 1940 temperature had no apparent relationship to terrestrial carbon storage. Since 1940, however, Houghton's analysis suggests that the accumulation of carbon in terrestrial ecosystems has been negatively correlated with global temperature; that is, a warming has released carbon from storage on land. The correlation shows a 3.4- to 6.4-Pg carbon release (or reduced uptake) per degree centigrade, with the release lagging behind temperature by up to 7 years. There may be several factors involved in explaining the difference in response during these two periods, but the observation is consistent with the recognition, on the basis of experience with studies of vegetation (Chapters 1 and 19 above), that the carbon balance of terrestrial ecosystems is sensitive to rates of change in temperature.

The terrestrial ecosystems that are most vulnerable to the changes in global climate are forests and peatlands of the middle and higher latitudes of the Northern Hemisphere, where the climatic changes will be most abrupt. The midlatitude forests of the temperate zone have been considered on the basis of modeling studies to be absorbing a large amount of the otherwise unaccounted-for carbon being released by human activities. These forests, already heavily stressed by exposures to ozone, acid deposition, and photochemical smog, can be expected to respond rapidly to further temperature changes. The probability is high that the response will be an increase in total respiration relative to gross photosynthesis, a change that will produce rapid, further impoverishment with the release of additional carbon as CO_2 and, if soils are sufficiently wet, CH_4 (see Chapters 8, 9, and 10) into the atmosphere. If this region has been serving as a net sink for atmospheric CO_2, as suggested by Tans et al. (1990) (see Chapters 20 and 21), the strength of that sink can be expected to weaken rapidly and the region can be expected to become a source of additional CO_2.

Support for this analysis comes from the global record of temperature and the record of temperature and CO_2 anomalies following the Mt. Pinatubo eruption in June and July, 1991. Its emissions of volcanic dust and sulfur aerosols have reduced the temperature of the earth by approximately 0.5°C, and the rate of CO_2 accumulation in the atmosphere has declined abruptly as well (see Figure 1.2). Such triggering events as the Pinatubo eruption may have caused the abrupt reversals of climatic trends observed in records such as that of the Vostok Core (Figure 1.1). The assumption at present is that concentrations of the heat-trapping gases, including especially CO_2, have reached the point at which their influence, despite the feedback effect of the drop in temperature, will remain large as the atmosphere clears and the temperature will return to its global upward trend. Evidence of the emergence of dominance by heat-trapping gases is especially important because of the positive feedback now recognized as characteristic of past climatic changes. The feedback has always existed. When temperature was dominated by other factors, such as dust in the upper atmosphere, the feedback was present but the influence of heat-trapping gases was not strong enough to dominate in modulating temperature globally. Now that these gases are emerging as the dominant influence, the importance of the feedback becomes conspicuous and the potential exists for a far more rapid and significant excursion of global temperature.

HOW LARGE A BIOTIC FEEDBACK?

The question of the amount of carbon that might be mobilized as CO_2 and CH_4 is the key to the scale of the net feedback. Although the feedback may be positive overall, as indicated previously, it is far from clear that the negative feedbacks are insignificant. There is a considerable body of evidence to suggest that at present the world carbon budget is being influenced by the combination of successional changes in forests and a stimulation of net primary production from increased concentrations of CO_2 in the atmosphere (Chapters 3, 4 and 6). What does seem clear is that as the total warming increases and environmental conditions drift farther from the slower changes of recent centuries, the effectiveness of the negative relative to the positive feedbacks will diminish. A 10% decrease in the total pool of carbon held on land in plants, soils, and peats (Table 23.2) seems possible over a few decades. Such a change could involve the release of as much as 200 Pg carbon at a rate of 4 Pg/year over 50 years. Using similar logic and the assumption that a 1°C increase in temperature might increase rates of respiration by 10%, Woodwell (1989) suggested that a warming of forests and tundra in northern latitudes could produce from that source alone as much as 3 Pg carbon per year as CO_2 and, under certain circumstances, CH_4, through the increased decay of organic matter in soils.

Models of changes in vegetation in response to a warming of the earth often, but not uniformly, predict increased carbon storage in terrestrial ecosystems under future, warmer climates (see Prentice and Sykes, Chapter 17, for a discussion; see also Solomon and Shugart 1993). The predictions generally assume a new equilibrium climate; the increased storage of carbon results from an increased area of forest, especially in the far north. Will this transition occur in a time of human interest? At least two factors work against such an expectation. First, as outlined previously in several ways, accommodation of existing forests to the changes in climate is not likely to provide negative feedbacks in most instances. Second, the changes in ecosystem structure that must accompany the establishment of forest over large regions will proceed on time scales of centuries to millennia. Existing forests, adapted to present conditions, must disappear and be replaced in toto by new forests, complete with stocks of soil carbon and nutrients and with trees from other regions selected anew for a climatic and photoperiodic regime that is novel for them. If increased storage of carbon can be anticipated in such forests in a new "equilibrium," it is unlikely that the pathways by which the earth reaches this state will be marked by negative feedbacks of significance in the global carbon cycle in the next 100 years. Rates of forest development are too slow.

At the same time it is important to observe that a warmer ocean absorbs and holds less CO_2 from the atmosphere simply because the solubility of CO_2 in seawater declines as the temperature rises (Sarmiento 1993).

Table 23.2A. Estimates of Total Carbon (Pg) Held in Plants, Including Roots, Globally[a]

Estimate	Reference	Comment
450	Bolin (1970)	
833	Reiners (1973)	
1081	Bazilevich (1974)	Assuming carbon equals 45% dry mass
592	Duvigneaud (1972)	In Bolin et al. (1979)
827	Whittaker and Likens (1975)	
760	Baes et al. (1976)	
700	Bolin et al. (1979)	
560	Atjay et al. (1979)	
500	Brown and Lugo (1981)	
558	Olson et al. (1983)	
830	Bolin (1983)	
734	Matthews (1984)	Mean of four estimates
594	Goudriaan and Ketner (1984)	
657	Esser 1987	
737	Smith et al. (1992)	

[a]Mean ± standard deviation, coefficient of variation 23%: 694 ± 162. Median: 700.

Table 23.2B. Estimates of Total Carbon (Pg) Held in Soil Organic Matter Globally[a]

Estimate	Reference	Comment
700	Bolin (1970)	
1080	Yakushevskaya (1971)	Seen in Baes et al. (1976)
1051	Keeling (1973)	
1407[b]	Bazilevich (1974)	Assuming carbon equals 50% dry mass
1000	Baes et al. (1976)	
2946[b]	Bohn (1976)	
1456	W. H. Schlesinger (1977)	
2840	Duvigneaud (1979)	In Bolin et al. (1979)
1672[b]	Bolin et al. (1979)	
1636[b]	Atjay et al. (1979)	
2070	Atjay et al. (1979)	Alternate estimate
1457	Meentemeyer et al. (1981)	
1380	Brown and Lugo (1981)	
1395	Post et al. (1982)	
2200[b]	Bohn (1982)	
1900[b]	Bolin (1983)	
1532	Goudriaan and Ketner (1984)	
1551	W. H. Schlesinger (1984)	
1477	Buringh (1984)	
1272	Post et al. (1985)	
1576	Eswaran et al. (1993)	

[a]Mean ± standard deviation, coefficient of variation 35%: 1601 ± 565. Median: 1467.
[b]Includes a specific estimate for peat.

Table 23.2C. Estimates of Total Carbon (Pg) Held in
Peat Globally[a]

Estimate	Reference	Comment
114	Bazilevich (1974)	Assuming carbon equals 51.7% dry mass (Gorham 1991)
862	Bohn (1976)	From histosol maps
882	Bolin et al. (1979)	
225	Atjay et al. (1979)	Including swamp and marsh
150	Bramryd (1980)	
300	Sjörs (1980)	
124	Kivinen and Pakarinen (1981)	Calculated by E. Gorham
377	Bohn (1982)	From histosol maps
500	Bolin (1983)	
557	Gorham, new[b]	
446	Gorham, new[c]	
357	Eswaran et al. (1993)	For histosols from global soil data bases

[a]Mean ± deviation, coefficient of variation 66%: 412 ± 272. Median: 377.
[b]This new estimate is based on 133 kg carbon m^{-2} in 342 million ha of northern peatlands (Gorham 1991), multiplied by a new total area of world peatlands, 419 million ha, in which the world area estimated by Kivinen and Pakarinen (1981), 450 million ha, is reduced by a correction for Canada, where the area estimate of Gorham (1991) is substituted.
[c]This new estimate is calculated as above but using a different total area of world peatlands (bogs and fens), estimated by Aselmann and Crutzen (1989) at 335 million ha.

CONCLUSIONS: WHAT WE KNOW . . . AND SOME OF WHAT WE NEED TO KNOW

Significant uncertainties continue to plague our knowledge of the global carbon cycle. There is little question that the atmospheric burdens of CO_2, CH_4, and other radiatively important trace gases are increasing as a result of human activities. There is also little question as to the importance of terrestrial ecosystems, especially forests, and oceans in modulating the atmospheric increase. Between 1988 and 1991 the increase varied between about 2 and 5 Pg carbon without an obvious anthropogenic cause. The modulating factors are obviously inconstant and the variability is high. The details, however important, are not known.

The most serious questions have to do with the potential for just such surprises, especially surprises that lead to positive feedbacks. The potential appears significant—

Table 23.2D. Estimates of Total Carbon (Pg) Held in Plants
and Soil Globally[a]

Total pool	Comment
2295	Mean phytomass + mean soil organic carbon
2167	Median phytomass + median soil organic carbon

[a]It is not always clear whether these estimates include carbon in standing dead biomass and forest litter. Bazilevich (1974) and Atjay et al. (1979) estimate the former at 75 and 30 Pg, respectively. Litter estimates by Bazilevich (1974), Whittaker (1975), and Atjay et al. (1979) are 97, 56, and 60 Pg, respectively. Bolin et al. (1979) estimate standing dead biomass and litter together at 150 Pg; Bolin (1983) estimates 60 Pg.

unfortunately more significant than the potential for surprises involving negative feedbacks. If, for instance, a sufficient decline in the water table occurs in the boreal and tundra peatlands, subterranean fires could speed oxidation of the peat in the vast, remote peatlands of Canada and Russia, spewing forth smoke, CO_2, and CH_4 throughout the Northern Hemisphere for years. Alternatively, if water tables remain high, these peatlands might shift toward the production of CH_4 at high rates, as outlined by Gorham (Chapter 9). The issue is all the more serious because the positive feedbacks may be large enough to speed the warming and are coupled directly to some of the most difficult human problems, such as the growth of the human population and the expansion of demands for food, fiber, land, and energy. Urgency is attached to efforts to reduce the risks. Action hinges heavily on clear understanding of the problems and the consequences. The feedback issues, critical as they are, cannot be separated from basic understanding of the global climatic cycle. The following topics, slightly modified from the summary of the IPCC workshop, emerge at the moment as especially worthy of further consideration by the scientific community:

1. The implications of the feedbacks intrinsic in the data from the Vostok, Greenland, and other ice cores, which provide a record of atmospheric composition and climatic changes through glacial and interglacial periods. Do these data provide an envelope of experience within which the current world can be expected to operate? What are the causes of reversals of the trends of temperature change over glacial time?

2. The implications of the contemporary correlations between temperature and the CO_2 content of the atmosphere, in which changes in temperature precede changes in CO_2 level by about 5–7 months. Parallel correlations for CH_4 and N_2O raise additional questions worthy of detailed examination.

3. Improvement of the inventory of the carbon retained in terrestrial ecosystems globally. Satellite imagery offers the possibility of making direct measurements of changes in that inventory by quantifying changes in the area and distribution of forests and peatlands, changes in successional status, and changes in the structure of forests, all with continuously improving detail. Measurement of the frequency, area, and severity of fires bears heavily on calculation of carbon stocks on land. The data are to be used to define the role of the terrestrial vegetation, especially forests, in controlling the composition of the atmosphere. Satellite imagery and computer techniques (e.g., geographic information systems) for handling data make this realm a new frontier that offers great opportunities for innovations in scholarship and for improving management of resources in the public interest.

4. Improvement of global budgets of trace gases that affect climate, including CH_4, N_2O, dimethyl sulfide, and ozone.

5. The implications for the composition of the atmosphere, the energy budget, and the temperature of the earth of the systematic impoverishment of the biosphere through the cumulative effects of progressive chronic disturbance.

6. The implications of the changes in climate—including changes in temperature, precipitation, fire, and composition of the atmosphere—for the growth, selection, and survival of plants and especially for the storage of carbon as net ecosystem production in terrestrial ecosystems.

7. The development of methods sensitive enough for the direct monitoring of flows of carbon in its various forms between the atmosphere and the land, with errors reduced to the order of 10% or less.

8. The refinement of analyses of the mass transport of atmospheric components, especially CO_2, through global circulation patterns as the basis for the testing of data and hypotheses on stocks and flows of carbon among the major reservoirs of land, air, and oceans. These techniques have been heavily dependent on isotopic data and on modeling.

9. Continued exploration of the role of the oceans in absorbing and sequestering carbon transferred as atmospheric CO_2 and as fixed carbon through terrestrial runoff. Have we overlooked an oceanic sink for carbon of significant size? How will the warming of the earth, or other changes now in process or probable, influence the role of the oceans and their currents in affecting the energy budget globally?

10. Exploration of the potential for an increase in UV-B or other human disturbance to affect the current biotic fluxes of organic carbon in the world oceans and thereby upset current projections of the course of the global energy budget over the next decades.

11. Continued examination of various suggestions for improving the storage of carbon on land or in the sea by nutrient fertilization or other manipulations of biogeochemical cycles.

12. Continued monitoring of atmospheric trace gases globally, with increased attention to terrestrial sites where the mobilization of large pools of carbon may become conspicuous as the warming gains momentum.

REFERENCES

Alley, D. B., D. A. Meese, C. A. Shuman, A. J. Gow, K. C.Taylor, P. M. Grootes, J. W. C. White, M. Ram, E. D. Waddington, P. A. Mayewski, and G. A. Zielinski. 1993. Abrupt increase in Greenland snow accumulation at the end of the Younger Dryas event. *Nature* 362:527–529.

Aselmann, I., and P. J. Crutzen. 1989. Global distribution of natural freshwater wetlands and rice paddies, their net primary productivity, seasonality and possible methane emissions. *J. Atmos. Chem.* 8:307–358.

Atjay, G. L., P. Ketner, and P. Duvigneau. 1979. Terrestrial primary production and phytomass. In *The Global Carbon Cycle*, ed. B. Bolin, E. T. Degens, S. Kempe, and P. Ketner, pp. 129–181. Wiley, New York.

Bacastow, R., and C. D. Keeling. 1973. Atmospheric carbon dioxide and radio carbon in the natural carbon cycle. II. Changes from A.D. 1700 to 2070 as deduced from a geochemical model. In *Carbon and the Biosphere*, ed. G. M. Woodwell and E. V. Pecan, pp. 86–135. U.S. Atomic Energy Commission, Washington D.C.

Baes, C. F., H. E. Goeller, J. S. Olson, and R. M. Rotty. 1976. The global carbon dioxide problem. ORNL-5194. Oak Ridge National Laboratory, Oak Ridge, Tenn.

Barnola, J. M., D. Raynaud, Y. S. Korotkevich, and C. Lorius. 1987. Vostok ice core provides 160,000-year record of atmospheric CO_2. *Nature* 329(6138):408–414.

Bazilevich, N. I. 1974. Energy flow and the biological regularities of the world ecosystems. In *Proceedings of the First International Congress of Ecology*, pp. 172–186. The Hague, The Netherlands.

Bohn, H. L. 1976. Estimate of organic carbon in world soils. *Soil Sci. Soc. Am. J.* 40:468–470.

Bohn, H. L. 1982. Estimate of organic carbon in world soils II. *Soil Sci. Soc. Am. J.* 46:1118–1119.

Bolin, B. 1970. The carbon cycle. *Sci. Am.* 223(3):124–132.

Bolin, B. 1983. Changing global biogeochemistry. In *Oceanography: The Present and Future,* ed. P. G. Brewer, pp. 305–326. Springer Verlag, New York.

Bolin, B., E. T. Degens, P. Duvigneau, and S. Kempe. 1979. The global biogeochemical carbon cycle. In *The Global Carbon Cycle,* ed. B. Bolin, E. T. Degens, S. Kempe, and P. Ketner, pp. 1–56. Wiley, New York.

Bramryd, T. 1980. The role of peatlands for the global carbon dioxide balance. In *Proceedings of the 6th International Peat Congress,* pp. 9–11.

Broecker, W. S. 1987. Unpleasant surprises in the greenhouse? *Nature* 328:123–126.

Broecker, W. S., and T.-H. Peng. 1992. Interhemispheric transport of carbon dioxide by ocean circulation. *Nature* 356:587–589.

Brown, S., and A. E. Lugo. 1982. The storage and production of organic matter in tropical forests and their role in the global carbon cycle. *Biotropica.* 14:161–187.

Buringh, P. 1984. Organic carbon in soils of the world. In *The Role of Terrestrial Vegetation in the Global Carbon Cycle,* ed. G. M. Woodwell, pp. 91–109. Wiley, New York.

Cain, S. A. 1944. *Foundations of Plant Geography.* Harper and Brothers, New York.

Davis, M. B., and C. Zabinski. 1992. Changes in geographical range resulting from greenhouse warming: Effects on biodiversity in forests. In *Global Warming and Biological Diversity,* ed. R. L. Peters and T. E. Lovejoy, pp. 297–308. Yale University Press, New Haven, Conn.

Esser, G. 1987. Sensitivity of global carbon pools and fluxes to human and potential climatic impacts. *Tellus* 39B:245–260.

Eswaran, H., E. Van Den Berg, and P. Reich. 1993. Organic carbon in soils of the world. *J. Soil Sci. Soc. Am.* 57:192–194.

Gear, A. J., and B. Huntley. 1991. Rapid changes in the range limits of Scots pine 4000 years ago. *Science* 251:544–547.

Gorham, E. 1991. Northern peatlands: Role in the carbon cycle and probable responses to global warming. *Ecol. Appl.* 1:182–195.

Gorham, E., and J. A. Janssens. 1992. The paleorecord of geochemistry and hydrology in northern peatlands and its relation to global change. *Suo* 43:117–126.

Goudriaan, G., and P. Ketner. 1984. Are land biota a source or sink for CO_2? In *Interactions between Climate and Biosphere,* ed. H. Lieth, R. Fantechi, and H. Schnitzler, pp. 247–252. Swets and Zeitlinger, Lisse, Germany.

Houghton, J. T., G. J. Jenkins, and J. J. Ephraums. 1990. *Climate Change. The IPCC Scientific Assessment.* Cambridge University Press, Cambridge and New York.

Houghton. J. T., B. A. Callander, and S. K. Varney. 1992. *Climate Change 1992. The Supplementary Report to the IPCC Scientific Assessment.* Cambridge University Press, Cambridge and New York.

Idso, S. B., B. A. Kimball, and S. G. Allen. 1991. Net photosynthesis of sour orange trees maintained in atmospheres of ambient and elevated CO_2 concentration. *Agric. For. Meteorol.* 54:95–101.

Jenkinson, D. S., D. E. Adams, and A. Wild. 1991. Model estimates of carbon dioxide emissions from soil in response to global warming. *Nature* 351:304–306.

Kauppi, P. E., K. Mielikäinen, and K. Kuusela. 1992. Biomass and carbon budget of European forests, 1971–1990. *Science* 256:70–74.

Keeling, C. D. 1973. The carbon dioxide cycle: Reservoir models to depict the exchange of atmospheric carbon dioxide with the oceans and land plants. In *Chemistry of the Lower Atmosphere,* ed. S. I. Rasool, pp. 251–329. Plenum Press, New York.

Keeling, C. D., R. B. Bacastow, A. F. Carter, S. C. Piper, T. P. Whorf, M. Heimann, W. G. Mook, and H. Roeloffzen. 1989. A three dimensional model of atmospheric CO_2 transport based on observed winds: 1. Analysis of observational data, pp. 165–236. *Geophysical Monograph 55. American Geophysical Union,* Washington, D.C.

Kellogg, W. W. 1983. Feedback mechanisms in the climate systems affecting future levels of carbon dioxide. *J. Geophys. Res.* 88:1263–1269.

Khalil, M. A. K., and R. A. Rasmussen. 1989. Climate-induced feedbacks for the global cycles of methane and nitrous oxide. *Tellus* 41B:554–559.

Kivinen, E., and P. Pakarinen. 1981. Geographical distribution of peat resources and major peatland complex types in the world. *Ann. Acad. Sci. Fenn. Ser. A* 132:1–28.

Kuo, C., C. Lindberg, and D. J. Thompson. 1990. Coherence established between atmospheric carbon dioxide and global temperature. *Nature* 343:709–713.

Lashof, D. 1989. The dynamic greenhouse: Feedback processes that may influence future concentrations of atmospheric trace gases and climatic change. *Climatic Change* 14:213–242.

Lorius, C., N. I. Barkov, J. Jouzel, Y. S. Korotkevich, V. M. Kotlyakov, and D. Raynaud. 1988. Antarctic ice core: CO_2 and climatic change over the last climatic cycle. *Eos* 681–684.

Lugo, A. E., and S. Brown. 1986. Steady state terrestrial ecosystems and the global carbon cycle. *Vegetatio* 68:83–90.

MacDonald, G. J. 1990. Role of methane clathrates in past and future climates. *Climatic Change* 16:247–281.

MacDonald, G. M., T. W. D. Edwards, K. A. Moser, R. Pienitz, and J. P. Smol. 1993. Rapid response of treeline vegetation and lakes to past climatic warming. *Nature* 361:243–246.

Marston, J. B., M. Oppenheimer, R. M. Fujita, and S. R. Gaffin. 1991. Carbon dioxide and temperature. *Nature* 349:573–574.

Matthews, E. 1984. Global inventory of the pre-agricultural and present biomass. In *Interactions between Climate and Biosphere,* ed. H. Lieth, R. Fantechi, and H. Schnitzler, pp. 237–246. Swets and Zeitlinger, Lisse, Germany.

Meentemeyer, V., E. O. Box, M. Folkoff, and J. Gradner. 1981. Climatic estimation of soil properties: Soil pH, litter accumulation, and soil organic content. *Bull. Ecol. Soc. Am.* 62:104.

Meybeck, M. 1981. River transport of organic carbon to the ocean. In *Flux of Organic Carbon by Rivers to the Oceans,* ed. G. E. Likens et al., pp. 219–269. Conf.-8009140, National Technical Information Service, Springfield, Mo.

Olson, J. S., J. A. Watts, and L. J. Allison. 1983. Carbon in live vegetation of major world ecosystems. ORNL 5862. Environmental Sciences Division, Oak Ridge National Laboratory, Oak Ridge, Tenn.

Post, W. M., W. R. Emanuel, P. J. Zinke, and A. G. Stangenberger. 1982. Soil carbon pools and world life zones. *Nature* 298:156–159.

Post, W. M., J. Pastor, P. J. Zinke, and A. G. Stangenberger. 1985. Global patterns of soil nitrogen storage. *Nature* 317:613–616.

Post, W. M., J. Pastore, A. W. King, and W. R. Emanuel. 1992. Aspects of the interaction between vegetation and soil under global change. *Water Air Soil Pollut.* 64:345–363.

Raynaud, D., J. Jouzel, J. M. Barnola, J. Chappellaz, R. J. Delmas, and C. Lorius. 1993. The ice record of greenhouse gases. *Nature* 259:926–934.

Sarmiento, J. L. 1993. Ocean carbon cycle. *Chem. Eng. News* May 31, pp. 30–43.

Sarmiento, J., and E. Sundquist. 1992. Revised budget for the oceanic uptake of anthropogenic carbon dioxide. *Nature* 356:589–593.

Schindler, D. W., and S. E. Bayley. 1993. The biosphere as an increasing sink for atmospheric carbon: Estimates from increased nitrogen deposition. *Global Biogeochem. Cycles* 7:717–734.

Schlesinger, M., and J. F. B. Mitchell. 1985. Model projections of the equilibrium climatic response to increased carbon dioxide. In *DOE: The Potential Climatic Effects of Increasing*

Carbon Dioxide, eds. M. C. MacCracken and F. M. Luther, pp. 81–147. U. S. Department of Energy, Carbon Dioxide Research Division, Washington, D. C.

Schlesinger, W. H. 1977. Carbon balance in terrestrial detritus. *Annu. Rev. Ecol. Syst.* 8:51–81.

Schlesinger, W. H. 1984. Soil organic matter: A source of atmospheric CO_2. In *The Role of Terrestrial Vegetation in the Global Carbon Cycle,* ed. G. M. Woodwell, pp. 111–127. Wiley, New York.

Schlesinger, W. H., 1986. Changes in soil carbon storage and associated properties with disturbance and recovery. In *The Changing Carbon Cycle,* ed. J. R. Trabalka and D. E. Reichle, pp. 194–220. Springer-Verlag, New York.

Sjörs, H. 1980. Peat on earth: Multiple use or conservation? *Ambio* 9:303–308.

Smith, T. M., J. F. Weishampel, H. H. Shugart, and G. B. Bonan. 1992. The response of terrestrial C storage to climate change: Modeling C dynamics at varying temporal and spatial scales. *Water Air Soil Pollut.* 64:307–326.

Solomon, A. M., and W. P. Cramer. 1993. Biospheric implications of global environmental change. In *Vegetation Dynamics and Global Change,* ed. A. M. Solomon and H. H. Shugart, p. 25–52. Chapman and Hall, New York.

Solomon, A. M., and H. H. Shugart. 1993. *Vegetation Dynamics and Global Change.* Chapman and Hall, New York.

Tans, P. P., I. Y. Fung, and T. Takahashi. 1990. Observational constraints on the global atmospheric CO_2 budget. *Science* 247:1431–1438.

Whittaker, R. H. 1975. *Communities and Ecosystems,* 2nd ed. Macmillan, New York.

Whittaker, R. H., and G. E. Likens. 1975. In *Primary Productivity of the Biosphere,* ed. H. Lieth and R. H. Whittaker, pp. 305–328. Springer-Verlag, New York.

Wofsy, S. C., M. L. Goulden, J. W. Munger, S. M. Fan, P. S. Bakwin, B. C. Daube, S. L. Bassow, and F. A. Bazzaz. 1993. Net exchange of CO_2 in a mid-latitude forest. *Science* 260:1314–1317.

Woodwell, G. M. 1983. Biotic effects on the concentration of atmospheric carbon dioxide: A review and projection. In *Changing Climate,* pp. 216–241. National Academy of Science Press, Washington, D.C.

Woodwell, G. M. 1989. The warming of the industrialized middle latitudes 1985–2050: Causes and consequences. *Climatic Change* 15:31–50.

Woodwell, G. M. 1990. The earth under stress: A transition to climatic instability raises questions about biotic impoverishment. In *The Earth in Transition: Patterns and Processes of Biotic Impoverishment,* ed. G. M. Woodwell, pp. 3–7. Cambridge University Press, New York.

Woodwell, G. M., and R. H. Whittaker. 1968. Primary production in terrestrial ecosystems. *Am. Zool.* 8:19–30.

Index

Page numbers referring to entries in tables or figures are set in *italics*.

Printed in the United States
70394LV00001B/5

9 780195 086409